Sustainable Engineering Forum 2021

Held at the 2021 AIChE Annual Meeting

Boston, Massachusetts, USA and Online
7 - 11 November and 15 - 19 November 2021

ISBN: 978-1-7138-5716-7

Printed from e-media with permission by:

Curran Associates, Inc.
57 Morehouse Lane
Red Hook, NY 12571

CURRAN ASSOCIATES INC. proceedings.com

Some format issues inherent in the e-media version may also appear in this print version.

AIChE
120 Wall Street, FL 23
New York, NY 10005-4020

Phone: (800) 242-4363
Fax: (203) 775-5177

www.aiche.org

Additional copies of this publication are available from:

Curran Associates, Inc.
57 Morehouse Lane
Red Hook, NY 12571 USA
Phone: 845-758-0400
Fax: 845-758-2633
Email: curran@proceedings.com
Web: www.proceedings.com

TABLE OF CONTENTS

624954 Local Renewable Chemical Production for Combined Heat and Power in Remote Locations: Optimal Design and Operation

Abstract:

Generation of electrical power from intermittent sources of renewable energy such as the wind or the sun is a promising approach to improving energy sustainability. Renewable energy storage is necessary to balance this intermittent generation with electrical power demands. Batteries are the most used energy storage technology, but their capital cost and relatively low energy density makes them unsuitable for longer-term, high-capacity energy storage. Production of hydrogen (H_2) via water electrolysis or further processing to ammonia (NH_3) are alternative energy vectors more suitable to seasonal storage. Power generation via H_2-fed fuel cells is well-established whereas NH_3 has received worldwide attention [1-4] for power generation with either fuel cells or traditionally fossil-fueled technologies (e.g. internal combustion gensets) with minor modifications. Previously, we have investigated 1 MW residential electrical power supply systems with chemical energy storage in 15 American cities which represent various climate-demand regions throughout the continental United States [5]. We found that in every location, cost savings can be achieved by using both synergistically. While promising, these systems are not yet economically competitive with centralized fossil fuel-based electricity generation and transmission.

In this work, we optimize and analyze combined heat and power (CHP) systems in remote locations where the conventional supply paradigm is less economically favorable. At present, power and heat supply in these areas is expensive, entailing long distance electrical transmission and the import of fossil fuels for local heating or small-scale local generation. Combining heat and power demands allows for effective utilization of waste heat from power generation and furthermore, renewable-derived H_2 and NH_3 can be used to generate heat using boiler-type technologies [6]. Economically competitive renewable CHP design is complicated and highly unintuitive because these systems inherently operate at unsteady-state and hence their operating schedules should be accounted for at the design stage. To this end, we have developed a mixed integer linear programming combined optimal combined design and scheduling (CODS) model. The model minimizes the annualized system net present cost (NPC) by optimally selecting and sizing the units in the system while simultaneously scheduling the operation of these units (i.e. on/off, production rates, storage inventories) during each distinct period of a scheduling horizon which captures both diurnal and seasonal variation in intermittent renewable generation and power demand. Our CODS modeling framework generates variable length scheduling periods by clustering consecutive hours of full year time series data for wind capacity factor, solar capacity factor and power demand. This temporal aggregation makes the model computationally tractable, allowing for high throughput computational studies.

We specifically use CODS to determine optimal local CHP systems for (i) Mahaka, Hawaii, (ii) Nantucket, Massachusetts and (iii) Northwest Arctic Borough, Alaska. There are approximately 2,500 homes in each location; we consider residential power and heat demands arising from these. To elucidate the true economic potential of renewable H_2 and NH_3 production, we simultaneously allow continued business-as-usual power and fuel purchases while considering installation of PV arrays, wind turbines, batteries, electric boilers, and hot water thermal storage as well as H_2 and NH_3 production and storage to fuel local power generation and/or heat generation. We compare these optimal hybrid systems to business-as-usual only and electrification-only (i.e. batteries, electric boilers, and hot water heat storage) benchmarks. Overall, this analysis provides an understanding of the most economically competitive residential renewable deployments in today's energy landscape.

References

[1] Klerke, Christensen, Nørskov, & Vegge. (2008). Ammonia for hydrogen storage: challenges and opportunities. *J. Mater. Chem. 18*(20), 2304-2310.

[2] Zamfirescu & Dincer. (2008). Using ammonia as a sustainable fuel. *J. Power Sources 185*(1), 459-465.

[3] Smith, Hill, & Torrente-Murciano. (2020). Current and future role of Haber–Bosch ammonia in a carbon-free energy landscape. *Energy Environ. Sci. 13*(2), 331-344.

[4] Palys, Wang, Zhang, & Daoutidis. (2021). Renewable ammonia for sustainable energy and agriculture: vision and systems engineering opportunities. *Curr. Opin. Chem. Eng. 31*, 100667.

[5] Palys & Daoutidis. (2020). Using hydrogen and ammonia for renewable energy storage: A geographically comprehensive techno-economic study. *Comput. Chem. Eng. 136*, 106785.

[6] Samsatli & Samsatli. (2019). The role of renewable hydrogen and inter-seasonal storage in decarbonising heat – Comprehensive optimisation of future renewable energy value chains. *Appl. Energy 233*, 854-893.

629661 Distributed and on-Demand Ammonia Synthesis By Inorganic Membrane Reactor

Abstract:

The world's primary energy sources are currently fossil fuels, including coal, oil and natural gas. Although society has benefited from the steady use of energy, continuing to use fossil fuels at such rates could lead to dire consequences. This motivates the development of new processes and technologies to produce carbon-neutral energy that breaks linear scaling relations while critically addressing environmental needs.

Ammonia (NH_3) is an important raw material for fertilizer production and today it is gaining attention as a new energy carrier for hydrogen (H_2). Conventionally, ammonia is synthesized through the well-known Haber-Bosch process at 400-500 °C and P~150 bar. Both critical reaction conditions and massive production (176 million metric ton of NH_3 in 2019 globally) make it one of the most energy extensive processes, consuming 1-2% of the world's total energy.

Here, an intensified process using a 'smart' reactor (*aka* membrane reactor) is proposed to produce H_2 from methane steam reforming and NH_3 in the same system at lower operating conditions than the traditional process. In addition, the system allows for the separation of a highly concentrated stream of carbon dioxide, which can be captured to address environmental issues. This alternative system allows to synthesize NH_3 on demand with near zero emissions, without the necessity of centralized facilities.

The smart reactor consists of a stainless-steel module, equipped with two C-based gaskets at both ends of the tubular membrane to prevent the mixing of the permeate and retentate flow streams. The membrane has a thin Pd-Au layer deposited onto a porous ceramic support via electroless plating. The physical membrane characteristics are as follows: total length of 75 mm, active Pd length of 50 mm, and a 10 mm O.D.

Two commercial catalysts were packed into the smart reactor. The Ni-based catalyst – selective towards methane steam reforming reaction - was packed in the annulus region of the membrane reactor, while the Fe-based one - selective towards ammonia synthesis - was packed in the core of the membrane. Prior to the reaction tests, the Pd-based membrane was tested in pure gas permeation tests in order to evaluate the hydrogen/other gases ideal selectivity. Successively, natural gas steam reforming reaction tests have been performed at T = 400 °C, reaction pressure = 30 bar, and various permeate pressure and sweep-gas flow rate. Ammonia synthesis was then performed in the same system at the best operating conditions, in terms of hydrogen permeated flow rate, achieved in the previous experimental campaign.

Laboratory data suggests that this technology is promising for innovative NH_3 synthesis processes operating at reduced conditions while paying attention to the environment.

625095 Spatial Variation in Cost of Renewable Electricity-Driven Continuous Ammonia Production across the United States

Abstract:

Ammonia production contributes more than 1% of the global greenhouse gas emissions (GHG) while being used to serve a majority of the demand for nitrogen-containing fertilizer for agricultural use. The predominant route for ammonia production relies on natural gas as a source of energy and hydrogen for thermochemical Haber-Bosch (HB) synthesis, and is estimated to result in about 2.3 tonnes of CO_2 per tonne NH_3 ammonia produced.[1] Declining costs of variable renewable energy (VRE) based electricity and electrolyzers have raised interest in producing low-carbon H_2 via electrolysis, as well as its use in decarbonization of industrial ammonia production. This electrically-driven ammonia route could not only serve existing uses for fertilizer production, but also be deployed to service energy needs for other end-use sectors where ammonia is being contemplated (e.g. marine transport). Here, we study the spatial variations in cost of producing ammonia across the U.S. predominantly based on the temporal variations in using renewable electricity input in conjunction with thermochemical HB synthesis. The plant configuration targets continuous ammonia supply to an end-use customer throughout the year while rely on energy input as electricity, supplied from a combination of on-site VRE generation as well as grid imports. Production of feed H_2 and N_2 for HB synthesis is modeled using pressure swing adsorption (PSA) and low temperature proton exchange membrane (PEM) electrolysis. The energy inputs and capital cost for the HB synthesis loop, including compressors, heat recovery and reactor system, are parameterized based on offline Aspen simulations.

Our approach goes beyond prior techno-economic assessments of electricity-driven ammonia production by explicitly accounting for variability in electricity supply and its implications on plant design, cost and emissions. This is achieved by using a least-cost integrated design and operations modeling framework that treats as variables the relative sizing of various units (e.g. electrolyzer, PSA, renewables capacity), including possible on-site deployment of alternative forms of on-site storage (battery energy storage, gaseous H_2 and liquid N_2). The model accounts for intra-annual variability in inputs (VRE resource, grid electricity marginal price and emissions) at an hourly resolution and includes various constraints such as: a) such as inter-temporal constraints governing energy storage (H_2, battery) and electrolysis operation, b)inflexible ammonia production because of the relatively harsh operating conditions (500°C, 250 bar) , with possibility to go offline for maintenance with minimum up and downtime requirements, c) constraints enforcing hourly (constant) and annual production (>95% availability) requirements as well as any emissions intensity constraints. The overall mixed-integer linear programming (MILP) model is able to optimize for the minimum annualized cost of providing round-the-clock ammonia under the required system emission and flexibility constraints.

We used the model to evaluate the levelized cost of ammonia for multiple electricity source and location scenarios in the US. Our analysis shows that the levelized cost of a grid-connected ammonia is nearly 50% higher than current natural gas, HB-produced ammonia in the absence of carbon tariffs. Relying on a grid electricity mix with higher VRE penetration (e.g. New York) allows for producing ammonia with lower overall emission intensity (e.g. 1.6 CO_2 tonne/tonne) compared to conventional natural gas-based routes. We also evaluated dedicated VRE-based ammonia production for locations in close proximity to existing NH_3 production facilities and agricultural hubs in the US, to identify the cost-optimal VRE mix and storage requirements. Our analysis shows that a standalone renewable ammonia production facility makes use of storage of intermediate products (N_2, H_2) in the production process so as to be able to dispatch them during non-availability of renewable electricity. To meet the minimum power input necessary to operate the thermochemical HB process, electrochemical storage (e.g. Li-ion) is also needed. However, if the thermochemical HB process can be operated at less than nameplate feed flow rates, the need for Li-ion battery storage is minimized, allowing for more cost-effective production options.

[1] IEA (2019), *The Future of Hydrogen*, IEA, Paris https://www.iea.org/reports/the-future-of-hydrogen

623227 Paired Electrolysis of Furanic Compounds to Valuable Chemicals in Electrochemical Flow Cells

Abstract:

Organic electrosynthesis is emerging as a cost-effective and environmental-friendly chemical production strategy by utilizing renewable electricity. Paired electrolysis cogenerates valuable chemicals at both electrodes can optimize the energy efficiency and economic feasibility. We report pairing hydrogenation and oxidation of 5-(hydroxymethyl)furfural (HMF) or furfural to desired chemicals at a single electrolysis cell. Electrocatalytic hydrogenation of HMF to 2,5-bis(hydroxymethyl)furan (BHMF) and furfural to furfural alcohol (FA) with high selectivity of >90% can be operated at near-neutral pH on Ag-based and Pb-based catalysts, respectively. In addition, oxidizing HMF to 2,5-furandicarboxylic acid (FDCA) and furfural to furoic acid can both be realized at TEMPO mediated process by using carbon-based catalysts or at Ni-based catalyst in an alkaline medium. Taken together, HMF or furfural can be performed in a single electrolysis cell with a minimized cell voltage only around 1.6 V. Products selectivity and faradaic efficiency are highly related to the reaction conditions, including potential or current density, architectures of the reactor, type of catalysts. By optimizing the single flow reactor, a three-electrode system, two-electrode membrane assembly architecture, and pH-symmetric and pH-asymmetric structure can be designed to reduce the capital expense, minimize required energy, and simplify processing steps. Finally, a complete electrons economy can be achieved by pairing two electrochemical reactions, and the overall charge efficiency can attain over 170% without any crossover issue detected. As a result, the continuous cogeneration of high value-added BHMF or FA and FDCA or furoic acid can be performed in a single electrolyzer.

625620 An Integrated Reaction and Separation Process for "Green Ammonia" Production from Distributed and Intermittent Renewable Energy.

Abstract:

Industrial ammonia production is currently responsible for 50% of global food supply through fertilizer and 1-2% of CO_2 emissions through the conventional Haber-Bosch process. "Green ammonia" produced from water, air, and renewable electricity has garnered accelerating attention in recent years for both energy storage and sustainable fertilizer. However, the conventional Haber-Bosch process is incompatible with intermittent and isolated renewable energy because it is designed to operate continuously for months at a time using fossil fuels in centralized locations. An electric Haber-Bosch process, while benefiting from efficient electric compressors and modular electrolysers, still requires a high pressure reactor with a recycle-loop. A novel integrated process has been demonstrated at the lab scale using $Ru/Cs/CeO_2$ catalyst and $MnCl_2/SiO_2$ absorbent to synthesize and separate ammonia in a single vessel.[1] The catalyst was designed for low-temperature and high-conversion activity by removing hydrogen and ammonia inhibition, respectively. The absorbent utilized $MnCl_2$ rather than conventional $MgCl_2$ because it was resilient to decomposition after exposure to water. This unique absorbent was supported on silica to prevent sintering and improve cyclical capacity. The temperature, pressure, and N_2:H_2 ratio have been optimized with respect to the forecasted production rate of the process, and economic analyses have compared the integrated process with the conventional Haber-Bosch process.[2] Such an integrated process re-defines the Haber-Bosch process by removing the requirement for high pressures and a recycle. It is anticipated that an integrated "green ammonia" process will be the foundation for both sustainable fertilizer and long-term energy storage in the future when paired with distributed and intermittent renewable energy.

Figure | Exceeding single-pass equilibrium with a single or double layer integrated flow system. Modified from [1].

1. Smith, C. and L. Torrente-Murciano, Exceeding Single-Pass Equilibrium with Integrated Absorption Separation for Ammonia Synthesis Using Renewable Energy-Redefining the Haber-Bosch Loop. Advanced Energy Materials, 2021. 11.
2. Smith, C., A.K. Hill, and L. Torrente-Murciano, Current and future role of Haber-Bosch ammonia in a carbon-free energy landscape. Energy & Environmental Science, 2020. 13(2): p. 331-344.

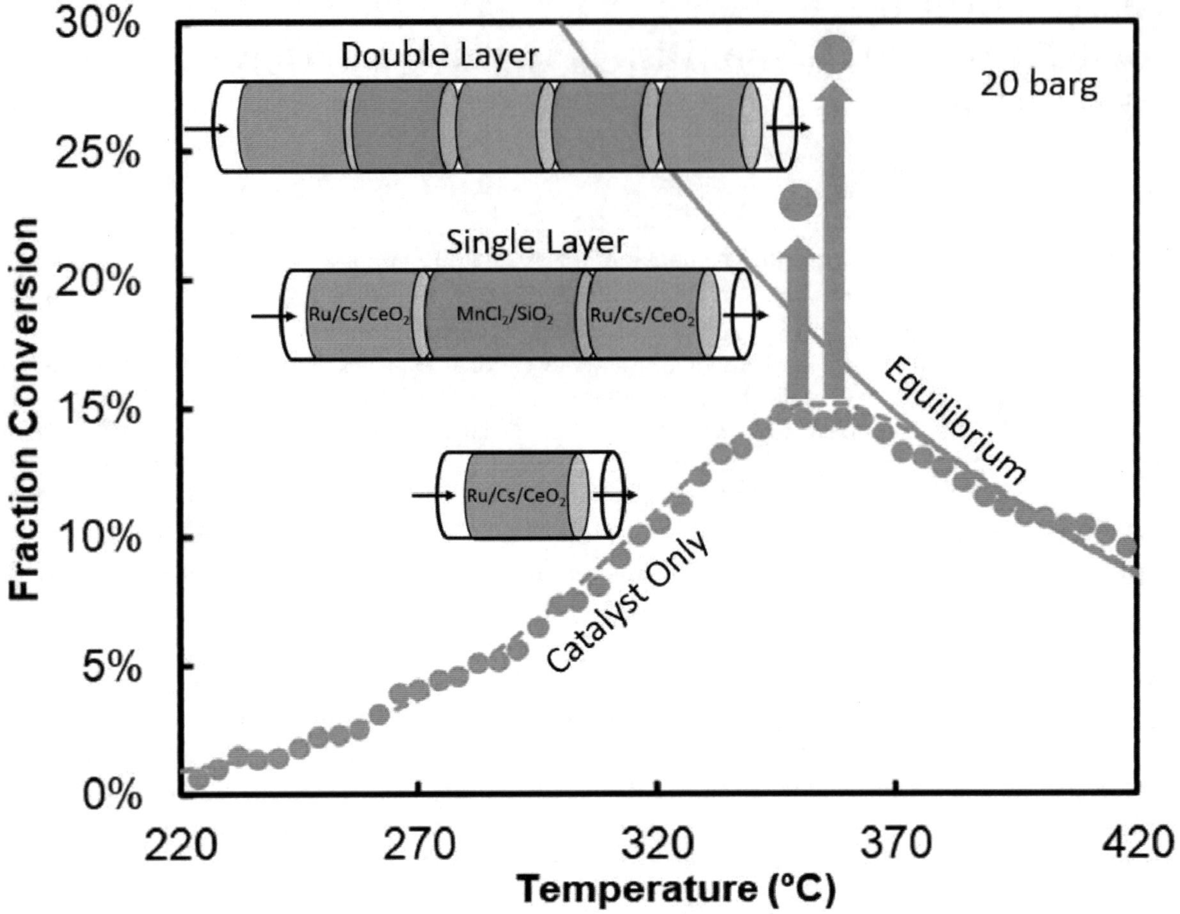

30%
25%
20%
15%
10%
5%
0%
Fraction Conversion
Double Layer
20 barg
Single Layer
Ru/Cs/CeO$_2$
MnCl$_2$/SiO$_2$
Ru/Cs/CeO$_2$
Ru/Cs/CeO$_2$
Equilibrium
Catalyst Only
220
270
320
370
420
Temperature (°C)

626807 Life Cycle GHG Emissions, Water and Fossil-Fuel Consumption for Polyethylene Furanoate (PEF) and Its Co-Products from Lignocellulosic Biomass *Via* Furanics Conversion

Abstract:

The growing environmental impact of fossil-fuel derived plastics has motivated researchers to look for alternative solutions. Biobased plastics are one of the promising solutions since it can potentially reduce greenhouse gas (GHG) emissions and fossil-fuel consumption as compared to their fossil-fuel counterparts. Polyethylene furanoate (PEF) is a bioplastic that has been studied as a potential replacement for fossil-based polyethylene terephthalate (PET) resin, one of the most commonly used resins for packaging and synthetic fiber applications. Recent studies show a technical and economic viability of producing PEF and other potential biofuel co-products (i.e., furfuryl ethyl ether (FEE), methyl levulinate (ML), and dimethyl ether (DME)) from a lignocellulosic feedstock via furanics conversion pathways. In this presentation, we investigate three different PEF production pathways and assess cradle-to-gate GHG emissions, fossil-fuel consumption, and water consumption using the Greenhouse Gases, Regulated Emissions and Energy Use in Technologies (GREET) life-cycle analysis (LCA) tool. The system boundary analyzed in this study includes all of the processes involved in the three PEF pathways from cradle to plant gate: wheat straw cultivation & farming, organosolv (or furanosolv) pretreatment, enzymatic hydrolysis (or alcoholysis), glucose isomerization to fructose, furanics production, furanics recovery, furanics oxidation & polycondensation into PEF, furfural recovery, furfuryl ethyl ether (FEE) production, and on-site combined heat and power (CHP) generation. For the co-product allocation methods, mass-allocation is used as a baseline method for both allocations between wheat straw and grain production and among the final products of PEF production (i.e., PEF, FEE, ML, and DME). Electricity net export from the CHP generation is assumed to displace the U.S. average mix grid electricity. Different sensitivity scenarios are also addressed in this study to see the impact of the upstream production of the material inputs and allocation methods on the LCA results. The results showed that all three PEF pathways are net carbon-negative ranging from -0.77 kgCO_2e/kg-PEF (-124% reduction from fossil-based PET) to -1.44 kgCO_2e/kg-PEF (-145% reduction). Fossil-fuel consumption for the three different PEF pathways was also significantly lower than its fossil-fuel counterpart: 13 – 28 MJ/kg-PEF vs. 74 MJ/kg-PET. The GHG emissions and fossil-fuel consumption metrics for the biobased PEF were also better than those for biobased PET (-0.61 kgCO_2e/kg-PET, 51 MJ/kg-PET). However, the water consumption for the biobased PEF production was higher for the two pathways than its fossil-based counterpart (i.e., 10 – 16 L/kg-PEF vs. 6.4L/kg-PET) while only the third pathway for PEF (i.e., 1.2L/kg-PEF) outperformed its fossil-fuel counterpart. These LCA results show that biobased PEF has the potential to reduce fossil energy use and GHG emissions compared with conventional fossil-derived PET, with the tradeoff of increased water consumption for some of the pathways.

628467 Techno-Economic Greenhouse Gas Life Cycle Assessment of the CO_2-EOR Operations on Farnsworth Unit

Abstract:

Objectives/Scope

The processes involved in the capture, transportation and injection of CO_2 for use as an enhanced oil recovery (EOR) technique and for storage have operational units/processors which may generate GHG emissions which are not captured or accounted for. This paper will present a full life cycle assessment (LCA) of the GHG emissions of the CO_2-EOR operations on Farnsworth Unit located in Ochiltree county, Texas, USA.

Methods, Procedures, Process

Two scenarios would be presented in this paper; the paper would first consider LCA of current operational conditions of the FWU (utilizing field data provided by the operators). Here CO_2-EOR processes with and without gas separation would be considered. secondly, a number of optimized operational conditions based on the LCA results from the first scenario will also be employed. Based on the operator's new operational strategy and number of years to continue injection, a forecast of the injection and production activity through a simulation work would be conducted. The second Scenario would be applied to this forecast and LCA of the GHG conducted. Finally, economic analysis for all scenarios with respect to technology and operational units would be assessed.

Results, Observations, Conclusions

Net Emission factors (in $kgCO_{2e}$/ bbl. of crude produced) and mass emissions ($kgCO_{2e}$) of GHG would be presented for the various operational units within the FWU boundary. Comparison with current FWU operational conditions and the optimized cases would be presented. This would be in terms of net emission factors, net mass emissions and net storage. Results of Economic analysis of all scenarios would also be emphasized to best describe the feasibility of all proposed optimized cases. The idea here will be to optimize conditions to minimize to the barest minimum net emissions and maximize net storage of CO_2. Thus, units including but not limited to energy sources, compressors and pumps, artificial lifts, land use/construction activities, produced fluid processing and storage facilities would be factored.

Novel/Additive Information

This paper in its entirety would provide an insight to operators of CO_2-EOR (more specifically to FWU) on key unit processes which contributes to GHG emissions, and possible areas of optimization to increase net CO_2 storage with minimal/reduced net GHG emissions.

630159 Life Cycle Analysis and Techno-Economic Assessment of Emerging Sustainable Polymer Processes

Abstract:

Of the 350 million tonnes of plastic waste generated worldwide in 2018, only 18% was recycled, while a majority 58% was landfilled or lost to environment.[1,2] The remaining 24% was incinerated, which emits greenhouse gases. In this context, the design of sustainable polymers that contribute less solid waste and less greenhouse gas emissions have arisen as a priority. Design of sustainable polymers can entail the use of renewable raw materials (like biomass), incorporating recyclability and/or degradability to benign compounds, and reducing energy and water consumption in the production of polymers, among other strategies. We evaluated three polymers designed to incorporate sustainability attributes including production from biomass and biodegradability with techno-economic and life cycle analyses. The goal was to analyze benefits/disadvantages of implementing these processes at different scales, compared to the similar baseline systems in use today.

The first process explores the synthesis of caprolactone monomers from biomass via the production on an intermediate p-cresol.[3] This was compared to the conventional method of producing petroleum derived caprolactone monomers. Caprolactone monomers are essential as they are the building blocks of polycaprolactones which are subsequently employed in the production of polyurethanes. A preliminary cradle-to-gate life cycle analysis revealed the comparative advantage of such a process in terms of greenhouse gas emissions, water use and energy use. The second process employs biomass-derived cellulose nanocrystals (CNCs)[4] as nano-fillers to reinforce polylactic acid (PLA), which is a biodegradable polymer. The biodegradability of these nanofillers is important to ensure the biodegradability of the polymer composite. The process of producing a piece of cutlery (fork) from this entirely biodegradable material was compared against two scenarios: (1) where the polymer is PLA with a non-biodegradable nanofiller (like graphene, titanium dioxide, calcium carbonate, silica etc.) and (2) where the polymer itself is non-biodegradable, like polypropylene, along with a non-biodegradable nanofiller. In the third process, the production of b-cyclodextrin polymers[5] (which can be sourced from biomass cellulose) for the adsorption of micropollutants (especially PFAS and PFOA) in primary treated wastewater was explored for their environmental and economic footprint. This was compared with the production of conventional adsorbents like activated carbon for similar functional performance of micropollutant removal.

Impact categories explored in all the life cycle analysis included greenhouse gas emissions, water use and energy intensity. Argonne National Laboratory's GREET model was employed to conduct the life cycle analyses. In the case of the cyclodextrin adsorbents, aquatic toxicity was also investigated as an impact category and the U.S. Environmental Protection Agency's AQUATOX model was used as a means to establish the toxicity of baseline PFAS/PFAO emissions. For the techno-economic analysis, a discounted cash flow approach method was employed to evaluate a minimum selling price of the concerned product.

For large-scale applications of these emerging processes, these environmental and economic analyses can serve as an integral tool to identify possible bottlenecks and subsequently guide the direction of future research efforts.

References:

1. Geyer, R., Jambeck, J. R. & Law, K. L. Production, use, and fate of all plastics ever made. *Sci. Adv.* **3**, e1700782 (2017).
2. *A circular economy for plastics – Insights from research and innovation to inform policy and funding decisions, European Commission.* https://op.europa.eu/en/publication-detail/-/publication/33251cf9-3b0b-11e9-8d04-01aa75ed71a1/language-en (2019).
3. Lundberg, D. J., Lundberg, D. J., Hillmyer, M. A. & Dauenhauer, P. J. Techno-economic Analysis of a Chemical Process To Manufacture Methyl-ε-caprolactone from Cresols. *ACS Sustain. Chem. Eng.* **6**, 15316–15324 (2018).
4. Calvino, C., Macke, N., Kato, R. & Rowan, S. J. Development, processing and applications of bio-sourced cellulose nanocrystal composites. *Prog. Polym. Sci.* **103**, 101221 (2020).
5. Ling, Y., Alzate-Sánchez, D. M., Klemes, M. J., Dichtel, W. R. & Helbling, D. E. Evaluating the effects of water matrix constituents on micropollutant removal by activated carbon and β-cyclodextrin polymer adsorbents. *Water Res.* **173**, 115551 (2020).

624490 Carbon Footprint and Land Use Reductions of Electromicrobial Production Systems Quantified By Process Modelling and Comparative Life Cycle Assessment

Abstract:

The reliance of most industrial bioprocesses on heterotrophic organisms such as *Escherichia coli* has generated concerns around the environmental impacts of feedstock production, including the impact of land use on ecosystems, increased pollution from fertilizer production and application, and competition with food production. To circumvent these challenges, researchers have proposed various electromicrobial production (EMP) systems, a broad strategy in which electrochemically generated mediator molecules provide energy for microbes to fix carbon dioxide and produce value-added products. Based on previous work in modelling bioreactors for use in EMP systems, we have developed process models and performed cradle-to-gate life cycle impact assessments (LCIA) of three proposed electro-microbial production systems: a Knallgas bacteria-based system, a formatotroph-based system, and an acetogen-based system. The generation of various products, including biomass, industrial enzymes, and lactic acid is studied. The life cycle impact assessment demonstrates that all three systems will have a lower global warming potential than a traditional glucose-fed bioproduction system if the electricity is provided by a renewable source such as solar or wind energy. Additionally, the LCIA shows that the EMP systems have less than 5% of the land occupation footprint compared to the traditional bioproduction system analyzed. When comparing the systems to each other, we found that the hydrogen-fed Knallgas bacteria-based system uses the least amount of energy and has the lowest carbon footprint compared to the other two EMP systems. We also identified the environmental hotspots of each proposed EMP system and discuss strategies to overcome these bottlenecks. This analysis demonstrates the potential of electro-microbial production systems as a sustainable bioproduction strategy and identifies conditions under which they become attractive alternatives.

623058 Integrated Biorefinery Design Under Multi-Scale Uncertainties

Abstract:

With the increasing awareness to reduce the carbon oxide emission in the chemical industry, using biomass as the feedstock for chemical production has gained significant interest in recent years. Promising processes have been designed to produce p-xylene from cellulose[1], furfural[2] and other value-added products[3] from hemicellulose, as well as phenols from lignin[4]. The concept of the biorefinery is developed to enable efficient utilization of all components in the biomass feedstock to increase the revenue and thus the process economic viability. The challenge of biorefinery design lies in rationally choosing different feedstocks, potential products and technology alternatives under different demand and supply constraints. Consequently, a superstructure optimization framework is often utilized to solve this problem. Yuan and Eden applied superstructure optimization to an integrated fast pyrolysis and gasification biorefinery to compare different conversion routes.[5] Superstructure model is also developed by Gong and You for an algal biorefinery to determine the optimal configuration and operating conditions.[6]

Many parameters used in the biorefinery design come with considerable uncertainties. For instance, experimental yields and separation efficiencies are often utilized for industrial-scale plant design. However, there are inevitably subject to intrinsic variations and inaccurate measurement. Moreover, the performance of scaled-up processes may differ from its bench-scale counterpart.[7] The feedstock supply and composition provide another source of uncertainties to be considered in process design. Meanwhile, the prices of chemicals are volatile and unpredictable, affecting the plant's profitability.[8] Incorporating uncertainties in optimization has gained growing attention over recent years.[9] Currently, the number of uncertain parameters considered in the process design is limited by computational complexity. Two commonly used optimization methods under uncertainty are robust optimization and stochastic programming. Stochastic programming uses probability distributions of uncertain parameters to build the scenario tree and calculate the expected values as objective functions. Proper sampling and scenario reduction techniques are crucial to its implementation.[10] When ensuring process feasibility is the top priority, or only ranges of parameters are known, the robust optimization problem should be adopted to provide a conservative process design that is feasible even in the worst case.[11]

In practice, uncertainties from various sources may have different impacts on the process outcome.[12] In the multistage optimizations under uncertainty, wait-and-see decisions are taken after the uncertainty's realization, which are captured by the affine decision rule in the adaptive robust optimization.[13] On the other hand, decisions may affect the endogenous uncertainty's distribution or the time it realizes.[14] Furthermore, uncertainties associated with process operation or product quality need to be treated with more care to ensure the process's feasibility in all scenarios. Other uncertainties, including market prices, are less restricted, for which optimizing expected profit is more appropriate.[12] Yue and You also formulated a stochastic robust optimization that distinguishes uncertainties associated with strategic and operational decisions.[7]

In this work, we consider a unified framework to include different levels of uncertainties in the biorefinery design. Various feedstocks, including lignocellulosic biomass, food waste and technical lignin, introduce the flexibility for supply but also composition uncertainty. Based on experimental data, processing block flowsheets are developed in Aspen Plus to synthesize a variety of chemicals from biomass.[15] The input, output, operating and capital cost information of each block is extracted and scaled to formulate the optimization problem's objective functions and constraints. Both the profitability and environmental benefits are main driven forces for biorefinery development. Thus, a multi-objective optimization problem is formulated to simultaneously minimize the greenhouse gas emissions based on life cycle assessment (LCA) and maximize profit. We first identify the primary sources of uncertainties and classify them based on their main effects of various decision making including process design, operation and LCA. The LCA calculation is independent of the previous stages, meaning that its uncertainty analysis could be performed after finalizing the plant configuration. Its uncertainty provides a reference for the accuracy of our objective values obtained from the optimization problem. Next, uncertainties on the prices and demands are considered in the strategic decisions that maximize the expected profit, while those associated with feedstock and operating conditions are more restrictive so that the feasibility is satisfied throughout the entire uncertainty range. This biorefinery design problem is solved by a combination of robust optimization and stochastic programming techniques.

Reference

1. Athaley, A.; Annam, P.; Saha, B.; Ierapetritou, M., Techno-economic and life cycle analysis of different types of hydrolysis process for the production of p-Xylene. *Comput. Chem. Eng.* **2019,** *121*, 685-695.

2. Alonso, D. M.; Hakim, S. H.; Zhou, S.; Won, W.; Hosseinaei, O.; Tao, J.; Garcia-Negron, V.; Motagamwala, A. H.; Mellmer, M. A.; Huang, K.; Houtman, C. J.; Labbé, N.; Harper, D. P.; Maravelias, C. T.; Runge, T.; Dumesic, J. A., Increasing the revenue from lignocellulosic biomass: Maximizing feedstock utilization. *Science Advances* **2017,** *3* (5), e1603301.

3. Athaley, A.; Saha, B.; Ierapetritou, M., Biomass-based chemical production using techno-economic and life cycle analysis. *AIChE J.* **2019,** *65* (9), e16660.

4. Liao, Y.; Koelewijn, S.-F.; Van den Bossche, G.; Van Aelst, J.; Van den Bosch, S.; Renders, T.; Navare, K.; Nicolaï, T.; Van Aelst, K.; Maesen, M.; Matsushima, H.; Thevelein, J.; Van Acker, K.; Lagrain, B.; Verboekend, D.; Sels, B. F., A sustainable wood biorefinery for low–carbon footprint chemicals production. *Science* **2020**, eaau1567.

5. Yuan, Z.; Eden, M. R., Superstructure optimization of integrated fast pyrolysis-gasification for production of liquid fuels and propylene. *AIChE J.* **2016,** *62* (9), 3155-3176.

6. Gong, J.; You, F., Optimal Design and Synthesis of Algal Biorefinery Processes for Biological Carbon Sequestration and Utilization with Zero Direct Greenhouse Gas Emissions: MINLP Model and Global Optimization Algorithm. *Ind. Eng. Chem. Res.* **2014,** *53* (4), 1563-1579.

7. Yue, D.; You, F., Optimal supply chain design and operations under multi-scale uncertainties: Nested stochastic robust optimization modeling framework and solution algorithm. *AIChE J.* **2016,** *62* (9), 3041-3055.

8. Noh, N. M.; Bahar, A.; Zainuddin, Z. M., Scenario Based Two-Stage Stochastic Programming Approach for the Midterm Production Planning of Oil Refinery. *MATEMATIKA: Malaysian Journal of Industrial and Applied Mathematics* **2018,** *34* (3), 45-55.

9. Sahinidis, N. V., Optimization under uncertainty: state-of-the-art and opportunities. *Comput. Chem. Eng.* **2004,** *28* (6), 971-983.

10. Martín, M.; Martínez, A., Addressing Uncertainty in Formulated Products and Process Design. *Ind. Eng. Chem. Res.* **2015,** *54* (22), 5990-6001.

11. Grossmann, I. E.; Apap, R. M.; Calfa, B. A.; Garcia-Herreros, P.; Zhang, Q., Mathematical Programming Techniques for Optimization under Uncertainty and Their Application in Process Systems Engineering. *Theor. Found. Chem. Eng.* **2017,** *51* (6), 893-909.

12. Chen, Y.; Lin, M.; Jiang, H.; Yuan, Z.; Chen, B., Optimal design and operation of refinery hydrogen systems under multi-scale uncertainties. *Comput. Chem. Eng.* **2020,** *138*, 106822.

13. Zhao, L.; Ning, C.; You, F., Operational optimization of industrial steam systems under uncertainty using data-Driven adaptive robust optimization. *AIChE J.* **2018,** *65* (7), e16500.

14. Zhang, Q.; Feng, W., A unified framework for adjustable robust optimization with endogenous uncertainty. *AIChE J.* **2020,** *66* (12), e17047.

15. Ibarra-Gonzalez, P.; Rong, B.-G., Integrated Methodology for the Optimal Synthesis of Lignocellulosic Biomass-to-Liquid Fuel Production Processes: 1. Simulation-Based Superstructure Synthesis and Development. *Ind. Eng. Chem. Res.* **2020**.

622660 Towards a New Understanding of the Retro-Aldol Reaction for Oxidative Conversion of Lignin to Aromatic Aldehydes and Acids

Abstract:

Lignin is the most abundant natural biopolymer after cellulose and accounts for 30% of non-fossil carbon on earth. The retro-aldol reaction is one of the key steps involved in the oxidative conversion of lignin to aromatic aldehydes and acids. In principle, the retro-aldol reaction can proceed in the absence of oxygen. In this work, a new approach based on the influence of oxygen on the oxidation of lignin was investigated. The effect of reaction chemistry, time, temperature, and lignin feedstock play a key role on the yield of aromatic aldehydes and acids. At 140°C, oxidation of softwood Lignoboost kraft lignin for 40 minutes results in combined maximum yield of 5.17% w/w of vanillin and vanillic acid. In comparison, using the new approach in which oxygen was charged for only 20 minutes during the 40 minute reaction improved this yield considerably to 6.95%. Further, yield improvement was obtained when applying this approach to different lignin feedstocks. Oxidation also increased the carboxyl content of lignin from 0.49 mmol/g to 1.41 mmol/g which represents consequential improvement. The current study provides further evidence showing that the oxidation reaction is a crucial pathway for lignin valorization.

Effect of temperature on the yield of vanillin and vanillic acid

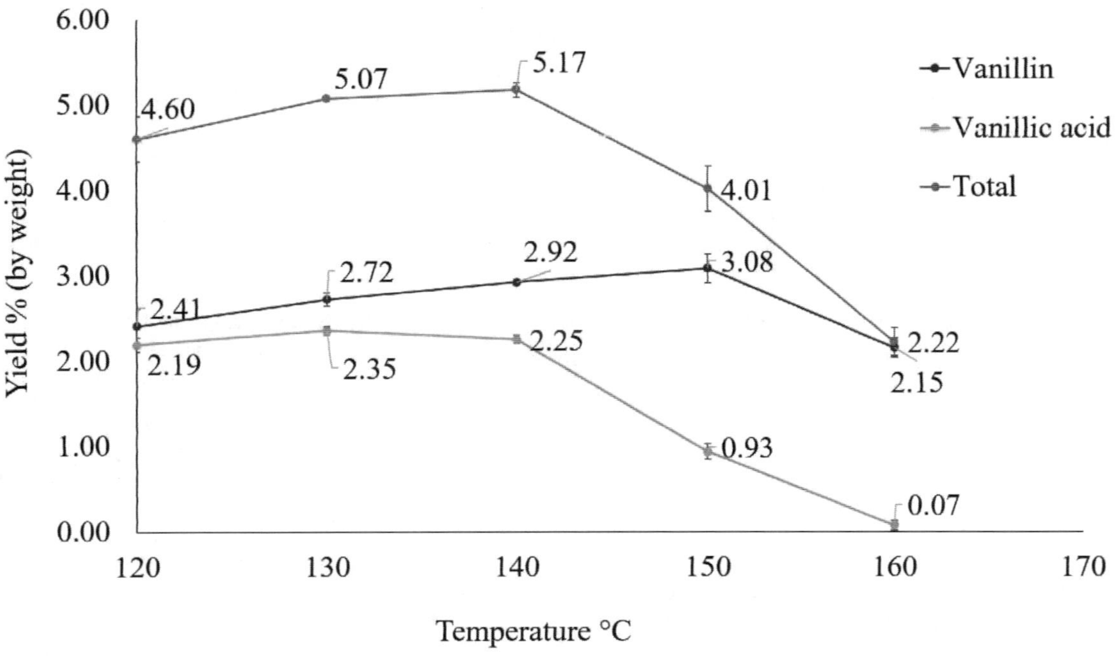

Effect of temperature on carboxyl content of LBoostS

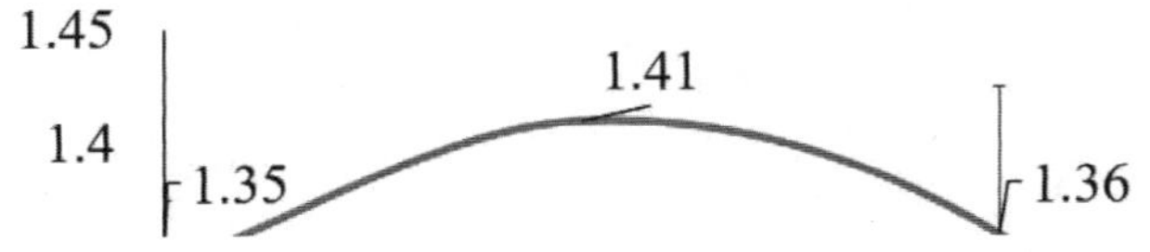

***** Formatting Issue - Best Available Paper/Graphic *****

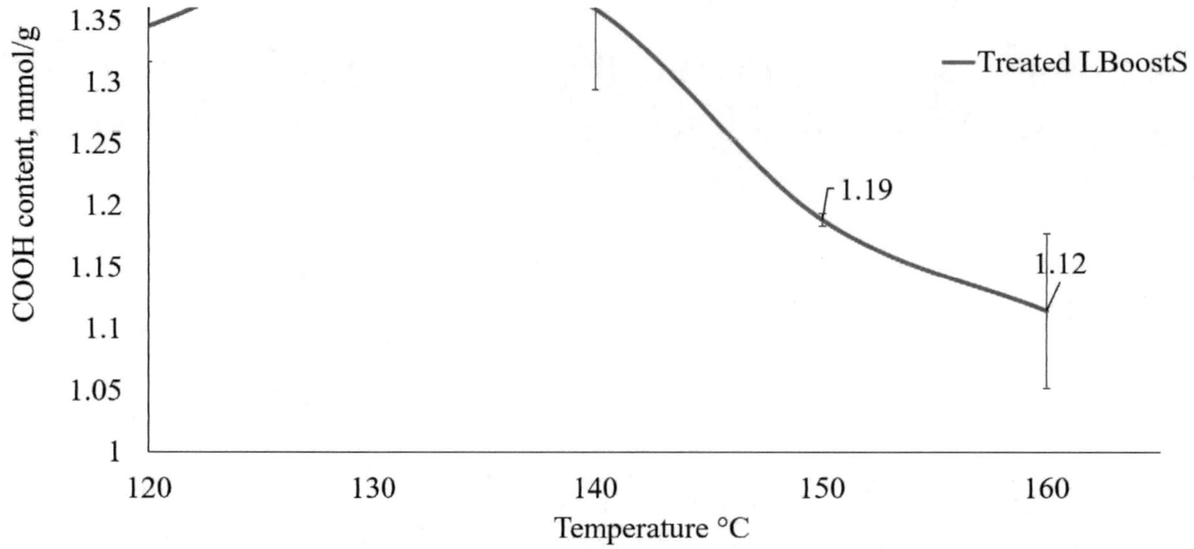

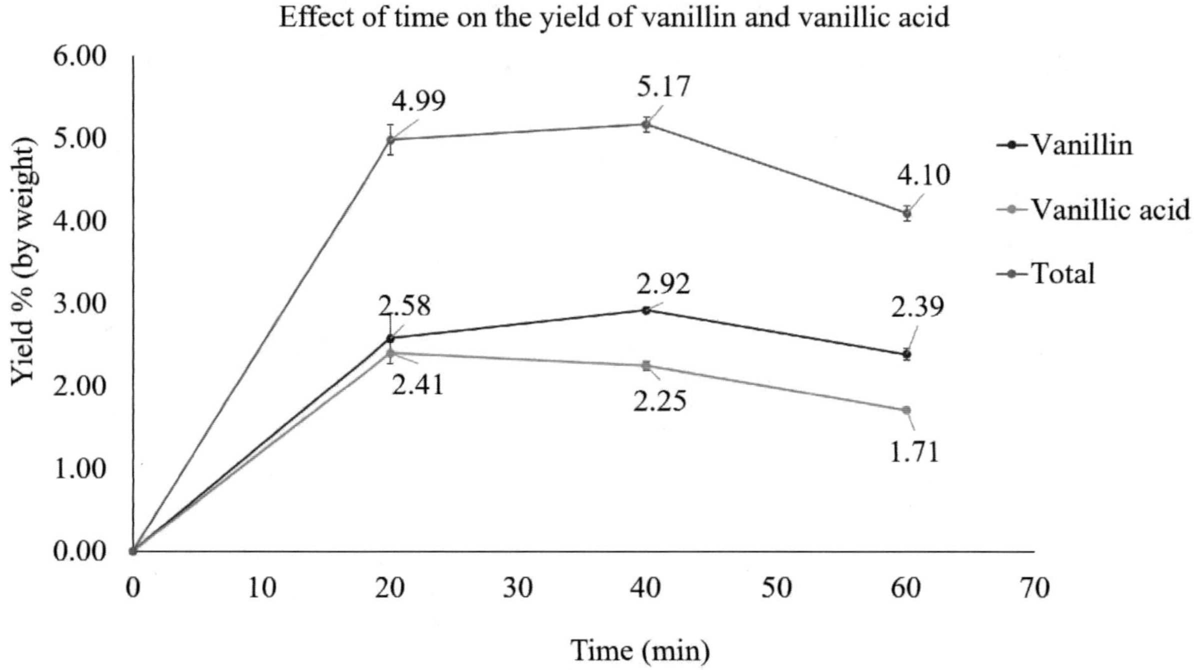

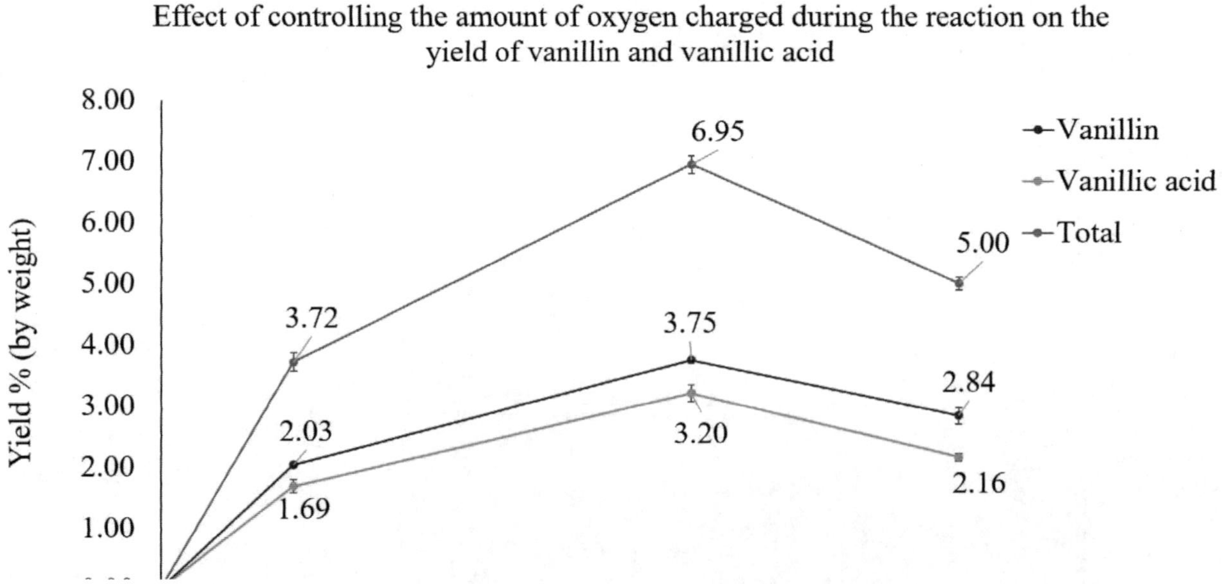

*** Formatting Issue - Best Available Paper/Graphic ***

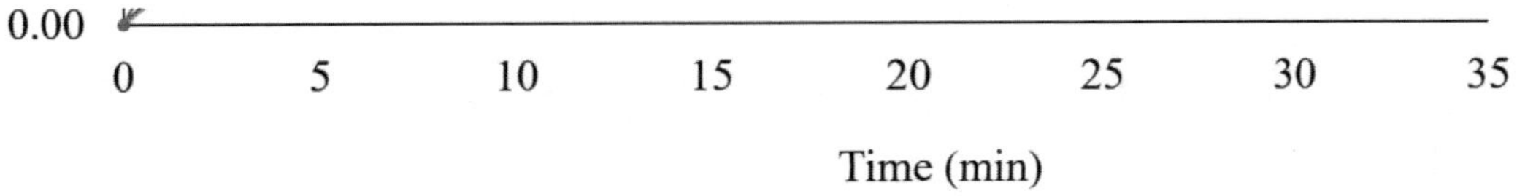

Effect of controlling the amount of oxygen charged during the reaction on the yield of vanillin and vanillic acid from different lignin feedstocks

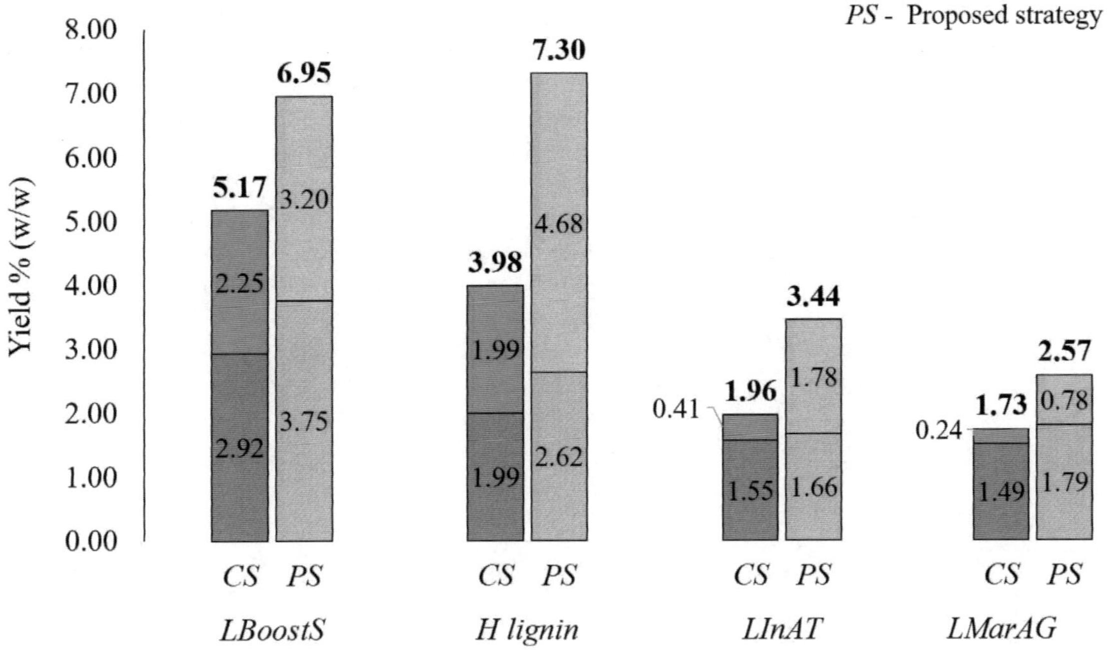

628460 Lignin Depolymerization in a Molten Salt Hydrate (lithium bromide trihydrate)

Abstract:

A molten salt hydrates (MSH) is a concentrated aqueous solution of an inorganic salt, in which the molar ratio of water to salt is close to the coordination number of the salt cation. Because of their unique properties, in particular the ability to dissolve cellulose, some MSHs are promising solvents and reaction media for biomass processing. For example, LiBr molten salt hydrate (~61 wt% LiBr aqueous solution) has been successfully used in hydrolysis of cellulose and lignocellulosic biomass, lignin quantitation, isomerization of glucose, furan-based chemicals from biomass, synthesis of oligosaccharides, and fabrication of nanocrystals, films, and aerogels from cellulose and biomass in our previous studies. It was also observed that lignin was extensively depolymerized when biomass was treated in the LiBr system. This presentation is to report our recent investigations into the depolymerization mechanisms of lignin in the LiBr system. The studies revealed that all the ether bonds of lignin in methoxyl, β-*O*-4, β-5, and β-β structures except for the 4-*O*-5 bond in diphenyl structure could be cleaved in the LiBr system. The ether bonds were cleaved via the S_N2 substitution pathway except for β-*O*-4 bond. The β-*O*-4 bond was primarily cleaved via the benzyl cation and enol ether intermediates, leading to Hibbert's ketones. Besides, some β-*O*-4 structures were transformed into benzodioxane structures, which were stable in the LiBr system.

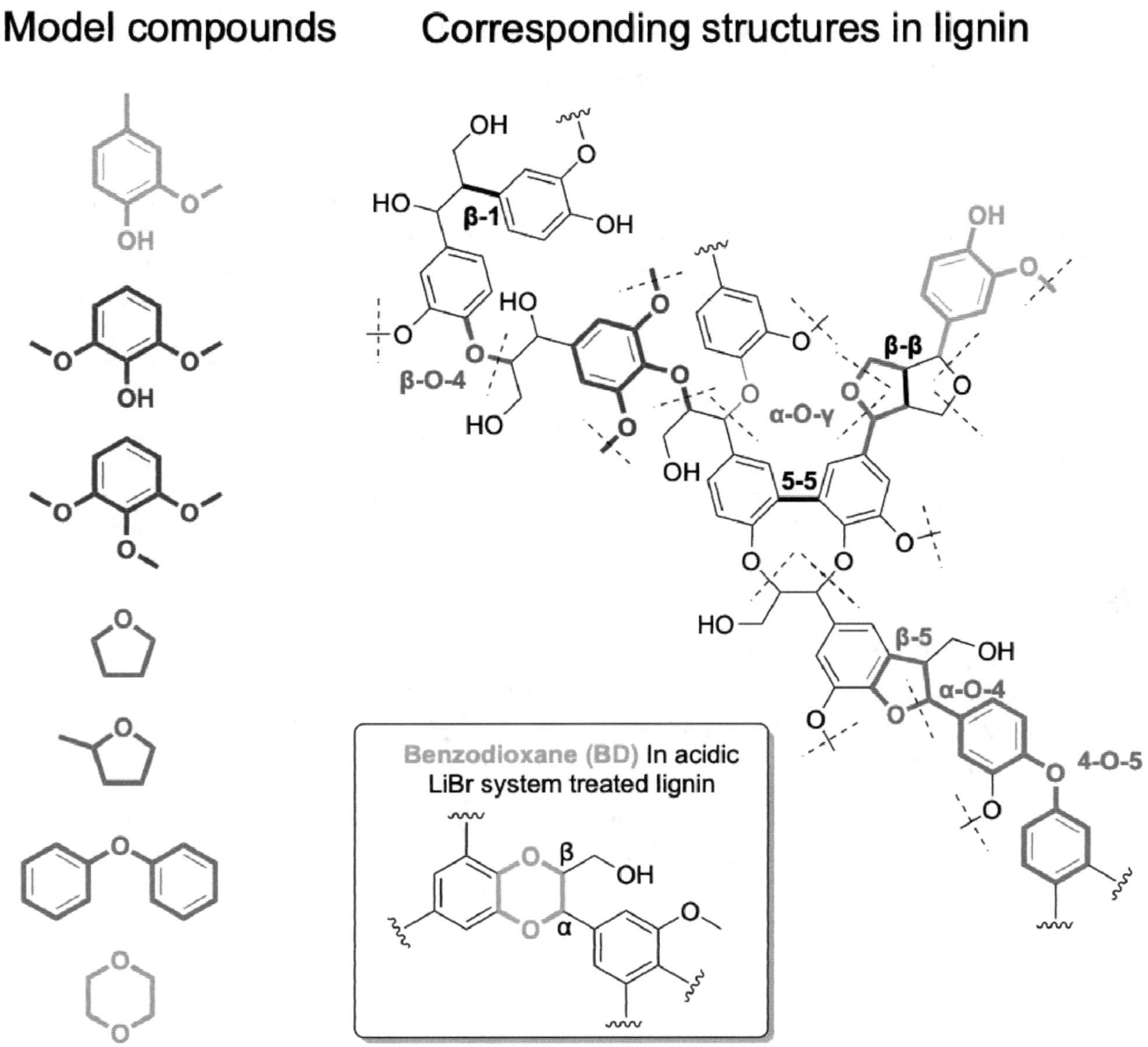

Model compounds
Corresponding structures in lignin
OH
β-1
β-O-4
HO
α-O-γ
β-β
5-5
β-5
α-O-4
4-O-5
Benzodioxane (BD) In acidic
LiBr system treated lignin
β
α

631322 Direct Catalytic Conversion of Lignocellulosic Biomass into Liquid Paraffins and Aromatics

Abstract:

Today, gasoline is the major economic driver for petroleum refineries and its production accounts for nearly 20% of the total volume of products derived from a barrel of crude oil. However, the market share occupied by gasoline consumption is expected to decline 1% annually through 2050 and the petrochemical industry will become more dependent on the production of non-combustible petrochemicals, jet fuel and marine diesel due to increased car fuel efficiency, emerging electric and hydrogen fuel cell vehicle utilization. Thus, according to the BP Energy Outlook 2020, biofuels are expected to play a major role for the aviation and marine industries, but they also can play a role in accelerating the displacement of petroleum-based gasoline and non-combustible fuels, which could be substituted by the exact bio-based equivalents if there are biomass conversion technologies that enable such potential. Herein, a zeolite-supported metal catalyst in a presence of a hydrocarbon solvent was used to directly convert lignocellulosic biomass into a range of gaseous and liquid hydrocarbons at 280 Celsius via hydrodeoxygenation. Biomass conversion and carbon yields higher than 90% were obtained with a solvent to biomass ratio of 3:1 and 8h of residence time. C5-C7 alkanes (61% of total carbon) and C3-C4 alkanes (14% of total carbon) along with lower level of aromatics (10% of total carbon) were the main products generated. The produced alkanes can be fractionated and used directly in many applications today or subjected to further oxidative coupling followed by oligomerization to produce carbon range jet fuel and diesel.

624532 Energia - an Integrated Framework and Software Prototype for Multiscale Energy Systems Transition Modeling, Optimization and Scenario Analysis

Abstract:

De-carbonization of energy infrastructure, while meeting growing global energy demand, requires a shift in primary energy production and supply towards renewable technologies. Renewables, such as wind, solar and biofuels, are typically available intermittently, and are subject to seasonal variability and uneven geographical distribution. This necessitates the explicit consideration of spatiotemporal characteristics and variabilities in the analysis of proposed low-carbon energy scenarios. Future energy systems may further involve integrated value chains of multiple energy sectors with multiple technology options for sustainable energy generation, production of chemicals and synthetic fuels, and transportation [1]. A holistic, multiscale energy systems engineering approach [2] can provide a systematic framework to address these challenges toward assessing promising energy transition pathways, linking decisions at the process synthesis, scheduling, and supply chain level, and enabling trade-off analysis.

In this work, we present the developments of ENERGIA, a novel multi-scale energy systems transition modeling, optimization and scenario analysis framework and software prototype. ENERGIA integrates (i) energy supply chain and transportation considerations, (ii) detailed energy production aspects, and (iii) scheduling decisions for operation and inventory management of energy and resources storage. Its key features include (i) detailed data and models for the description of process alternatives and units and the corresponding supply chains, (ii) a detailed time-varying scheduling model [3], (iii) a library of surrogate modelling techniques, for both the nonlinear process models, as well as scheduling decisions, and (iv) an effective multi-period, multi-location mixed-integer optimization solution strategy, coupled with environmental, risk and uncertainty analysis. ENERGIA's python-based environment allows users to visualize resource availabilities and demands at various temporal and geographic scales and resolutions, and compare competing objectives and renewable-based energy strategies. A hydrogen-economy energy transition problem is presented to highlight the potential and key capabilities of the proposed framework.

[1] Demirhan, C. D., Tso, W. W., Powell, J. B. & Pistikopoulos, E. N., A multi-scale energy systems engineering approach towards integrated multi-product network optimization. *Applied Energy* 281, 116020 (2021).

[2] Kakodkar, R., He, G., Demirhan, C. D., Arbabzadeh, M., Baratsas, S.G., Avraamidou, S., Mallapragada, D., Miller, I., Allen, R. C., Gençer, E & Pistikopoulos, E.N., A review of analytical and optimization methodologies for transitions in multi-scale energy systems. [Manuscript submitted, 2021]

[3] Allen, R. C., Baratsas, S.G., Kakodkar, R., Avraamidou, S., Powell, J. B., Heuberger, C.F., Demirhan, C. D. & Pistikopoulos, E.N., An optimization framework for solving integrated planning and scheduling problems for dense energy carriers. *11th IFAC International Symposium on Advanced Control of Chemical Processes (ADCHEM)* (2021). [In press]

630180 Distributed and Sustainable Hydrogen Economy Via Renewable-Integrated and Intensified Process Design and Optimization

Abstract:

Hydrogen is an energy carrier and versatile feedstock with a wide range of applications in refining and petrochemicals, automotive and aerospace, power generation, and transportation sectors. There are several ongoing efforts to reduce the cost of manufacturing hydrogen and incorporate it in traditional chemical industries (e.g., steel, refining, ammonia, etc.) while ensuring process sustainability and reducing environmental footprint. Currently, steam methane reforming (SMR) is the predominant route for hydrogen production [1]. However, the major challenges include high energy intensity, low methane conversions and large environmental footprint. To counter these challenges, our objective is to develop a sustainable and efficient chemical process pathway for producing blue hydrogen from natural gas feedstock while simultaneously compressing emitted carbon for sequestration purposes. To this end, we develop a sustainable process that leverages a combination of dynamic process intensification (DPI) and renewable sources (solar and wind) to power the energy demands of the hydrogen production process and carbon capture. Specifically, the intensified hydrogen production process is based on sorption enhanced steam methane reforming (SE-SMR), which utilizes an admixture of Ni-based SMR catalyst and hydrotalcite adsorbent for in-situ carbon capture [2]. In comparison to conventional SMR process, the intensified SE-SMR process has several advantages including higher reaction conversion and product purity, lower energy demands (due to lower operating temperature) and easier product separation. However, the optimal design and operation of SE-SMR systems is especially challenging as they are typically represented by complex and detailed nonlinear algebraic and partial differential equation (NAPDE)-based first-principles models. In addition, the temporal variation in renewable availability and the SE-SMR process dynamics significantly increase the complexity of the overall operation.

To handle these complexities, we develop a computational framework for optimal design and scheduling of the intensified hydrogen production module while handling the process variabilities and intermittency in renewable availability. The framework is based on a large-scale mixed integer linear programming (MILP) model that minimizes hydrogen production cost while incorporating (i) SE-SMR process design and operation constraints, (ii) power flow from grid and renewables to process equipment, and (iii) process economics. For performing high-fidelity simulations of SE-SMR processes, a generalized reaction-adsorption modeling and simulation (GRAMS) platform is utilized [3]. To maintain the MILP nature of the problem, the complex nonlinearities in SE-SMR process dynamics and cost correlations have been adequately represented by artificial neural network (ANN)-based regression models. A nationwide analysis has been performed on more than 1000 locations in the United States to evaluate the potential of the developed intensified hydrogen production technology for localized manufacturing for hydrogen refueling stations. The results indicate that with future projected cost estimates of solar and wind energy, blue hydrogen can be manufactured to meet the key cost benchmark of $2.5 per kg for small-scale applications using the proposed hydrogen production pathway.

References

[1] M. Wieliczko and N. Stetson, "Hydrogen technologies for energy storage: A perspective," *MRS Energy & Sustainability*, vol. 7, 2020.

[2] A. Arora, I. Bajaj, S. S. Iyer, and M. M. F. Hasan, "Optimal Synthesis of Periodic Sorption Enhanced Reaction Processes with Application to Hydrogen Production," *Comput. Chem. Eng.*, vol. 115, pp. 89–111, 2018.

[3] A. Arora, S. S. Iyer, and M. M. F. Hasan, "GRAMS: A General Framework Describing Adsorption, Reaction and Sorption-Enhanced Reaction Processes," *Chem. Eng. Sci.*, vol. 192, pp. 335–358, 2018.

625511 Unlocking the Potential of Open-Ocean Wind Energy Via Green Offshore Ammonia Production

Abstract:

The world is transitioning toward renewable sources of energy, such as solar and wind, at an unprecedented pace. In particular, the offshore wind market grew nearly 30% per year between 2010 and 2018 and is expected to increase 15-fold over the next two decades [1]. There is vast untapped potential in offshore wind, especially at locations far from shore where wind velocities are significantly higher than close to or on shore. However, transmitting electricity generated on the open ocean to onshore demand points is a major challenge since using electrical cables for such long distances can be prohibitively expensive. In this work, we propose to address this challenge with green offshore ammonia, where the idea is to directly produce ammonia offshore using electricity generated from the wind; the ammonia then serves as an energy carrier that can be readily shipped to shore. This is feasible since, besides electricity, the production of green ammonia only requires air and water [2], which are available in abundance on the ocean.

While research has been conducted on ammonia production with offshore resources, most of them only consider transmitting offshore generated electricity to shore for ammonia production on land [3,4]. In this work, we assess the techno-economic viability of wind-powered offshore ammonia production by applying a superstructure-based optimization approach that determines the optimal design and operational schedule of the process. This approach allows a systematic selection among various means of chemical production, storage, and transportation as well as different energy transmission, conversion, and storage technologies. We optimize the design with regard to the levelized cost of ammonia (LCOA) for given demands at locations with different distances to shore, water depths, and wind profiles. Our results show that the LCOA depends heavily on the wind speed and means of chemical transportation or electricity transmission. When the distance to shore is relatively small, it is economically more favorable to transmit the generated electricity to shore and produce ammonia onshore. However, as the distance increases, shipping offshore produced ammonia becomes the better option. Also, as the distance increases, the economics favor transporting ammonia using ships rather than pipelines. Our analysis demonstrates the potential of green offshore ammonia and provides insights that can help identify suitable locations for realizing this new concept.

References:

[1] International Energy Agency (2019). Offshore Wind Outlook 2019.

[2] Palys, M. J., Wang, H., Zhang, Q., & Daoutidis, P. (2021). Renewable ammonia for sustainable energy and agriculture: vision and systems engineering opportunities. Current Opinion in Chemical Engineering, 31, 100667.

[3] Morgan, E. R., Manwell, J. F., & McGowan, J. G. (2017). Sustainable ammonia production from US offshore wind farms: a techno-economic review. ACS Sustainable Chemistry & Engineering, 5(11), 9554-9567.

[4] Serna, Á., & Tadeo, F. (2014). Offshore hydrogen production from wave energy. International journal of hydrogen energy, 39(3), 1549-1557.

632984 Evaluation of Ultrasonic Pretreatment on Catalytic Hydrothermal Liquefaction of Woody Biomass

Abstract:

In this study, ultrasonication pretreatment of a woody substrate (e.g., pinewood) was evaluated to understand its effectiveness on different products generated during catalytic hydrothermal liquefaction processing. It was anticipated that the ultrasonication would break open the cell walls by cleaving the chain linkages among the macromolecules and releasing polysaccharides. Ultrasonication of previously pulverized and milled *Pinus* ponderosa in water was performed using a 20 kHz probe and the effect of different ultrasonic parameters such as power input, time of sonication and slurry composition was evaluated. Ultrasonically pretreated woody biomass was treated with hydrothermal liquefaction in the presence of a Ca-salt catalyst in a bench scale 300 mL SS 316 Parr batch reactor at 250°C with some initial N_2 pressure for 1-2 hours. Hydrothermally processed slurry was filtered, and heavy bio-oil (HBO) was extracted with acetone. The products such as HBO and char were analyzed by elemental analyzer for CHNO and calorific value was calculated. The aqueous biocrude was also analyzed by GCMS to identify oxygenated hydrocarbons. The results obtained with catalytic hydrothermal processing of ultrasonically pretreated pinewood will be presented.

624219 Development of a Novel Nickel-Ceramic Filter for Hot Gas Removal of Tars and Particulates from Biomass Syngas

Abstract:

Gasification, the high temperature partial oxidation of carbonaceous fuels to produce syngas, is a promising route for converting biomass to energy. Syngas has many applicable uses including generation of electricity, production of specialty chemicals and fuels, and further separation for hydrogen production. Downstream syngas applications require contaminants to be below certain levels in order for the process to be effective. While particulates, responsible for equipment damage, can be separated through high temperature filtration, tar formation is the most cost-inducing problem whereby their condensation clogs downstream equipment presenting a major hurdle limiting biomass gasification commercialization. Catalytic tar reforming is a cost-effective strategy for tar abatement whereby conversion of tars into syngas improves the overall efficiency of the gasification process. By using a ceramic filter as a catalyst support, process intensification is achieved through reforming tars and filtering particulates in one unit. Nickel was chosen as a catalyst for its low cost and effectiveness towards tar reforming. The surface area of the ceramic filter (113 m^2/g) makes it an ideal support for catalyst material and its high porosity (85%) leads to a low pressure drop with minimal internal mass transfer limitations. Steam reforming evaluation tests using naphthalene as a tar simulant were performed with nickel loadings of 2-40 wt% at 750°C, steam/carbon = 5, 0.05 s residence time, and inlet naphthalene concentration of 9 g/m^3 to mimic typical fluidized bed gasifier outlet conditions. Peak naphthalene conversions of 80% were observed with a nickel loading of 15 wt% over 2 hours with no increase in pressure drop. PIXE analysis showed uneven dispersion of nickel on the filter using incipient wetness impregnation leading to investigations with vacuum urea impregnation method. This improved the surface area and achieved 81% naphthalene conversion with an activation energy of 123.3 kJ/mol, comparable with studied tar removal catalysts.

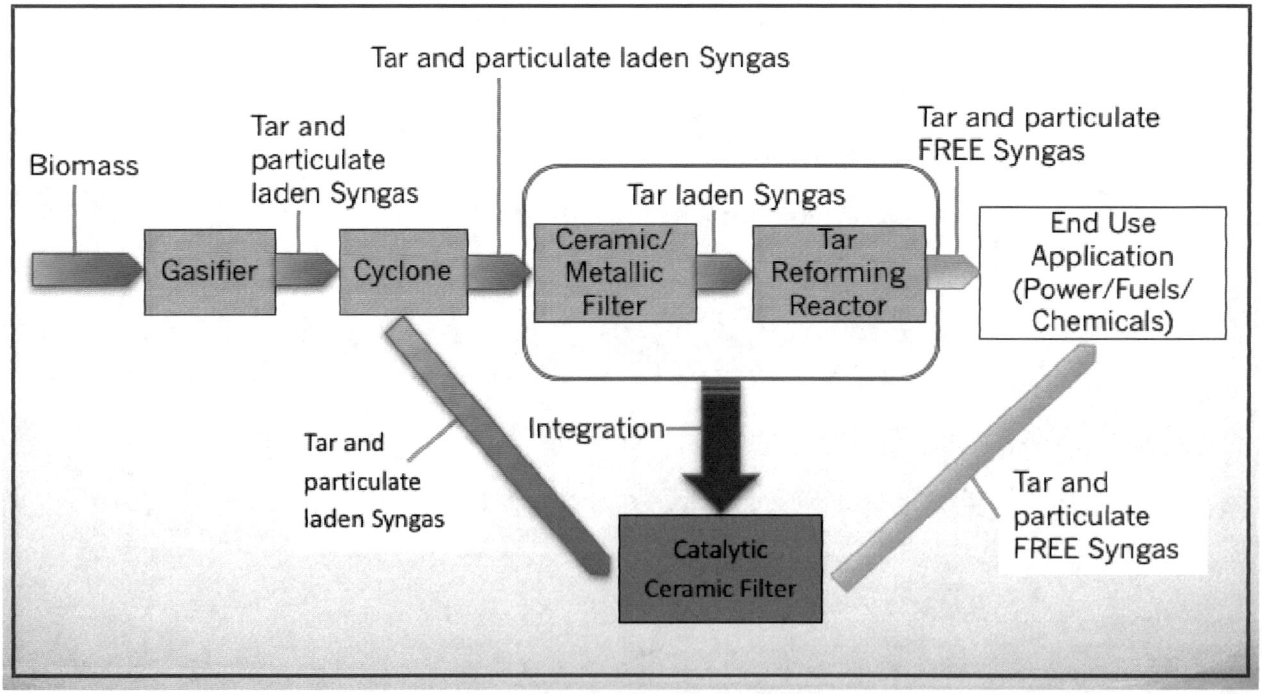

623090 One-Pot Conversion of Biomass in Aprotic Solvent for Chemicals and Quality Lignin Based on Plasma Electrolysis

Abstract:

Plasma electrolysis can occur when high-voltage electricity is applied to a solution. Due to the combined effect of joule heating and active species generated, plasma electrolysis of biomass demonstrates unique advantages in comparison to conventional thermal-based conversion. In this study, plasma electrolysis was employed as a novel method to convert red oak in γ-Valerolactone (GVL) and diluted sulfuric acid. It was observed that during the plasma electrolysis, the solution temperature could increase to 165 °C without external heating. Biomass was completely liquified by plasma electrolysis to produce liquid products containing furfural, levoglucosenone (LGO), levulinic acid (LA) and 1,4:3,6-dianhydro-α-D-glucopyranose (DGP). The monomer yields were affected by acid concentration, biomass mass loading, water co-solvent, electricity power parameters, as well as reaction time. With 4% biomass mass loading, LGO yield up to 40.6 mol% per cellulose fraction and furfural yield of 98.1 mol% per hemicellulose fraction were obtained in a GVL solution with 10.5 mM acid. With a lower biomass mass loading of 1.3%, the maximum LGO yield of 45.2 mol% was obtained using only 3.5 mM acid. A comparison study showed that thermally converting red oak at a 165 °C GVL with 10.5 Mm acid can only produce up to 13.1 mol% LGO and 84.7 mol% furfural using much longer reaction times. During plasma electrolysis, lignin dissolution in the solution was followed by dehydration of hemicellulose to furfural, and finally dehydration of cellulose to LGO, LA and DGP. In this study, the solubilized lignin was precipitated and pyrolyzed. As a result, up to 18.71% phenolic monomers were obtained from the plasma-derived lignin, compared to 6.32% from organosolv lignin, and 8.61% from milled lignin. The 2D-NMR analysis showed that α-carbons in β-O-4 and β-5 linkages were absent in the plasm-derived lignin.

628290 Modeling Green Diesel Hydroprocessing Via Molecule-Based Reactor

Abstract:

With the increasing demand to regulate carbon emission, the sustainable use of alternative feedstocks is a critical issue in next decades. Upgrading renewable oils via hydroprocessing is one promising technology. Triglycerides and their derivative fatty acids are the major molecular components of renewable oils. Green diesel is upgraded from renewable feedstocks by hydrodeoxygenation (HDO) and hydroisomerization (HI). Triglycerides and fatty acids are converted to a mixture of paraffinic molecules while limiting the production of CO2.

We developed a detailed HDO kinetic model of green diesel at the molecular level using the Aspen Molecule-Based (MB) reactor[1]. Starting from a list of triglycerides molecules, the detailed reaction paths of HDO were created and grouped to around 10 reaction families. As a result, more than 200 distinct reactions and around 100 molecules are included in that model. We applied a linear free energy relationship (LFER) to constrain the parameters of the model. Using one component of MB reactor: MB model builder, the HDO reaction network and kinetics were automatically converted to a dll that was loaded into the MB reactor unit op in Aspen HYSYS. Using the calibration function in MB reactor, we tuned the kinetic parameters of the HDO model based on literature data. The parity plots showed good agreement between the predicted results and experimental data.

We used Aspen MB hydrotreater to model the hydroisomerization of the effluent from MB HDO. By leveraging selected separation unit ops in HYSYS such as flash, column, etc., a waste cooking oil hydroprocessing case was created in Aspen HYSYS. The representative results of the case will be discussed.

Therefore, MB reactor in Aspen HYSYS gives users the opportunity to simulate and optimize the sustainable use of renewable feedstocks.

[1] Hou and Campbell "Molecule-Based EQUATION ORIENTED Reactor Simulation Infrastructure And Its Model Reduction",Docket No.: 1086.2062-000

631565 Production of Short-Chain Oligomers from Cellulose Via Selective Hydrolysis in Molten Salt Hydrates and Separation

Abstract:

Cellulose is the most abundant homopolysaccharide in nature which consists of D-glucose units connected through β(1→4) glycosidic bonds. Selective production of glucose from hydrolysis of cellulose is the key step for efficient conversion of lignocellulosic biomass. Cellulose can be dissolved and hydrolyzed into glucose with a high selectivity in specific solvents, including ionic liquids, deep eutectic solvents and molten salts hydrates (MSHs). However, the separation of the formed glucose is challenging due to its high solubility in these solvents.

To address this issue, a stepwise method is introduced, where cellulose is hydrolyzed into short-chain glucan oligomers in the MSH of LiBr.[1] It was found that cellulose can be efficiently hydrolyzed into short-chain oligomers with a degree of polymerization of 4-11 under mild conditions in the MSH of LiBr. We demonstrated that compared to glucose, the formed glucan oligomers can be efficiently separated from the MSH hydrolysate by an anti-solvent precipitation method.[2] Under optimized conditions, 90.3% of glucan oligomers can be produced from crystalline cellulose and separated from the MSH with the addition of methanol, and the precipitated glucan oligomer can be hydrolyzed into glucose with a yield of 99.7% using dilute sulfuric acid. The precipitation efficiency is influenced by the glucan oligomer chain-length, glycoside bond type and concentration. The obtained oligomer can be selectively converted into monosaccharide or might be used in other emerging applications such as healthcare and agriculture.

1. Liu, Q.; Ma, Q.; Sabnis, S.; Zheng, W.; Vlachos, D. G.; Fan, W.; Li, W.; Ma, L., Production of high-yield short-chain oligomers from cellulose via selective hydrolysis in molten salt hydrates and separation. Green Chemistry 2019, 21 (18), 5030-5038.

2. Liu, Q.; Luo S.; Fan, W.; Ouyang X.; Qiu X., Separation of short-chain glucan oligomers from molten salt hydrate and hydrolysis to glucose, submitted.

628662 Conversion of Food Waste to Levulinic Acid Using a Catalytic Membranemembrane Reactor

Abstract:

One-pot conversion of biomass to levulinic acid (LA), a promising platform chemical of bio-refinery, is an important technology for biomass conversion. Multiple biomass waste materials such as agriculture residues, forest residues and food wastes have been considered as promising feedstocks for bioprocessing and bio-refinery. Here, cellulose as well as food wastes from vegetables and other starch-based food materials have been investigated for their conversion to levulinic acid using a unique solid acid catalyst immobilized on a membrane substrate. Our reusable membrane catalyts is superior to the corrosive homogeneous acid or toxic metal-based catalyst. In addition, the membrane substrate enables the immediate separation of the levulinic acid from the rest of the feed stream driving the reaction to completion and improving the yield. Finally, hot water extraction as a pre-processing step was found to be effective in removing some of the proteins and other soluble carbohydrate. A levulinic acid yield of over 80% has been achieved for cellulose. For starch-based food waste such as rice and noodles, over 90% of levulinic acid yield was obtained. For the more recalcitrant cellulosic vegetables, up to 50% of levulinic acid yield has also been achieved indicating the promising potential of our technology for food waste utilization.

631216 Capturing Spatial, Temporal and Technological Detail in Hydrogen Supply Chains By a Bilevel Optimization Strategy

Abstract:

Related background and motivation

Hydrogen is one of the key enablers of the energy transition and the decarbonization of our societies as confirmed by the most recent energy roadmaps [1] and national plans aiming for carbon neutrality by 2050. For hydrogen to play a significant role in the future demand mix, its costs should yet go down implying improved infrastructure design and operations. This paper presents a methodological framework for the optimal design of Hydrogen Supply Chains (HSC) to provide effective support for the study of deployment scenarios. According to the energy source used for its production, hydrogen can be labeled as: green, if produced through Power-to-Gas (PtG) systems, converting renewable electrical energy into green hydrogen and offering a more flexible power grid, better management of the intermittency of renewable energy sources, and a low-carbon fuel supply: as grey if obtained from fossil energy sources with Steam Methane Reforming (SMR) which is the most common process used today and finally as blue h if produced by the SMR process combined with carbon capture, utilization, and storage (CCUS).

Due to the characteristics of the HSC network, economic and environmental assessment is key to its optimal design and deployment that requires the consideration of the set of all possible raw materials, energy sources, technology maturity levels, potential sites for location, transportation and distribution options, storage, end uses, as well as the dynamic change of hydrogen demand over time.

Several works have been identified in the scientific literature to address this complex optimization problem [2]. Most studies involve mathematical models with mixed-integer linear programming (MILP) formulation, associated with an epsilon-constraint technique for tackling the multi-objective issue, mainly based on the total daily cost of hydrogen (TDC) and the Greenhouse Gas (GHG) emissions of the whole supply chain [3,4]. Yet for more than one objective, the construction of the Pareto front by mathematical programming techniques is usually a time-consuming task, since each Pareto solution requires at least one scalar optimization. An alternative to the solution of such problems is to use metaheuristics, and specifically, Multi-objective Evolutionary Algorithms (MOEAs). MOEAs are stochastic population-based nature-inspired algorithms, able of handling multiple objectives, continuous and discrete variables, nonlinear and multimodal functions. However, as these search algorithms are designed to work over unconstrained search spaces, they generally exhibit poor performance when solving highly constrained problems, even if sophisticated constraint-handling techniques have been developed. The scientific objective of this work is thus to propose a hybrid methodology to solve the HSC problem, which decomposes the global problem in such a way that some advantages can be taken of the strengths of every single approach, mitigating simultaneously their disadvantages.

Methods

The original HSC problem [4] is decomposed into two parts (see Figure 1): the (master) sub-problem and the continuous (slave) sub-problem. The master sub-problem is treated by the multi-objective evolutionary algorithm, in which the integer variables (number of production units) and their size (continuous variables) are encoded in every individual in the population. For each individual, a linear programming sub-problem is solved to determine the corresponding optimum production rates and transportation flows. This Linear Programming (LP) problem is constructed according to the values of the integer variables provided by the MOEA in such a way that it is almost a canonical transportation problem. The integer variables produced by the MOEA can be repaired to ensure that this problem is feasible. Subsequently, the MOEA recovers the continuous variables from the LP solver, to compute the *TDC* and *GHG* objectives. Once evaluated, each individual has its fitness function value assigned by the MOEA, which evolves the population for the next generation.

The proposed model takes the economy of scale for production and storage facilities into account with the "six-tenth-factor rule" [5]. The technological maturity of the various units that is reflected in their installation cost, power efficiency, maximal production capacity and the cost evolution of the different energy feedstocks over time is considered. The resulting model can thus be formulated as a mixed-integer nonlinear programming problem (MINLP) with MATLAB environment. An MOA algorithm is a tailored made version used to solve the upper-level problem, whereas the solution of the lower-level problem is performed using CPLEX v12.8.0.

Computational experiments

The proposed model is applied to a case study, i.e., Midi-Pyrénées part of the Occitanie Region in France [6]. The available resources are renewable energy sources (wind, solar and hydro) and natural gas, which must be exported. The set of options for hydrogen production accounts for green hydrogen (alkaline electrolysis cells (AEL) or polymer electrolyte membranes (PEM) with renewable energy sources), blue (SMR with CCUS), and grey (SMR without CCUS). Hydrogen demand is related to the transport sector for use in fuel cell electric vehicles (FCEV).

The region under study is divided into eight subregions (grids) and the time horizon (2020-2050) is divided into seven periods of five years each. Figure 2 presents the obtained Pareto front for this instance. It is composed of 1000 points yielded by MOEA. Each point represents a complete solution, that is, the result of strategic decisions, i.e., size, type, location of production and storage technologies, as well as logistics related to hydrogen transport between grids, for each period.

The Pareto front is divided into segments that are colored according to the shade of hydrogen mostly used to meet the hydrogen demand. Segment 1 comprises non-inferior solutions with Levelized Cost of Hydrogen (LCOH) from 1.64 - 1.90 USD/kgH_2 ; segment 2 from 1.90 - 1.95 USD/kgH_2 TDC; segment 3 from 1.95 - 2.13 USD/kgH_2 TDC; segment 4 from 2.13 - 2.49 USD/kgH_2 TDC; and segment 5 from 2.49 - 2.86 USD/kgH_2 TDC. In this way, the solutions in the left part of the front mainly involve the SMR technology, whereas the solutions at the right part of the front mainly use PEM electrolysis powered by renewable (wind) energy to meet the hydrogen demand.

For the sake of illustration, let us examine the composition of the segment that is mainly constituted of solutions using water electrolysis technology to satisfy the hydrogen demand. These solutions can be considered as the most expensive alternatives for deploying the HSC, but involve the best green options. Figure 3 shows a detailed solution from this segment. The squares indicate the number of production units, the arrows between grids indicate transport flows, and the pie charts indicate demand satisfaction by production type. It can be observed that the solution presents an LCOH of 2.62 USD/kgH_2, and GHG emissions of 1.86 kgCO_2-eq/kgH_2. Alkaline electrolysis (AEL) technology is employed using electricity from wind. In the first period, 33% and 67% of H_2 demand are fulfilled by AEL/wind and PEM/wind, respectively. Note that the AEL demand satisfaction drops from 33% to only 6%, from the first to the second period, due to the decrease in PEM technology cost compared to AEL. In periods 5 and 6, the network shows the installation of PEM electrolysis coupled with photovoltaic systems; nevertheless, most of the hydrogen is still produced by PEM/wind technology. In the last two periods, the contribution of demand satisfaction by technology remains the same, even if two new electrolysis PEM/wind plants are installed in the last period.

Conclusions

This work proposed a comprehensive methodological framework for the optimal design, deployment and management of Hydrogen Supply Chains HSC focusing on eco-mobility applications and their spatio-temporal integration considering the potential of Renewable Energy Sources. This framework can be viewed as a decision support system for scenario analysis. to achieve the desired step-up across the hydrogen value chain. An innovative point of the methodology is to explore the wide range of capacity sizes for a range of new technologies that may be involved gradually on the market.

The optimization strategy could now be extended to address other objective functions such as risk, the inclusion of dynamic aspects associated with the seasonal variability of energy sources, and the incorporation of other hydrogen demands and interactions with systems (ammonia and methanol production, buildings and industrial sector, etc.).

Reference

(1) IEA - International Energy Agency. *The Future of Hydrogen*; 2019.

(2) Li, L.; Manier, H.; Manier, M. A. Hydrogen Supply Chain Network Design: An Optimization-Oriented Review. *Renew. Sustain. Energy Rev.* **2019**, *103* (January), 342–360. https://doi.org/10.1016/j.rser.2018.12.060.

(3) Almansoori, A.; Shah, N. Design and Operation of a Future Hydrogen Supply Chain: Multi-Period Model. *Int. J. Hydrogen Energy* **2009**, *34* (19), 7883–7897. https://doi.org/10.1016/j.ijhydene.2009.07.109.

(4) De-León Almaraz, S.; Azzaro-Pantel, C.; Montastruc, L.; Boix, M. Deployment of a Hydrogen Supply Chain by Multi-Objective / Multi-Period Optimisation at Regional and National Scales. *Chem. Eng. Res. Des.* **2015**, *104*, 11–31. https://doi.org/10.1016/j.cherd.2015.07.005.

(5) Böhm, H.; Zauner, A.; Rosenfeld, D. C.; Tichler, R. Projecting Cost Development for Future Large-Scale Power-to-Gas Implementations by Scaling Effects. *Appl. Energy* **2020**, *264* (December 2019), 114780. https://doi.org/10.1016/j.apenergy.2020.114780.

(6) Région Occitanie / Pyrénées-Méditerranée. *Scénario Région à Énergie Positive de La Région Occitanie / Pyrénées-Méditerranée*; 2017.

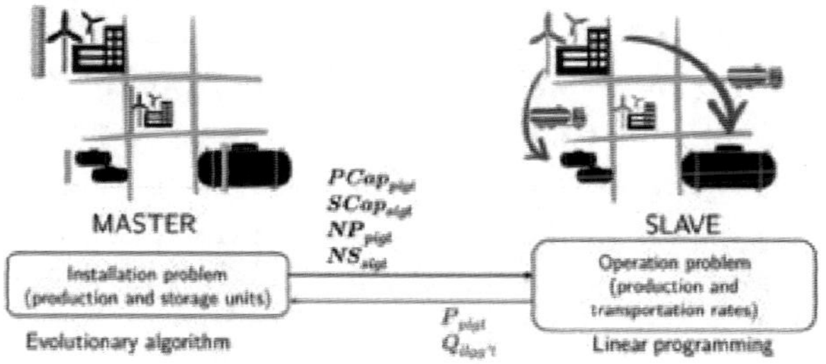

Figure 1. A bilevel solution approach.

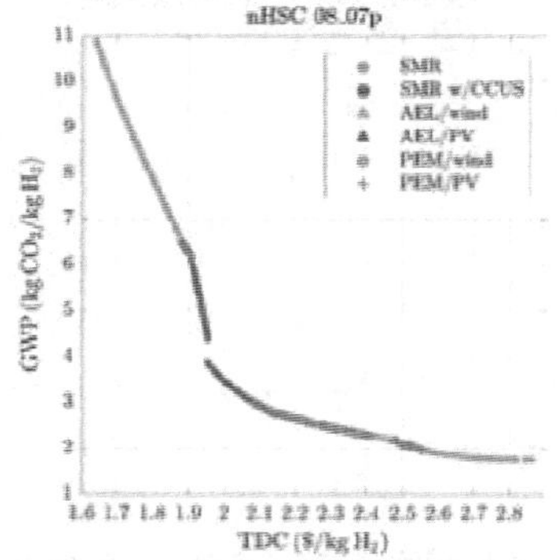

Figure 2. Pareto front approximation of HSC08g07p instance (8 grids-7 periods). The color code indicates the majority of production technology used for hydrogen demand.

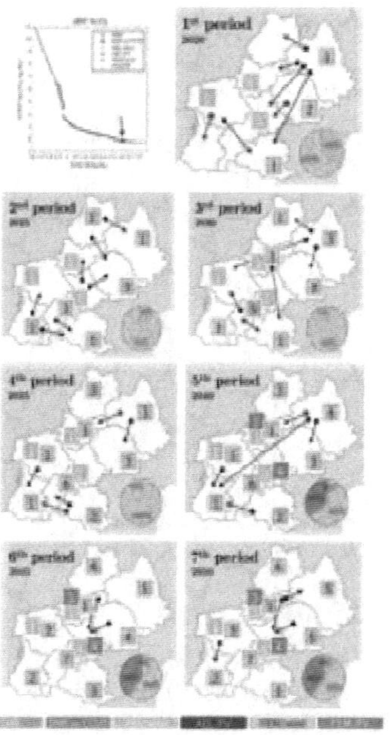

Figure 3. Presentation of a HSC solution from segment 5 for nHSC08g07p instance.

630571 Spatially-Explicit Techno-Ecological Design for Sustainable Manufacturing and Minimized Air Pollution Health Impacts

Abstract:

Sustainable design is increasingly in focus for academia, industry, and policy makers. With technology taking prioritization in the pursuit of sustainability, nature's solutions to regulating pollutants and environmental hazards are forgotten. Although technology is necessary to maintain the current state of pollution control, both technology and ecology should both be considered as solutions moving forward. Because ecological solutions require time to grow to mature states, it can be argued that these should have immediate priority to reap greater benefits in the future. Previous work in techno-ecological process design has shown this along with economic analyses that reveal competitive feasibility. In some cases, ecological solutions have proved to be cheaper than technological alternatives. Along with decreasing negative environmental impacts, including ecosystems in sustainable design can lead to innovative solutions which utilize the functions and co-benefits of nature as we search for sustainable solutions amid increasing population and demand for these services.

One challenge that is characteristic of ecosystems and ecological models, is the spatial and temporal variation of ecosystem services based on land-cover heterogeneity, meteorological patterns, seasonality of ecological functions, and other variables. It is important to understand how, where, and when mass and energy flow across components of techno-ecological systems to inform design opportunities and impacts. This work focuses on gaseous pollutants, air quality regulation, and exposure to local communities. We incorporate a detailed geophysical and meteorological dispersion model to simulate atmospheric dispersion of air pollution from a given site and the capacity of local ecosystems to uptake those pollutants through dry deposition. This research introduces a spatially-explicit techno-ecological design framework, enabling design of both the manufacturing process and the surrounding landscape. An additional layer of spatially-informed population and exposure to health risks is included within the framework to add a design objective based on health and safety.

Both the framework and a case study will be presented. The case focuses on criteria air pollutants (SO_2, NO_2, and PM_{10}) and carbon dioxide emissions for a power generation station near Cincinnati, OH. The case will compare pollution removal between technological equipment and land-use change opportunities in the region, optimizing minimum-cost and minimum-health risk solutions that meet given pollution removal requirements. Different scenarios will be demonstrated, showing the impacts of different model spatial scales, different ecosystem service accounting methods, budget constraints, and others. The results yield both long-term operating conditions of technology and maps of land changes over time. Further, we will discuss sensitivity to critical variables such as conditional costs of conducting land changes. Our results show opportunities and limitations for land management as a solution to air quality regulation, specifically for a power station on the border between a rural and urban region. In presenting future opportunities, we will also show initial work in integrating spatial landscape design with high-resolution temporal process operations of pollution control technology. Exploring spatially-explicit design options for techno-ecological systems enables smarter industrial site design and is one step closer towards bridging ecological knowledge with engineering practice.

627240 Decision Tree-Based Optimisation for Flexible Energy Storage Dispatch

Abstract:

Electrical energy storage (EES) devices have shown a number of advantages in both front-of-meter and behind-the-meter applications, especially with the increased penetration of intermittent renewable energy generators (Ma et al., 2018). Some of these merits are the improvements in power reliability, flexibility, and quality, and reduction in the final electricity bill of end users. However, owing to problems of a prolonged pay-back period in large scale deployment of these devices, focus has been drawn to revenue stacking – where a collection of tasks are performed by the EES device in order to generate more revenue (Fisher et al., 2019; Roberts & Brown, 2020). In behind-the-meter applications, for example, the ability to stack the revenue streams of price arbitrage, surplus PV value uplift and ancillary service provision (via aggregation) has the potential to improve the pay-back period. In such situations, the scheduling of the EES device plays a key role in determining the total revenue generated as well as the lifetime of the device.

A number of mathematical models have been developed to optimise the operation of these devices under differing conditions of energy demand, energy generation sources, electricity prices and/or accessible revenue streams (García Vera et al., 2019), however the final results obtained are often a single set of time-dependent decision variables which must be followed to maximise revenue. In reality, these solutions require strict implementation (mostly with the help of computer systems), and do not always perform well when conditions of demand, price or energy generation deviate from assumptions/predictions, even when uncertainty is taken into consideration. A more flexible approach towards energy dispatch involves identifying key metrics/parameters which characterise the optimal operation of the EES device and from which rules can be generated. This has the advantage of being easily implementable and applicable over a wider range of system variability.

To this end, we propose a flexible approach to energy dispatch for EES devices using a decision tree-based optimisation approach. Using output from a two-stage optimisation-based approach, we train a decision tree to obtain highly optimal and feasible dispatch results with no prior knowledge of future load profile and market data. The first stage of the optimisation approach solves an optimal dispatch optimisation model to obtain a set of distinct optimal solutions for training data, and the second stage is a feature extraction stage which finds the optimal price range where charge and discharge actions are executed in order to minimise the total cost. These results are then mapped to a decision tree for obtaining real-time dispatch actions for the EES device. The approach is applied to a microgrid with an EES asset having access to the UK day-ahead energy market. Given a load demand which it must satisfy, this approach proffers the optimal dispatch actions for the EES device in order to ensure minimal total costs.

References

Fisher, M., Apt, J., & Whitacre, J. F. (2019). Can flow batteries scale in the behind-the-meter commercial and industrial market? A techno-economic comparison of storage technologies in California. *Journal of Power Sources*, *420*, 1–8. https://doi.org/10.1016/j.jpowsour.2019.02.051

García Vera, Y. E., Dufo-López, R., & Bernal-Agustín, J. L. (2019). Energy Management in Microgrids with Renewable Energy Sources: A Literature Review. *Applied Sciences*, *9*(18), 3854.

https://doi.org/10.3390/app9183854

Ma, T., Shen, L., & Li, M. (2018). Electrical Energy Storage for Buildings. In *Handbook of Energy Systems in Green Buildings* (pp. 1079–1107). Springer Berlin Heidelberg. https://doi.org/10.1007/978-3-662-49120-1_44

Roberts, D., & Brown, S. F. (2020). Identifying calendar-correlated day-ahead price profile clusters for enhanced energy storage scheduling. *Energy Reports*, *6*, 35–42. https://doi.org/10.1016/j.egyr.2020.02.025

630889 Tracking Embodied Co, Li, Mn, and Ni Material Flows in the US through Economic Input-Output-Based Network Analysis

Abstract:

The transition toward cleaner energy and electric vehicles (EVs) is creating unprecedented demand for mineral resources on which these technologies rely. For many of these minerals, the US is currently significantly dependent on imports (between 50 – 100%), often from countries that are geopolitically sensitive. Securing the supply of these so-called 'critical materials' is therefore crucial for a successful transition to cleaner energy and EVs. An essential step in securing the supply of critical materials is tracking their flows within the economy and understanding their usage to elucidate opportunities for efficiency, substitution and recycling. Established methods for material flow tracking depend heavily on empirical data that are resource intensive to procure, thereby limiting the reproducibility of many material flow analyses and their application to supply chain modeling. This study presents an economic input-output (IO)-based framework for characterizing the national trade structure of embodied critical materials. We generate national IO models for select lithium-ion battery materials using annual industry consumption data and the US national IO accounts at the detailed sector level. The focus in this study is on characterizing cobalt, lithium, manganese, and nickel flows to demonstrate the use of the modeling framework, which could then be applied to other materials for which annual industry consumption data are available. To uncover competing uses and potential secondary sources for these materials, we further use network analysis to identify significant sectors in the domestic metals supply chains. Initial estimates of the potential recoverable secondary supply of cobalt, lithium, manganese, and nickel from the sectors highlighted by the network analysis using multiple metrics are presented as a precursor to more detailed analyses of the sector-specific uses of these materials in future work.

632024 Dynamic Material Flow Analysis of Crystalline Silicon Solar Cells for Circular Economy Transition

Abstract:

Solar photovoltaics (PV) are the fastest growing renewable energy technology for clean, inexpensive, and sustainable electricity generation. Along with numerous technical roadmaps to improve system cost, performance and reliability, the PV industry should also plan to handle large volumes of silicon panel waste, which is initially estimated to be ~13 million metric tons (MT) by 2050 in the U.S. alone. Understanding the magnitude of material needs and how material flows throughout the PV panel life cycle could change with different design, operational, and end-of-life (EOL) choices will help transition into a circular, resource-conserving economy. Herein, we introduce a computational framework for dynamic material flow analysis (DMFA) based on electricity generation to quantify time-series stocks and flows of bulk PV materials (e.g., solar glass and aluminum frames). We apply this PV DMFA model to evaluate cradle-to-cradle life cycles of utility-scale silicon PV systems in the United States in the years 2000 - 2100. We modeled a range of scenarios to understand how material demands depend on selected PV-related parameters, different material circularity strategies, and recent module design trends (e.g., bifacial or high power). We found that cumulative float glass and aluminum needed for PV installations would likely reach 100 million MT and 12 million MT by 2100, respectively, in the baseline scenario. The most influential parameters for PV installation and subsequent waste reduction include module lifetime, module efficiency, annual degradation rate and material intensity. Ongoing trends to produce larger power frameless modules could save 10 million MT of glass and ~9 million MT of aluminum. Module recycling and component remanufacturing were found to be potentially the most effective material circularity strategies for waste minimization if challenges to their implementation can be overcome. Standardization of module size, minimally intrusive recycling, and careful scrap handling and cleaning could facilitate improved material circularity and reduced waste. The model will be publicly available, and the framework serves as a sustainability support tool for efficient material management of PV energy systems in the circular economy.

630189 Developing Innovation Roadmaps for Plastics Value-Chain Using Sustainable Circular Economy Framework

Abstract:

The plastics value-chain is leading to large amounts of nutrient pile up in the environment through oceanic gyres, over-filled landfills, and micro-plastics. Grocery carrier bags, films and beverage bottles form a major chunk of this pollution, an estimated 45%. Since this plastics problem is very pressing, it has inspired researchers from multiple disciplines to propose innovations in various parts of the value-chain. It is crucial to re-evaluate our current value-chain operation for minimizing this pile-up while also keeping a check on the life-cycle impact of circularizing the value-chain. Researchers at The Global Kaiteki Center have systematically built a list of innovations in the plastics value-chain and categorized them based on the sectors which they are likely to displace [1]. This includes information ranging from novel AI based separation techniques and chemical recycling to household collection schemes and incentives. This list of innovations is an exhaustive set of emerging technologies, novel policy action and behavioral strategies. In collaboration, we at The Ohio State University have developed a general Sustainable Circular Economy framework which can design cradle-to-cradle life-cycle value chains for multiple stakeholders using multi-objective optimization [2]. This work aims to systematically rank the list of innovations and construct a roadmap towards a Sustainable Circular Economy (SCE), using the developed optimization framework in conjunction with hotspot and sensitivity analyses.

In our previous work [2,3], we have found optimal value-chains within the design space of conventional alternatives (called 'superstructure' networks) for the grocery sacks value-chain. This included decision variables ranging from type of bags, collection methods, primary and secondary waste treatment methods, etc. We also found optimal value-chain designs for the three objectives: circularity, global warming potential and life-cycle cost. Further, we constructed a pareto surface containing trade-off or compromise solutions. As mentioned earlier, the focus of this work is to study each innovation tabulated in the developed list based on enhancement of Sustainable Circular Economy. We also broaden the scope of our previous work [2] to also include beverage bottles, since a major chunk of emerging technologies are aimed at dealing with PET bottle waste.

We use life-cycle allocation and displacement methods to find hotspots of emissions, loss of circularity and cost in the PET beverage bottles value-chain. This yields information on sectors of the value-chain which have the highest scope of improvement. Additionally, we have categorized the list of innovations based on stage of adoption, namely: Innovation gap, Theoretical, Experimentation, Commercialization and Diffusion. Information from both hotspot and adoption stage are used to select which technologies to introduce in the value-chain network of alternatives, or 'superstructure'. This is followed by data collection and generation for introducing life-cycle of the chosen 'emerging technology' within the design space. This requires generation of life-cycle inventory for novel technologies using prospective LCA [4,5]. In order to fill several data voids and gaps, simulation of chemical plants and physical processes are performed to find parameters of the life-cycle model. For example, within the PET bottles case study we explore novel technologies to make bio-based compostable polyhydroxy-alkanoate and bio-PET bottles; and chemical recycling using glycolysis. Following this, the SCE framework is implemented using multi-objective optimization to construct a new pareto surface. Ultimately, the presence of the innovation within the pareto optimal set is investigated to determine if the innovation is worthy of being included within the roadmap. This process, starting from hotspot analysis is repeated, until every sector of the value-chain is covered, thereby developing a roadmap for innovation. As a result, the generated innovation roadmap contains novel technologies in each sector ranked according to the potential of enhancing the SCE of the system. In conclusion, this talk will contain a systematic method to convert a list of innovations into a Roadmap of most promising innovations using systemic multi-objective optimization approaches on cradle-to-cradle life-cycle superstructures. The method will be demonstrated for products such as grocery sacks and beverage bottle value-chains, thereby aiding corporate strategy and policy action towards most effective SCE innovations.

References

1. Hafsa, F., Dooley, K., Basile, G., and Buch, R. (2020), "Innovations for more circular plastic packaging", under review at *Journal of Cleaner Production*.
2. Thakker, Vyom, and Bhavik R. Bakshi. "Toward sustainable circular economies: A computational framework for assessment and design." *Journal of Cleaner Production*295 (2021): 126353.
3. Thakker, Vyom, and Bhavik R. Bakshi. "Towards Sustainable Circular Economy: Design Framework and Application to Grocery Sacks". *2020 AIChE Annual Meeting*. Video available at: https://youtu.be/dWj6R0UXMjM
4. Yao, Yuan, and Eric Masanet. "Life-cycle modeling framework for generating energy and greenhouse gas emissions inventory of emerging technologies in the chemical industry." *Journal of Cleaner Production* 172 (2018): 768-777.
5. Thonemann, Nils, Anna Schulte, and Daniel Maga. "How to conduct prospective life cycle assessment for emerging technologies? A systematic review and methodological guidance." *Sustainability*3 (2020): 1192.

623627 Systems Analysis Approach to PET and Olefin Plastics Supply Chains in the Circular Economy

Abstract:

Plastic waste has become a global crisis on our planet. Most of the plastic waste generated in the U.S. was landfilled (~76%) and only a small fraction was recycled (~9%) in 2018 [1]. Of this total plastic waste generated, polyethylene terephthalate (PET) and polyolefins such as high-density polyethylene (HDPE), low-density polyethylene (LDPE), and polypropylene (PP) represent a significant portion (~80%) of valuable plastic waste stream. The current plastic linear take-make-dispose system in the U.S. represents not just environmental threat but also an economic loss of ≈ $10 billion/year of embodied value ending up in landfills [2]. Therefore, a systems approach is needed to tackle this plastic waste crisis and move towards circular economy. However, the environmental, economic, and social impacts of implementing a circular economy in plastic waste supply chains are still lacking, and trade-offs among these impacts at the systems level are still unknown.

To address this research gap, our study presents a novel framework called a 'systems analysis framework for PET and olefin plastics in a circular economy' [3]. This framework identifies different processes involved in a plastic waste supply chain and quantifies the environmental, economic and social impacts of plastic waste in a circular economy. The presented framework can address many different research questions such as, but are not limited to:

- How would the optimized material flow of plastics look like in a future circular economy with minimum environmental impacts, when compared with the linear economy?
- What combinations of recycling technologies are needed to achieve a system with minimum energy consumption and environmental impacts?
- What are the environmental and economic trade-offs of implementing a circular economy at systems level, when compared with linear economy?
- What unit operations within the various chemical recycling technologies for PET and olefin plastics consume the most energy and emit most pollution?
- How would process level changes (e.g. process intensification, energy integration) affect the economic and environmental performance of the chemical recycling processes?

The first research objective of this study was to identify the availability and quality of datasets, process models, and knowledge gaps associated with plastic waste supply chain processes. The second research objective of this study was to showcase an application of the presented framework by conducting a preliminary systems analysis of PET bottles in a circular economy to address some of the research questions highlighted above. For this research purpose, we developed a simple model that utilized a linear programming optimization method to minimize environmental impacts of implementing a circular economy in a PET bottle supply chain. This model enables system-level understanding of the causes of greenhouse gas (GHG) emissions and interaction among the system processes. The key result from our optimization model indicated that a combination of both mechanical and chemical recycling technologies is needed to achieve a system with least GHG emissions, specifically reductions of **24%** when compared with the linear economy [4].

References

1. USEPA. National Overview: Facts and Figures on Materials, Wastes and Recycling. 2019. Available at: https://www.epa.gov/sites/production/files/2020-11/documents/2018_tables_and_figures_fnl_508.pdf (Last accessed: March 29, 2020).

2. Sherwin, G. Closed Loop Partners- The Circular Shift: Four Key Drivers of Circularity in North

America. 2020. Available at: https://www.closedlooppartners.com/closed-loop-partners-launches-report-on-unprecedented-shifts-in-the-circular-economy-in-north-america/ (Last accessed: March 29, 2020).

3. Shonnard, D. R.; Tipaldo, E.; Thompson, V.; Pearce, J. M.; Caneba, G. T.; Handler, R., Systems analysis for PET and olefin polymers in a circular economy. *Procedia CIRP* **2019,** *80*, 602-606, DOI 10.1016/j.procir.2019.01.072.

4. Chaudhari, U.; Lin Y.; Thompson V.; Hander R.; Joshua J.M.; Caneba G.; Muhuri P.; Watkins D.; Shonnard, D., Systems analysis approach to PET and olefin plastics supply chains in the circular economy: A review of Datasets and Models. ACS Sustainable Chemistry and Engineering *(Awaiting final decision),* ***2021.***

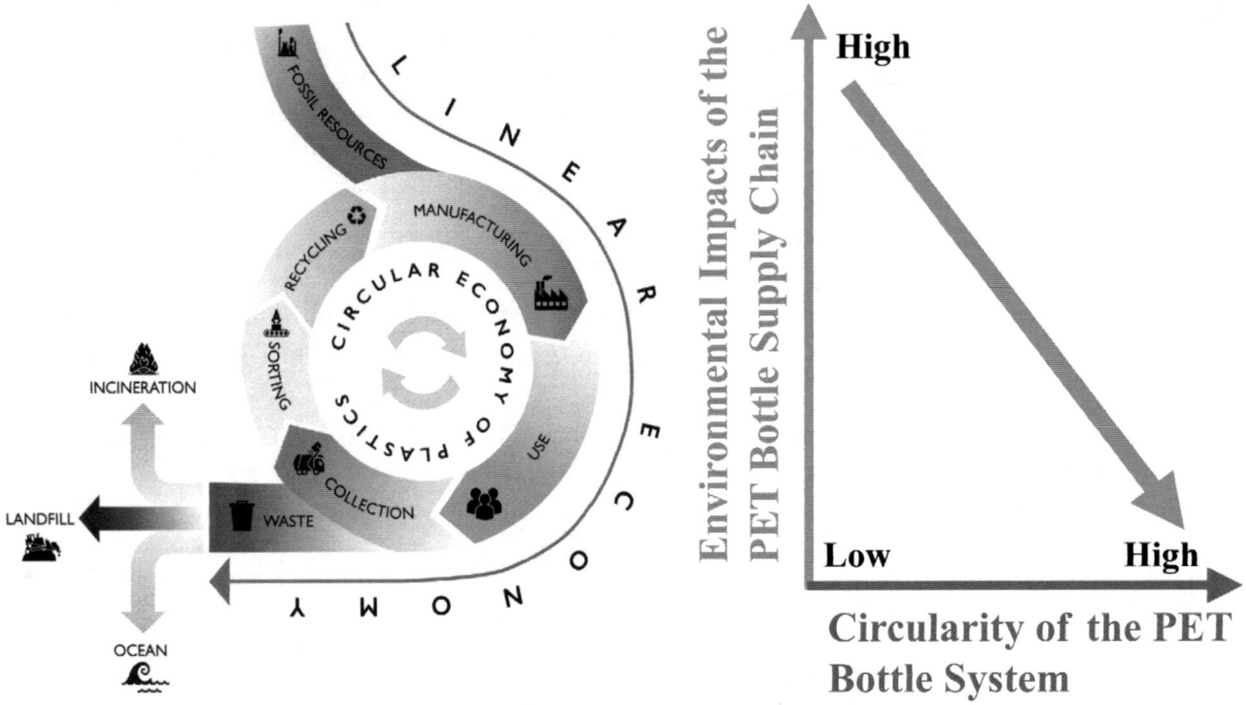

624874 Micron: A Quantitative, Holistic and Robust Circular Economy Assessment Framework at the Micro Level

Abstract:

Circular Economy (CE) aims to solve resource, waste, and emission challenges confronting society by creating a production-to-consumption total supply chain that is restorative, regenerative, and environmentally benign (a). A method for evaluating and comparing the circularity of different companies or scenarios is vital for effective decision making (b). A variety of metrics has been developed for measuring different aspects of CE, however the focus is mainly on the macro and meso levels (c-f). This is attributed to some of the intrinsic characteristics of circularity assessment at micro level, where CE indicators frequently concentrate on just a subset of CE principles or have a narrow focus to specific products and materials, therefore missing the bigger picture and leading to sub-optimal results (g).

This work aims to address this research gap by introducing "MICRON" (MIcro CirculaR ecOnomy iNdex), a GRI (Global Reporting Initiative) based, quantitative web-tool that takes into consideration all goals of CE holistically. First, principal categories are designed based on the goals and key characteristics of CE. Then, clearly specified indicators and readily measurable metrics are selected for each of these principal categories. The weighted average of metrics in each principal category constitutes the Categorical Circularity Indexes, while their linear average constitutes the Overall Circularity Index of each company. Using the proposed web-tool, companies are able to track their transition towards CE, conduct temporal analysis, compare and benchmark their performance against their peers and industry's standards. The applicability and the capabilities of the developed CE methodology is demonstrated through case studies in the Energy & Utilities, Manufacturing and Automotive sectors, where Categorical and Overall Circularity Indices are calculated over a period from 4 up to 10 years. The results demonstrate a clear trend towards circularity, however the implementation of CE is still premature and there are tremendous opportunities for improvement.

References:

a. Avraamidou S, Baratsas SG, Tian Y, Pistikopoulos EN. Circular Economy - A challenge and an opportunity for Process Systems Engineering. Computers & Chemical Engineering. 2020;133:106629.
b. Baratsas SG, Masoud N, Pappa VA, Pistikopoulos EN, Avraamidou S. Towards a Circular Economy Calculator for Measuring the "Circularity" of Companies. 31st European Symposium of Computer Aided Process Engineering: ESCAPE-31: Elsevier; 2021.
c. Cayzer S, Griffiths P, Beghetto V. Design of indicators for measuring product performance in the circular economy. International Journal of Sustainable Engineering. 2017;10(4-5):289-98.
d. Elia V, Gnoni MG, Tornese F. Measuring circular economy strategies through index methods: A critical analysis. Journal of Cleaner Production. 2017;142:2741-51.
e. Parchomenko A, Nelen D, Gillabel J, Rechberger H. Measuring the circular economy-A Multiple Correspondence Analysis of 63 metrics. Journal of cleaner production. 2019;210:200-16.
f. Saidani M, Yannou B, Leroy Y, Cluzel F. How to assess product performance in the circular economy? Proposed requirements for the design of a circularity measurement framework. Recycling. 2017;2(1):6.
g. Kristensen HS, Mosgaard MA. A review of micro level indicators for a circular economy–moving away from the three dimensions of sustainability? Journal of Cleaner Production. 2020;243:118531.

626667 Integrated Framework for the Design of Intensive Cattle and Agricultural Operations Towards Circular Economy.

Abstract:

Nowadays, the specialization of agriculture and the intensification of livestock have separated both sectors, so that dairy and meat production are located in one place, and crops are grown in another. This leads to a dependence on mineral fertilizers in crop areas and an excess concentration of nutrients in livestock areas (Rick Kersbergen,2021). On the one hand, livestock is one of the main generators of anthropogenic CO_2 (IPCC,2014), in addition to other environmental impacts such as contamination of soils, eutrophication of water resources, and the generation of bad odors (FAO,2006). On the other hand, the increase in the population worldwide has pushed agricultural systems to increase their production and cultivation area increasing the total consumption of mineral fertilizers. Its excessive and uncontrolled use does not only imply an increase in the carbon footprint due to the energy used to obtain the mineral fertilizer, but also contamination of soils and water resources due to the excess of nutrients (Lammel, 2001). Organic waste can be treated through a system that integrates anaerobic digestion (Kafle,2016), struvite production (Martín-Hernández,2020), and ammonia stripping (Lei,2007) to produce biogas and organic fertilizers, which can be used to reduce the use of mineral fertilizers in crops.

Integrated farming and livestock systems simulate natural energy and nutrient cycles by converting cellulosic ruminant feed into protein and recycling nutrients from livestock manure into the cell structure of crops (Oltjen and Beckett, 1996). Nowadays, these integrated systems refer to small organic farms (extensive livestock) where such integration consists of a grazing activity where manure is deposited naturally (Reddy,2016). To extend this integration to intensive systems of thousands of animals, it is necessary to control and understand how the nutrients are transformed in the digestion process of the animals. The target is to maintain the balance between the nutrients necessary for the crops and those supplied because a small imbalance in nutrients can lead to soil depletion or over-fertility.

In this work, an integrated system comprising intensive livestock and agriculture is compared with traditional systems, from various points of view, economic and environmental. A mathematical optimization framework is developed including models for estimating energy and nutritional requirements of beef cattle, for waste treatment (anaerobic digestion), to recover nutrients, and for the management of crops. This integrated model allows relating the formulation of the diet of the animals with the composition of their feces, the necessary cultivation area (13 different crops are considered), required raw materials (water, chemicals, additional fertilizer, energy, supplements, etc.) as well as carrying out the economic and environmental evaluation of the entire system. Through the application of the model to a representative case study where the complete 6-year life cycle of 1000 calves is studied, a **49% reduction** in the environmental impact of the combined agriculture-livestock system has been achieved with an economic benefit **4.7 times larger** than the decentralized case, showing that not only integrated systems are economically viable but also that the environmental impact of two of the most polluting economic activities carried out by human beings can be reduced, making a correct formulation of the feed and optimal crop selection. Besides, the nutrients recovered represent **17% of Nitrogen, 5% of Phosphorous, and 26% of Potassium** with respect to the necessary nutrients.

Acknowledgement

The authors acknowledge JCyL for a Ph.D. fellowship. M.L.C.

References

Edgar Martín-Hernández, Gerardo J. Ruiz-Mercado, Mariano Martín (2020). Model-driven spatial evaluation of nutrient recovery from livestock leachate for struvite production, Journal of Environmental Management, Volume 271,2020,110967,

FAO. (2006). La ganadería amenaza el medio ambiente. http://www.fao.org/newsroom/es/news/2006/1000448/index.html (accessed March 18,2020)

Gopi Krishna Kafle, Lide Chen, Comparison on batch anaerobic digestion of five different livestock manures and prediction of biochemical methane potential (BMP) using different statistical models. Waste Management, Volume 48,2016, Pages 492-502.

IPCC (2014): Climate Change 2014: Synthesis Report. Contribution of Working Groups I, II and III to the Fifth Assessment Report of the Intergovernmental Panel on Climate Change. [Core Writing Team, R.K. Pachauri and L.A. Meyer (eds.)]. IPCC, Geneva, Switzerland, 151 pp.

Kersbergen, R. Integrating Livestock with Crop Production Yields Benefits for Both. https://mosesorganic.org/farming/farming-topics/livestock/integrating-livestock-with-crop-%20production%20/ (accessed March 18,2020)

Kersberguen,R. Integrating Livestock with Crop Production Yields Benefits for Both. https://mosesorganic.org/farming/farming-topics/livestock/integrating-livestock-with-crop-%20production%20/ (accessed March 18,2020)

Lammel,J.(2001). Mineral Fertilizer in the Future– Sustainable Farming. https://arablefarmer.net/fileadmin/_migrated/content_uploads/mineral_fertil.pdf (accessed March 18,2020)

Oltjen, J. W., & Beckett, J. L. (1996). Role of ruminant livestock in sustainable agricultural systems. Journal of Animal Science, 74(6), 1406. doi:10.2527/1996.7461406x

Reddy P.P. (2016). Integrated Crop–Livestock Farming Systems. In: Sustainable Intensification of Crop Production. Springer, Singapore. https://doi.org/10.1007/978-981-10-2702-4_23

Xiaohui Lei, Norio Sugiura, Chuanping Feng, Takaaki Maekawa (2007).Pretreatment of anaerobic digestion effluent with ammonia stripping and biogas purification. In: Journal of Hazardous Materials, Volume 145, Issue 3, Pages 391-397,

636455 Use of Environmental Sustainability Principles to Help Solve Environmental Problems of Concern Today

Abstract:

In 1987, the concept of sustainable development was first put forward in the Bruntland Report which was prepared for the World Commission on Environment and Development. The links between inequality, poverty, and environmental degradation were recognized; the report sought a way for a fairer world with less damage to the environment that supports communities. The report defined sustainable development as "Development that meets the needs of the present without compromising the ability of future generations to meet their own needs". As a consequence, environmental sustainability ensures that in meeting our needs for water, food, and shelter as well as engaging in activities that make our lives enjoyable, we don't damage the environment, deplete our natural resources, and protect global ecosystems. Important aspects tied in with address: (1) Behavioral economics; (2) Intergenerational equity; and (3) What to do with the developing world. As engineers, we will increasingly need to be considering these concepts.

Recent research activities aimed toward environmental sustainability have involved: condensate recovery in air handlers; improved building energy conservation; use of green roofs combined with cistern collection and rain gardens for storm water mitigation on campus buildings; modeling of pollutants fate and transport in the Nile River tributaries; use of electrochemical sensors in storm drains to characterize storm water runoff in urban roadways; use of vertical gardening systems and community gardens to address urban food deserts; lead contamination in drinking water systems physical/chemical treatment of hazardous wastes, particularly in the area of sonication (acoustic cavitation) and photocatalysts to treat organic contaminants; and waste plastics in the marine environment. Highlights of several of these research topics will be described in the context of environmental sustainability.

636472 Advancing Methodologies for Sustainability assessment of transition to Low Carbon and Circular Economy via integration of process systems engineering and macroeconomics

Abstract:

Decarbonization and implementing Circular Economy are two major goals for sustainable development to mitigate the challenges of climate change and increasing anthropogenic wastes. Among numerous pathways proposed for this transition, it is important to evaluate the systems level economic and environmental impacts of these pathways for selection of the most sustainable pathways. Industrial Ecology (IE) is a field of study that focuses on developing methodologies and computational tools to assess large scale impact of these interventions. One key method in IE utilizes the macroeconomic Input-Output model for evaluating the impact of planned transitions such as shifts to renewable energy, bio-based production etc. However, the existing techniques do not efficiently integrate process engineering and thermodynamics knowledge in sustainability assessments, thus limiting the automation in evaluation of new technologies under development at scale. Further, lack of engineering integration and reliance on only monetary based model introduces uncertainty in analysis due to price variation and obsolete models along with missing feedback in considering sustainability criteria in design. In this talk, I will present the advancement in systems level sustainability assessment techniques by addressing these limitations through integrating process systems engineering and thermodynamics with macroeconomic Input-Output framework. Specifically, I will present novel methodologies and computational tools that will enable fast, and reproducible quantitative evaluation of systems level environmental and economic impacts of adopting emerging technologies for decarbonization and circular economy pathways.

628891 DFT-Informed Energetics of Plasma-Enabled Reactions Pathways and Microkinetic Modeling for Ammonia Synthesis on Transition and Low-Melting Point Metals

Abstract:

Ammonia is a crucial chemical (> 200 millions tons a year) due to its widespread use in fertilizer production, and its potential as a hydrogen carrier in a future "hydrogen economy." Currently synthesized using thermal catalysis, ammonia synthesis is currently a fossil energy-powered, high temperature-high pressure, energy-intensive process that is responsible for ~2% of the world's CO2 emissions. Ideally, as the world transitions to cleaner, renewable energy, the net CO2 emissions with ammonia synthesis would decrease. However, the harsh conditions required for thermally catalyzed ammonia synthesis demands a highly centralized, continuously-on process that is incompatible with renewable energy. Thus, currently, alternative methods to synthesize ammonia under milder conditions are actively sought, including plasma catalysis in plasma reactors.

Optimizing plasma-assisted ammonia synthesis, including optimizing the catalyst to be loaded on the reactor, is currently challenging due to an incomplete understanding of the most relevant reaction mechanisms involved in the process. One snippet of this debate include the role of vibrational versus radical species on ammonia formation. Some first principles-informed models have attempted to shed light on the relevant mechanism, but centered on vibrational species and only the typical Langmuir-Hinshelwood reactions associated with thermal catalysis were considered [1]. More comprehensive models including plasma radicals also attempted similar insights, but used generic, metal-agnostic parameters, and while including Eley-Rideal reactions, it only considered NH_X species [2]. These models thus may be at odds with experiments reporting the appearance of N2Hx species on the metal [3].

To enable first principles-informed comprehensive modeling of all plausible Langmuir-Hinshelwood (LH) and Eley-Rideal (ER) reactions involving N_YH_X surface species and plasma radicals we performed density functional theory (DFT) to obtain reaction energies for 50+ reaction events across 9 metal surfaces of different degrees of nitrophilicity (Fe, Co, Ni, Pd, Cu, Au, Ag, Ga, Sn), and used scaling relationships (both from the literature and new ones generated from our own select DFT calculations) to provide corresponding activation energies. Hydrogen dissolution events were also included in our calculations, motivated by previous observation of a hydrogen sink effect in experiments by some of us [4]. Ultimately, the obtained data was input into a microkinetic model, which we used to explore the effect of plasma composition on catalyst state and dominant reaction pathways. Our results indicate that plasma radical species such as N(g) and H(g) only need minimal concentration in the plasma phase (even several orders of magnitude lower than that of vibrational species) to make ER pathways dominant and the formation of N_YH_X species plausible. However, N_YH_X seem more likely to hydrogenate until saturation and spontaneous dissociation via ER reactions with H(g) radicals than to dissociate early via LH reactions. The dominant role of ER reactions makes for a crucial factor blurring differences in activities across different metals. Finally, we present relationships between energetic descriptors and experimentally measured catalyst activities in DBD and RF reactors.

1. Metha et al. Nature Catal. 2018, 1, 269-265

2. Van't Veer et al. *J. Phys. Chem. C* 2020, 124, 42, 22871–22883

3. Lea et al. *ACS Catal.* 2020, 10, 24, 14763–14774

4. Shah et al. ChemCatChem 2020, 12, 1200-1211

626387 Plasma-Assisted Catalytic Approaches for the Production of Chemicals

Abstract:

Plasma-assisted catalysis has grown rapidly in the past few years as a nonthermal, ambient pressure approach to activate stable molecules (e.g., N2). Utilizing renewable energy, plasma processes have the potential to reduce the CO2 footprint compared to thermal processes, and convert reactants at point sources owing to the modular nature of these systems. Our research in the area of plasma-assisted catalysis is focused on the ability to control the catalyst performance by tuning the plasma environment or plasma-catalyst interactions. We have been particularly interested in the use of nonthermal plasmas for C-H and N2 activation via dry reforming of methane and ammonia synthesis reactions, respectively. In recent publications from our groups, we observed remarkable synergy between the plasma and catalyst for NH3 synthesis from N2/H2: 1) the rate enhancements with plasma stimulation were more sensitive to N2 concentration, 2) the ideal catalyst was not the same in a plasma environment as in traditional thermal catalysis, and 3) NH3 yields above the thermal equilibrium limits were observed and sensitive to the catalyst composition. Further advances in plasma-assisted catalysis requires the fundamental knowledge of plasma interactions with catalyst surfaces. However, the consequences of plasma activation for the interaction of gas species with a catalyst surface are currently speculative, and observation of these species on the surface is very challenging. This presentation will focus on exploiting catalyst-plasma interactions for nitrogen and hydrocarbon activation, and the role of the plasma in enhancing catalytic performance.

627791 Models and Observations of Plasma-Catalytic Nitrogen Oxidation

Abstract:

Coupled plasma and catalyst enables reaction yields that exceed the contributions of the individual components. Reliable kinetic models are essential to rationalize observations and to guide the identification of appropriate plasma-catalyst combinations. Kinetic models of plasma- and of heterogeneous-catalytic-chemistries are individually well developed. Models to couple the two are still primitive. Here, we demonstrate DFT-parametrized microkinetic models to couple plasma and surface reactions. The models are applied to N_2 oxidation, an alternative nitrogen fixation process, enabled only with coupled non-thermal plasma and catalysts. We compare the intrinsic activity of Pt and Au under plasma conditions. We build integral reactor models coupling simplified plasma reaction and surface reaction networks to predict product yield. We first show plasma excited species can enhance turnover frequencies of both Pt and Au. The enhancing potential of catalysts to chemical yield from plasma, however, are condition- and material- dependent. Plasma/Pt generates more or comparable nitric oxide than plasma/Au under most thermal and plasma parametric conditions. Plasma/Au outperforms plasma/Pt under low thermal but high plasma excitations. Results are compared against experimental observations, highlighting that material selection must consider plasma properties, thermal conditions and reactor configurations.

630666 Methane-Products Process Intensification through a Nanosecond Plasma Discharge

Abstract:

The energy industry, such as other sections, moves toward more efficient systems and newer technologies hires by companies. Due to excellent energy-engineering setups that we have developed in many years, we could revolute the energy industry in the 20th century. The usages of natural gas, a clean feedstock, increase every year, and the majority of people in civilized countries have decided to move toward less polluted earth. Among all products that are produced at different plants, hydrogen plays a vital role. Producing hydrogen more efficiently at the existing plants is crucial. Moreover, hydrogen is a clean energy source and can help move toward a sustainable energy economy.

Most of (about 95%) hydrogen is produced from fossil fuels by steam reforming of natural gas, partial oxidation of methane, and coal gasification as of 2020. All these methods of hydrogen production require a considerable amount of energy due to high temperate operations. Therefore, the first question is how to produce hydrogen and light hydrocarbons cheaper with more eco-friendly technologies.

In this work, a nanosecond pulsed plasma reactor was designed to produce hydrogen and liquid hydrocarbons. The conversion rate, efficiency, and cost of the designed-made reactor show a promising method for converting hydrogen. Currently, hydrogen is priced between $3 and $8 per kilogram. We are presently producing hydrogen below $1.

Most of the researches that have been done on converting methane to hydrogen or other products have focused on the reactor's efficiency. This study attempts to present a novel reactor design and perform process intensifications for future industrial uses. Moving toward hydrogen as a source of energy can diversify our energy sources and enhance energy resiliency and security. Moreover, to avoid constructing gigantic plants to make them profitable, new-more efficient technologies are required to make the production in smaller-scale plants profitable. Also, having more distributed-smaller production plants would create opportunities to reach stranded reservoirs.

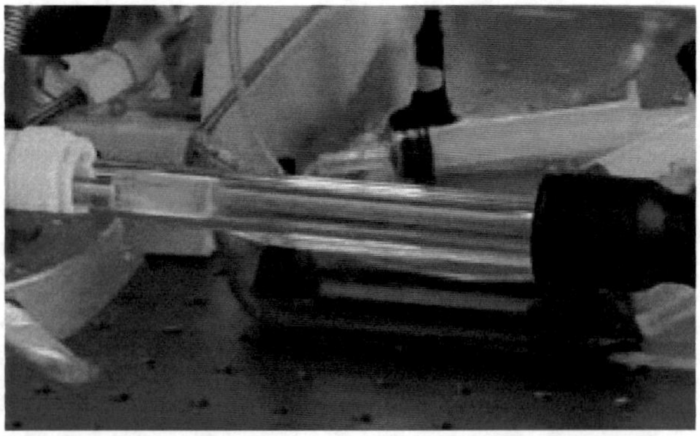

631226 Plasma-Assisted Upgrading of Methane: Mechanistic Insights from in-Situ PM-IRAS and OES Spectroscopy

Abstract:

Non-thermal plasmas (NTPs) produce reactive chemical environments including electrons, ions, radicals, and vibrationally-excited molecules and drive chemistry at temperatures at which such species are thermally inaccessible. The integration of a catalyst with reactive NTPs can promote novel chemical transformations. In particular, we are interested in upgrading of light hydrocarbons (e.g. methane) by direct coupling with nitrogen to produce value-added liquid chemicals (e.g. pyrrole and pyridine) in a plasma-assisted catalytic process. To design effective catalysts and plasma-catalytic processes, we need comprehensive understanding of the plasma catalytic processes including characterization of plasma-catalyst interactions. However, to date, the fundamental understanding of plasma-catalyst interactions is still in its infancy. In this work, we use a newly developed multi-modal polarization-modulation infrared-reflection absorption spectroscopy (PM-IRAS) tool with optical emission spectroscopy (OES) to probe plasma-catalyst interactions of CH_4/N_2 plasma species with different model catalytic surfaces (Ni, Pd, and Au). CH_4/N_2 feed gas mixture is introduced in three different ways: (1) CH_4-N_2 in sequence, (2) N_2-CH_4 in sequence, and (3) simultaneous with varying CH_4/N_2 ratio. We characterize excited species in the plasma-phase using OES, detect surface-adsorbed species using PM-IRAS, and detect gas-phase products using mass spectrometry. We find that different sequences in the feed gas introduction result in different surface-adsorbed species, and the presence of different surfaces dictates the surface-adsorbed species present. We also find that varying CH_4/N_2 feed ratio affects the macroscopic properties of the plasma and the production of intermediates such as cyanide (CN).

628510 Computational Evaluation of Plasmonic Photocatalysis in Au and Ag Nanoparticles

Abstract:

Many industrial chemical processes require high operating temperatures to drive catalytic reactions which are crucial to human development. These high temperatures are both energy demanding and detrimental to the long term life of the catalyst. Plasmonic metal nanoparticles offer an interesting alternative to traditional heterogenous catalytic processes due to their ability to be activated by light. One potential mechanism of action is the localized surface plasmon resonance, which involves oscillations in surface electron density that decay into so called "hot" electrons. These hot electrons can then transfer to unoccupied orbitals of surface intermediates to break chemical bonds or accelerate reactions. While plasmonic photocatalysis is a well-known event, the exact mechanism of these reactions is still debated. Understanding the precise workings of plasmon transfer is important for rational design of novel catalytic structures. Here, we utilize real-time, time-dependent density functional theory (RT-TD-DFT) to excite a system with oscillating electric fields and track the subsequent excited state dynamics in real time. This method models light of varying frequency in order to observe the mechanism of plasmon-mediated reactions in Au and Ag nanoparticles. Our results indicate that adsorbates can have qualitatively different charge transfer behavior. Specifically, O_2 loses electrons while N_2 gains electrons when exposed to identical electric fields on both Au and Ag nanoparticles. Additionally, charge density calculations indicate that these excitations are $\pi \rightarrow \pi^*$ on short time scales and a mixture of π, $\sigma \rightarrow \pi^*$, σ^* as time evolves. We intend to further explain the differences between O_2 and N_2 via careful inspection of the electronic structure.

632268 Bio-Project "Derisking" through Development of Systematic Methodologies and Frameworks for Risk Assessment

Abstract:

One of the primary hindrances to producing a viable, sustainable domestic biomass industry for renewable biofuels, bio-products and bio-power is the lack of understanding and quantification of the risks associated with both the biomass supply chain and preprocessing and conversion technologies. Currently a consistent method for assessing, comparing, and quantifying risks in biomass supply chains does not exist, creating a major investment barrier to bioenergy projects in the U.S. The lack of a standardized approach has resulted in bioenergy stakeholders independently using inconsistent approaches and evaluation criteria, leading to unreliable and incomparable assessments of risks and financing barriers to bio-project development. Along with the challenges of inconsistent risk assessment for supply chain risk, technology specific risks based on variability in biomass properties are not fully understood and can pose significant unforeseen challenges for bioenergy projects. In many cases these properties have not yet been identified and the impacts on the proposed technology and products unquantified. This is particularly challenging for emerging preprocessing and conversion technologies. Without a firm understanding of the preprocessing/conversion technology-specific critical properties, the risk of a proposed bio-project cannot be fully evaluated.

To address inconsistent risk evaluation in the biomass supply chain supporting project financing, a Biomass Supply Chain Risk Standards (BSCRS) framework was developed. The BSCRS framework includes a comprehensive list of known and perceived risks (Risk Indicators) to the supply chain developed through 100's of interviews with bioenergy industry experts spanning from feedstock growers and suppliers to representatives from the financial sector. These risks have been organized into a manageable hierarchy of Risk Categories and Risk Factors that can be practically assessed. This BSCRS framework also provides mitigation strategies for multiple Risk Indicators from best available industry practices and research findings. Additionally, a risk quantification methodology for each Risk Factor, Risk Category, and the bio-project as a whole was developed to enable capital markets to assess feedstock risk more efficiently and more accurately. Multiple case studies representing existing bio-projects have been used to evaluate and verify the BSCRS framework and scoring methodology.

To address technological risk *along* with the supply chain risk captured in the developed BSCRS framework, this work also focuses on development of a systematic criticality assessment tool using well-accepted, quantitative risk analysis methods to evaluate bioenergy feedstock critical properties impacting system unit operations. The proposed Failure Mode and Effect Analysis (FMEA) approach uses a team of subject area experts (SAEs) for each targeted unit operation within a system. Collectively, the team will develop and use a quantitative scoring system to assess the material attributes, process parameters, and quality attributes for key unit operations that have already been identified. The FMEA process generates Risk Priority Numbers (RPNs) for the various failures and predominant causes for each material/process unit/product combination resulting in a semi-quantitative, standardized methodology for assessing technological risk and biomass properties contributing to that risk.

629512 Design and Optimization of a Sustainable Beccs Supply Chain in the European Union to Reduce the Stress on Core Planetary Boundaries

Abstract:

The raising interest gained by negative emission technologies and practices (NETPs) in academia, industry and international organizations is motivated by the ambitious goal set by the Paris Agreement[1] to reduce the global temperature rise by 2100 well below 2°C above pre-industrial levels. In fact, since the current scenario still features massive deployment of fossil fuels in hard-to-abate or costly-to-abate sectors, such as cement production or aviation, the removal of greenhouse gases (GHG) has become necessary to meet the target. In particular, research has been focusing on carbon dioxide removal (CDR).

In recent years, diverse NETPs have been studied and developed, from nature-based to more engineering options. Among them all, bioenergy with carbon capture and storage (BECCS) has been identified as the most promising one, because of its potential to remove CO_2 from the atmosphere and provide a sustainable and reliable source of energy. In a nutshell, BECCS involves the capture of the CO_2 from the atmosphere *via* photosynthesis during the plants growth; the biomass is then processed and transported to be converted into products, e.g., bioenergy or biofuels. The CO_2 emitted during the process is captured and stored permanently underground, creating a net carbon sink.

A highly detailed BECCS supply chain (SC) model has been designed and analyzed in the context of the European Union (EU-28 referring to 2018 data) to provide insights on the impact on the planetary boundaries (PBs)[2], [3] of the infrastructure associated with the carbon removal and electricity production on a multi-regional level. The model is based on a six-type feedstock biomass chosen among energy crops and residues types. The energy crops are cultivated on marginal land to avoid competition with food production and the amount available is based on the regional yield and land availability. The model includes five echelons (cultivation, processing, pelleting, combustion with carbon capture and storage) linked by transportation stages with three different transportation means. The CO_2 is capture employing a post-combustion technology with a monoethanolamine (MEA) aqueous solution and it can be stored in three different sites.

The goal of this work is to determine the configuration of an SC across all the EU members that meets a minimum CDR requirement in a given year and minimizes the transgression on the PBs. The CDR is shared among the countries of the EU-28 according to a cooperative approach. A linear programming (LP) optimization model has been implemented and solved in GAMS[4] interfacing with CPLEX. The analysis follows a life cycle optimization approach, considering the CO_2 removal as the functional unit, and focuses on nine PBs from cradle to grave. The solution has been enforced to be cost optimal using an epsilon-constrained method.

The results of the optimization is driven by the CO_2 removal: 473 Mt of biomass is cultivated in the EU in addition to 511 Mt of residues collected. The amount of biomass determines the amount of CO_2 removed from the atmosphere and it is constrained by the storage capacity in the EU-28. The solution to the minimum transgression scenario favors the deployment of biomass residues available in each country with the assumption that they are not held accountable for impacts from cultivation such as land and water usage – the main source of controversy of BECCS. At the end of a year the negative emissions provided by the SC amount to 0.95 Gt CO_2 with a total cost of the processing and infrastructure of 156 billion EUR in 2018. The supply chain is shared among all the countries of the EU-28 (except for Malta, which does not contribute at any echelon), however transportation of biomass across national borders only occurs from the land to the pelleting sites, while the combustion takes place in the same country of pelleting. The CO_2 is then distributed *via* a pipeline network across the EU to be stored. The cooperation among the countries allows taking advantage of the land of each country to grow or collect the biomass to meet the CDR target and to compensate for those countries which do not have any storage capacity.

After the minimization of the PBs transgression, we take a step forward and analyze the results in the context of the current transgression scenario in the EU focusing on three core PBs linked to CO_2: climate change CO_2 concentration, climate change energy imbalance and ocean acidification. We calculate the current transgression using the consumption-based emissions from the EORA database. The results show that the CDR provided by the SC can reduce the stress on the three core PBs by 20% where credits for the electricity produced are not even included.

In this work, we demonstrated that BECCS is a NETP viable option to meet the goal set by the Paris agreement while accounting for all the infrastructures involved in the SC in a very detailed optimization model. Additionally, the use of biomass on marginal land and residues allows for a sustainable feedstock which does not compete with food production. We proved that the collaboration among the states makes possible almost a 1 Gt CO_2 removed in a given year and can decrease the pressure on three fundamental PBs by 20%. The cost associated with the removal is shared among all the countries and they amount to about 11% of the GDP in the whole EU in 2018.

[1] "The Paris Agreement | UNFCCC." https://unfccc.int/process-and-meetings/the-paris-agreement/the-paris-agreement.

[2] J. Rockström *et al.*, "Planetary boundaries: Exploring the safe operating space for humanity," *Ecol. Soc.*, vol. 14, no. 2, 2009, doi: 10.5751/ES-03180-140232.

[3] W. Steffen *et al.*, "Planetary boundaries: Guiding human development on a changing planet," *Science (80-.).*, vol. 347, no. 6223, 2015, doi: 10.1126/science.1259855.

[4] "GAMS - Cutting Edge Modeling." https://www.gams.com/.

624535 Techno-Economic Evaluation of Manure Pre-Treatment Strategies for thc Production of Biogas

Abstract:

Sustainable resources are playing an increasingly important role in energy generation. Previous research has shown that biomass is suitable for producing electricity through technologies such as anaerobic digestion (AD) [1]. AD is currently being applied to stabilize organic matter in animal manure, to recover energy (in the form of biogas), and to reduce environmental impact of manure application (e.g., reduction of greenhouse gas emissions) [2,3]. On the other hand, AD systems in the US face financial barriers, mainly due to high uncertainty in investment and operations, as well as the relatively low price of output products in current markets [4,5]. Studies have presented various ways to strengthen the competitiveness of AD, including co-digestion [6,7], policy incentives [8], addition of biochar [9], and pretreatment of waste [10]. Recent experimental results have shown that proper AD pretreatment methods can enhance the fermentation process of dairy manure and double the biogas yield [11]. To identify the economic feasibility of the novel pretreatment methods, techno-economic analysis needs to be conducted to quantify the trade-off between increased revenue and potential additional equipment and operational expenses.

In this work, we considered 5 different pretreatment scenarios of dairy manure, including (a) no pretreatment, (b) 2% HCl (acid), (c) 10% NaOH (alkaline), (d) 5% NaOH and 10% H2O2 (AHP), and (e) 2.5% H2SO4 and 6.0% Na2SO3 with irradiation (SPORL). For each of these scenarios, a conceptual process design and techno-economic analysis was conducted, and the corresponding mass and energy balance was estimated using a combination of experimental and simulation data. Our analysis computed capital investment and annual production cost of AD, desulphurization, combined heat and power unit, and biogas upgrading unit [12,13]. We also used a supply chain optimization framework to test the impact of logistical issues and calculated the breakeven price of derived products (electricity and biomethane) in the State of Wisconsin [14]. It was found that, while pretreatment yields a significant increase in biogas yields, the AD systems are in general not profitable. The pretreatment method (b) and (c) have the best economic performance and can reduce the breakeven price of derived products by 20-30%, while the pretreatment method (d) and (e) increase the breakeven price due to high chemical and equipment costs. A series of sensitivity analyzes were conducted to inspect process bottlenecks; we found that, surprisingly, the biogas cleaning unit is the main economic bottleneck. This cleaning unit is responsible for removing trace contaminants generated during the AD process (such as H2S). We thus conclude that research should be centered not only on increasing biogas yield but also on suppressing trace contaminants.

References:

[1] Surendra, K. C., et al. "Biogas as a sustainable energy source for developing countries: Opportunities and challenges." Renewable and Sustainable Energy Reviews 31 (2014): 846-859.

[2] MacDonald, James M. Manure use for fertilizer and for energy: report to congress. DIANE Publishing, (2009).

[3] US DA, US EPA, & US DOE. "Biogas opportunities roadmap: Voluntary actions to reduce methane emissions and increase energy independence." (2014).

[4] Gloy, Brent A., and Jonathan B. Dressler. Financial barriers to the adoption of anaerobic digestion on US livestock operations. Agricultural Finance Review (2010).

[5] Yousuf, Abu, et al. "Financial sustainability of biogas technology: Barriers, opportunities, and solutions." Energy Sources, Part B: Economics, Planning, and Policy 11.9 (2016): 841-848.

[6] Angelidaki, Irini, and L. Ellegaard. "Codigestion of manure and organic wastes in centralized biogas plants." Applied biochemistry and biotechnology 109.1 (2003): 95-105.

[7] Gelegenis, John, et al. "Optimization of biogas production by co-digesting whey with diluted poultry manure." Renewable Energy 32.13 (2007): 2147-2160.

[8] Sampat, Apoorva M., Gerardo J. Ruiz-Mercado, and Victor M. Zavala. "Economic and environmental analysis for advancing sustainable management of livestock waste: A Wisconsin Case Study." ACS sustainable chemistry & engineering 6.5 (2018): 6018-6031.

[9] Masebinu, S. O., et al. "A review of biochar properties and their roles in mitigating challenges with anaerobic digestion." Renewable and Sustainable Energy Reviews 103 (2019): 291-307.

[10] Ardİc, Ilker, and Fadime Taner. "Effects of thermal, chemical and thermochemical pretreatments to increase biogas production yield of chicken manure." Fresenius Environmental Bulletin 14.5 (2005): 373-380.

[11] Kim, Joonrae Roger, and K. G. Karthikeyan. "Solubilization of Lignocellulosic Biomass Using Pretreatments for Enhanced Methane Production during Anaerobic Digestion of Manure." ACS ES&T Engineering (2021).

[12] Beddoes, Jenifer C., et al. "An analysis of energy production costs from anaerobic digestion systems on US livestock production facilities." USDA NRCS Technical Note 1.1 (2007).

[13] Okoro, Oseweuba Valentine, and Zhifa Sun. "Desulphurisation of biogas: a systematic qualitative and economic-based quantitative review of alternative strategies." ChemEngineering 3.3 (2019): 76.

[14] Sampat, Apoorva M., et al. "Optimization formulations for multi-product supply chain networks." Computers & Chemical Engineering 104 (2017): 296-310.

625088 Modeling and Designing Renewable Energy Systems in Rural Areas with Flexible Operating Units

Abstract:

Designing energy systems with flexible inputs is increasingly important because distributed and renewable energy production forms an increasing fraction of the energy supply (Marsden, 2011; Tabassum, Kashem and Mathew, 2018). Distributed energy systems are thought to be less prone wide-spread service interruptions, and renewable energy production tends to be intermittent. For these reasons, distributed energy systems have quite different challenges compared to centralized ones, and novel design and modeling methods are in demand and have been proposed. We present a new approach to this problem focused on the structure of the system itself which not always considered in adequate detail.

In addition, the growing public interest in making energy generation more sustainable (Kutscher, Milford and Kreith, 2018) has created a rising interest in the use of renewable energy sources like wood, grass, manure, etc. Because these energy sources depend on agriculture, they are making necessary the design of energy systems able to manage flexible inputs. In addition, because these sources are low value commodities, their transportation for long distances is impractical. Sustainability, however, has a social and economic component consistent which makes the revitalization of local and regional economies also a priority. This is important because it creates stakeholder with a vested interest in sustainable energy generation.

Previous work applied the P-graph methodology to design the structure for an energy system based on biomass in rural conditions (Vance et al., 2015) typical for central Europe. But flexible inputs to the energy generation process where not explicitly considered. A more recent extension of the P-graph framework makes it possible to have flexible inputs to a production process, and this also makes the use of more accurate equipment models possible (Éles, Heckl and Cabezas, 2020). This flexibility is important because as already mentioned, the availability of the sources of renewable energy, e.g. biomass, is not steady but variable with seasons, weather, and other conditions. Hence, in this work a method based on the aforementioned extension of the P-graph methodology is proposed to design the structure of a renewable energy system in a rural area. The study uses the data collected for the region around Bad Cell in Upper Austria, Europe. Then the notion of flexible operating units is applied to the new case study, giving us the possibility of identifying optimal composition of biomass feeds and energy plant sizing given the particular geography of the region. Our model includes the collection and transportation of local renewable resources, fermenters and combined heat and power (CHP) plants, and the operation of all of these is optimize for profit from generated heat and electricity. This is again important because it has the potential of creating local stakeholders with a vested interest in the success of renewable energy generation.

The results of this work highlight the importance of thoroughly considering the effect of system structure in the design and optimization of renewable energy systems. They also offer some concrete examples where profitable energy generation systems can be established and operated in a central European setting. Future work will include a more careful consideration of system dynamics and more detailed analysis of the environmental sustainability of such processes.

References

Éles, A., Heckl, I. and Cabezas, H., 2020. Modeling technique in the P-Graph framework for operating units with flexible input ratios. Central European Journal of Operations Research, pp.1-27.

- Kutscher, C.F., Milford, J.B. and Kreith, F., 2018. Principles of sustainable energy systems. CRC Press, Boca Raton, Florida, USA.
- Marsden, J., 2011, April. Distributed generation systems: A new paradigm for sustainable energy. In 2011 IEEE Green Technologies Conference (IEEE-Green) (pp. 1-4). IEEE.
- Tabassum M., Kashem S.B.A., Mathew K. (2018) Distributed Energy Generation—Is It the Way of the Future?. In: Garg A., Bhoi A., Sanjeevikumar P., Kamani K. (eds) Advances in Power Systems and Energy Management. Lecture Notes in Electrical Engineering, vol 436. Springer, Singapore. https://doi.org/10.1007/978-981-10-4394-9_61
- Vance, L., Heckl, I., Bertok, B., Cabezas, H. and Friedler, F., 2015. Designing sustainable energy supply chains by the P-graph method for minimal cost, environmental burden, energy resources input. Journal of Cleaner Production, 94, pp.144-154.

625885 Analysis of Marginal Land Definition on Integrated Landscape Design and Biofuel Supply Chain Network Design Optimization

Abstract:

Land available for lignocellulosic feedstock production is unevenly distributed across the landscape and depends heavily on the criteria used to classify it. Additionally, the landscape-scale decisions of where to plant bioenergy crops and how to manage them may have a large effect on the efficiency of the integrated biofuel supply system. Therefore, a detailed and integrated analysis of the landscape design and biofuel supply chain network design is needed to better understand the trade-offs among different types of marginal land and how the supply chain can be designed alongside the landscape to find holistically sustainable and economically beneficial outcomes for the large-scale production of biofuels.

In this study, we use the SALUS (systems approach to land-use sustainability) process-based model to generate spatially explicit, high-resolution estimates of biomass yields and soil organic carbon (C) changes under switchgrass (*Panicum virgatum*). We then use an integrated mixed-integer linear programming approach to simultaneously optimize the landscape design and biofuel supply chain network design for switchgrass grown on alternative definitions of marginal land. These include non-arable lands (i.e., Land Capability Classifications V-VIII) and parcels identified as historically abandoned and recently abandoned in previous studies.

We find that the differences in yield distribution, soil organic C sequestration distribution, and overall available land area, lead to very different optimal supply chain configurations and policies for landscape design (establishment and management), and that C market valuation (\$/MgCO_2e) significantly influences what lands are considered favorable to establish bioenergy crops. With a higher cost of C, biomass may be planted on lands better suited to sequestering C located farther from a biorefinery despite the increase in transportation costs. Furthermore, by designing the landscape and supply chain simultaneously and at a high resolution, the model finds that integrated 'win-win' solutions are possible such that a significant amount of C negative biofuel can be produced for only a modest increase in biofuel price when compared to coarse resolution models or independently optimized landscapes and supply chains.

626370 Economic Analysis Linking Fuel Property to Potential Value of Diesel-Boiling-Range Bio-Blendstocks to Petroleum Refiners

Abstract:

Research and development of diesel-boiling-range bio-blendstocks has gained increasing interest as they exhibit beneficial fuel properties and simultaneously reduce greenhouse gas (GHG) emissions in the medium- and heavy-duty transportation sector. Recently, several bio-blendstocks have been identified with superior properties as diesel blendstocks in sooting potential, cetane number and cold weather operability, as well as reasonable potential production costs and scalability. Competitive bio-blendstocks should have either lower price or superior properties or both relative to petroleum-derived diesel blendstocks. Additionally, the potential economic values of bio-blendstocks is expected to vary with refinery configuration. In this work, we developed full-refinery linear programming (LP) models based on open-source data to identify the key fuel properties of diesel-boiling-range bio-blendstocks that can bring value to petroleum refiners and to assess the potential economic impact of blending them into the diesel pool of different petroleum refinery configurations. Eight bio-blendstocks in the diesel boiling-range and three representative refinery configurations were investigated. This presentation describes the overall modeling approach, data collection, non-linear property blending models, fuel properties as key economic drivers, and the potential economic value of bio-blendstock fuel properties to refiners. This analysis determined sulfur content and cetane number are the primary fuel properties with the potential to provide value to refiners. Note that, these refinery models will later-on be coupled with life cycle assessment to identify the potential environmental benefits and tradeoffs to refiners.

624247 Replacement of Fluorinated Aqueous Fire-Fighting Foams (AFFF)

Abstract:

Fluorinated aqueous fire-fighting foams were developed to fight flammable liquid (Class B) fires. These foams contain per- and poly-fluoro alkyl substances such as perfluorooctanesulfonic acid (PFOS). The United States military was directed to replace these foams with non-fluorinated firefighting agents by October 1, 2023 as part of the 2020 National Defense Appropriations Act (NDAA). This presentation will provide an overview of the characteristics and performance of foams used to fight Class B fires under different conditions such as pooled liquids or tank fires, along with obstacles to the development of replacement foams matching the performance of fluorinated AFFF. Lastly, issues related to the environmental performance of the replacement foams will be discussed.

623036 Green and Sustainable Processes for Critical Metal Recovery Using Oxalate Chemistry

Abstract:

Green chemistry is the design of energy-efficient chemical processes and products with environmentally benign input materials while preventing waste generation. This presentation will describe sustainable metal extraction processes from three different types of mixed metal oxide materials: spent lithium-ion battery cathodes, bauxite, and ilmenite. Recycling of spent lithium-ion battery cathodes have the potential to meet the growing demand of two critical metals: lithium (Li) and cobalt (Co). The demand for both of these metals are exponentially increasing with the rapid growth in electric vehicles. A green, closed-loop, and sustainable process for recovery of both of these metals from lithium cobalt oxide ($LiCoO_2$) has been developed utilizing oxalate chemistry. In this process, the recovery and recycle of the aqueous acid through a novel ion-exchange process minimizes waste generation. The Li is separated in the form of lithium carbonate (Li_2CO_3) and the Co is separated either as cobalt oxide (Co_3O_4) or cobalt hydroxide ($Co(OH)_2$). The oxalate-based process has also been applied for refining of ores such as bauxite and ilmenite and recovery of commercially valuable metals such as aluminum (Al), iron (Fe) and titanium (Ti). These metals can be efficiently separated and recovered in the form of their metal hydroxides. This presentation will cover the experimental and mechanistic details along with a techno-economic analysis of our closed-loop processes with existing technologies.

626392 Evaluating Green Solvent Nanoparticle Production Via Supercritical Fluid Synthesis

Abstract:

Nanoparticles, materials with at least one dimension less than 200nm, have been the subject of increasing research interest due to unique nanoscale properties for a variety of applications. In addition to high surface-area-to-volume ratio, nanoparticle shape and nature of exposed surfaces impacts functionality with much research dedicated to controlling these nanoparticle features. Supercritical fluid (SCF) synthesis can utilize “green” (e.g., non-toxic, abundant, and renewably sourced) solvents, such as water or ethanol, by applying the unique properties of a solvent as it transitions from subcritical to supercritical state to specifically manipulate nanoparticle formation, with potential to control for size and shape. It is a relatively simple process with few steps and has been successfully scaled with supercritical water, enabling manufacture of 1000 tons of nanoparticles per year. While this water synthesis has shown to be successful, scaled synthesis with supercritical ethanol is nascent and may have potential benefits over water synthesis. For example, ethanol has a lower critical point than water so potentially lower energy requirements for similar SCF synthesis benefits. In addition, ethanol can act as a reducing agent at SCF conditions, enabling synthesis of pure metal nanoparticles, while water is oxidizing and mainly creates metal oxide nanoparticles unless substantial amounts of reducing agents are added. In this study, these potential benefits are evaluated with a life cycle perspective to weigh considerations with solvents, synthesis energy consumption, and nanoparticle quality. Here, knowledge of the full-scale supercritical water plant along with Aspen process models are applied to evaluate the life cycle assessment impacts of supercritical ethanol synthesis versus supercritical water synthesis at the lab scale and full production scale. This study informs solvent selection for supercritical fluid nanoparticle synthesis to make the best decision regarding environmental impacts and quality of nanoparticle product.

624346 Green Solvent Selection Using End-of-Life Metrics

Abstract:

Solvents are an often overlooked yet indispensable aspect of chemical reactions and processing. Improving solvent consumption and reuse/recycle has the opportunity to greatly decrease overall process waste production. Current green solvent selection guides either completely ignore solvent process impacts or lack a comprehensive metric for determining the role of solvent end-of-life options in their solvent ranking. End-of-life options include solvent recycle using separations to purify and reuse the solvent or disposal/incineration. In previous works the determination of solvent end-of-life impacts are either overly simplified or require rigorous computational analysis and/or significant process and thermophysical property data. The goal of this work is to provide a comprehensive yet simple way to compare solvent end-of-life options solely relying on basic chemical properties.

The end-of-life options that are compared in this work include recycling via distillation and membrane separations as well as incineration. For each option, the minimum energy of separation is determined and compared to the energy recovered via incineration as well as energy demand for solvent manufacture. This approach is used across a large range of solvents and solvent mixtures in order to help determine the lowest impact options for each solvent and help to evaluate green solvents for various process considerations.

624493 Emergy and SPI Assessment of Solvent Recovery Pathways

Abstract:

Introduction

Conventional waste solvent handling techniques such as incineration, onsite, and offsite disposal increase the process's ecological burden and carbon footprint (Cseri et al., 2018; Das et al., 2012; Oppelt, 1990; Riber, 2007). The solvent waste generation will continue to increase as global solvent consumption expands (Chea et al., 2020). Sustainability assessment has become fundamentally important to decision-makers due to the increased environmental concerns associated with waste solvent disposal. Therefore, the recovery and reuse of solvents from waste streams is paramount to improving the overall sustainability of industrial processes (Cavanagh et al., 2014; Slater et al., 2010, 2007).

Emergy and Sustainable Process Index (SPI) are sustainability metrics that have can be used to evaluate the energy efficiency and greenness of industrial processes (Brown and Ulgiati, 2002; Krotscheck and Narodoslawsky, 1996; Narodoslawsky, 2015; Ulgiati et al., 1994). Emergy, defined as the amount of available energy of one kind that is directly and indirectly used up in transformations to make a product, can help identify the major energy-based hotspots within a process. Emergy helps to account for all the work done by nature and human-economy using solar-emjoules (sej) as the basis for calculation. However, SPI is an ecological footprint that evaluates the environmental burden of processes. It estimates the total area needed by a process to produce a unit product or service. In sustainability assessment, it is important to have multiple indices, as a single metric is insufficient in capturing and addressing the various dimensions encompassing a process. Therefore, to offer a holistic sustainability analysis for solvent recovery processes, both SPI and Emergy, need to be integrated in the evaluations.

Methodology

In this work, the Emergy and SPI of solvent recovery framework has been evaluated using a superstructure-based optimization approach that captures all the complexities of the problem. Individual technologies were categorized into stages. The Emergy models include contents of the waste solvent, material of construction of technologies, labor, purchase cost of technologies, utilities, maintenance and overhead costs. The SPI models consist of areas needed for raw material consumption, energy consumption, infrastructure, staff accommodation, embedded air, water, and soil emissions into the ecosphere. The optimization was formulated as a mixed-integer non-linear programming (MINLP) problem. The objective was to simultaneously minimize the SPI, Emergy and Economics of the recovery pathway while meeting the exit stream's purity requirements.

Results & Discussion

The recovery framework was applied to an illustrative case study. In this case study the goal is to recover dimethoxyethane (DME) at a purity of 95% from a process waste stream consisting of 21.3% DME, 35.3% water, 41.3% toluene, and 1.3% 1-ethoxy-1-methoxy ethane (EME). The combined capital and operating cost of the recovery process is $330,000 per annum (Chea et al., 2020). Figure 1 shows the results for the SPI and Emergy analysis for this representative case study. It can be observed that the major contribution to Emergy is the maintenance and overhead cost while the area needed to accommodate water emissions is the dominant footprint for the SPI analysis. The results support the need to have multiple metrics to help identify hotspots and implement corrective measures to reduce the impacts of the major contributing footprints. This was observed based on the representative case study

References

Brown, M.T., Ulgiati, S., 2002. Emergy evaluations and environmental loading of electricity production systems. J. Clean. Prod. 10, 321–334. https://doi.org/10.1016/S0959-6526(01)00043-9

Cavanagh, E.J., Savelski, M.J., Slater, C.S., 2014. Optimization of environmental impact reduction and economic feasibility of solvent waste recovery using a new software tool. Chem. Eng. Res. Des. 92, 1942–1954. https://doi.org/10.1016/j.cherd.2014.02.022

Chea, J.D., Lehr, A.L., Stengel, J.P., Savelski, M.J., Slater, C.S., Yenkie, K.M., 2020. Evaluation of Solvent Recovery Options for Economic Feasibility through a Superstructure-Based Optimization Framework. Ind. Eng. Chem. Res. 59, 5931–5944. https://doi.org/10.1021/acs.iecr.9b06725

Cseri, L., Razali, M., Pogany, P., Szekely, G., 2018. Organic Solvents in Sustainable Synthesis and Engineering, in: Green Chemistry. Elsevier, pp. 513–553. https://doi.org/10.1016/B978-0-12-809270-5.00020-0

Das, A., Gupta, A.K., Mazumder, T.N., 2012. A comprehensive risk assessment framework for offsite transportation of inflammable hazardous waste. J. Hazard. Mater. 227–228, 88–96. https://doi.org/10.1016/j.jhazmat.2012.05.014

Krotscheck, C., Narodoslawsky, M., 1996. The Sustainable Process Index a new dimension in ecological evaluation. Ecol. Eng. 6, 241–258. https://doi.org/10.1016/0925-8574(95)00060-7

Narodoslawsky, M., 2015. Chapter 3 - Sustainable process index, in: Klemeš, J.J. (Ed.), Assessing and Measuring Environmental Impact and Sustainability. Butterworth-Heinemann, Oxford, pp. 73–86. https://doi.org/10.1016/B978-0-12-799968-5.00003-8

Oppelt, E.T., 1990. Air emissions from the incineration of hazardous waste. Toxicol. Ind. Health 6, 23–51.

Riber, C., 2007. Evaluation of waste specific environmental impacts from incineration. Kgs. Lyngby.

Slater, C.S., Savelski, M.J., Hounsell, G., Urbanski, F., Geiger, J., Knoechel, D., 2010. Green Engineering Analysis of a Multi-use Solvent Recovery System for Small Volume Waste Steams in the Pharmaceutical Industry. Presented at the 2010 Meeting of the American Institute of Chemical Engineers, Salt Lake City, UT.

Slater, C.S., Savelski, M.J., Taylor, S., Kiang, S., LaPorte, T., Spangler, L., 2007. Pervaporation as a green drying process for solvent recovery in pharmaceutical production, in: Amer. Institute Chem. Engng. Annual Meeting.

Ulgiati, S., Odum, H.T., Bastianoni, S., 1994. Emergy use, environmental loading and sustainability an emergy analysis of Italy. Ecol. Model. 73, 215–268. https://doi.org/10.1016/0304-3800(94)90064-7

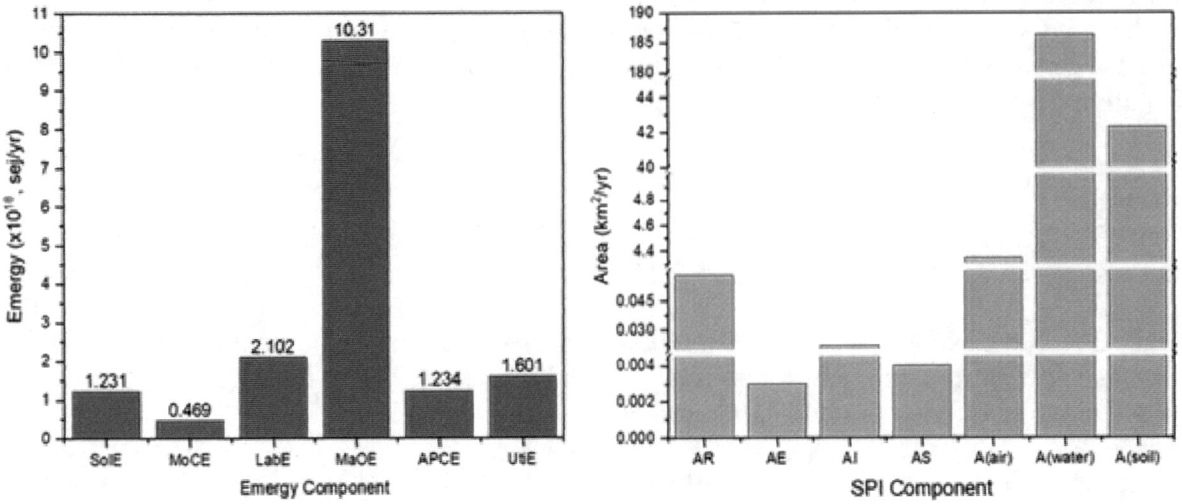

Figure 1. SPI and Emergy analysis for optimal pathway of sedimentation-pervaporation-ultrafiltration. [AR - Area for raw material consumption; AE - Area for energy consumption; AI - Area for installation; AS - Area for staff usage; A(air) - Area needed to embed air emissions; A(water) - Area needed to embed water emissions; A(soil) - Area needed to embed soil emissions; SolE – emergy content of waste solvent; MoCE – emergy associated with material of construction of selected technologies; LabE – emergy associated with labor; MaOE – emergy due to maintenance and overhead cost; APCE – emergy due to annualized purchase cost of technologies; UtiE – emergy associated with utilities]

633459 First-Principles Understanding and Design of Metal Oxides for Photocatalytic Water Splitting

Abstract:

Renewable sources of energy are increasingly needed and solar production of hydrogen fuel from water offers significant potential to contribute to these needs if new photocatalyst materials and/or more efficient photocatalytic systems can be identified for visible-light-driven water photoelectrolysis. A variety of materials have been synthesized and tested for use as electrocatalysts. Among others, transition metal oxides are considered as one of the most promising electrocatalysts as they are earth abundant, eco-friendly, and stable in aqueous solution. Particularly, Bismuth Vanadate (BiVO4) has recently been identified as an inexpensive, promising photoanode for use in water-splitting photoelectrochemical cells (PECs). However, pristine BiVO4 exhibits poor electron-hole separation and sluggish water-oxidation kinetics, limiting its solar energy conversion efficiency. Significant research efforts have been made to enhance the PEC performance of BiVO4 through doping with metal/nonmetal elements and loading of oxygen evolution co-catalysts. However, the design and successful realization of commercially viable BiVO4-based photoanodes are still rather elusive. This talk will highlight the strong effects of crystal phase and doping on the photocatalytic performance of BiVO4 that have been found from combined first-principles and experimental studies. The improved understanding offers important guidance for the rational design of metal oxide-based photocatalysts.

632359 Structural and Stability Trends in Single (ABO_3) Perovskite Oxides from DFT-Optimized Bond Valence Structures

Abstract:

Ternary perovskite oxides (ABO_3) are a diverse class of technically important materials that exhibit exceptional properties for a variety of applications; as such, ternary perovskite material properties have been characterized in several high-throughput density functional theory (DFT) investigations. To ensure computational tractability, however, these previous investigations employ structural and stability approximations that are not rigorously supported. Neglecting these assumptions, we report a comprehensive investigation of 200 experimental ABO_3 compositions generated in all possible ternary perovskite geometries— as dictated by the 11 unique Glazer tilts— and optimized with DFT. Glazer tilt enumeration was performed with the Structure Prediction and Diagnostic Software (SPuDS) optimization package— which uses a bond valence method (BVM) based global instability index (GII) minimization procedure to predict octahedral (BO_6) tilting— and relaxed with GGA+U parameters compatible with the Materials Project (MP) database. We report a general linear correlation between GII and DFT energy that enables GII minimized structures to predict DFT optimized structures, particularly those that exhibit significant octahedral tilting. ~95% of DFT predicted ground states arise from the heavily distorted a-b+a- (orthorhombic) Glazer tilt, verifying that BO_6 distortions are the primary mode of stabilization in these compositions. Furthermore, we define the "energetic well depth" to be the difference between the undistorted a0a0a0 (cubic) geometry and the DFT predicted ground state, and show that linear scaling of the well depth energy predicts the decomposition enthalpy of the ground state— computed relative to MP tabulated competing phases— with RMSE = 40 meV/atom. By leveraging this relationship and slightly modifying the GII minimization procedure used by SPuDS to improve cubic lattice predictions, stability can be rapidly and accurately predicted from a single DFT optimization of the 5-atom cubic geometry. This study thereby constitutes the first time an integrated SPuDS+DFT scheme has been used to examine perovskite oxides, and represents a valuable addition to the growing arsenal of high-throughput investigative tools for targeted ABO_3 applications.

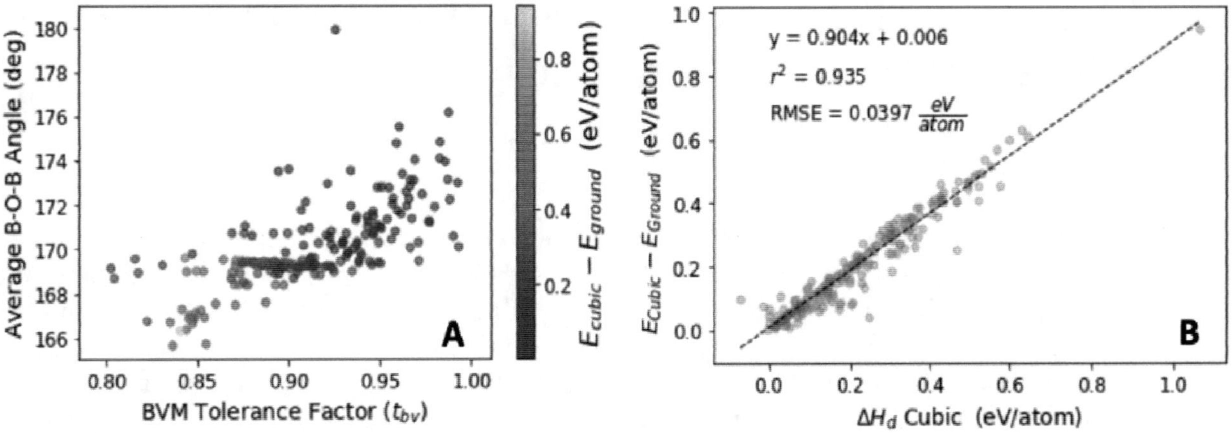

Shown for the 200 experimental compositions considered in our dataset: **A)** the relationship between the average B-O-B angle of the minimum DFT energy polymorph (i.e., the DFT predicted ground state, y_1-axis), the DFT energy difference between the undistorted cubic polymorph and the minimum DFT energy (distorted ground state, y_2-axis) polymorph, and the bond valence Goldschmidt tolerance factor (t_{bv}, x-axis). In the region $t_{bv} < 1.0$, the average B-O-B angle of the DFT ground state polymorph and the depth of the DFT predicted energetic well both increase as t_{bv} decreases, in agreement with SPuDS. In the region $t_{bv} > 1.0$, A and B site displacements supersede BO_6 tilting as the dominant form of distortion; this region is therefore beyond the scope of the current investigation. **B)** the stability of the cubic polymorph with respect to the convex hull (ΔH_d Cubic, x-axis) is linearly related to the depth of the energetic well (E_cubic - E_ground, y-axis). Linear regression shows that the magnitude of the energetic well depth is ~90% that of the cubic phase stability ($r^2 = 0.935$). Therefore, the DFT predicted ground state stability with respect to the Materials Project (MP) computed convex hull can be estimated as $\Delta H_{d,\,ground} = 0.1 \cdot \Delta H_{d,\,cubic}$ with reasonable accuracy (RMSE = 0.0397 eV/atom).

633322 Extracting the RedOx Thermodynamics of Perovskites via Combined Experiment and Theory

Abstract:

The analysis and design of thermochemical looping systems for renewable fuel production and energy storage requires detailed knowledge of the partial molar enthalpies and entropies of reduction of the active metal oxides. Experimental data normally consists of measured non-stoichiometries at a series of temperatures and oxygen partial pressures. Thermodynamic properties are then extracted via the van't Hoft or other analysis, such as the compound energy formula. These techniques often allow for trade off of enthalpy or entropy, providing non-unique solutions. In contrast computational methods such as density functional theory struggles to reproduce the complex temperature dependent behavior due to the large number of potential configurations. Here we present a new method for extracting thermodynamic properties which leverages a combination of DFT and experiment. This method removes the ambiguity between enthalpy and entropy, while providing temperature dependent behavior. The solution does not require pre-made assumptions.

633448 Materials Discovery and Development for Lower Temperature and Near Isothermal Thermochemical H2 Production

Abstract:

Materials discovery and development related to thermochemical materials for H2 production has historically been focused on materials that have large entropy changes and moderate enthalpies, in order that temperature swings between reduction and oxidation steps are minimized and thermal reduction temperatures are reduced below 1400 °C. Isothermal or near-isothermal cycling has been proposed as a means to decrease sensible heating requirements to the solid phase, but theoretical and experimental work has yielded mixed predictions and results. Based on our recent theoretical and experimental work, we postulate that the prior mixed results reported in literature stem from a combination of materials and conditions for isothermal operation (e.g., CeO2-δ at pO2 ~ 10-4 atm and 1400°C-1500 °C), that has been far from those desired for optimal performance. In this work, we implement a materials design concept that identifies and investigates new material classes, compositions, and operating conditions that are ideal for isothermal thermochemical H2 production. Starting from La-Sr-Mn perovskites, which have been theoretically and experimentally identified as having more favorable properties than ceria, we are utilizing high throughput screening coupled to direct DFT computations of down-selected materials for key performance properties such as oxide formation energies and oxygen vacancy formation and migration energies. DFT computation and screening is coupled closely with experimental synthesis of powdered, replica and direct foam compositions that afford enhanced heat and mass transfer properties. Compositions are experimentally characterized through thermogravimetric analysis and relaxation experiments at a larger scale to test the stability and performance of the replica and foamed compositions.

629134 Thermodynamic Characterization of Doped Spinels for Thermochemical Fuel Production

Abstract:

Technologies such as solar thermochemical hydrogen (STCH) production, which leverage the oxygen-exchange capacity of metal oxides to split water, are thermodynamically attractive, as the entire solar spectrum can be exploited to supply process heat. Here, a metal oxide is subjected to cyclic reduction-oxidation reactions that facilitate either oxygen or hydrogen production, depending on the operating conditions. Chemical thermodynamics suggest that the forward direction of each redox half-reaction is favorable in different temperature and/or oxygen partial pressure regions, and as a result, in practice, large irreversibilities generally arise from sensible heating between redox regimes. To address this issue, recent research has primarily focused on: (1) investigating novel perovskite materials with tunable thermodynamic properties that enable greater reduction extents at more moderate temperatures; and (2) integrating auxiliary techniques, such as mechanical or thermochemical pumping, that reduce the operating oxygen partial pressure (and thus temperature) in the endothermic reduction step. However, until highly effective and reliable solid-solid heat recovery methods are devised, inefficiencies inherent to temperature-swing operation will continue to negatively impact performance. Alternatively, at the expense of lower hydrogen productivity, the water-splitting (oxidation) step can be operated at temperatures where reduction is favorable, thereby minimizing thermal stresses and avoiding complexities associated with solid-solid heat recuperation. Herein, motivated by the immediate practicality of isothermal operation, the behavior of low-cost cobalt-iron aluminate spinels was characterized under conditions relevant to isothermal STCH. Using thermogravimetry, equilibrated changes in oxygen content were quantified at different oxygen partial pressures and temperatures to extract thermodynamic properties, namely the standard molar enthalpy and entropy of relevant defect reactions, and thus enable performance predictions. Results indicate that this class of redox intermediates is capable of higher hydrogen productivities under isothermal operation than previously established benchmark materials (e.g., ceria and lanthanum manganite-based perovskites) evaluated under temperature-swing. The favorable thermodynamic properties of spinels at conditions better suited for industrialization present a step forward towards sustainable hydrogen production with solar thermal energy.

622622 Sustainable High-Purity Nitrogen Production Via a Coupled PSA-Thermochemical Process for the Ammonia Industry

Abstract:

In order to achieve the Paris agreement's target of maximum global warming of 1.5 °C, a sustainable and CO_2 lean production, not only of fuels, but also of key commodities like ammonia, is necessary. Ammonia is a fundamental chemical commodity for fertilizer production, and becomes increasingly important as an energy vector, especially considered as fuel for maritime applications. Since not only the production of H_2, but also the production of the feedstock N_2 itself causes high CO_2 emissions, its sustainable production is mandatory. Economic analysis reveals that the combination of a solar-energy-powered pressure-swing absorption (PSA) and a redox-oxide-based, solar thermochemical cycle decreases the thermal energy input of high-purity N_2-production significantly. The viability of thermochemical air separation and the benefit of using pre-purified air has already been proven in laboratory scale experiments, but the approach of coupling PSA with such a thermochemical cycle has not yet been demonstrated.

In this contribution we present an overall approach including both the redox oxide material selection as well as the solar reactor design. The characteristics of the selected redox materials are analyzed by in-situ high-temperature x-ray diffraction (XRD) in a temperature range of 500-1100°C and thermogravimetric analysis (TGA). Various oxygen partial pressures (0-83 % O_2) are applied in both characterization methods.

The proposed air separation process is demonstrated with a PSA unit coupled to an indirectly heated, pressurized packed bed reactor (Fig. 1). The influence of the process parameters such as pressure, reduction and oxidation temperatures on the performance criteria are shown and compared to results gained by simulations. Moreover, we illustrate how the process parameters can be influenced by tuning the composition of a representative redox material $Ca_{1-x}Sr_xMnO_3$. Therefore, the impact of the Sr-amount on the structure, structural changes and the enthalpy are presented. The performance of the overall system is evaluated on the basis of the achieved purity of the outlet gases, the volume flow rates and the efficiency.

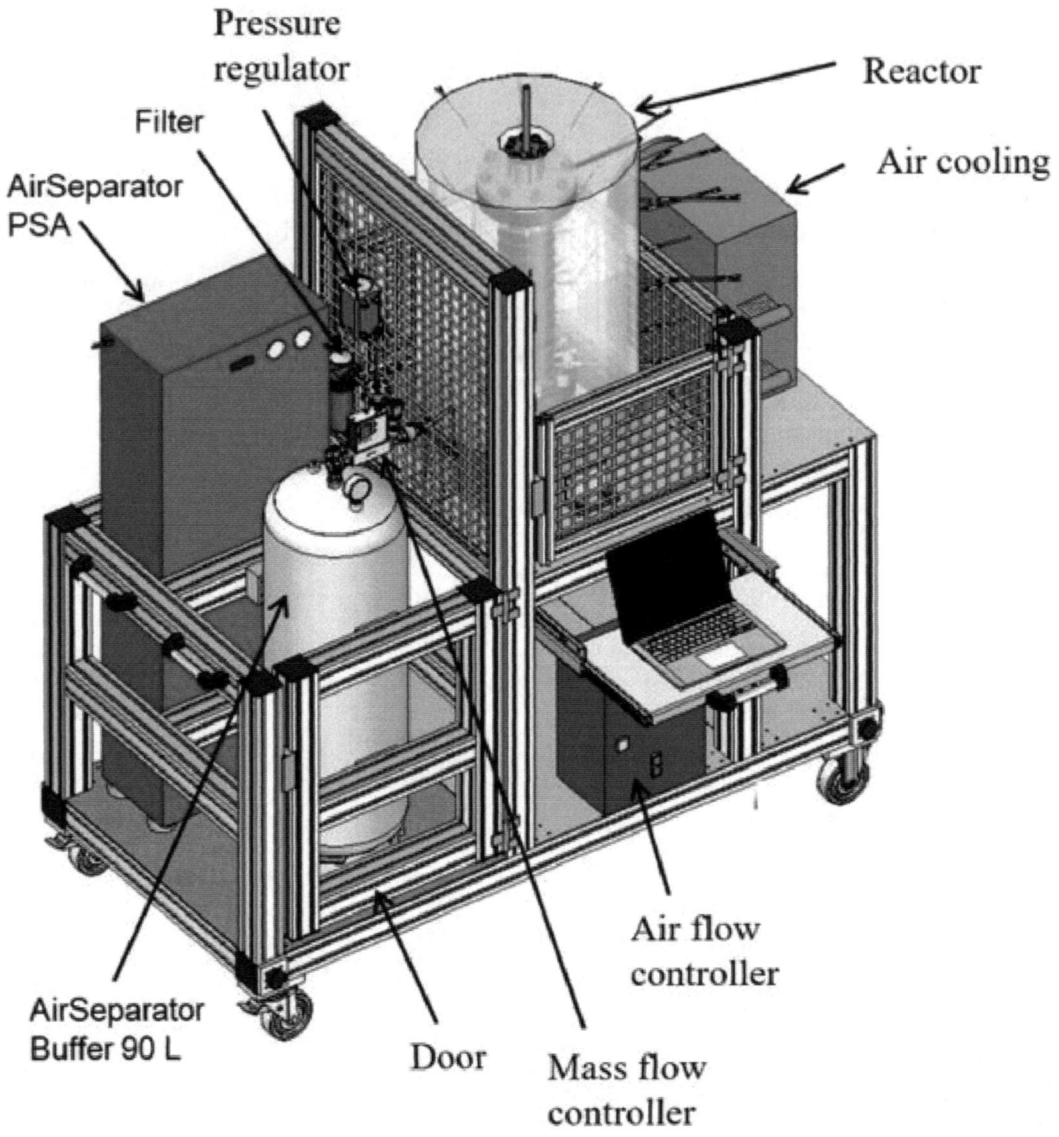

Fig. 1. Reactor design for a two-step air separation process

633022 High-Purity Nitrogen Production from Air By Pressure Swing Adsorption Combined with $SrFeO_3$ Redox Chemical Looping for Trace Oxygen Removal

Abstract:

This work focuses on the production of high purity nitrogen from air via a combination of pressure swing adsorption (PSA) with a downstream redox chemical looping cycle used as a deoxygenation unit. The non-stoichiometric perovskite $SrFeO_{3-\delta}$ was selected for the redox chemical looping cycle because of its favorable thermodynamics, rapid oxidation kinetics and intermediate reduction temperatures. A first principles analysis of the system showed that this combined process could improve upon both the energy balance and production capacity of a PSA system. The chemical stability of the material was demonstrated over 250 redox cycles via thermogravimetry with less than 1 % degradation in the chemical storage capacity. The oxidation kinetics were also measured at 523 K over a range of oxygen concentrations, and fit with a suitable kinetic model. The kinetics were then incorporated into a 1D convection-diffusion model of a packed bed reactor. The model indicates that, for a targeted oxygen impurity of $x_{O2} < 3 \times 10^{-6}$, a chemical looping unit added downstream to a PSA system could approximately triple the capacity and reduce the energy demand to 14 kJ/mol of N_2 [Bulfin *et al.* 2020]. The results are promising for further study, with the goal of bridging the gap between PSA and cryogenic systems in terms of scale, purity and cost.

Bulfin, B., Buttsworth, L., Lidor, A., & Steinfeld, A. (2020). High-purity nitrogen production from air by pressure swing adsorption combined with SrFeO3 redox chemical looping. Chemical Engineering Journal, 127734.

.

633348 Development of low-cost Receivers for Parabolic Trough Assembly for developing countries like India: Optical and Thermal optimization

Abstract:

India's energy consumption has seen an exponential growth of 91% between 1990 to 2008. As per International Energy Agency (IEA), India's energy demand will increase by 50% by 2030, and by 2040 India will have the largest increase in energy demand. This will indeed lead to an increase in pollution and carbon footprint. For sustainable development, India has to develop renewable energy technology at a higher pace. Amongst renewable sources, solar energy has seen a massive development in the past couple of years. Installed capacity has increased from 12.3 GW to 37 GW from 2017 to 2020 which is roughly 4% of total power generation. Based on India's ambitious policy, more than 40% of the total power generated will be solar power. Presently, PV and CSP contribute 34.62 GW and 500 MW respectively. PV is the cheaper and more developed form of solar technology as compared to CSP. With advances in thermal energy storage (TES) systems, CSP can be made cheaper than PV. The worldwide installed capacity of CSP has increased from 1.3 GW to 6.3 GW in the last decade.

Among CSP, Parabolic Trough Collector (PTC) is the most mature technology. The worldwide installed capacity of PTC has increased from 632 MWe to 4.7 GWe in the last decade. Receiver tubes constitute nearly 30% of total solar field cost and hence play a vital role in deciding the cost of power production. The minimum cost of power production in India is INR 10.5 at Diwakar, Rajasthan. This cost is incompetent concerning PV and hence low-cost receivers have to be deployed to bring down the production cost.

A patented novel PTC design has been fabricated to harness solar energy efficiently. The cost of assembly is $75/m^2-aperture. To bring down the cost of assembly, a single parabolic reflector has been replaced by multiple linear mirror strips of 30 mm width and 1 m length placed on a laser-cut parabolic frame. This provides operational flexibility and easy replacement in case of breakage. The proposed square cavity receiver is optimized by an in-house solar ray-tracing code. The maximum optical efficiency attained by the system is 83% for highly reflecting glasses and a highly absorbing solar selective coat on the absorber tube. Based on the optimization, a cavity of 60 mm width (aperture width) and 71 mm height is used for further thermal optimization. The center of the absorber tube is positioned at a height of 1.6.45 m from the vertex of the parabolic frame.

CFD technique has been applied to predict the thermal performance of the proposed air-filled square cavity receiver. Heat loss value at various absorber tube temperatures was: 44.75 W(200 ^{0}C), 64.27 W(250 ^{0}C), 87.89 W(300 ^{0}C), 116.36 W(250 ^{0}C), 150.5 W(400 ^{0}C), 191.24 W(450 ^{0}C) and 239.42 W (500 ^{0}C). Simulations have been carried out to study the effect of key performance parameters like emissivity of solar selective coat, orientational of receivers, the reflectivity of optical cavity and casing of receiver, wind flow around the receiver, and insulation material. To compare the performance of the proposed receiver, data has been compared with commercialized SCHOTT PTR® 70 Receivers. Heat loss from proposed receiver against SCHOTT at respectively

627897 Solar Thermochemical Water Splitting Using Iron-Aluminate Spinels

Abstract:

Technologies like solar thermochemical hydrogen (STCH) production, which leverage the oxygen-exchange properties of metal oxides to split water, are attractive because they can harness the entire solar spectrum, whereas photovoltaic-driven alternatives rely only on specific wavelength ranges. Generally, this approach involves alternating between two steps namely, (1) the endothermic removal of oxygen from an oxide and (2) the exothermic oxidation of the reduced oxide with water to produce hydrogen. The endo/exothermic nature of each reaction implies that opposite changes in temperature are required to drive the hydrogen yield per cycle. However, in practice, implementing a temperature swing between redox regimes imposes significant irreversibilities that arise from sensible heating of the solid, thus resulting in inefficient operation unless heat is recovered via ultra-high temperature thermal energy storage (a technology still in its infancy). Therefore, in this work, a scalable prototype reactor – situated within a well-insulated cavity receiver – was designed and operated such that the water-splitting (oxidation) step was intentionally initiated at temperatures where reduction is favorable, thereby avoiding the complications associated with solid-solid heat recuperation. To mitigate the undesired thermodynamic consequence of operating isothermally, iron aluminate spinels, an earth-abundant class of materials with an exceptional capacity for hydrogen production, were considered for this reactor demonstration. The stable experimental results, obtained under concentrated radiation from a 45 kW_e solar simulator, provide compelling motivation for the commercial viability of solar thermochemical routes for hydrogen production.

627942 Numerical Simulation of Concentrated Solar Energy Absorption in Packed and Fluidized Bed Systems

Abstract:

Concentrated solar energy (CSE) is an excellent source for thermal energy storage medium in packed bed of near-blackbody particles. In addition, CSE is capable of directly providing the required clean energy for many chemical, pharmaceutical, and biological processes. Thus, understanding and searching for the most efficient way to store concentrated solar radiation and transfer the stored energy is very important. In this study, we attempt to gain a better understanding of the radiation energy by CSE received by packed and fluidized bed systems, using our Computational Fluid Dynamic (CFD) model to simulated radiative energy received and distributed in our fluidized and packed bed systems. In the fluidized bed system, gas enters the bottom of the fluidized through a distributor, specially designed to eliminate the large bubbles that cause gas bypassing and loss of particles by entrainment. Inside the fluidized bed, particles receive concentrated solar power, while the gas directly removes the heat from the high-temperature particles. Drag forced modification was developed based on the Clark sub-gird model for fluidized beds. The radiative heat transfer was modeled as an effective conductivity based on the Breitbach and Barthels study and the conductive heat transfer was modeled using Zehner-Bauer-Schlünder (ZBS) conductivity model. The numerical simulation results were compared with experimental data of Tregambi et al (2016). The result of our two-dimensional simulation agreed well with thermal measurement by infrared cameral on a directly irradiated packed bed and fluidized bed solar absorber experiments by Tregambi et al.

Tregambi, C., Chirone, R., Montagnaro, F., Salatino, P., Solimene, R., 2016. Heat transfer in directly irradiated fluidized beds. Solar Energy, 129, pp.85-100.

631755 Operational Limits of Redox Metal Oxides Performing Thermochemical Water Splitting

Abstract:

Abstract:

Solar thermochemical hydrogen production is an attractive technology that enables the storage of intermittent solar energy in the form of chemical bonds. Efficient operation requires the identification of a redox-active metal oxide (MO_x) material that can achieve high conversion of water to hydrogen at minimal energy input. The water splitting occurs by consecutive reduction and re-oxidation reactions. The MO_x is reduced to $MO_{x-\delta}$ and, in the second step, is re-oxidized by water to recover the initial MO_x. The material must reduce to $MO_{x-\delta}$ at temperatures achievable in concentrated solar receiver/reactors, while maintaining a thermodynamic driving force to split water. At equilibrium, extent of reduction depends on temperature and oxygen partial pressure, and in this analysis, a set of thermodynamic properties, namely enthalpy, and entropy of oxygen vacancy formation, is sufficient to represent the MO_x. The work presented here is a method to easily classify materials based on these thermodynamic properties under any condition of oxygen partial pressure and temperature.

Methodology:

This work extends the thermodynamic analysis performed by Bayon et al. [1] and Li et al. [2]. The chemical equilibrium of the thermal reduction reaction at T_{red} and $pO_{2,red}$ is given by:

$\Delta h_{o,red} - \Delta s_{o,red}\, T_{red} = -1/2RT_{red} \ln((pO_{2,red})/(p°))$ (1)

And the equilibrium for the reoxidation reaction is analogous and described by the conversion of H_2O into H_2 (θ) as well as the equilibrium constant K_{ws}:

$\Delta h_{o,ox} - \Delta s_{o,ox}\, T_{ox} = -RT_{ox} \ln((1-\theta)/\theta\, K_{WS})$ (2)

This work is based on the countercurrent reactor and the definition of the thermochemical properties at the outlet of the reduction and inlet of the oxidation provides the operational limits in terms of T_{red}, $pO_{2,red}$, T_{ox} and θ. Using Equations 1 and 2, the enthalpy of oxygen vacancy formation can be obtained from the entropy and temperatures of reduction and re-oxidation separately. To study the effect of these system variables, we hold fixed the reduction oxygen partial pressure at the inlet of the reduction reactor and the conversion of water into hydrogen leaving the re-oxidation reactor.

Advance of results:

An example of the application of the thermodynamic method, this section reports the system effects of for reduction oxygen partial pressure of 10^{-5} and a value of conversion from H_2O to H_2 of 10%. By overlapping both equilibriums lines described in equation 1 and 2, we can obtain Figure 1. Generally, the higher the enthalpy of oxygen vacancy formation and the lower the entropy of oxygen vacancy formation, results in lower temperature differences between reduction and re-oxidation (note the

-1

difference of the height in the dashed and solid lines for 300, 400 and 500 kJ mol). Only with high reduction temperatures is it possible to split water at high re-oxidation temperatures. Lowering the reduction temperature results in increasing the temperature difference between reduction and re-oxidation (see distance between both dashed and solid lines in Figure 1).

The intersection of both lines is the isothermal equilibrium temperature. Isothermal water splitting will only be possible, for a conversion of 10% of water, at a temperature of 2030 K and for unique pair of values of enthalpy and entropy. The pairs of and are: 300 kJ mol^{-1} and 100 J mol^{-1} K^{-1}; 400 kJ mol^{-1} and 149 J mol^{-1} K^{-1}; and 500 kJ mol^{-1} and 200 J mol^{-1} K^{-1}. We cannot find materials in the literature that can meet the isothermal criteria for oxygen partial pressure of 10^{-5} in reduction and water conversion of 10% in re-oxidation. Either the values of entropy at a given enthalpy are too high or the values of enthalpy at a given entropy are too low. Above that intersection of oxygen partial pressures, we could potentially re-oxidize the material at a higher temperature than the reduction.

The work presented in this abstract will be extended to cover a wide range of conditions of oxygen partial pressure and temperatures. In addition, we will be presenting a classification of redox materials in operational ranges using the method briefly presented here. Despite the simplicity of the method, we believe this analysis will support future research in targeting thermodynamic properties of redox active metal oxides.

References:

[1] A. Bayon, A. de la Calle, K. K. Ghose, A. J. Page, R. McNaughton. *Int. J. Hydrogen Energy* **2020**, *45*, 12653–12679.

[2] S. Li, V. M. Wheeler, A. Kumar, M. B. Venkataraman, C. L. Muhich, Y. Hao, W. Lipiński. *Energy Technol.* **2021**, *2000925* (19), 1–18, DOI: 10.1002/ente.202000925.

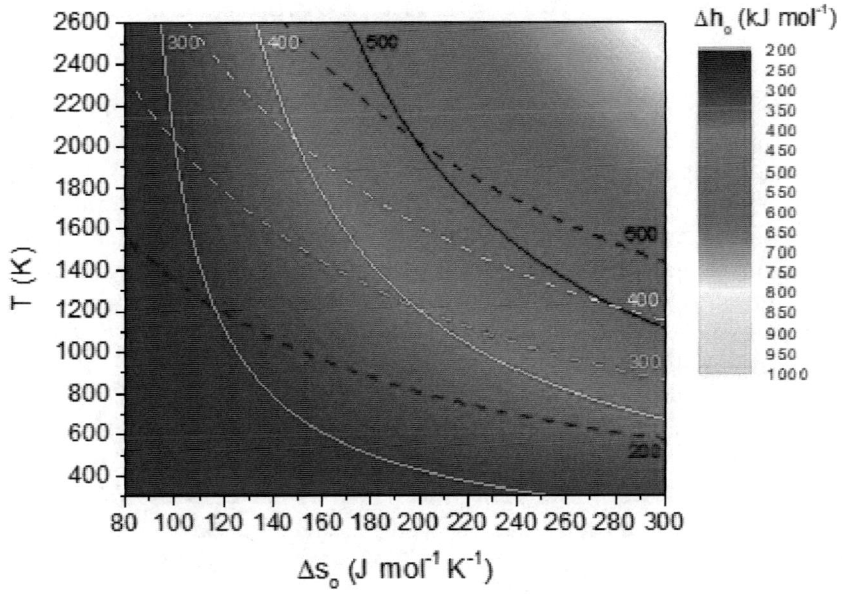

Figure 1: Overlapping of both equilibriums for several values of Δh_o and Δs_o. The thermal reduction is represented by dashed lines of $\frac{p_{O_2,red,in}}{p^o} = 10^{-5}$ and re-oxidation is represented by solid lines ($\theta_{out} = 10\%$). Lines highlighted shows Δh_o values of 200, 300, 400 and 500 kJ mol^{-1}.

632868 Towards Chemical Equilibrium in Thermochemical Water Splitting

Abstract:

The efficiency of many processes strongly depends on limiting thermodynamic irreversibilities, i.e., how close to equilibrium the system is throughout the process. In thermochemical cycles for water and/or carbon dioxide splitting, operating near equilibrium means that the chemical potential of oxygen in the solid and gas phases must not differ significantly. Here, we show that in co- and counter-current reactors for two-step redox-active metal oxide thermochemical cycles, and in the absence of multiple gas or solid injection points, approaching this ideal reversible path is possible only if the reactions occur at specific, material-dependent, thermodynamic conditions.

The method presented determines the chemical equilibrium at each infinitesimal point along the reaction and flow path, leading to a variable temperature, thermal reduction and re-oxidation reaction, based on both mass conservation and the Gibbs chemical equilibrium criterion. The mass balance imposes a linear relationship between the extent of reduction (δ) and the O_2/inert gas partial pressure ratio ($p^*=pO_2/p_{inert}$) in thermal reduction or δ and the H_2 molar fraction ($\theta=pH_2/(pH_2+pH_2O)$) in re-oxidation. The Gibbs criterion limits the temperature (T) for which the reaction can take place spontaneously at each infinitesimal point along the flow path. The reaction Gibbs free energy depends on three variables for each reaction: T, pO_2 and δ for thermal reduction, and, T, θ and $\delta$for re-oxidation. Since at equilibrium the reaction Gibbs free energy is zero, knowing any two of the three variables, (T, pO_2, δ) or (T, θ, δ) determines the third. For a co- and counter-current reaction, the mass balance line further constrains (pO_2, δ) for thermal reduction and (θ, δ) for re-oxidation. The equilibrium reaction temperature profile is therefore not arbitrary but follows uniquely from the mass balance and the Gibbs criterion and the boundary conditions.

A common simplification used for thermodynamic system analyses is to assume that each reaction occurs at a single temperature. However, a single-temperature (or isothermal) reacting path leads to the reaction having to always be out of equilibrium at all but one point along the reaction path. Although such an isothermal flow path is theoretically possible, engineering for constant temperature is challenging given expected fast kinetics, as a heat source or heat sink would struggle to supply the energy balance in response. In both cases, the isothermal and the equilibrium path will be compared in this presentation.

The design of the reactor for each reaction depends on four operating variables which fully determines both ends of the mass balance line. Although, the most obvious selection could be the coordinates of both ends at the mass balance line, using these boundary conditions are not the most convenient to specify, since the outlet extent of reduction in both reactions and also the outlet oxygen partial pressure in thermal reduction are more typically reactor outputs. Instead of these four variables, we find it more convenient to set and control: a) the inlet extent of reduction (δ_{OX}), the inlet oxygen partial pressure ($pO_{2,in}$), the maximum reduction temperature ($T_{TR,max}$), and the inert sweep ratio ($\lambda=n_{inert}/2nO_2$) for thermal reduction and b) the inlet extent of reduction (δ_{TR}), the inlet and outlet hydrogen molar fraction (θ_{in} and θ_{out}, respectively) and the minimum temperature ($T_{OX,min}$) for re-oxidation.

Since the efficiency of two-step thermochemical water and carbon dioxide splitting cycles are critically dependent on the selection of the operating conditions, we have analyzed these conditions in detail. As an example, we choose ceria for the redox active metal oxide, the Zinkevich, et al. [1] model for the ceria thermodynamic properties, and the Coolprop library [2] for the oxygen, hydrogen and water thermodynamic properties.

For thermal reduction, we observe that for a given peak reduction temperature ($T_{TR,max}$) and inert sweep ratio (λ), the purity of the inert gas ($pO_{2,in}$) has a limited influence on the outlet extent of reduction, the relevance decreasing with decreasing $pO_{2,in}$. The reaction is almost complete at the peak temperature, and after that the primary effect is a trade-off between T and pO_2, not substantially furthering the reaction extent. This result suggests a potential process optimization avenue, in terms of resources expended for inert gas purification. Something similar happens with the re-oxidation. For a given minimum re-oxidation temperature and outlet H_2 molar fraction the inlet hydrogen molar fraction has a limited influence. Again, this result suggests a potential process optimization avenue, in terms of resources expended for the H_2/water separation at the gas outlet/ solid inlet of the reactor.

Increasing the inert sweep ratio makes more relevant the purity of the sweep gas in the reduction reaction, further increasing the cycle extent of reduction ($\Delta\delta$). However, the feasibility of high sweep ratios has limitations. In practical reactors, flows may not be arbitrarily fast and must allow sufficient interaction time for the mass and heat transfer processes to approach equilibrium. Similarly, reducing the outlet H_2 molar fraction makes more relevant the purity of the water stream in the re-oxidation reaction, also increasing $\Delta\delta$. However, the feasibility of reducing the outlet H_2 fraction also has limitations since the fraction is indicative of the amount of excess steam in the reactor.

In conclusion, the temperature path, i.e., increasing the temperature difference between the peak reduction temperature and the minimum re-oxidation temperature, is the less restrictive approach than the presented before (i.e. reduce the $pO_{2,in}$ and/or θ_{in}, and increase λ and/or reduce the θ_{out}) to high thermochemical water and carbon dioxide splitting efficiencies assuming feasibility of efficient high-temperature heat recovery and high temperature structural materials.

References:

[1] M. Zinkevich, D. Djurovic, and F. Aldinger, “Thermodynamic modelling of the cerium-oxygen system,” *Solid State Ionics*, vol. 177, no. 11–12, pp. 989–1001, 2006.

[2] I. H. Bell, J. Wronski, S. Quoilin, and V. Lemort, “Pure andPseudo-pure Fluid Thermophysical Property Evaluation and the Open-SourceThermophysical Property Library CoolProp.,” *Ind. Eng. Chem. Res.*, vol. 53, no. 6, pp. 2498–2508, Feb. 2014.

[3] A. de la Calle, I. Ermanoski, and E. B. Stechel, “Towards Chemical Equilibrium in Thermochemical Water Splitting. Part 1: Thermal Reduction,” In preparation.

[4] A. de la Calle, I. Ermanoski, and E. B. Stechel, “Towards Chemical Equilibrium in Thermochemical Water Splitting. Part 2: Re-oxidation,” In preparation.

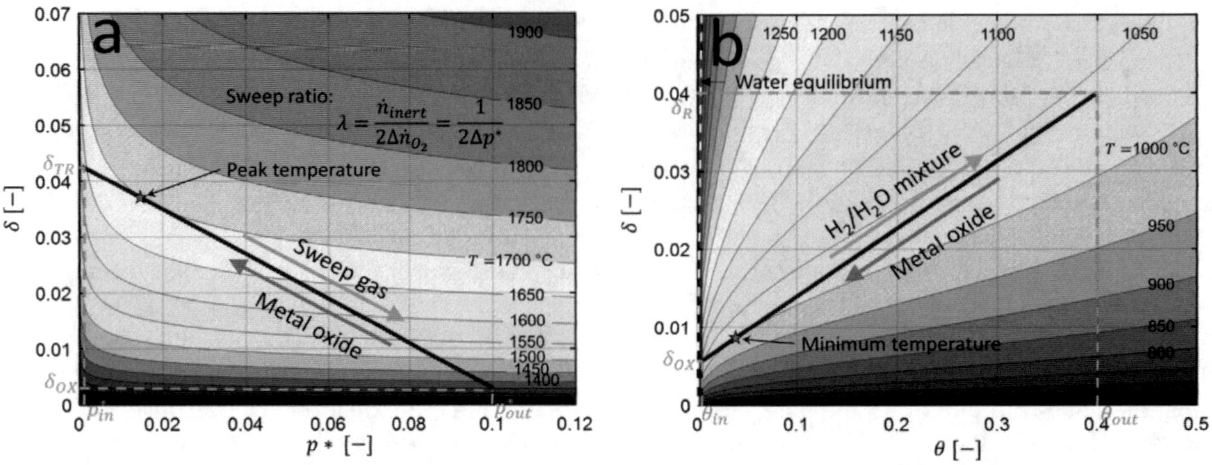

Figure 1: a) Thermal reduction reaction: mass balance representation in $\delta - p^*$ coordinates
b) Re-oxidation reaction: mass balance representation in $\delta - \theta$ coordinates

627357 Kinetic Investigation of Iron Aluminate-Based Materials for Fuel Production

Abstract:

Thermochemical gas splitting of water and/or carbon dioxide via redox is a promising technology to renewably produce syngas for the diversification of gas-to-liquid (GTL) fuels. Here, a metal oxide intermediate undergoes a two-step redox cycle, where either oxygen or hydrogen and/or carbon monoxide is produced. To date, experimental demonstrations have yet to report efficiencies necessary for competing with conventional technologies, primarily due to the insufficient properties of the active materials considered and the requirement for large temperature swings between redox regimes. As a result, recent research has largely focused on discovering new materials with thermodynamic properties that permit lower temperature operation without sacrificing product yields. However, such efforts have proven difficult, as materials that are reducible at low temperatures often become harder to oxidize (and produce fuel).

In terms of efficiency, performing each reaction at the same temperature is appealing due to the elimination of the significant sensible heating penalties that are associated with conventional temperature swing strategies. Although isothermal operation implies lower hydrogen/carbon monoxide yields due to operating the exothermic oxidation reaction under unfavorable conditions, material selection for isothermal shifts to focus on materials that undergo large extents of reaction within the attainable range of oxygen chemical potential.

Recent research has shown that iron aluminate-based materials exhibit uniquely large yields under isothermal conditions, making them ideal candidates for efficient fuel production. Here, a kinetic investigation of these materials is conducted in a stagnation flow reactor and thermogravimetric analyzer to gain insight on how these materials split water and/or carbon dioxide. Results indicate these materials can provide a path forward to a clean, efficient generation of fuels.

624128 An Optimization-Based Framework for Material Selection and System Design for Integrating Thermochemical Energy Storage in Solar Power Systems

Abstract:

A promising approach to manage intermittency in concentrating solar power (CSP) plants is to integrate them with thermal energy storage (TES). It is estimated that CSP plants integrated with TES can contribute to 11% of the global electricity generation by 2050.[1] The commercially deployed CSP plants, which are typically integrated with two-tank molten-salt sensible heat storage unit, however, are less efficient because they deliver heat to the power block at low temperature (565 ^{o}C). Furthermore, the energy density of molten-salt systems is low and it can become too expensive to be used in large-scale systems. Thermochemical energy storage (TCES), which is based on converting solar-thermal energy to chemical energy, is a promising option because it allows high operating temperatures, high storage density, and low heat loss over long periods.

Today, TCES is in its infancy due to, among other reasons, the system complexity. It remains unclear which configuration and reaction should be used for the integrated CSP-TCES system. Towards this goal, we make contributions in three aspects. First, we identify nine major TCES strategies by combining three representative types of reactions (oxides, hydroxides, and carbonates) and six configurations. Second, we develop a novel two-stage stochastic programming optimization model for the design and operation of CSP plants under seasonal solar variability[2,3]. Third, using this model, we obtain the optimal performance of all TCES strategies. This analysis allows us to develop general guidelines for the selection of optimal process configurations.

In the second part of the talk, we discuss an extensive reaction screening to generate a ranked list of promising reaction candidates. To do this, we first generate a comprehensive list of reaction candidates. Then, we perform a preliminary screening, and choose the reactions with positive reaction enthalpy and equilibrium temperatures between 400 ^{o}C and 1500 ^{o}C. Finally, the model is solved for the reaction candidates obtained after preliminary screening and the reactions are ranked based on the cost of electricity production. The analysis identifies available promising TCES materials. Finally, we formulate a "material optimization" problem by treating key material properties as decision variables. This analysis provides insights in the material and reaction properties that are most suitable for CSP-TCES plants and serve as guidelines for the development of novel TCES systems.

References

1 IEA publications, *Energy Technology Perspectives 2017 - Catalysing Energy Technology Transformations*, 2017.

2 X. Peng, T. W. Root and C. T. Maravelias, *AIChE J.*, 2019, **65**, e16458.

3 X. Peng, M. Yao, T. W. Root and C. T. Maravelias, *Appl. Energy*, 2020, **262**, 114543.

625062 Economic Analysis and Environmental Impact of a Novel Solar Parabolic Trough Plant Used for Industrial Process Heat That Utilizes Flexible Heat Integration

Abstract:

Despite industrial process heat representing a large portion of the world's energy consumption, solar industrial process heat (SIPH) provides a very small percentage of supplied process heat worldwide. SIPH plants have found to vary if they are financially viable for different cases and applications [1-2], but have an environmental benefit by reducing fossil fuel emissions from traditional energy sources. Because of this, more research is needed in analyzing the environmental benefits of SIPH in tandem with their economic viability to enable ways for SIPH to become more common.

This study focuses on an economic analysis and environmental impact of a novel SIPH plant that utilizes flexible heat integration (FHI) (see Figure 1). FHI has been studied in solar thermal power plant applications [3-5], but its application is novel for SIPH. Flexible heat integration allows for more degrees of freedoms in solar thermal plants, allowing the plant to respond optimally to changing weather conditions. In the novel SIPH plant proposed, the heat transfer fluid (HTF) setpoint and the ratio of HTF flow going to high and low temperature heat exchangers are allowed to be manipulated based on a heuristic optimization scheme. Economic analysis of this SIPH plant is conducted for two cases, a plant that doesn't include FHI as well as a plant that utilizes optimized FHI. Results show a lower levelized cost of heat (LCOH) by implementing FHI and a quicker return on investment by offsetting natural gas usage. An environmental impact is also conducted showing a large reduction in emissions for both SIPH cases, with a larger reduction coming from the optimized FHI case. This highlights that through an economic and environmental lens, SIPH plants utilizing FHI can help to push more renewable energy in industrial process heat applications.

[1] A. K. Sharma, C. Sharma, S. C. Mullick, and T. C. Kandpal, "Financial viability of solar industrial process heating and cost of carbon mitigation: A case of dairy industry in India," *Sustain. Energy Technol. Assessments*, vol. 27, no. March, pp. 1–8, 2018, doi: 10.1016/j.seta.2018.03.007.

[2] G. Quiñones, C. Felbol, C. Valenzuela, J. M. Cardemil, and R. A. Escobar, "Analyzing the potential for solar thermal energy utilization in the Chilean copper mining industry," *Sol. Energy*, vol. 197, pp. 292–310, Feb. 2020, doi: 10.1016/j.solener.2020.01.009.

[3] K. Ellingwood, K. Mohammadi, and K. Powell, "A novel means to flexibly operate a hybrid concentrated solar power plant and improve operation during non-ideal direct normal irradiation conditions," *Energy Convers. Manag.*, vol. 203, no. August 2019, p. 112275, 2020, doi: 10.1016/j.enconman.2019.112275.

[4] K. Rashid, S. M. Safdarnejad, and K. M. Powell, "Process intensification of solar thermal power using hybridization, flexible heat integration, and real-time optimization," *Chem. Eng. Process. - Process Intensif.*, vol. 139, no. January, pp. 155–171, 2019, doi: 10.1016/j.cep.2019.04.004.

[5] K. Ellingwood, K. Mohammadi, and K. Powell, "Dynamic optimization and economic evaluation of flexible heat integration in a hybrid concentrated solar power plant," *Appl. Energy*, vol. 276, no. April, p. 115513, 2020, doi: 10.1016/j.apenergy.2020.115513.

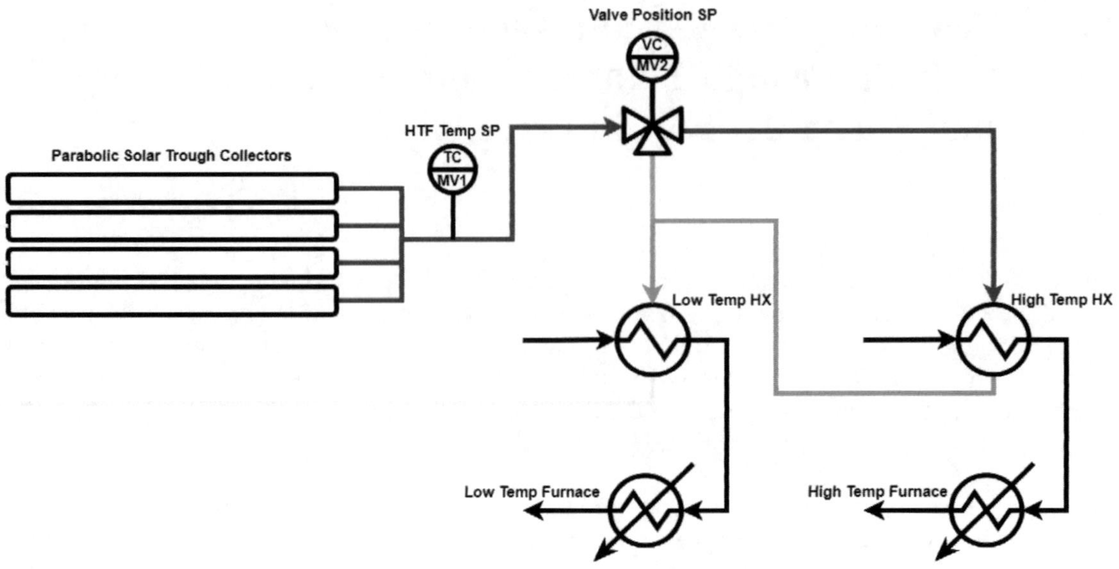

Figure 1. SIPH Plant Process Flow Diagram

632564 Interlayer Vibrational and Thermal Transport Phenomena in 2D Layered Perovskites

Abstract:

2D lead halide perovskites offer a versatile platform for the creation of atomically precise interfaces between organic and inorganic layers. In this talk, I will describe the vibrational, thermal, and phase change properties of 2D alkylammonium lead halide perovskites. Highlights will include 1) selective order/disorder transitions of the organic subphase, 2) observation of zone-folded longitudinal acoustic (zf-LA) phonons in 2D superlattices, and 3) evidence for ballistic phonon transport in interlayer heat transfer.

634247 Boosting Hybrid 2D Perovskite Stability through Computational Screening of Ligand Chemistries

Abstract:

2D hybrid perovskite materials are attractive from the perspective of providing a much larger design space of organic ligand chemistries in comparison with bulk 3D materials. Likewise, many recent reports have highlighted how tuning the ligand chemistry can impact the opto-electronic properties and stability profiles of the resulting 2D materials. In this talk, we'll describe how we have implemented a high-throughput molecular simulation framework to screen prospective ligand chemistries for synthesizability and stabilization of the perovskite lattice. Using this framework, we have been able to characterize tens of thousands of perovskite materials and establish some novel design rules for ligand chemistries. We will also present a few case studies of materials that have been synthesized and tested in response to this effort.

625598 Engineering Entangled Photon Pairs with Metal–Organic Frameworks

Abstract:

The discovery and design of new materials with competitive optical frequency conversion efficiencies can accelerate the development of scalable photonic quantum technologies. Metal–organic framework (MOF) crystals without inversion symmetry have shown potential for these applications, given their nonlinear optical properties and the combinatorial number of possibilities for MOF self-assembly. In order to accelerate the discovery of MOF materials for quantum optical technologies, scalable computational assessment tools are needed. We develop a multi-scale methodology to study the wavefunction of entangled photon pairs generated by selected non-centrosymmetric MOF crystals *via* spontaneous parametric down-conversion (SPDC). Starting from an optimized crystal structure, we predict the shape of the $G^{(2)}$ intensity correlation function for coincidence detection of the entangled pairs, produced under conditions of collinear type-I phase matching. The effective nonlinearities and photon pair correlation times obtained are comparable to those available with inorganic crystal standards. Our work thus provides fundamental insights into the structure–property relationships for entangled photon generation with metal–organic frameworks, paving the way for the automated discovery of molecular materials for optical quantum technology.

634315 Invited Talk: A Charge Transfer Framework That Describes Supramolecular Interactions Governing Structure and Properties of 2D Perovskites

Abstract:

Organometal halide perovskites can efficiently absorb light; perovskite solar cells can generate electricity with power conversion efficiencies now exceeding 25%. However, halide perovskites decompose when exposed to light, heat, or moisture, making them unsuitable for use as active components in solar cells. Two-dimensional perovskites with organic ammonium spacer ligands offer improved environmental stability, but the power conversion efficiencies of the solar cells containing these two-dimensional perovskites are low, reflecting inefficient charge transport. As the interactions between organic ligands and the perovskite octahedra affect the octahedral tilts, directly affecting charge transport, there is an opportunity for a comprehensive molecular design strategy that can manipulate organic-inorganic interactions, control octahedral tilt, enhance charge transfer, and impact device performance. In this talk, I will discuss our approach to molecular design, and using both theoretical and experimental support, elucidate the role of the organic cation in controlling the macroscopic properties of the resulting two-dimensional perovskites. Importantly, we learned that attractive point interactions within the organic bilayer result in a withdrawal of the organic cation from the interstitial site, and less penetration of the organic cation correlates to better performing solar devices. This design paradigm can guide future innovations in solution processable perovskite optoelectronics, with the goal of creating both high performing and stable next generation solar cells.

627921 Molecular Modeling of Halide Diffusion in 2D Organic-Inorganic Hybrid Perovskites

Abstract:

Organic-inorganic hybrid perovskites are promising semiconductor materials for a wide range of optoelectronic applications due to their unique properties, including high carrier mobility, solution processibility, and strong light absorption. However, facile halide diffusion is a source of instability for 3D perovskites that leads to rapid degradation of device performance. In contrast, 2D perovskites exhibit relatively suppressed halide diffusion that is hypothetically further improvable via surface cation design. Nonetheless, the halide diffusion mechanism in 2D perovskites remains largely unknown and is challenging to experimentally characterize. To understand the microscopic origin of the diffusion and the role of organics cations, we have employed the molecular dynamics (MD) simulation to explore different halide migration paths at the atomic scale and the role of surface cation in suppressing diffusion. Steered MD and thermodynamic integration were used to characterize the free energy profiles of each migration path at surface/bulk layers for various systems. In comparison to 3D perovskites, we observe that 2D perovskites exhibit relatively high diffusion activation energies and a large the asymmetry across the accessible channels, which accounts for the suppression of halide diffusion. In addition, the change in the energy barrier when applying different organic cations and the number of inorganic layers offers guidelines for molecular engineering of perovskite materials toward long-term stability.

627960 Understanding the Impact of Pore-Polymer Interactions on the Mobility of Poly(ethyleneimine) Confined in Mesoporous SBA-15: Quasi-Elastic Neutron Scattering Studies

Abstract:

Supported amines are considered a promising class of CO_2 sorbents based on their capability of capturing CO_2 from various sources such as flue gas stream from power plants (5-15% CO_2) as well as ambient air (~400 ppm). Among supported amines, poly(amines) physically impregnated within the mesopores of solid oxide supports have been extensively used for achieving high amine contents (i.e., high capacity) and physical/chemical stability (i.e., limited loss via evaporation/oxidative degradation of amine compounds).[1] Within supported amine systems, the physical properties of the poly(amines) play key roles in determining CO_2 uptake performance. In particular, the chain dynamics of poly(amines) govern diffusion rates of CO_2, which ultimately determine CO_2 sorption/desorption dynamics.[2] Therefore, understanding the dynamics of poly(amine) species confined in solid supports will promote designing optimal supported amines for CO_2 capture.

Herein, we study the impact of the interaction modes between the pore surface and poly(amines) with a model system comprised of branched-poly(ethyleneimine) (BPEI) supported in mesoporous SBA-15 silica. Chain mobilities of BPEI are quantitatively analyzed using quasi-elastic neutron scattering (QENS) in tandem with coarse-grained molecular dynamics (MD) simulations. The pore-polymer interactions are tuned by grafting organosilanes having different end groups to the silica surface, leading to different levels of attraction between BPEI and the pore walls of silica. Chain mobilities of BPEI undergoing four different interaction modes with pore walls are investigated. The first mode brings covalent tethering of amine sites of BPEI to pore surface via halide-amine reaction, where alkyl halides are grafted onto pore surfaces. The second interaction is acid-base and hydrogen bonding interactions based on silanols decorating the native SBA-15 silica, which strongly interact with amine groups of BPEI. The third interaction considers relatively weak interaction with pore surface and BPEI by substituting acidic silanols into basic aminoalkyl groups on the pore walls, so that pore walls interact with BPEI mainly by inter-amino hydrogen bonding. Lastly, repulsive, van der Waals interaction (via grafting aliphatic chains on pore walls) yields the weakest interaction between pore walls and BPEI. QENS (High-Flux Backscattering Spectrometer; HFBS at NIST) and MD simulation revealed that the chain mobilities of BPEI are significantly influenced by different surface groups appearing on the pore walls. The strong/transient interaction (acid-base, hydrogen bonding) due to silanol-BPEI interaction yielded the slowest motions, covalent bonding and repulsive interactions resulted in slightly increased mobility, and hydrogen bonding (aminoalkyl-BPEI; without acid-base interaction) brought significantly recovered mobility as compared to bulk PEI. We hypothesize that PEI's mobility depends not only on the attraction level with the pore walls but also on its chain conformation in the presence of surface capping groups on the pore walls. To understand further, these results are currently being coupled with additional QENS experiments, alongside insights developed via other techniques.

References:

(1) Choi, S.; Drese, J. H.; Jones, C. W. Adsorbent Materials for Carbon Dioxide Capture from Large Anthropogenic Point Sources. *ChemSusChem* **2009**, *2* (9), 796–854.

(2) Holewinski, A.; Sakwa-Novak, M. A.; Carrillo, J.-M. Y.; Potter, M. E.; Ellebracht, N.; Rother, G.; Sumpter, B. G.; Jones, C. W. Aminopolymer Mobility and Support Interactions in Silica-PEI Composites for CO2 Capture Applications: A Quasielastic Neutron Scattering Study. *J. Phys. Chem. B* **2017**, *121* (27), 6721–6731.

622996 Overcoming the Permeability-Selectivity Tradeoff Via the Design of Heterogeneous Pore Wall Chemistries

Abstract:

Developing next-generation membranes for water treatment demands system-specific materials with precisely tuned functionality. In particular, the purification and reuse of highly contaminated water calls for solute-specific designs where a key challenge is the separation of small neutral solutes. One such solute is boric acid, which is strictly regulated in water reuse for agricultural irrigation. Precise tuning of membrane pore chemistry requires a fundamental understanding of the mechanism by which functional moieties at the pore wall modulate the transport of water and solutes. Simulations enable the nanoscopic investigation of the mechanism by which a chemically heterogeneous pore modulates hydration water and thus its water-mediated interactions with solute molecules. Here, we use molecular dynamics simulations to investigate water and boric acid structure, dynamics, and transport in a cylindrical nanopore patterned with nonpolar methyl and polar hydroxyl groups. We find that changes in the arrangement of the methyl and hydroxyl groups, in addition to the density of each group, affect the transport of both water and boric acid in the pore. These changes not only affect the selectivity of water over boric acid, but also the permeability of water through the pore, leading to improvements in the permeability-selectivity behavior, a pernicious trade-off in membrane technology. We thus apply an optimization procedure based on a genetic algorithm to design the chemical patterning of the pore wall to simultaneously optimize water selectivity and permeability. The optimization identifies non-intuitive patterns that significantly hinder the transport of boric acid through the pore, simply by altering the functional group patterning. This inverse design procedure and the accompanying molecular-level understanding of the effect of chemical heterogeneity on solute selectivity in a pore will inform the design of more efficient water purification membranes, as well as the modulation of water-mediated surface-solute interactions in other systems.

624568 Hydrogen Bond Patterns of Deep Eutectic Solvents and Non-Deep Eutectic Solvents

Abstract:

Deep eutectic solvents (DESs) have shown promise as sustainable solvents to extract substrates from aqueous solutions as they are composed of natural compounds. Currently, a trial-and-error method is used to find new DESs experimentally. This process is inefficient and time consuming as all permutations of binary systems must be tested and then ruled out if no DES is formed. Hydrogen bond interactions between the binary compounds are considered the prevailing reason for the eutectic point. We investigated the relationship between hydrogen bond features and the potential of forming DESs for a binary system using a computational approach. We conducted molecular dynamics simulations to analyze the structural and dynamic properties of hydrogen bonds in 50 DESs and 100 non-DES systems and investigated their pattern using a statistical approach. Our results could be used to develop molecular principles for designing new DES based on natural hydrogen bond donors and acceptors.

620632 Silica-Kaolinite Interface for Shale Gas Application: A Molecular Modeling Study

Abstract:

Methane mostly makes up the constituent of shale gas and it is currently being exploited in fulfilling the world's energy demands. Molecular modeling techniques including Density Functional Theory (DFT) and Molecular Dynamics (MD) techniques are employed to understand methane transport in the pores at typical reservoir conditions. Herein, we study the interfacial phenomenon of methane adsorption on Silica-Kaolinite interface. This interface is created as it simulates a shale model which in reality is made up of clay (kaolinite) and quartz-like (silica) materials. The simulations revealed that the interface is formed by a chemical bond between a silicon atom from the silica surface and two oxygen atoms from the kaolinite surface. The adsorption of methane was studied at three different positions on the Silica-Kaolinite interface namely, silica-dominated region, interface region, and kaolinite-dominated region. The mode of adsorption of methane irrespective of its position on the interface was found to be physisorption with the kaolinite region having a greater adsorption. This is in agreement to previous studies which used experimental techniques for study methane adsorption. The results found herein which will be useful in shale estimating the original gas in place for shale reservoir application.

This work is funded by NPRP grant 12S-0130-190023 from from the Qatar National Research Fund (a member of Qatar Foundation). The statements made herein are solely the responsibility of the authors.

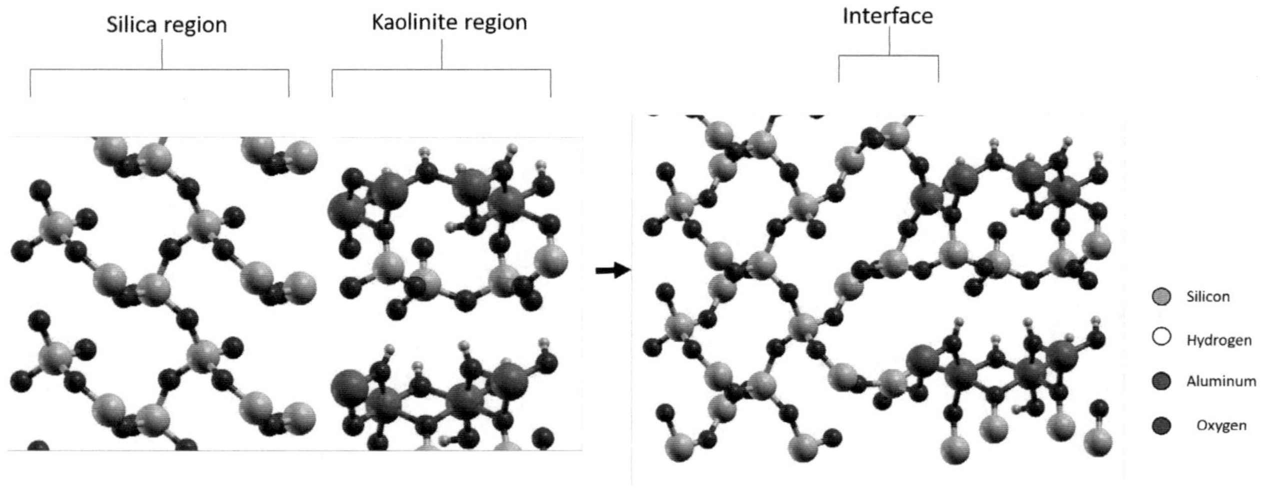

629486 Investigation on the Sodiation of Selenium-Graphene for Na-Ion Batteries: A First-Principles Study

Abstract:

Sodium–selenium (Na–Se) batteries have recently attracted attention as an emerging alternative for the traditional Li-ion batteries owing to their high theoretical volumetric capacity (~ 3250 mAh/cm^3) and electronic conductivity of Se (~ 10^{-3} S/m). Carbon composites are proposed as promising host materials to buffer the volume fluctuation of Se during charge/discharge, provide conductive channels, and physically trap the high-order polyselenides. In this talk, we evaluate the role of carbon host in establishing the reaction kinetics as the fundamental understanding of the sodiation of the Se-C composite is still limited. Based on first-principles simulations, we first present interfacial structure, charge transfer, and diffusion characteristics of Na atoms at the interface between Se and graphene (C). Our calculations demonstrate that there is a substantial enrichment of Na atoms at the interface and corresponding electron transfer to both Se and C nearby. The interactive strength of Se with graphene is found to increase with sodiation, which could lead to mechanical failures as compressive stress build-up at the interface. The presence of C host results in a more populated distribution of longer Se chains, as less reduced Se (due to electron transfer toward graphene) tends to be clustered. We observed a faster mobility of interfacial Na atoms along the graphene surface when compared to the diffusion of Na atoms in Se. Our calculations also show that the Na–Se/C system exhibits n-type metallic characteristic, which could provide fast electron flux for the sodiation reaction.

630087 Molecular Insights into the Electrowetting Behavior of Aqueous Ionic Liquid

Abstract:

Electro-wetting is the modulation in the wettability of a surface prompted by an external voltage [1, 2] across the solid/liquid interface. Ionic liquids (ILs) which have unique physicochemical properties,[3] gain industrial significance, due to its development as a promising alternative medium for electrowetting. This phenomenon is exploited in various applications such as variable fluid lens systems, electrowetting displays, programmable optical filters, paint drying, micro motors, electronic microreactors, liquid optics, or lab-on-a-chip devices.[4-7]

Molecular dynamics simulations were applied to study the wetting behavior of nanoscale aqueous hydrophilic and hydrophobic Imidazolium-based ionic liquid (IL) droplets on a solid graphite substrate subjected to the perpendicular electric field. Imminent transformation in the droplet configuration was observed at $E = 0.08$ V/Å both for hydrophobic ILs [EMIM][NTF_2] and SPC/E water droplets. However, for hydrophilic, [EMIM][BF_4], the droplet was completely stretched to be column-shaped under the electric field $E = 0.09$ V/Å for lower weight percentages of ILs and $E > 0.1$ V/Å at a higher weight percentage of ILs (> 30 wt%), because of the hydrophilic nature of [EMIM][BF_4]. We explore the impact of electric-field through various parameters such as mass density distribution, contact angle, orientation, and hydrogen bonds.

[1] A. Quinn, R. Sedev, J. Ralston, Contact Angle Saturation in Electrowetting, The Journal of Physical Chemistry B, 109 (2005) 6268-6275.

[2] J. Restolho, J.L. Mata, B. Saramago, Electrowetting of ionic liquids: Contact angle saturation and irreversibility, The Journal of Physical Chemistry C, 113 (2009) 9321-9327.

[3] J.T. Hirvi, T.A. Pakkanen, Wetting of Nanogrooved Polymer Surfaces, Langmuir, 23 (2007) 7724-7729.

[4] D. Bratko, C.D. Daub, K. Leung, A. Luzar, Effect of field direction on electrowetting in a nanopore, Journal of the American Chemical Society, 129 (2007) 2504-2510.

[5] Y.S. Nanayakkara, H. Moon, D.W. Armstrong, A tunable ionic liquid based rc filter using electrowetting: A new concept, ACS Applied Materials & Interfaces, 2 (2010) 1785-1787.

[6] Y.S. Nanayakkara, H. Moon, T. Payagala, A.B. Wijeratne, J.A. Crank, P.S. Sharma, D.W. Armstrong, A fundamental study on electrowetting by traditional and multifunctional ionic liquids: possible use in electrowetting on dielectric-based microfluidic applications, Analytical chemistry, 80 (2008) 7690-7698.

[7] Y. Wang, G.A. Voth, Unique spatial heterogeneity in ionic liquids, Journal of the American Chemical Society, 127 (2005) 12192-12193.

629294 Methanol Synthesis Processes

Abstract:

Hydrogenation of carbon dioxide may contribute to the reduction of industrial carbon dioxide emissions. The bulk chemical methanol is a suitable hydrogenation product. It is used for paints, plastics, and adhesives, as feedstock for the production of chemicals such as formaldehyde, ethylene, propylene, methyl tertiary-butyl ether, and acetic acid, and as fuel additive or alternative fuel. It provides promising properties for energy storage, as it can raise the energy density of hydrogen-based energy carriers by one order of magnitude. Low temperature level in reforming also suggests methanol for fuel cell applications. State-of-the-art methanol production is based on syngas, a mixture of carbon monoxide, carbon dioxide, and hydrogen, and preferably $Cu/ZnO/Al_2O_3$ catalysts. State of the art reactor- and process design suggest looping systems with sophisticated internal heat energy management. Huge efforts have meanwhile been made to directly process carbon dioxide instead of syngas. To our knowledge neither of the carbon dioxide based processes has established on the market, mainly because of catalyst life time limitations. To sum up the present status, the establishment of direct methanol synthesis from carbon dioxide suffers from limitations in feed gas preparation and/or appropriate catalysts. Appropriate high carbon dioxide concentration sources are available in huge quantities in the iron and steel industry or cement industry, but appropriate solutions are still missing.

When processing carbonate based iron ores (siderite) in hydrogen atmosphere our group registered that within a temperature window of 600 to 800 °C and ambient pressure it is possible to alter the carbon monoxide/carbon dioxide ratio of the exiting gas. The final solid product from ore reduction still maintains this "catalytic" property, suggesting it for syngas preparation from carbon dioxide with a cheap and robust catalyst which can easily be processed after deactivation. The process was tested and validated in a lab size plug flow reactor setup.

For experimental simulation/validation of the methanol synthesis in lab scale a CSTR loop reactor system with external cooling loop for condensing the gaseous products and pressure drop controlled gas feed was constructed. The operation window of the setup is 500 °C at maximum and 10 MPa. At present the reactor setup is equipped with two gas supply lines equipped with mass flow controllers, limiting operation flexibility to fixed stoichiometry of the carbon dioxide/carbon monoxide source. The methanol synthesis process was validated with a Cu/ZnO/MgO catalyst prepared by wet incipient impregnation and stoichiometric CO_2/H_2 feed at 250 °C and 5 MPa. Whether this catalyst can withstand long term operation with CO_2/H_2 feed with acceptable loss of activity, or whether an upgrade of the feed gas by pretreatment in the syngas reactor is needed, cannot be concluded from present experience. However, either route, dual step synthesis with feed gas pretreatment or direct synthesis shows promising features, since catalyst life span is not a crucial issue.

630139 Systematic Approaches for Discovering Innovations to Enable a Sustainable Circular Economy

Abstract:

The recent push towards Sustainable Circular Economy (SCE) needs to be substantiated with modelling and systems design of value-chains. In our previous work, we have built a general framework to assess and design life-cycle value chains within a superstructure network for establishing SCE [1]. This SCE framework was applied to a carrier bags case study with various plastic and paper alternatives and multiple End-of-Life (EoL) options in the superstructure [2]. Due to the possibility of multiple stakeholders, multi-objective optimization was performed within the framework for objectives from the three domains: Environmental, Economic and Circularity. The trade-offs between these three objective types were quantified using pareto surfaces, and compromise solutions were proposed from the pareto optimal set. This study led to insights about currently available alternatives of the value-chain, e.g., quantifying the paper-plastic dilemma, probing effect of re-use, selecting best EoL strategy etc.

However, this approach is limited by the expanse of the constructed superstructure network. The proposed designs can only be derived from the set of currently available technologies and product systems in the life cycle of products. Evidently, win-win solutions are likely to be obtained by substituting conventional technologies with innovations and emerging technologies. This work is aimed at (i) guiding innovation in value chains and (ii) discovering novel products and pathways. For the first goal, we use hotspot and sensitivity analysis, while the second goal relies on superstructure-free design methods.

First, we identify activities in the carrier bags value-chain which act as hotspots of emissions, cost and 'loss of circularity' using life-cycle allocation and displacement methods. For instance, the highest amount of circularity loss in the optimal carrier bags value-chain occurs at the segregation process. These hotspots are obtained for various scenarios of technological advancements and recycling consciousness in society. Through these hotspots, we gain knowledge about where the highest scope of improvement in all the three objective domains lie. This is followed by sensitivity analysis using Monte-Carlo simulations to find effect of perturbations in the parameter space on the pareto surface. We also explore prospects of performing more comprehensive yet tractable sensitivity analysis using (i) analytical approaches: contingent derivatives of perturbation maps [3]; and (ii) stochastic approaches: multi-objective evolutionary algorithms (MOEA)[4]. Sensitivity analysis allows us to rank value-chain activities based on pareto dominance that can be achieved by enhancing their respective efficiencies through innovation . In turn, this helps in guiding innovations towards most promising sectors for SCE. For example, such an approach can point toward the need for innovation in chemical recycling technologies of plastic carrier bags.

For discovering innovations, we explore the use of superstructure-free methods. As stated earlier, the SCE framework only evaluates technologies and connections specified within the superstructure. We expect that novel products and pathways can be found if a more flexible design space is used. In order to do so, we evaluate various PSE methods such as State Task Network (STN), General Modular Framework (GMF) and Unit, Port, Conditioning Stream (UPCS) representation [5] to best describe the nodes and connections within the cradle-to-cradle value chain. We scan the method pool for prevalent analogies between units and life-cycle activities, possibility of directed multigraph required for circular flows and flexibility in designing node connections. However , we do not define or generate the superstructures beforehand. Instead, we let the connections and amounts transferred be decision variables of the design problem. UPCS defines the 'minimal' and 'feasible' rule s for nodes which ensure that no redundant flows enter a unit, and the necessary flows for transformations do enter, respectively. In addition to these, we add the following rules: (i) each product flow is associated with both input and output ports of UPCS, and (ii) there must exist a unit between every output and input port. The streams linking the ports are captured through an adjacency matrix with binary decision variables in each cell. Each cell of this matrix corresponds to a transformation and refers to a sub-matrix of tuples containing (i) scaling factor of transformation (decision variable), (ii) enthalpy and exergy changes (parameters), (iii) yield per unit input (parameter). Within the manufacturing sector, these transformations can be modelled as reaction-separation networks (e.g., from Reaxys), whereas value-chain activities can be modelled using exergy changes. Physico-chemical constraints such as flow conservation are added as a function of adjacency matrix and scaling factors. We demonstrate this method using an illustrative example and find optimal nodes and connections for a small set of products and elementary flows (to and from nature). We also probe the scalability of this method for larger number of products, and to this effect, consider using techniques like Bender's decomposition and evolutionary algorithms (like NSGA-II) for superstructure-free design.

Through this presentation we aim to convey further developments in application of our SCE framework [1,2] for guiding innovation and emerging technologies towards activities with most scope and highest impact, using hotspot and sensitivity analyses, respectively. Additionally, we create a highly flexible design space for value chains using superstructure and superstructure-free methods. With a large enough sub-matrix for underlying transformations, we expect that this new methodology can yield optimal solutions which haven't yet been considered in academia or industry. We also probe the scalability of these methods for large and practically relevant problems using advanced mathematical modeling techniques.

References

1. Thakker, Vyom, and Bhavik R. Bakshi. "Toward sustainable circular economies: A computational framework for assessment and design." *Journal of Cleaner Production*295 (2021): 126353.
2. Thakker, Vyom, and Bhavik R. Bakshi. "Towards Sustainable Circular Economy: Design Framework and Application to Grocery Sacks". *2020 AIChE Annual Meeting*. Video available at: https://youtu.be/dWj6R0UXMjM
3. Tanino, Tetsuzo. "Sensitivity analysis in multiobjective optimization." *Journal of Optimization Theory and Applications*3 (1988): 479-499.
4. Deb, K., Pratap, A., Agarwal, S., and Meyarivan, T. A. M. T. "A fast and elitist multiobjective genetic algorithm: NSGA-II". *IEEE transactions on evolutionary computation*, *6*(2) (2002): 182-197.
5. Wu, WenZhao, Carlos A. Henao, and Christos T. Maravelias. "A superstructure representation, generation, and modeling framework for chemical process synthesis." *AIChE Journal*9 (2016): 3199-3214.

SHAPING THE FUTURE WITH A HYDROGEN VALUE CHAIN SIMULATION PLATFORM

Ian Willetts, AVEVA Software LLC, Carlsbad, CA
Cal Depew, AVEVA Software LLC, Lake Forest, CA

Abstract

Today's top international oil companies (IOCs) are responding to the climate crisis and setting goals to be net zero carbon emission companies by 2050 or sooner. They are on a journey to transition from being international oil companies to becoming international energy companies (IECs). These companies will implement near term, medium term, and longer-term initiatives for decarbonization as they progress along their energy transformation journey. These steps will include:

- Reducing methane emissions
- Reducing carbon dioxide emissions
- Carbon Capture, Utilization and Storage (CCUS)
- Accelerating decarbonization

A key accelerator all companies are embracing as part of their strategy to drive to their net-zero goals is digital transformation which involves identifying and implementing methods to bring digital tools and work processes closer together to drive more informed business decisions. Many industry stakeholders see the Digital Twin as the most important building block for the digital transformation of the process industries and digital transformation of standard work processes is required to accelerate today's energy transition.

While legacy simulators are well-suited to accurately simulate conventional processes, their decades-old architectures mean they are not ideally suited to serve the entire plant lifecycle nor to support the digital transformation journey for the nascent hydrogen or low-carbon economy.

Digital transformation is undoubtedly a critical lever that both EPCs and Owner Operators can use to deliver both short term sustainability gains and to elevate the quality of capital project execution while minimizing the environmental impact of new capital investments for traditional oil and gas processes, CCUS projects or low or zero carbon energy projects.

Process simulation is accepted as a required tool for the process design of traditional oil and gas and chemical plant facilities, from conceptual design, through engineering and on to operations. However, with the energy transition underway, the landscape of the process industries will change dramatically over the next three decades. As such, a process simulator must be able to simulate both existing and new energy processes as well as support new and improved process engineering practices. The paper discusses the challenges of the current simulation tools from a process modeling perspective when it comes to the need for:

- Improved process equipment models suitable, for example, for modeling gray and blue hydrogen generation including steam methane reforming, absorption processes to capture carbon dioxide and adsorption processes to create high purity hydrogen

- Environmental and economic KPI calculations that allow process engineers to design for production, and quality goals while balancing economic and environmental objectives.
- New equipment models to support the modeling and optimization of the full hydrogen value chain including green hydrogen generation including, for example, renewable power sources, electrical grid models and electrolyzers
- Accurate thermodynamics for key sustainable processes including, for example, carbon capture and hydrogen liquefaction
- Capabilities to simulate liquid organic hydrogen carrier (LOHC) processes that increase the energy density of hydrogen to facilitate easier transportation

The paper also discusses the challenges of today's tools from a platform perspective when it comes to updating the tools for the hydrogen value chain:

- Legacy process simulators are overloaded with tightly woven niche features and functions only used by specialists. The complexity of the code makes it difficult for software vendors to adapt them fast enough for rapidly changing industry trends.
- Legacy process simulators are typically poorly integrated into engineering workflows beyond the process world, and if so, only with a single directional information flow. This leads to poor FEED quality and subsequent capital project cost and schedule overruns.

To address these challenges, the next generation process simulator needs the following, and other functionality, to be true enablers of digital transformation for the new lower carbon economy:

- Emission calculations linked with capital and operating cost models to support multi-constraint optimization of new process designs
- An open modeling framework to be able to quickly add new process models as they are rapidly developed to address the evolving hydrogen economy
- The ability to take Hydrogen value chain models used for process design and reuse them for dynamic simulation, control strategy design and for operator training simulators thereby shortening project schedules and lowering the total cost of simulation
- The ability to put those same models on-line for real time performance monitoring, process optimization and predictive analytics for incremental profitability, all in a cost-effective manner

The authors will show that the above identified obstacles can be readily overcome today with a next generation process simulator - AVEVA Process Simulation – that has been prepared with all the necessary capabilities and architecture needed to be able to support EPCs and Owner Operators as they go through their energy transition journey.

Beginning an Energy Transition Journey

Today's top international oil companies (IOCs) are responding to the climate crisis by setting goals to be net zero Carbon companies by 2050 or sooner. They are inevitably on a journey to transition from being international oil companies to being international energy companies (IECs).

This journey will include deploying near term sustainability solutions for current assets as well as investing in lower carbon energy projects in the future. TotalEnergies, for example, plans to ramp its capital investment in clean energy projects from 3% of total capital investment today to 20% over the next 20 years.

A Deloitte Insights article titled "The future of work in oil and gas" [1] published in Oct 2020 outlined one potential energy transition pathway that oil and gas companies will likely undertake to achieve their net-zero goals. This is shown in Table 1.

Table 1. Energy Transition Pathways for Oil and Gas Companies

Energy Transition Pathway	Theme
First Steps Foundational sustainability measures and HSE requirements	Water, safety, health, land etc.
Building Blocks Advanced emissions, electrification, and energy efficient measures	Asset electrification
	Energy efficiency and management
	Fugitive emissions
New Blueprints Portfolio of lower-carbon energy sources, greener fuel mix, and carbon capture solutions	Energy from low-carbon sources
	Greener end-product mix (natural gas)
	New revenue models (renewables, biofuels, EV stations etc.)
	Carbon capture solutions and services
Winning the Future License to lead in a clean-energy economy	Net zero emissions from purchases and operated assets
	End-of-life treatment of sold products to create a circular economy
	Shape and influence paths to net-zero of consumers and societies

While Deloitte is looking at the world from the Owner Operator perspective, the building blocks and new blueprints above will rely on EPCs to align with Owner Operators to deliver on those initiatives. As such this transition will equally impact both EPCs and OO's and both will have to rethink the way they do business today.

This view by Deloitte is well aligned with the viewpoint of the Oil and Gas Climate Initiative (OGCI) who published a progress report "Delivering on a Low Carbon Future"[2] in December 2020. In the report the OGCI argue that it is a commercial imperative for oil and gas companies to implement a broad and ambitious portfolio of near, mid, and long-term low carbon solutions to meet their net-Carbon zero goals. The focal areas they outline include:

- Reducing methane emissions
- Reducing carbon dioxide emissions
- Carbon Capture, Utilization and Storage (CCUS)
- Accelerating decarbonization

Table 2 captures some of the recommended levers that oil and gas companies should deploy in these four areas:

Table 2. Levers that Oil and Gas Companies Should Deploy

Focus	Primary Levers	Additional Comment
Reduce methane emissions	Detect, measure, and mitigate fugitive emissions	Must be cost effective and scalable to be deliver substantive emissions reductions
	Replace or upgrade high-emitting equipment	
Reduce carbon dioxide emissions	Improve energy efficiency	Focus on reduction in LNG and Refineries. Consider CO_2 emissions from production facilities as well as from emissions associated with the required imports of electricity and steam
	Reduce and ultimately eliminate flaring	
	Electrify operations with renewables where possible	
	Cogenerate electricity and useful heat	
Carbon Capture, Utilization and Storage (CCUS)	Captured CO_2 can be stored underground with or without enhanced oil recovery. Captured CO_2 can also be used for producing chemicals and concrete	Scaleup is the biggest challenge. Per the IEA CCUS in Clean Energy Transitions report (Sept 2020) a 38-fold increase in CCS is needed to meet the 2020 1.5 Gt goals.
Accelerate decarbonization	Requires investment in low and zero carbon energy sources including green hydrogen	11% of GHG emissions come from energy production including oil and gas. The percentage is much higher when considering Scope 1,2 and 3 emissions

Confluence of Energy Transition and Digital Transformation

A key accelerator all companies are embracing as part of their strategy to drive to a net-zero goal is digital transformation. Digital transformation involves bringing together today's digital tools and work processes to drive more informed business decisions.

Figure 1. Digital Transformation Merges Digital Tools and Work Processes

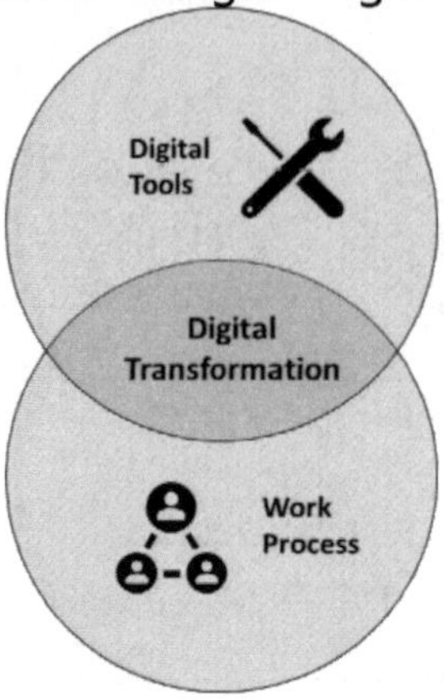

Per Covestro[3], new process technology must outperform existing technology to be acceptive in a competitive environment. But the existing technology and processes have been optimized for decades. We now have only a short time for the introduction of new sustainable processes. To achieve the same level of maturity, we need innovation for both the processes and the design workflows to accelerate the new process technology. This need for innovation applies to both chemical companies addressing the circular economy and oil and companies in their energy transition to become energy companies.

As oil and gas companies begin to execute their energy transition strategies, they must still deliver profitability in what remains a highly competitive industry. More informed business decisions can drive many traditional key performance indices (KPIs) of both EPCs and Owner Operators, across greenfield and brownfield facilities including:

- Higher quality FEED design leading to lower capital cost and shorter schedules
- Engineering efficiency improvements across disparate multi-disciplinary teams
- Improved safety
- Less unplanned downtime
- Higher margins and profitability

With the drive for a sustainable future, additional metrics come into consideration, including but not limited to:

- Project emission factors
- Overall CO2 intensity of products produced
- Resource-energy efficiency
- Specific Energy Intensity
- Carbon efficiency

One key accelerator of digital transformation is the Digital Twin, which is a digital representation of a real-world entity, a model that mirrors an actual asset to be built or an existing brownfield asset. In the Process Industries there are many different views of what represents a Digital Twin. One common viewpoint mirrored by many of the largest oil and gas companies is shown in the following image from AVEVA Group Plc, a leading provider of digital transformation solutions.

Figure 2. The Digital Twin

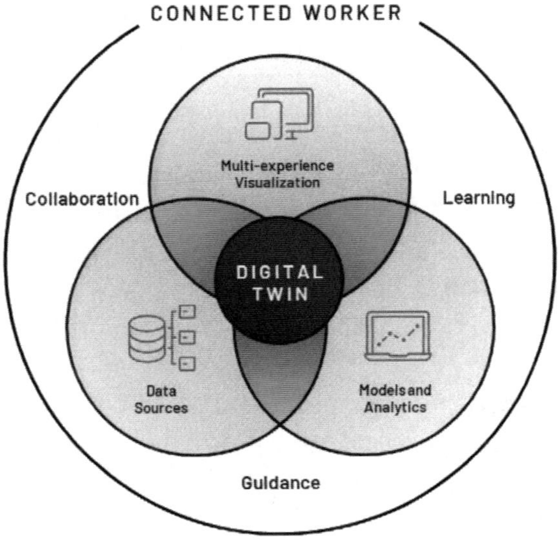

The Digital Twin includes:

- Data sources: the engineering design data and real time plant data
- Simulation models: first principles, physics-based models, data analytic models and hybrid models
- Visualization layer: the tools necessary for the people to be able to access and visualize data so that they can make appropriate and informed decisions to drives KPIs

The simulation model is often referred to as the behavioral digital twin as it can provide knowledge and insight into how the process will behave. Knowledge and insight provided by such a model cannot be gained in any other way other than by operating the plant. However, in the nascent hydrogen economy there are many projects planned but almost all are in the early conceptual or the early FEED project phase where there is no opportunity to gather plant operating data.

Simulation Throughout the Capital Project Lifecycle

Incorporating a process simulation model as part of the digital twin at every phase of a CAPEX project reduces risk and ensures projects are much more likely to be delivered on schedule and within budget. Successfully developing and leveraging the Process Digital Twin throughout a project lifecycle requires both the EPC and the Owner Operator to think differently and change the traditional business models. The following highlight the value a digitally transformed workflow can provide a Capital project from initial concept through design, commissioning, and startup and into on-going operations:

From concept to FEED

The Construction Industry Institute (CII) report number RR331-11[4] that a project with a high FEED maturity and accuracy outperformed projects with a low FEED maturity and accuracy by 24 percent in terms of cost growth, and 12 percent in terms of change order performance.

A process simulation model is a key enabler to achieve a highly mature FEED by using the insight and knowledge provided by the simulation model to eliminate errors and mistakes in the design before capital is committed and when the cost to correct an error is the lowest.

Detailed design

As a project moves toward detailed design some of the equipment in the FEED phase will change. It is important to keep the simulation in lockstep with design changes to ensure that as the physical asset takes shape the digital twin remains an exact mirror image. The simulation can be used continually to ensure that the latest detailed design changes do not negatively impact the controllability and operability of the asset3.

Commissioning and start-up

Armed with a highly responsive digital twin, the EPC and the Owner-Operator can turn their attention to preparing for start-up and ensuring their teams have the right skills to safely start up and operate the plant. Training operations staff using the behavioral digital twin allows operations staff to practice startup repeatedly and even fail repeatedly in a completely safe environment. Practice makes perfect and the operations staff, over time, develop the required skills and competency to ensure a smooth and error free startup.

Ongoing operations

The value of a simulation does not end with startup. It can be used to support on-going operations and drive efficiencies. This can be in the form of connecting to plant data to provide inferential measurements that can support decision making for operational excellence. Models can also be used for troubleshooting to investigate the cause of abnormal operations. Staff turnover is also inevitable, and the Process Digital twin can be a backbone of a plant's workforce competency development program.

Simulation Challenges for Today's Industry

Legacy Code

Perhaps the most known simulation platforms in the Process Engineering domain are PRO/II, Aspen Plus, HYSYS and UniSim.

- PRO/II is the flagship steady state simulation package of AVEVA, formerly Simulation Sciences Inc who were incorporated in 1967. Today the product is known as AVEVA PRO/II Simulation.
- The Aspen Project started in 1976 and this project in turn led to the creation of Aspen Technology in 1981. Today Aspen Plus is one of the two the flagship steady state simulation software packages of Aspen Technology.
- HYSYS was first conceived and created by the Canadian company Hyprotech. The HYSYS Version 1.1 Reference Volume was published in 1996. In 2002 Aspen Technology acquired Hyprotech and today this product is known as Aspen HYSYS.
- Following a 2004 ruling by the United States FTC, Aspen Technology was forced to divest its Hyprotech assets, including HYSYS source code, selling these to Honeywell who

ultimately produced the UniSim product. UniSim is built on the same foundation as Aspen HYSYS.

These software programs have evolved and morphed over the years as they have been used by customers who require new features and updates as requirements change. Inevitably, such legacy software becomes more challenging to adapt as industry needs change. Today we are in an unprecedented industrial transformation and in need of a new and modern simulation platform built from the ground up, to serve the needs of the next generation, digitally native, process engineers.

Modeling Requirements

Whilst the existing legacy software platforms are very capable and certainly road tested, the energy transformation will drive a new set of modeling requirements to be fit for purpose including but not limited to:

- Improved process equipment models for gray and blue hydrogen generation such as reactor models for steam methane reforming, absorption processes to capture carbon dioxide and adsorption processes to create high purity hydrogen.
- Environmental calculations to allow process engineers to design processes that meet production and quality goals with minimum environmental impact.
- Equipment models for green hydrogen generation including renewable power sources, electrical grid models and electrolyzers to support the modeling and optimization of the full hydrogen value chain.
- Accurate thermodynamics for modeling hydrogen liquefaction and other sustainable processes.
- Capabilities to simulate liquid organic hydrogen carrier (LOHC) processes that increase the energy density of hydrogen and facilitate easier transportation.

Platform Requirements

Today's workers are digital natives and have very different expectations when it comes to technology. Unfortunately, legacy process simulators are overloaded with niche features and functions that have been added over the years and are often only usable by experts. This in turn makes it difficult for users to adapt them fast enough for today's rapidly changing industry trends.

A standalone simulation package is not an enabler of digital transformation. Legacy process simulators are typically poorly integrated into the engineering workflows beyond the process world. Even where some integration exists today, information exchange is only a single directional information flow. This leads to poor FEED quality and subsequent capital project cost and schedule overruns.

Hydrogen Value Chain Simulation Platform

As of August 2020, a Wood MacKenzie research report[5] highlights that the green hydrogen project pipeline has swelled from 3.5 GW to more than 15 GW just over the previous 10 months. The success of these and other future multibillion dollar capital projects will be measured against two key metrics, cost, and schedule. The traditional capital project execution methodologies of today result in projects being delivered, on average, 20 months late and 80% over budget.

It is paramount that this does not happen with these new renewable energy capital projects and a Digital Twin strategy involving simulation can materially impact these looming investments, provided there is a suitable Hydrogen value chain simulation platform available.

AVEVA Process Simulation

Unlike the legacy simulation platforms discussed earlier, AVEVA Process Simulation, conceived less than ten years ago, was developed from the ground up to deliver the process digital twin to the next generation of process engineers. AVEVA Process Simulation is differentiated from legacy platforms in three ways:

1. One process model can be used throughout the project lifecycle fully supporting a company's digital transformation goals.
2. It is built with next generation capabilities reflecting its conception and development into today's computing environment.
3. It is applicable to the full hydrogen value chain including gray, blue and green hydrogen processes.

One Model for the Process Lifecycle

Unlike any other simulator in the market today, AVEVA Process Simulation supports multiple simulation modes without the need to reconfigure the simulation and allows an engineer to simply toggle from one mode to another depending upon the task at hand. The following table highlights different ways that AVEVA Process Simulation may be used throughout the process lifecycle and the associated benefits for each use.

Table 3. Uses of Simulation throughout the Process Lifecycle

Project Phase	Requirements / Goal	Simulation Mode	Use Case	Benefit
Conceptual Design	Efficient and accurate optioneering	Steady State & Rating	Heat and Material Balance	Rapid optioneering on conceptual design supporting economic assessment for FID
			De-pressuring	Correct material selection for future integrity
			Safe Design	Ensure License to Operate will be granted and maintained
			Reactive Hazards Assessment	Design the process to mitigate the risk of reactive hazards to As Low As Reasonably Practicable (ALARP)
			Minimum Environmental Footprint	Includes GHG calculations to quantify environmental impact and allow most sustainable design to be built
FEED	High quality & maturity	Steady State and Dynamic Simulation	Most accurate design	Use the simulation to ensure a high quality or mature FEED which on average leads to 12% fewer change orders and 30% schedule reduction
			Equipment Sizing	Up to 3% saving in Total Installed Cost by eliminating overdesign or incorrect sizing leading to startup and commissioning delays
			Process Safeguarding	Implement a design that reduces the likelihood of loss of containment and that minimizes consequences should one occur
Detailed Engineering	Efficient and accurate design	Rating and Dynamics	Engineering Efficiency	Improved workflows from the digital transformation of the engineering processes can lead up to 30% fewer man-hours of engineering & 50% reduction in simulation effort
			Control and Operability Analysis	Ensure final design is easy to build and operate eliminating need for overly complex control strategies impacting safety and operating costs of the asset
			HAZOP Analysis	Support the HAZOP team to identify risk events or potential caused by deviations from design or operating intentions
			Flaring	Design so that flaring is not required during startup and CO2 emissions are minimized, e.g., include flare gas recovery in the initial design and not as a retrofit

Project Phase	Requirements / Goal	Simulation Mode	Use Case	Benefit
Detailed Engineering	Efficient and accurate design	Rating and Dynamics	Minimum Environmental Footprint	Includes GHG calculations to quantify environmental impact of standard operating procedures and abnormal operating conditions
			Energy Efficiency	Process intensification for lowest operating costs and lowest environmental impact
Commissioning and Startup	Controls Checkout and Operator Training	Rating and Dynamics	Controls Checkout	Verify and Validate controls database on the simulation model to eliminate issues and shorten commissioning time
			Development of Standard Operating Procedures	Establish SOPs on the simulator prior to live commissioning
			Start up and Shut Down	Practice start-up and shut-down procedures repeatedly in a safe environment to minimize commissioning and start up times
			Operator Training	Ensure the operations staff are ready for safe operations on Day One and ensure on-going
Operations	Process Improvement	Data Reconciliation, Steady State and Dynamics	Identify Integrity Operating Window Limits	Identify and eliminate conditions where equipment degradation is likely to occur ensuring minimum downtime
			Environmental Compliance	Maintain License to Operate
			Flaring	Reduced or zero flaring and lower environmental impact
			Process Upsets	Prepare operators to be able to manage abnormal operating conditions safely
			Troubleshoot Operations	Diagnose and correct operational issues using insight provided by the model
			Energy Efficiency	Identify and implement ways to minimize energy consumption
			Prescriptive Maintenance	intelligently schedule and plan asset maintenance with insight provided by the model
			Equipment and Asset Performance Monitoring	Monitor the technical and economic performance of the asset to identify opportunities for improvement

Next Generation Capabilities

The following table highlight some additional key features that differentiates AVEVA Process Simulation from other legacy systems.

Table 4. Key Features of AVEVA Process Simulation

<table>
<tr><th>Key Features</th><th colspan="2">Advantages / Benefits</th></tr>
<tr><td>Modern Graphical User Interface</td><td colspan="2">Contemporary user interface in line with the expectations of today's younger process engineers</td></tr>
<tr><td rowspan="12">Flexible Open Modeling</td><td rowspan="12">Easy to Customize and build libraries for new industries

Supports traditional process modeling as well as the modeling needs of the lower carbon future process industries</td><td>Process Library</td></tr>
<tr><td>Electrical Library</td></tr>
<tr><td>Utilities Library</td></tr>
<tr><td>Flare Library</td></tr>
<tr><td>Renewables Library</td></tr>
<tr><td>Economics Library</td></tr>
<tr><td>Controls Library</td></tr>
<tr><td>Pressure Swing Adsorption (PSA) Library</td></tr>
<tr><td>Rate Based Library</td></tr>
<tr><td>Membrane Library</td></tr>
<tr><td>Environmental KPI calculations</td></tr>
<tr><td>Others</td></tr>
<tr><td>Equation Oriented</td><td colspan="2">Fast to solve and handles complex recycles very efficiently</td></tr>
<tr><td>Cloud Enabled</td><td colspan="2">Supports collaboration across multiple disciplines</td></tr>
<tr><td>Optimization</td><td colspan="2">Incorporates the industry leading AVEVA Process Optimizer (ROMeo) optimizer which provides data reconciliation and multi-constraint optimization capabilities</td></tr>
<tr><td>Multi-Core/Threaded</td><td colspan="2">Speed and performance factored into the original design</td></tr>
<tr><td>Integrated with OSI PI</td><td colspan="2">Supports real time performance monitoring and is integrated with the number one historian in the process industries today</td></tr>
<tr><td>Integrated with AVEVA Engineering</td><td colspan="2">Improves engineering efficiency by keeping process and data digital twin in lock step through the process lifecycle</td></tr>
<tr><td>Integrated with AVEVA PRISM and Microsoft Bonai AI Platforms</td><td colspan="2">Full support for hybrid modeling combining first principles and data-based analytical models</td></tr>
<tr><td>Python Scripting Interface</td><td colspan="2">Digitally native engineers can enhance and integrate third party applications with contemporary programming languages</td></tr>
</table>

Hydrogen Value Chain Models

The following flowsheets demonstrate the usage of AVEVA Process Simulation for modeling gray, blue and green hydrogen production.

Figure 3. Steam Methane Reforming (Gray Hydrogen)

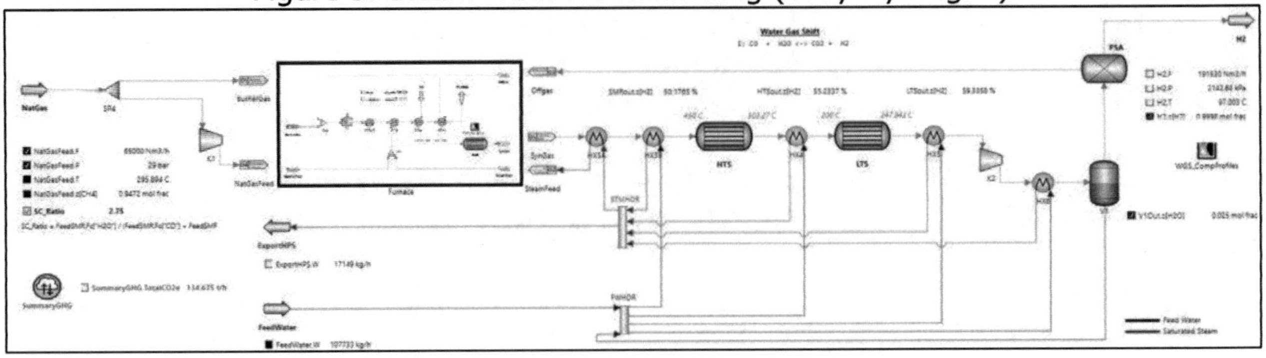

This example demonstrates a typical process to produce Hydrogen via Steam Methane Reforming (SMR). Natural Gas and high-pressure steam are reacted over a Ni / MgAl2O4 catalyst to produce H2, CO, and CO2. The CO and H2O in the outlet stream are then further refined in high-temperature and low-temperature water gas shift reactors. These reactors convert much of the remining CO and H2O into H2 and CO2. The final step of the process (not modeled here) separates H2 from the remaining components using Pressure Swing Adsorption (PSA).

Figure 4. Carbon Capture, Utilization and Storage (Blue Hydrogen)

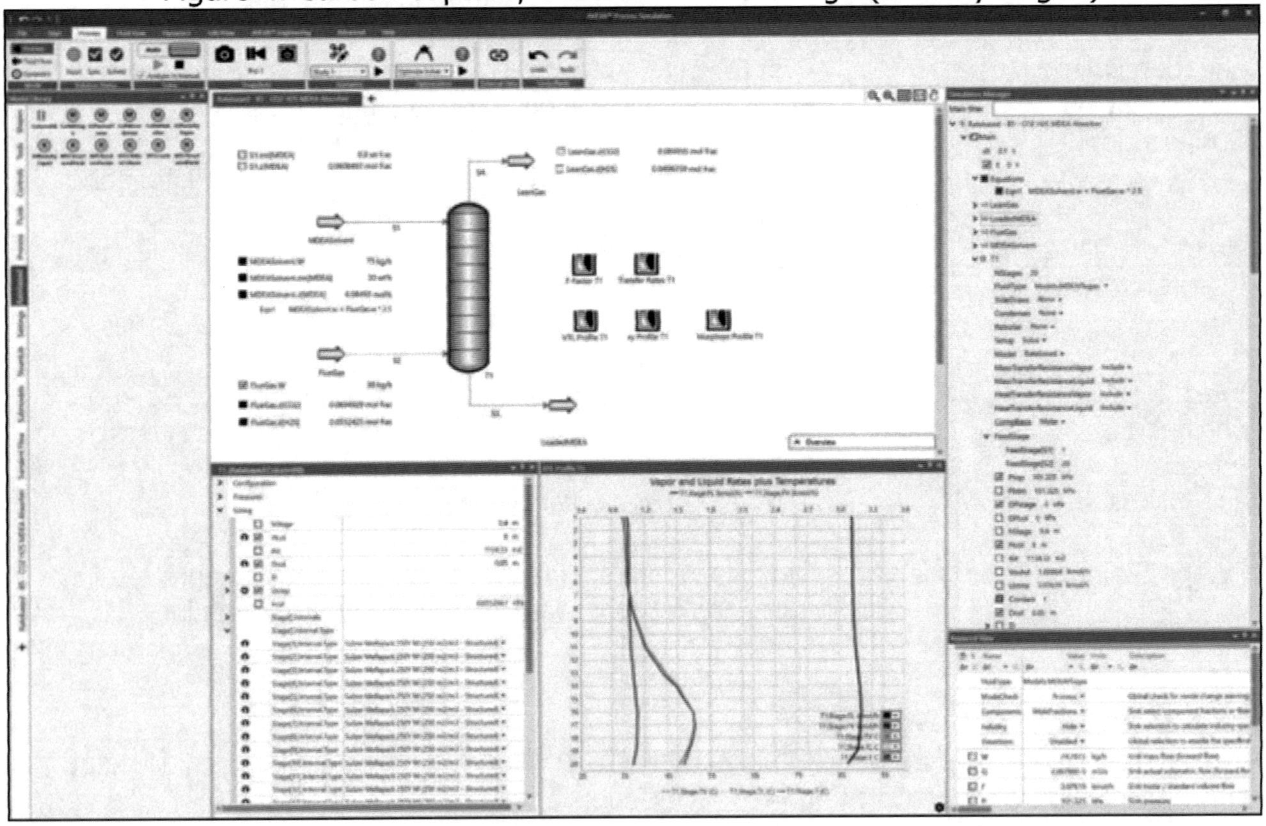

CO2 capture in absorbers is a not an equilibrium process but is mass transfer limited process. AVEVA Process Simulation includes a rate-based distillation column accurately modeling diffusion fluxes, diffusivity in the vapor and mass transfer calculations through the film between the vapor and liquid phases.

Figure 5. Sustainable Hydrogen Production (Green Hydrogen)

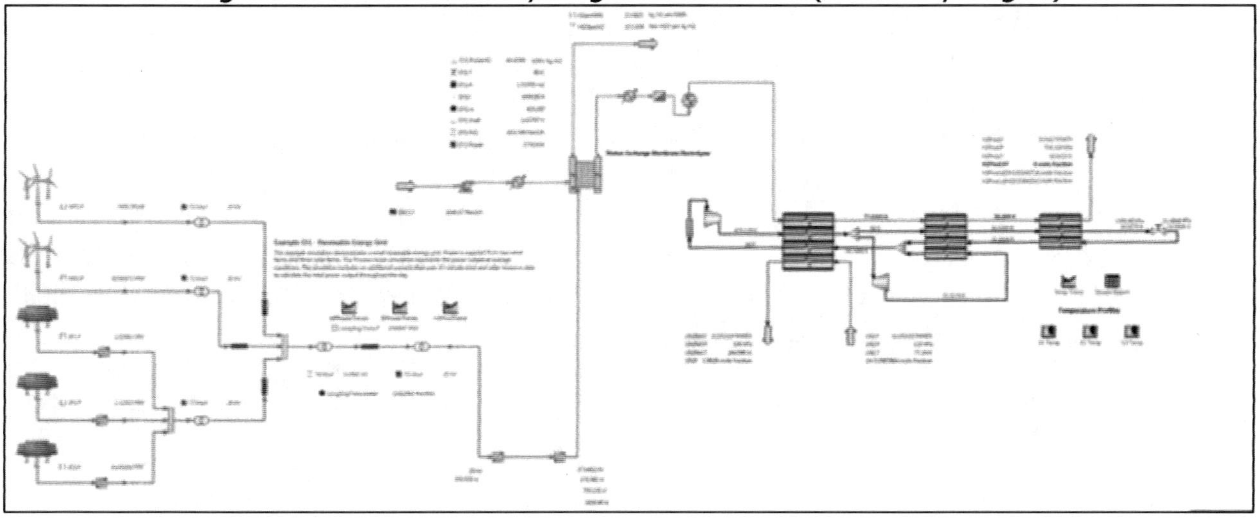

This simulation demonstrates a small renewable energy grid including renewable energy supplied from two wind farms and three solar farms. The model also includes both the process elements (pumps, heat exchangers, electrolyzers etc.), the electrical elements (transformers, power lines etc.) and a hydrogen liquefaction facility. The simulation runs at steady state to calculate average power output. The simulation also runs dynamically to predict the variability of power produced during the day as wind conditions change and the sun rises and falls. The model also looks at electricity supply and demand considering importing or exporting to the grid, spot prices and hydrogen requirements.

Conclusions

In response to climate change the process industries are today and will continue tomorrow to adapt to ensure a sustainable future. At the same time, oil and gas companies are catching up with the other industries and digitally transforming their businesses to operate smarter, more efficiently and more competitively. This digital transformation will be one key enabling building block for a successful energy transition.

AVEVA is strongly committed to a new sustainable future and has invested heavily over the last several years to develop AVEVA Process Simulation, a next generation simulation tool. AVEVA Process Simulation is ready to be the foundation of the digital twin, throughout the project lifecycle, for the new lower carbon projects that will deliver a more sustainable future.

Today's digitally native process engineers will be the front line when it comes to driving this change and the authors of this paper strongly encourage these change agents not to look at the simulation tools of the past but to look towards the future with AVEVA Process Simulation.

References

1. Dickson, D, Tilghman, N, Bonny T, "The Future of Work in Oil, Gas, and Chemicals", Deloitte (https://www2.deloitte.com/us/en/insights/industry/oil-and-gas/future-of-work-oil-and-gas-chemicals.html) 20 October 2020.
2. Oil and Gas Climate Initiative (OGCI) Report "Delivering on a Low Carbon Future" https://www.ogci.com/wp-content/uploads/2020/12/OGCI-Progress-Report-2020.pdf Dec 2020
3. Redepenning, C, Hanak, D. Broeker, S, Depew, C, "Building a Sustainable Chemical Industry Through Process Simulation" Chemweek Webinar, 11 May 2021
4. Construction Industry Institute Report No. RR331-11 - The Maturity and Accuracy of Front End Engineering Design (FEED) and Its Impact on Project Performance https://www.construction-institute.org/resources/knowledgebase/knowledge-areas/general-cii-information/topics/rt-331/pubs/rr331-11 2018
5. Hydrogen production costs to 2040: Is a tipping point on the horizon?", Wood MacKensie (https://www.woodmac.com/our-expertise/focus/transition/hydrogen-production-costs-to-2040-is-a-tipping-point-on-the-horizon/) 2020

622681 Nutrient Circularity to Abate Nitrogen Pollution: Techno-Economic Assessment of Nitrogen Recovery Systems for Livestock Facilities

Abstract:

Since the XIX century, the agricultural sector has experienced an accelerated industrialization progression pursuing its intensification and satisfy the demand of food production and agricultural products, i.e., increasing the agricultural production per unit of input resources (land, labor, time, fertilizer, seeds, and investment) (FAO, 2004). During the last decades, the agricultural intensification is driven by a sustained population increment, a worldwide average income growth, and the trade liberalization and logistics advancements leading the transnational trade of products (Baker and Da Silva, 2014). However, multiple environmental challenges must be faced as a consequence of the industrialization of the agricultural and farming activities: depletion of soil nutrients and organic matter due to mono-cropping practices, excessive and inefficient use of synthetic fertilizers to maintain high cropping yields, spatial concentration and inappropriate management of livestock manure, biodiversity loss, ecosystems degradation, etc. Focusing in the context of current nutrient management, the unbalanced use of synthetic fertilizers and the excessive organic waste generation from livestock facilities have decoupled a previous balanced scenario where the organic waste from livestock activities was used as a nutrient and organic matter supply for growing crops. This has created a dependency on mineral (phosphorus) and synthetic (nitrogen) fertilizers, and the efficient management of animal manure has become a critical need for livestock operations. Therefore, there is a disconnection between locations with excessive nutrient content from organic waste (livestock facilities), and croplands demanding nutrients to keep high yield rates. Also, due to the sparse location between facilities with in-excess nutrient content organic waste (including manure and digestate) and croplands with nutrient needs, the transportation of these wastes for nutrient redistribution is challenging and expensive. Therefore, the implementation of technologies and processes for the recovery of nutrients in a cost-effective form suitable for easy transportation and use in croplands is of utmost importance. Also, this would aid to restore the circularity of agricultural nutrients usage, disrupted by industrial practices (van der Wiel et al., 2020).

In this work, a systematic review and techno-economic analysis of the state-of-the-art technologies for nitrogen recovery from livestock waste is carried out through the development of modules for the calculation of the energy and mass balances, nutrient recovery yield, and techno-economic evaluation. The aim is to identify the most adequate technology to close the nutrient loop from livestock to crops, and the optimal design and operating conditions for its implementation. The processing and management of livestock waste involve all stages from waste collection to the production of the nutrient-rich product: manure preconditioning, optional anaerobic digestion of manure and biogas valorization, solid-liquid separation of manure or digestate, and nutrients recovery. The assessment of technologies is carried out with special emphasis on the resiliency of the technology selected for each scenario. This is because of the existence of multiple variable factors affecting the performance of nutrient recovery facilities, e.g., the variability of waste composition, the prices of the products obtained, and the cost of the utilities required by each unit. Therefore, a flexible framework capable of assessing changes in input parameters is developed, allowing the study of the impact of the fluctuation of these variables on the results, the techno-economic performance of the different nutrient recovery technologies, and the selection of the optimal configuration. In addition, the application of potential government incentives is evaluated, for designing and determining the most cost-effective incentive policies to stimulate the implementation of nitrogen recovery systems at livestock facilities.

The obtained information from the systematic design and evaluation of nitrogen recovery systems is essential for performing a multicriteria decision-making approach to deploy the optimal systems for nitrogen recovery considering the livestock facility size, type of livestock, its economic performance, and the availability of effective incentive policies to encourage the deployment of these facilities. Additionally, the obtained results are very valuable for the development of supply-chain studies to determine the optimal spatial distribution of nutrient recovery facilities, and the potential exchanges of waste and products across a multi-product coordinated network (Sampat et al., 2017).

References

Baker, D., and Da Silva, C., Trends in agri-food systems: Drivers, changes, impacts and overall assessment, FAO Policy Learning Programme (2014).

FAO, The ethics of sustainable agricultural intensification. FAO, (2004)

Sampat, A.M., Martin, E., Martin, M., and Zavala, V.M., Optimization Formulations for Multi-Product Supply-Chain Networks, Computers & Chemical Engineering, 104C, 296-310 (2017)

van der Wiel, B. Z., Weijma, J., van Middelaar, C. E., Kleinke, M., Buisman, C. J. N., and Wichern, F., Restoring nutrient circularity: A review of nutrient stock and flow analyses of local agro-food-waste systems. Resources, Conservation and Recycling, 160, 104901, (2020)

622575 Optimal Design of an Open-Cycle Ocean Thermal Energy Conversion System for Energy and Water Supply Considering Multiple-Objectives

Abstract:

Ocean thermal energy conversion systems (OTEC) represent an attractive economic alternative in communities where a high percentage of electricity production is based on oil, where the water is scarce and, where the communities are close to the ocean. This power generation technology uses the temperature difference between the deep cold water and warm surface water of the ocean to produce electricity and desalinated water using the principles of the Rankine cycle. The objective of this work is to present a rigorous model for the optimal design of the open cycle ocean thermal energy conversion system. This work addresses a multi-objective strategy for defining the relationships between design and utility policy prices in water-energy off-grid systems. The formulation allows obtaining the design and operation of the system. In addition, levels of subsidies and impacts social, energy, and economic policies are considered for defining the utility price for the end-user. The solution to the problem consists of minimizing the total annual cost while social welfare is maximized.

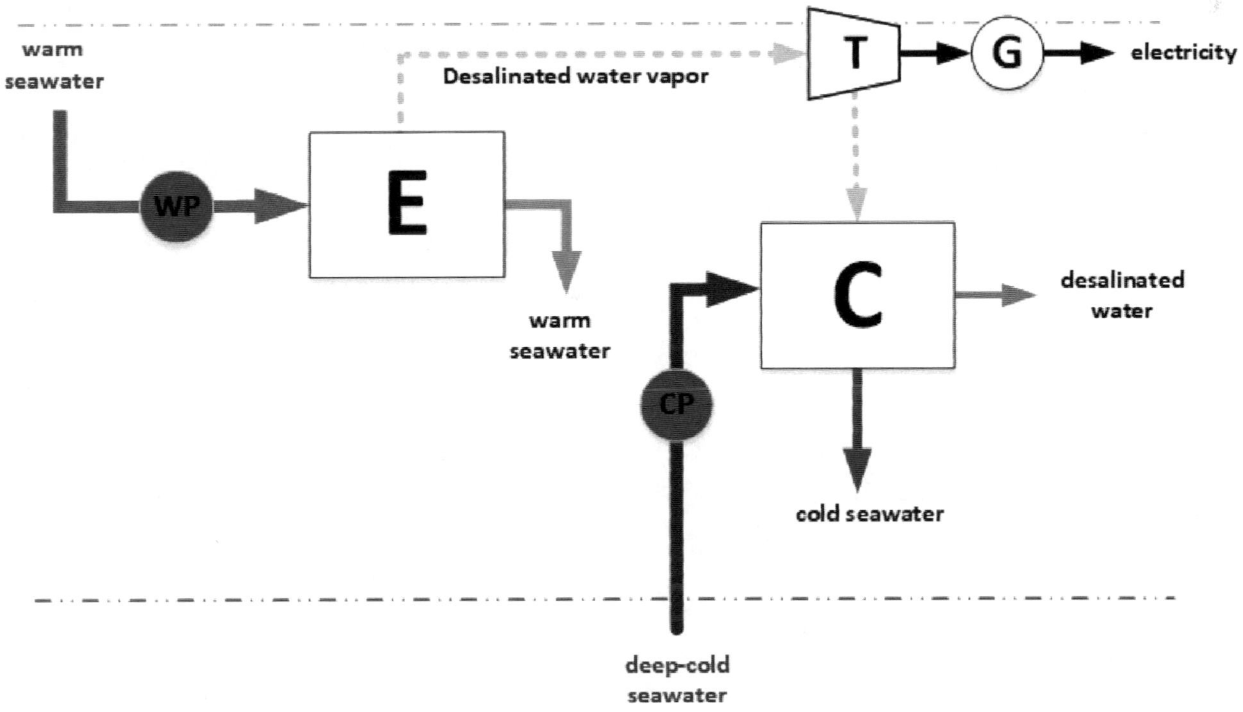

622627 Experiment Investigation of a Solar Reactor for Thermochemical Syngas Production Via the CeO_2-CH_4-CO_2 Hybrid Redox Cycle in a Concentrating Solar Tower

Abstract:

Solar thermochemical processing offers an efficient pathway for producing sustainable fuels using concentrated solar energy as the source of high-temperature process heat. The targeted fuel is syngas – a specific mixture of H_2 and CO that can be further processed to drop-in transportation fuels. Within this context, a 2-step redox cyclic process for syngas production is considered using CeO_2 as the redox material. This cycle has been extensively studied for splitting H_2O and CO_2 [1] and its on-sun feasibility was demonstrated as part of the EU project SUN-to-LIQUID [2][3]. An alternative process is a hybrid redox cycle that uses CH_4 as a reducing agent, thus significantly lowering the temperature of the endothermic reduction step. Furthermore, it can operate isothermally in the range 800-1200°C and achieve higher non-stoichiometries (δ), indicating a higher proportion of O_2 exchange and therefore a higher specific fuel output. The hybrid redox cycle is represented by two steps [4]:

Endothermic reduction:

$$CeO_2 + \delta CH_4 \leftrightarrow CeO_{2-\delta} + \delta CO + 2\delta H_2 \text{ (eq. 1)}$$

Exothermic oxidation:

$$CeO_{2-\delta} + \delta CO_2 \leftrightarrow CeO_2 + \delta CO \text{ (eq. 2a)}$$

$$CeO_{2-\delta} + \delta H_2O \leftrightarrow CeO_2 + \delta H_2 \text{ (eq. 2b)}$$

Combining the two redox steps yields the net reforming reactions:

Dry reforming:

$$CH_4 + CO_2 \leftrightarrow 2CO + 2H_2 \text{ (eq. 3a)}$$

Wet reforming:

$$CH_4 + H_2O \leftrightarrow CO + 3H_2 \text{ (eq. 3b)}$$

We report on a joint collaboration of Synhelion, IMDEA Energía, and ETH Zurich, to test the dry (CO_2-based) hybrid redox cycle and reforming reaction at the concentrating solar tower facility of IMDEA Energía in Spain. The solar reactor consisted of a cavity-receiver with a windowed aperture and lined with a reticulated porous ceramic (RPC) structure made of CeO_2, totaling 21 kg, which featured dual-scale porosity for enhanced heat and mass transfer. With this arrangement, the RPC is directly exposed to concentrated solar irradiation entering through the aperture. The main operating conditions were: solar radiative power input of 10 kW at a mean solar concentration ratio of 500 suns, nominal cavity temperatures between 800-1200°C, mass flowrates of 100 L/min (L: normal liters), and up to 40% reactive gas concentrations diluted in Ar. The peak molar conversions obtained during hybrid redox cycling (eq. 1 and eq. 2a) were CH_4 = 70% and CO_2 = 40%, which increased with temperature as predicted by thermodynamics, resulting in higher non-stoichiometries. Reaction selectivity to CO = 50% and to H_2 = 30% was reached at 930°C and at a low reactive gas concentration of 3%. The dry reforming operation mode (eq. 3a) is dominated by an endothermic reaction, hence CH_4 conversions increased with temperature independent of feeding ratio (CO_2:CH_4) to a maximum of 50% at 955°C. Conversions and selectivity were improved by incorporating a second tubular (non-solar) reactor downstream of the solar cavity [5]. Overall, the project proved the feasibility of operating the hybrid redox cycle with a directly-irradiated solar reactor in a solar tower configuration.

References

[1] Marxer, D., et al., *Solar thermochemical splitting of CO2 into separate streams of CO and O2 with high selectivity, stability, conversion, and efficiency.* Energy & Environmental Science, 2017. **10**(5): p. 1142-1149.

[2] Koepf, E., et al., *Liquid fuels from concentrated sunlight: An overview on development and integration of a 50 kW solar thermochemical reactor and high concentration solar field for the SUN-to-LIQUID project*, in *SOLARPACES 2018: International Conference on Concentrating Solar Power and Chemical Energy Systems*. 2019.

[3] Romero, M., et al., *Solar-Driven Thermochemical Production of Sustainable Liquid Fuels from H2O and CO2 in a Heliostat Field*, in *Proceedings of the ISES Solar World Congress 2019*. 2019. p. 1-12.

[4] Bulfin, B., et al., *Thermodynamic comparison of solar methane reforming via catalytic and redox cycle routes.* Solar Energy, 2021. **215**: p. 169-178.

[5] Ackermann, S., et al., *Process for the Production of Syngas*, EP18195213.6, filed 18.9.2018.

625622 Solar Syngas Production from H_2o and CO_2 Applicable for Methanol or Fischer-Tropsch Synthesis

Abstract:

We report on the experimental demonstration of the solar production of syngas – a specific mixture of CO and H_2 – from CO_2 and H_2O using a modular solar concentrator-reactor system. This system delivers concentrated solar process heat to drive the simultaneous co-splitting of CO_2 and H_2O via a ceria-based thermochemical redox cycle. We present representative on-sun runs with fully-automated consecutive redox cycles and show experimental parameter analysis for control strategies optimising syngas quality, yield, efficiency and conversion. We further present a dynamic grey-box model of the system that allows additional insights to the system's performance.

The solar reactor for effecting this redox cycle consists of a cavity-receiver containing a reticulated porous ceramic (RPC) foam structure made of pure CeO_2. Two identical solar reactors are mounted side-by-side on the focus of a solar dish concentrator that enables the operation of both solar reactors simultaneously by alternating the solar radiative input between them. Thus, while one solar reactor is performing the endothermic reduction step on sun, the second solar reactor is performing the exothermic oxidation step off sun. During the reduction step, the solar reactor is heated with concentrated sunlight up to 1500 °C and the total pressure is lowered to 4 mbar by a vacuum pump to evolve a pure stream of O_2 from CeO_2. During the oxidation step, CO_2 and H_2O are co-injected into the reactor's cavity, react with the reduced ceria, and are transformed into syngas - a specific desired mixture of CO and H_2.

The syngas composition can be tailored by changing H_2O:CO_2 feed ratios as well as choosing adequate oxidation start/end conditions to meet the desired quality and stoichiometry suitable for Fischer-Tropsch or methanol synthesis. Thus, the need for additional downstream refining of the syngas, e.g. via the energy intensive reverse water-gas shift reaction, is eliminated.

The entire system is controlled to perform fully automated consecutive redox cycles while continuously updating cycle parameters based on real time analysis and feedback loops. Changing process parameters such as reduction/oxidation temperatures, gas flow rates, or oxidation start/end conditions allows optimising the cycles towards maximising either efficiency, quality, yield or conversion. Full day on-sun consecutive cycles demonstrate the stability and robustness of the system.

A dynamic model of the reactor system allows further parameter analysis and offers additional insight to the system performance. The grey-box model is based on energy and mass balances and is using a lumped-parameter approach where parameter identification and validation is achieved through comparing numerically simulated results to experimental data collected over multiple characteristic cycles.

626086 Rotary Kilns: A Powerful and Effective Tool for Running Particle-Based Solar-Chemical Processes

Abstract:

Different concepts exist for coupling solar energy to an endothermal chemical process. One of the most effective option to use directly irradiated solar receiver-reactors, where the radiation is directly absorbed by the reactants; e.g. in a volumetric receiver-reactor or a particle receiver.

Considering that for heat storage applications a high mass flow rate and large amounts of active material are important, particle receivers are an attractive option. Rotary kilns combine the advantages of an effective homogenisation of the materials, effective heat transfer between the gas and solid phase, a high mass flow rate, and a high reactive surface area. Beyond this, the particles motion should decrease their sintering. Rotary kilns are a common and established device for diverse applications dealing with the thermal treatment of particulate material, with advantages of easy control of operating parameters, such as residence time or temperature, together with a simple design. They are well established devices for numerous thermochemical applications, e.g. the high temperature treatment of wastes, calcination processes, and the recovery of other non-ferrous metals.

The suitability of solar heated rotary kilns for thermochemical application was shown in several cases. For example, solar heated rotary kilns were tested for treatment of inert material, they were used to effect the endothermic calcination reactions (Moumin et al., 2019; Meier et al., 2006) or were used for the reduction of volatile metal oxides such as ZnO (Charvin et al., 2008), and for aluminum remelting (Glasmacher-Remberg et al., 2001).

Based on these experiences, the present study analyses the use of directly solar irradiated rotary kilns exemplarily for thermochemical heat storage, through the reduction and oxidation of non-volatile metal oxides, and for solar calcination processes. The main tasks and issues of such devices, i.e. effective mixing and heat transfer, managing particle flows and avoiding particles exiting the reactor, and ensuring the transparency of a window, are addressed based on experimental investigations in lab and pilot scale and numerical models.

In the first example the reactant is a metal oxide, precisely a mix of manganese and iron oxide, which undergoes consecutive reduction (endothermic) and oxidation (exothermic) reaction and thus serves as a thermochemical storage material. The metal oxide is shaped in mm-size granules. The reduction takes place in a solar rotary kiln which directly transfers the concentrated solar power to the reactive particles, which get heated and react. The system can treat several tenths of kg per hour at temperatures up to 1100°C, allowing to store a power of about 3kW.

The second example addresses the solar calcination of limestone. The experimental part addresses the testing of rotary kiln pilot calciner (up to 30 kg capacity and 5-10 kW solar power) in a solar simulator. A numerical model was developed to identify and verify the heat losses measured in the experiments. The parameters of the model were changed to best approach the experimental temperature distribution.

It became evident from experiments and models that beds exhibiting small size particles, like cement raw meal, show very little mixing in a cylinder without built-ins and the measured heat transfer coefficient was much lower than the model predictions. Using longitudinal strips as built-ins along the wall caused the material bed to collapse regularly, increasing its heat transfer by a factor of 2–3, up to the range of predictions from theoretical models.

Finally, an electrostatic precipitation system that can work at high temperature has been developed and adapted to the rotary kiln to keep the window free of dust. Significant improvement of window performance could be reached by this measure.

References:

Moumin, G., Tescari, S., Sundarraj, P., de Oliveira, L., Roeb, M., Sattler, C. 2019. Solar treatment of cohesive particles in a directly irradiated rotary kiln, Solar Energy 182, 480-490.

Meier, A., Bonaldi, E., Cella, G.M., Lipinski, W., Wuillemin, D., 2006. Solar chemical reactor technology for industrial production of lime. Solar Energy 80, 1355–1362.

Charvin, P., Abanades, S., Neveu, P., Lemont, F., Flamant, G., 2008. Dynamic modeling of a volumetric solar reactor for volatile metal oxide reduction. Chemical Engineering Research and Design 86, 1216–1222.

Glasmacher-Remberg, C., Roeb, M., Dersch, J., Schäfer, R., Funken, K.-H., 2001. Solar thermal recycling of aluminium. AL Aluminium and Its Alloys 135, 73–77.

623296 Performance of a Concentrated Solar–Gliding Arc Plasma Reactor for CO_2 Decomposition: Effect of Catalyst Monolith

Abstract:

Among approaches to reduce the increasing amount of atmospheric CO_2 [1], such as carbon capture and sequestration [2], the conversion of CO_2 into value-added fuels and chemicals using renewable energy does not only mitigate environmental emissions, but also contributes to sustainable development. Solar thermochemical CO_2 conversion approaches [3], by their direct use of the most abundant form of renewable energy, are among the most sustainable; however, their viability is limited by the inherent intermittency of solar irradiation. In contrast, plasmachemical methods can be very efficient and operate continuously, yet their sustainability depends on their use of electricity from renewable sources. The combined use of solar thermochemical and plasmachemical methods has the potential to lead to more sustainable and viable CO_2 conversion approaches. Computational studies have shown that CO_2 in nonequilibrium plasma state absorbs drastically more solar energy than CO_2 in thermodynamic equilibrium [4]. Moreover, experimental studies have demonstrated that greater CO_2 conversion efficiency is obtained by the combined use of concentrated solar radiation and plasma as compared to plasma or solar energy alone [5-7]. Prior studies of the combined use of concentrated solar radiation and a gliding arc discharge plasma for CO_2 conversion [5, 6] did not include the incorporation of catalytic monoliths in order to more clearly discern the interaction between solar photons and plasma species. The incorporation of catalytic monoliths, essential in solar thermochemical processes [8], has the potential to lead to significantly improved CO_2 conversion performance. The design, operation, and performance of a concentrated solar – gliding arc plasma reactor for CO_2 decomposition fitted with zirconia (Zr) foam catalytic monoliths are presented. A 6.5 kW high-flux solar simulator is used as the concentrated solar energy source. The focal point of the concentrated radiative flux is located at the point where a three-prong electric arc discharge, powered by two 120 W high-voltage power supplies, forms. The arc discharge plasma interacts with the incident solar radiation as it glides along three divergent electrodes to finally impinge on the Zr monolith. Optical visualization together with dynamic current, voltage, and power signals are used to characterize the process. Outflow gas samples are collected as function of process time and analyzed by Gas Chromatography to determine gas products. Conversion and processes efficiency as function of process parameters, including solar power, electric power, and catalytic monolith arrangement, are analyzed and presented.

Keywords: Energy – solar energy; Energy - Alternative Energy

Acknowledgements: This work has been supported by the U.S. National Science Foundation through award CBET-1552037.

References:

[1] Bhowmik, D. Global Carbon Dioxide Emissions in Hamilton Filter Model. *International Journal of Environment, Agriculture and Biotechnology*, *5*, 5.

[2] Rackley, S. A. (2017). Carbon capture and storage. Butterworth-Heinemann.

[3] Steinfeld, A. (2019). Concentrated solar energy – the path for efficient thermal conversion to power and fuels. Science Bulletin, 64:8, 485- 486.

[4] Elahi, R., Nagassou, D., Mohsenian S., & Trelles J. P. (2020). Enhanced Solar Radiation Absorption by Carbon Dioxide in Thermodynamic Nonequilibrium: A Computational Study, *Solar Energy*, 195, 369-381.

[5] Nagassou, D., Mohsenian, S., Nallar, M., Yu, P., Wong, H. W., & Trelles, J. P. (2020). Decomposition of CO_2 in a solar-gliding arc plasma reactor: Effects of water, nitrogen, methane, and process optimization. *Journal of CO_2 Utilization*, 38, 39-48.

[6] Nagassou, D., Mohsenian, S., Bhatta, S., Elahi, R., & Trelles, J. P. (2019). Solar–gliding arc plasma reactor for carbon dioxide decomposition: Design and characterization. *Solar Energy*, 180, 678-689.

[7] Mohsenian, S., Nagassou, D., Bhatta, S., Elahi, R., & Trelles J. P. (2019). Design and Characterization of a Solar-Enhanced Microwave Plasma Reactor for Atmospheric Pressure Carbon Dioxide Decomposition. *Plasma Sources Science and Technology*, 28, 065001.

[8] Bhatta, S., Nagassou, D., Mohsenian, S., & Trelles, J. P. (2019). Photo-thermochemical decomposition of carbon-dioxide in a direct solar receiver-reactor. *Solar Energy*, 178, 201-214.

623248 High-Temperature Heat Recovery System Coupled to a Solar Reactor for Thermochemical Redox Splitting H_2O and CO_2

Abstract:

The long-haul aviation and maritime sectors are strongly dependent on liquid transportation fluids, which are characterized by their high specific gravimetric and volumetric energy capacity. The production of drop-in fuels from H_2O and CO_2 using solar energy can eliminate greenhouse gas emissions derived from fossil fuels and provide the path to sustainable aviation and shipping.

One such promising pathway to solar fuels is via a thermochemical redox cycle for splitting H_2O and CO_2, driven by concentrated sunlight. The product is a specific mixture of H_2 and CO – syngas – which can be further processed downstream by established Fischer-Tropsch synthesis to liquid hydrocarbons such as kerosene (jet fuel) and methanol. The entire solar fuel process chain has been successfully demonstrated in concentrating solar tower and solar dish configurations. The solar reactor concept is based on a cavity-receiver lined containing reticulated porous ceramic (RPC) ceria structures. The solar-to-fuel energy efficiency achieved to date was 5-6% without the application of heat recovery. The relatively low value obtained is mainly due to the sensible heat rejected during the temperature-swing redox cycling. The reactor operates between a reduction temperature of 1500°C and an oxidation temperature of about 800°C, with the cooling losses accounting for up to 43% of the total solar energy input. Recovering this heat lost can significantly increase the solar-to-fuel energy efficiency.

We have developed a heat recovery method based on the thermocline energy storage (TES) concept. The solar reactor is coupled with two TES units. One TES unit operates between ambient and oxidation temperature, while the other operates between reduction and ambient temperatures. N_2 is used as an inert heat transfer fluid (HTF) to shift the thermoclines by charging and discharging the TES units. With this arrangement, the heat rejected when cooling the solar reactor from the reduction to the oxidation temperature is recovered and stored in the TES units, and delivered back to the solar reactor when heating from the oxidation to the reduction temperature. Recovering half of the sensible heat in the RPC can in principle boost the solar-to-fuel energy efficiency to values exceeding 20%.

We have designed and constructed an experimental setup, consisting of a lab-scale solar reactor and two thermocline-based TES units, each made of a packed-bed of alumina spheres. Preliminary testing at 1150°C in a high-flux solar simulator demonstrated the concept feasibility. The measured heat recovery effectiveness – defined as the ratio between the sensible heat extracted/charged from/to the solar reactor – was 50%. The experimental campaign included the study of several key parameters on the performance: the HTF flow rate, oxidation start temperature, oxidation end temperature, HTF inlet temperature during the heat recovery step, and stability during consecutive redox cycles.

627948 Modelling and Chemical Compatibility of Super-Critical Carbon Dioxide Flow in Micro-Channels of Microvascular Carbon/Carbon Composites

Abstract:

Solar energy generation in the United States has greatly increased in the last decade, with a tenfold increase since 2010 [1]. Concentrating solar power (CSP) systems use focused sunlight to convert thermal energy to electricity, usually through the use of conventional steam turbines. In a central tower configuration, heliostats are mirrors that follow the sun's position, directing the light into a receiver located at the top of a central tower. A major breakthrough in CSP systems is the use of thermal storage systems, whereby using molten salt, thermal energy can be stored for multiple hours, giving CSP systems on-demand energy generation capabilities. In light of that, central tower receivers have been developed initially to use molten salt as the heat transfer fluid.

Recently, a U.S. Department of Energy roadmap [2] has identified super-critical carbon dioxide (sCO_2) Brayton-cycles as an integral part of the next generation of CSP plants. By replacing the conventional steam turbines, higher efficiencies can be achieved, further contributing to lowering generation costs of CSP systems. Super-critical fluids have properties between gas and liquids, such as providing a lower pressure drop than liquids, while having a higher heat capacity when compared to gases. Carbon dioxide, whose critical point is at 73.9 bar and 304 K, has a higher density when supercritical, meaning that process equipment can be much smaller.

Recent research has focused on central gas receivers using sCO_2 as the heat transfer fluid [3] in place of molten salt. With this approach, the thermal losses from the heat transfer between molten salt and the sCO_2 Brayton-cycle are lowered. Additionally, the receiver can operate at higher temperatures, which generally correlates to higher global efficiencies. Molten salt receivers have to operate above 260 °C to avoid crystallization, and below 575 °C to avoid thermal decomposition of the salt. On the other hand, gas receivers operating with sCO_2 are currently being proposed to operate between 530 °C and 750 °C.

In this work, the use of a modular central receiver (Figure 1(a)) is proposed, where each module (Figure 1(b)) will consist of a novel Carbon/Carbon composite with an embedded microchannel network, where sCO_2 can flow as the heat transfer fluid. Traditional gas receivers use metallic or ceramic materials, which are known to be susceptible to thermal fatigue due to the daily startup operations of CSP plants. On the other hand, the use of C/C composites, which are strong, lightweight and highly heat conductive materials, have a much lower coefficient of thermal expansion, which greatly decreases problems with thermal fatigue. Modules with a micro-channel network are made by combining C/C composites fabrication techniques with the use of vaporization of sacrificial component (VaSC) technology, as previously presented [4].

However, the use of sCO_2 as the heat transfer fluid with C/C composites can cause mass loss of the module through a reaction that occurs between CO_2 and carbon materials above 600 °C. This effect may be significant, especially when considering the years of operation that the modules are expected to operate at. There are multiple proposed mechanisms for this reaction, however when it relates to the mass loss of a composite material, it is usually simplified to Equation (1). In addition, at high pressures, carbon monoxide (CO) can compete with CO_2 for active sites, changing the reaction mechanism to include a reversible step [5].

$CO_{2\,(g)} + C_{(s)} \rightarrow 2CO_{(g)}$ (1)

To predict the mass loss effects on the C/C composite modules, a Computational Fluid Dynamics (CFD) approach will be used to model the flow of sCO_2 through microchannels with a reacting surface. To perform the CFD modelling, the software Ansys Fluent is used. Preliminary results were obtained by modelling the flow of sCO_2 at high pressure (200 bar) through a straight microchannel using a 2-D axisymmetric geometry. The models and boundary conditions used for this case are as follows: axisymmetric (2D geometry), turbulent standard k-ε (viscous model), Peng-Robinson equation of state, inlet velocity as a parabolic profile with mean velocity of 5 m/s, inlet temperature of 600 °C, outlet pressure of 200 bar, constant wall temperature of 727 °C, radius of 0.25 mm and length of 10 mm.

Results from the modelling obtained thus far excludes the reacting surface, which will be used as a baseline for the development of an accurate model that describes the flow of sCO_2. The next step involves the inclusion of a reacting surface to the current model, as well as the evaluation of different geometries and operating conditions. To include the reaction surface into the model, the rate of mass loss of the carbon material can be described by Equation (2), where k is the kinetic constant, n is the pressure exponent and P_{CO2} is the partial pressure of carbon dioxide.

$$r_{CO2} = k\, P^{n}{}_{CO2} \quad (2)$$

Therefore, Equation (2) can be used to describe the mass loss of the reacting surface (the C/C composite), following a similar approach from works on surface recession of graphite rocket nozzles [6]. In order to implement a reacting surface into the model, the use of UDFs (User Defined Functions) will be necessary, which consists of a function written in C programming language, that can be used to couple the Fluent solver to complex equations.

The constants k and n from Equation (2) have to be determined experimentally, and since available data on carbon materials is usually limited to graphite, they will be initially used for baseline model results. Future kinetic experiments involve the use of TGA (thermogravimetric analysis) in order to obtain the rate constants for the composites fabricated in this work [4], thus more accurately representing the conditions (rate of mss loss) at which the microvascular C/C composites will be subjected to.

An experimental analysis of the flow of sCO_2 through microchannels of the fabricated C/C composite will be done next. By combining a setup under the expected temperature conditions (600 – 1000 °C) with a GC (gas chromatograph) analysis, the developed model can be validated by comparing the outlet gas composition.

By accurately predicting the effect of mass loss through the use of CFD modelling, such as changes in channel geometry, operational conditions (upper temperature limit) can be set, as well the evaluation of the use of protective coatings (such as silicon carbide (SiC)) that can slow the rate of reaction.

References:

[1] *U.S. Energy Information Administration (EIA)*. November 9, 2020, https://www.eia.gov/totalenergy/data/browser/

[2] Mehos, M., Turchi, C., Vidal, J., Wagner, M., Ma, Z., Ho, C., Kolb, W., Andraka, C., & Kruizenga, A. (2017). Concentrating Solar Power Gen3 Demonstration Roadmap.

[3] Zada, K. R., Hyder, M. B., Kevin Drost, M., & Fronk, B. M. (2016). Numbering-Up of Microscale Devices for Megawatt-Scale Supercritical Carbon Dioxide Concentrating Solar Power Receivers. *Journal of Solar Energy Engineering, Transactions of the ASME*, *138*(6), 1–9.

[4] Cordeiro, J. C.; Argot, J., Ramsurn, H., Crunkleton, D. W., Otanicar, T., Keller, M. (2020, November). "Microvascular Carbon-Carbon Composite for Concentrated Solar Power Gas Receivers". In 2020 Virtual AIChE Annual Meeting.

[5] E. S. Golovina. The gasification of carbon by carbon dioxide at high temperatures and pressures".

Carbon 18.3 (Jan. 1980), pp. 197-201.

[6] Piyush Thakre and Vigor Yang. Chemical Erosion of Carbon-Carbon/Graphite Nozzles in Solid-Propellant Rocket Motors". Journal of Propulsion and Power 24.4 (July 2008), pp. 822-833.

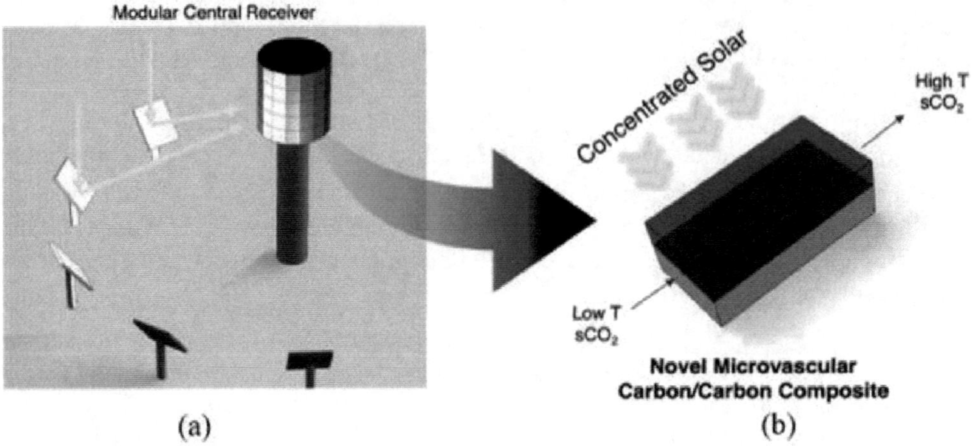

Figure 1. Modular (a) central tower receiver concept with a (b) novel microvascular Carbon/Carbon composite with sCO_2 as the heat transfer fluid.

631612 Life Cycle Assessment of Resource Recovery Technologies on Large-Scale Dairy Farms

Abstract:

Large-scale production of livestock for dairy products has a significant impact on the environment through greenhouse gas emissions, water pollution, and resource consumption. In particular, manure management at large-scale concentrated animal feeding operations (CAFOs) poses risks to air, water, and soil quality.

However, manure is also a rich source of organic matter, nutrients (nitrogen and phosphorus), and water. Resource recovery technologies offer a way to recycle valuable compounds in manure into beneficial products while simultaneously reducing harmful emissions. In this research, a novel Nutrients, Energy, and Water Integrated Recovery (NEWIR) system is compared to a typical California dairy CAFO with life cycle assessment (LCA) methodology. NEWIR consists of three innovative processes for integrated resource recovery from dairy manure: 1) hydrothermal carbonization (HTC) for energy, 2) algal raceway ponds for nutrients, and 3) membrane distillation (MD) for high-quality water. HTC creates a carbon-rich hydrochar and a nutrient-rich aqueous stream. Hydrochar can be burned in conventional coal power plants, replacing a fossil fuel with a renewable form of bio-energy. The aqueous HTC effluent serves as a water and nutrient source for microalgae that are used as a protein-rich cattle feed supplement to offset soybean production. Finally, MD uses waste heat from HTC to recover high-quality water from the algae effluent that is reused on-farm for cleaning and sterilization, replacing conventional reverse osmosis. LCA results are used to evaluate how resource recovery technologies affect relevant environmental impact categories such as carbon footprint, water footprint, and eutrophication. The life cycle inventory is developed from experimental data and modeling for a full-scale NEWIR system and analyzed using the ReCiPe impact assessment method. Results will be presented for a full-scale model of the NEWIR system, including avoided products due to resource recovery. By creating useful products from a biowaste stream, NEWIR paves the way towards sustainable food production on a large scale.

632629 Consumption-Based Accounting for Tracing Nutrient Footprints of Beef Supply Chains in the United States

Abstract:

Food supply chains span multiple spatial and temporal scales and have evolved to become increasingly complex. Food products go through many different processing steps before being ready for consumption. While highly connected food chains provide numerous benefits, they lack traceability. As such, understanding their resiliency, security, and associated environmental burdens of consumption become difficult. One specific class of impacts is the nutrient footprint of food consumption. Application of synthetic fertilizers and release of reactive nitrogen and phosphorus has been linked to degradation in water quality and loss of biodiversity across the globe. Beef production involves consumption of large quantities of animal feed, the production of which is heavily dependent on fertilizer application. Although there is a rich body of work on nutrient footprint estimation for beef production, there is a gap in the literature regarding the spatial distribution of the nutrient releases throughout the supply chain. Additionally, bulk of the studies focus on production related impacts making it difficult to trace environmental impacts of beef consumption that typically occur far from the point of consumption.

We develop an optimization-based framework to model supply chain networks of beef products at the county level. Using publicly available data, we construct a weighted network of nutrient flows based on nutrient mass balance, including synthetic fertilizers, manure production, and crop uptake and residues. Preliminary results show that the beef supply chain is highly concentrated in the central region of the United States, with a majority of slaughtering facilities owned by a small number of companies. Additionally, there is a large inflow of cattle to the central region for feeding and slaughtering. Feed is sourced from several different counties and states depending on the specific crop type. Analysis reveals that most states consume far more meat than they produce within their boundaries, and highly populated areas are reliant on the inflow of products from the Midwest. Our results also highlight the disparity between production and consumption-based impacts and spatial heterogeneity in nutrient footprint of beef. Implications of these findings for reducing the environmental impacts of beef supply chain, consumer food choice, and policymaking will be discussed.

623757 Assessment of Global Water Stress Using the Generalized Global Sustainability Model (GSM)

Abstract:

Water is essential for human survival and critical for other spheres of human civilization, namely, agriculture, industrial production, and energy production. Many regions in the world are expected to face water scarcity issues in the near future. However, the water sector cannot be studied in isolation. Agriculture, responsible for food and feed production, is the largest consumer of water as compared to other anthropocentric consumption routes such as municipal, industrial, livestock, and energy [1]. Energy is an important contributor to human development and well-being, and it is also a key input to agriculture. With electricity-powered pumps, water can be drawn from deeper wells or transferred for long distances, highlighting the role of energy in improving water accessibility. However, energy production itself requires water, and some energy production technologies are highly water-intensive. Biopower and biofuels have been shown to have high water footprints [2]. Additionally, excessive water withdrawal due to cheap energy may further deplete freshwater resources. Thus, ensuring adequate water availability for everyone is a problem that cannot be addressed in isolation. Instead, it is a part of the food-energy-water (FEW) nexus.

The inter-dependencies of these three sectors are complex. Scanlon et al. [4] highlighted the criticality of understanding the FEW nexus. Several researchers have studied the FEW nexus with different objectives. For example, D'Odorico et al. [5] closely examined the individual links between food-energy, energy-water, and food-water and identified the critical areas for action. However, this study did not capture the integrated nature of the FEW nexus. The scale of the FEW nexus is also an essential factor. However, Nie et al. [6] studied the optimization problem involving the FEW nexus on a small scale. Franz et al. [7] studied the FEW nexus at the regional scale, which again represents a limited spatial scale. However, as global systems are large, inter-dimensional influences may take time to manifest. Hence, integrated cross-disciplinary models exploring the interactions between these various components has become increasingly important. Such models and their applications can improve our understanding of the complex relationships between these sectors and provide prescriptive recommendations using scenario analysis. Development of such global models began with World3 by Meadows et al. [8]. The World3 model included subsystems such as food production, industry, population, pollution, and non-renewable resources. Some of the other notable examples of global models are GUMBO [9], Earth3 [10], and HANDY [11]. More recently, our group [12] developed the generalized Global Sustainability Model (GSM). This is conceptually similar to the earlier version developed by Kotecha et al. [13]. However, we revised the model by selecting parameters as well as variable ranges based on global historical data. As compared to other global models, its ecological dimension is represented with features such as multiple trophic levels and intra-trophic level diversity. Moreover, the explicit integration between the economic and ecological dimensions is another strong point.

One key limitation of the revised model was that it did not explicitly capture water as a resource. It was assumed to be a part of the generalized resource pool. Water, though, is a crucial resource and is a part of a nexus, as highlighted previously. The critical novel contribution of this work is the incorporation of a global water network in GSM. The revised model with the water network is then used to study scenarios and policy alternatives. The analysis work tackles the following questions using the GSM: with business-as-usual conditions, can we have a sustainable water supply for all? How would the situation change if there is a population explosion? Can the improvements under Sustainable Development Goal 6 help in preventing undesirable situations?

Figure 1 shows GSM with the water network system in place. The GSM consists of three ecological trophic levels and human society. There are eight ecological compartments, human households, and their inter-dependencies give rise to a complex food web. Lotka-Volterra type equations are used to represent these relationships. The economic dimension is represented in the form of four sectors: agriculture, livestock, industry, and energy. Typically, economic state variables, prices, and production targets for the four sectors are governed by their products' inventory and the wage rate. As a price-setting mechanism is used in this model in lieu of the Computable General Equilibrium Approach, the prices are used to compute human society's demands for each industry. The supply is constrained by the raw material availability limited by the ecological dimension. This is where the interaction of ecological and economic dimensions comes into the picture.

The water system in the GSM is developed in two steps: first, the global water reservoirs and flows are established, and then, the demands by different sectors are characterized. The water reservoir represents the shared pool of global exploitable water. Inaccessible water whose utility cannot be harvested owing to technological or economic limitations or pollution is shown by the non-exploitable water reservoir. The water recycling compartment represents the mass of water undergoing water recycling. Agriculture, industrial, energy and municipal consumption of water is shown by flows to P1, IS, EE and HH compartments. In order to model the water demands, we define sectoral water intensity as the quantity of water withdrawn per unit of the desired variable. Agriculture and livestock are part of the ecological system, and, hence, would require water for their survival. Considering this constraint, P1 and H1 masses are used as shadow variables. On the other hand, the industrial and energy sectors do not need continuous replenishment of water like the ecological sectors. These sectors withdraw water during the production process only. Hence, IS and EE production terms are used as shadow variables. For municipal water withdrawal, the human population is used as a shadow variable. To summarize, P1 and H1 mass for the agricultural and livestock sector, IS and EE production for the industrial and energy sector and population for the municipal withdrawal are used as shadow variables. A sectoral intensity trend is a relationship between the historical data and corresponding shadow variable. These trends are used for projecting the future global water demands for those sectors.

The region greatly influences the demand-supply dynamics of water under consideration. To study this aspect, countries are categorized into six groups based on the continent where they are located: Africa, Asia, Europe, North America, South America, and Oceania. As exact details of water demand and supply distributions for various sectors are not available, the shadow variables approach is used to obtain the regional distribution of the water variables for each sector. The availability of exploitable water in a region is relatively stable and can be considered proportional to its area. Thus, the area of the country is considered as a shadow variable here. Again, shadow variables represent the water demand for various sectors: municipal water demand - population; agricultural water demand - area under agriculture; industrial and energy sectors - GDP (measured in international $ 2011 and adjusted for inflation). This historical country-level data is aggregated, and the contribution of each of the six regions towards the total value is computed. The approach of the piece-wise linear fit [14] is used to obtain the linear regression fits. Global and regional demand-availability dynamics, the potential adverse effect of the population explosion, and the positive influence of water resource management are analyzed.

The model was parameterized using historical data until 2013, and simulations were performed until the year 2120. Simulation results indicate that overall global water resources are adequate to meet the demands of human society. Here, the metric used to quantify the relation between water demand and availability is net water stress. It is the ratio of total water demand and exploitable water resources. This metric is more realistic as it considers the techno-socioeconomic feasibility of water utilization. Accordingly, global net water stress is close to 60%. However, deeper investigation considering distribution across various regions in the world reveals a disparity in water availability. Asia is the most water-stressed region in the world. Its water demand is expected to exceed availability by 2036. The population explosion scenario also affects water stress in the long run. For Asia, the peak water stress for the population explosion scenario increases from 109% to 124%. Water use efficiency improvement offers a way to prevent or at least delay this impending crisis. Scenario analysis indicated that at least a 30% reduction in water demand in the agricultural sector is necessary to avoid the water crisis in Asia altogether.

To summarize, the global FEW nexus's sustainability analysis identifies Asia as a flashpoint of the impending water crisis and advocates aggressive reforms in water use agricultural patterns resulting in substantial efficiency improvement to prevent the same. The model will be used to identify specific technology improvement targets and policy alternatives to ensure global water sustainability.

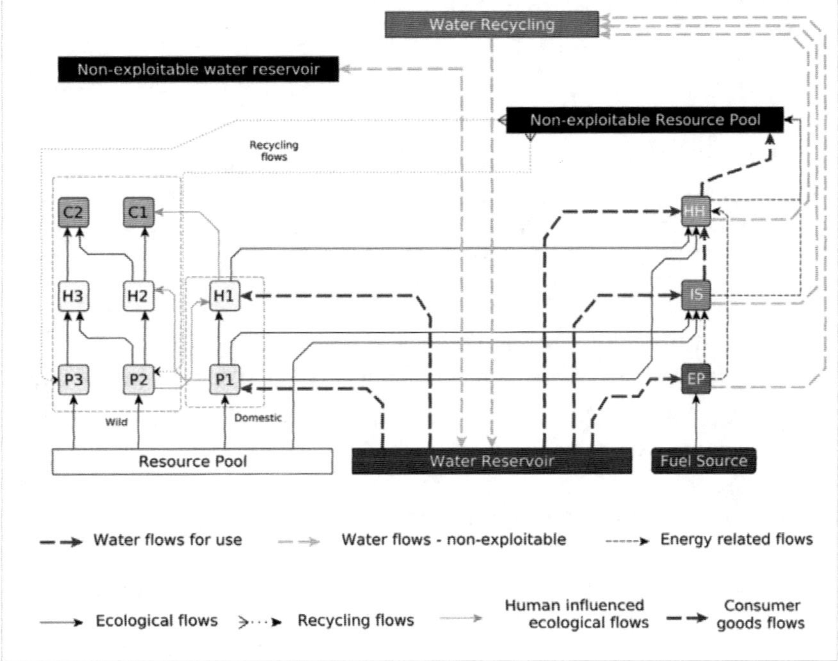

Figure 1: The Generalized Global Sustainability Model (GSM) is shown here. It has a trophic level structure. In the ecological dimension shown on the left side, higher trophic level depends on the lower trophic level. These are three ecological trophic levels are plants, herbivores and carnivores. These are classified into domesticated/human controlled and wild. Domesticated ecological compartments, that is, P1 and H1 represent the agricultural sector and the livestock respectively. P2 which is a part of the uncontrolled part of the ecosystem and represents grasslands is consumed by H1. Such multiple dependencies among the compartments give rise a complex food web. On the right side, HH represents human households, and IS and EP represent industrial sector and energy production sector respectively. The water dimension related components are shown in the center. Different types of flows are summarized below main figure.

624215 Bacteria Containment in Filtration

Abstract:

Drinking water systems contain a range of planktonic bacteria and bacterial communities that are most stable in the filtration system. Filters can have a negative impact if potential pathogens escape into distribution systems and contribute to infrastructure decay due to corrosion. Previous studies have determined that drinking water biofilms, especially those in biologically active sand filers, are stable communities that are able to persist the fluctuations in the source water community, and to develop resistance to disinfectants commonly used in drinking water. Focusing on the understanding of how bacteria biofilms can be contained further in filters by varying the filter material, and mechanical aspects of water. The effects of varying filter media and varying flow rates on bacteria biofilms by running trials in a plug flow reactor were measured.

627378 Use of the Aqueous Phase and Char from Hydrothermal Liquefaction of Biomass in Compost and Heavy Metal Adsorption Applications

Abstract:

Hydrothermal liquefaction (HTL) has shown much promise as a technique for converting waste biomass into bio-crude oil without the need for drying. Implementation of HTL at the large scale has been hindered by the limited options for value-added use of the aqueous phase (HTL-AP) and char (HTL-char) co-products.

This study investigated the applications of HTL-char as a mesoporous material for adsorption of heavy metals from water and of HTL-AP as an additive for compost. Food waste HTL-char, and a magnetic modified HTL-char prepared through co-HTL of food waste and red mud, were characterized and tested for removal of Pb and Cu from wastewater. Batch experiments were used to study the effects of magnetization, initial wastewater pH, contact time, and initial metal concentration on adsorption performance. Preliminary results from isotherm analysis showed that HTL-char derived from food waste has a higher adsorption capacity for Pb than for Cu. The maximum adsorption capacity of Pb (32.9 mg Pb g^{-1} dry char) was obtained at pH = 3.5, T = 23.1°C, and initial PB concentration of 80 mg/L. The feasibility of utilizing HTL-AP to add moisture (and some carbon/nutrients) to compost was assessed in terms of the influence on microbial activity, nutrient availability (C/N ratio), humidity of the compost pile over the course of production, and quality of the finished compost.

628007 Analysis of the Water-Energy-Food Nexus in Agriculture Dependent Rural Communities By Using Mathematical Programming

Abstract:

The food-energy-water (FEW) nexus has become a useful tool to address the complex and interrelated nature of global resource systems, on which society depends to achieve different social, economical and environmental goals. A rapidly growing world population is causing an unsustainable pressure on resources. The demand for water, food and energy is expected to increase by 30-50% in the next two decades. An effective connection between the elements of the nexus can be achieved through the use of new technologies or forms of energy production. For example, the use of agricultural and food biomass wastes for the production of biomaterials and bioenergy. The application of optimization tools in the analysis of the FEW nexus represents a useful alternative for decision-making as well as for the implementation of public policies when managing the elements of the nexus.

This work presents a mixed integer linear programming model to approach the food-energy-water nexus of agriculture based rural communities. The model considers the optimal management of water, food production involving different irrigation and cultivation schemes, and the use of biomass for the production of biofuels. The formulation includes the estimation of a socioeconomic development indicator and the potential implementation of public policies, such as government financing and incentive programs to agricultural activities. The effect of climate spatial variations on physical and biological resources and phenomena are also considered by using a multi-period approach. Three objectives are included in the model: maximization of annual net profit, minimization of water consumption and minimization of unsatisfied demand for food. The model was applied to a case study of an agricultural population, which has been solved through the GAMS modeling environment. The model applied to such case study consists of 23810 equations, 22054 continuous variables and 698 binary variables. The results show that the proposed methodology is capable of obtaining the optimal configuration of the food and biofuel production system with desirable results in terms of the sustainable management of resources and the application of financing programs and government incentives. The following are some of the conclusions obtained: *i)* it is shown that the application of government incentives has a positive impact on the modernization of agricultural activities, *ii)* the most important program for socio-economic development is government financing, *iii)* the modernization of irrigation allows a significant reduction in water consumption, *iv)* the installation of a biorefinery brings economic benefits and favors the production of some foods, *v)* it is possible to distribute food production, and therefore economic resources, in such a way that the same per capita income is obtained for each of the rural regions under analysis.

627509 Flue Gas (CO_2) Extraction and Bio-Sequestration (FGXB) for Increased Crop Yields and Water Utilization Efficiency (WUE)

Abstract:

There are no existing technologies and markets that can use the billions of tons or gigatons per year (GT/Y) of CO_2 produced by stationary industrial and power plant, and mobile transportation sources, until now. The Black sWan Cycle (BWC) proposed in this presentation uses stationary and mobile sources of CO_2 bearing exhaust to make food (agricultural products), increase biomass, and make water/reduce water consumption with Flue Gas eXtraction and Biosequestration (FGXB.) The increased biomass is used for fuel. BWC with Membrane Air Enrichment (MAE) performs Direct Air Capture (DAC) of CO_2, water, and oxygen and uses Oxy-Combustion to decrease fuel consumption, increase capacity, increase the heating value of low-quality biofuels, such as solid biomass and landfill/sanitation anaerobic digester biogas, and concentrate the CO_2 in the flue gas for more economic use and capture.

The first step of BWC utilizes the conventional science of Crop Carbon (dioxide) Enrichment (CCE) that has been practiced in greenhouses for almost a hundred years, and researched in crop Open Top Chambers (OTC) and Free Air Carbon (dioxide) Enrichment (FACE- meaning without any enclosures) systems for decades by the Brookhaven National Laboratory and USDA. CCE is the agricultural/ horticultural practice of raising a plant's biosphere from ambient 400 ppmv to under 1000 ppmv CO_2 for the purpose of increasing biomass, yield, growth rate, and water utilization efficiency of plants by minimum 40%. Dr. Bruce Kimball et al of USDA has produced dozens of studies for wheat, cotton, citrus, grapes, and soybeans validating this science and its performance, including a 17 year study showing increase of citrus yield by average 70% by using 550 ppm CO2. Conventionally the CO_2 is provided at a high operating cost ($300/Ton) by fit-for-purpose fuel fired flue gas (at 5 to 10% $CO_{2)}$ generators, or supplied from trucked-in/ locally stored near pure (90+%) CO_2 liquid or dry ice, and then diluted down to 1000 ppmv CO_2.

The Kolodji Corporation, an energy carbon management engineering company, utilizes patented and patent pending BWC technologies to commercialize and monetize the use of CO_2 in emissions with BWC FGXB by making food and water, saving water, and producing biomass for fuel. Another crucial aspect is the low energy/cost BWC Wig Membrane which uses Direct Air Capture (DAC) to recycle the fugitive CO_2 in a carbon enriched orchard by producing an enriched oxygen and water, and further CO_2 enriched permeate stream for substitution of combustion air in a conventional fired system by using the BWC Membrane Air Enrichment Oxy-Combustion (MAE) process. The benefits of MAE are reduced firing fuel consumption, increased firing efficiency, temperature, and heating value, as well as water recovery, Direct Air Capture of Water and CO_2, and increased concentration of the CO_2 in the flue gas for ease and more economic use and capture from the flue gas.

Four pilot scale applications of under 1000 Ton CO_2 per year BWC FGXB and MAE have operated in orchards in Kern County, California since 2018. The largest pilot facility is in a 50-acre almond orchard in the midst of 100s of acres of crops, across from a refinery emitting 200,000 tons per year

CO_2, and is funded by the California Department of Food and Agriculture (CDFA.) A demonstration scale MAE application that increases the sustainability and CO_2 capture potential of a biogas/natural gas fired 2.5 MW power plant, and a second smaller (under 1 MW) in two California sanitation districts are planned. These projects emphasize the key benefit of MAE to eliminate the need for supplemental firing of the 100% sustainable renewable biogas fuel with un-sustainable non-renewable natural gas by raising the heating value of the biogas using oxy-combustion. Further, the flue gas is concentrated from 5 to 15% CO_2 making it easier to capture or use the flue gas. Funding potential is for $4.4 Million in grants pending award by the California Energy Commission in April/June 2021.

Shared in this presentation are pilot plant results, and demonstration unit plans including a Bryan Research and Engineering (BR&E) PROMAX(TM) process simulation models, heat and material balances and process flow diagrams, as well as utility consumption benefits and a capital/ operating cost-benefit analysis to profit (return on investment) from fuel savings, increased agricultural yields, and water make/ savings, and as well, highlighted carbon (dioxide) balance. With wide acceptance of this BWC technology, Kolodji Corp will justify targets of carbon neutrality in the state of California (negative 300 Million Tons per year CO_2 production) by 2025, twenty years earlier than the 2045 mandated statutory requirement. As well, there is the goal of US wide CO_2 deceleration (-0.05 Tons/year/year or negative 5 GT/Y) of CO_2 by 2030, and global neutrality by 2035.

632638 *Electrosprayed Scalable 3D Graphene-CNT Electrodes for Li-Ion and Fuel Cell Applications*

Abstract:

Lithium-sulfur batteries (LSB) and advanced lithium-ion batteries (LIB) demonstrate promise as the next generation energy storage and conversion (ESC) technology especially as it pertains to wearable technology and electric vehicles. Although LIB dominate the battery market (>90%) due its reliability, long cycle life, and market maturity, alternative means of renewable energy must be discovered in order to reduce the cost of materials, dendritic issues at the solid-electrolyte interphase (SEI) layer, and limited charge capacity (~250 mAH h g^{-1}). The next generation of future batteries is the lithium-sulfur battery due its inexpensive cost ($0.25/kg sulfur), high theoretical energy density (2600 Wh kg^{-1}), and safety. How LIB and LSB systems incorporate into military applications such as drones and battery backs is what the interests the Army. However, LSB face many challenges before commercialization into wearable technology and electric vehicles due its fast capacity degradation, polysulfide "shuttling" effect, and short cycle life (<300 cycles). Carbon nanomaterials constitute an ideal miniaturized composite platform for integration with other solid-state electrode materials. Air-controlled electrospray has been recognized as an effective means of fabricating cathodes using a convective air jet flow that accelerates the drying and deposition process. This technique provides a uniform and well-dispersed coating onto the current collector while using low-cost commercial carbon nanomaterials in a matter of minutes. Currently, self-assembly is conducting using the bottom–up synthesis of 3D macrostructures using graphene and carbon nanotubes (CNTs) as building blocks. We demonstrate a novel, all-aqueous, and scalable 3D platform materials design process for the synthesis of carbon nanocomposite aerogels to serve as porous ultralightweight electrodes for the next generation energy storage and conversion applications. Specifically, the negatively charged surface functional groups on graphene oxide and oxidized CNTs will be the sites for electrostatic coordination of positively charged noble metal cations and polyelectrolytes from aqueous solutions. This enables the incorporation of carbon nanomaterials with any polyelectrolyte and noble metal nanostructures with precise connection of each of the individual components while maintaining their unique properties in both LIB and LSB systems. Through a scalable air-controlled electrospray approach, we can develop hierarchical porous 3D nanocomposite cathodes comprised of carbon nanomaterials with any precious or noble metal nanostructures (Pt, Pd, Au, Ru, Ag, and Ru) and polyelectrolyte that demonstrate a synergistic network between the active materials, carbon materials, and metal nanostructures. Electrochemical techniques such as cyclic voltammetry, linear sweep voltammetry, galvanostatic charge-discharge, and impedance spectroscopy are used to determine the electrical and ionic conductivity, electrocatalytic activity, and supercapacitor performance of the aerogels as lightweight high-power and energy density electrodes.

624456 Modeling the Brønsted Acidity of Lanthanum Exchanged Faujasite

Abstract:

Lanthanum stabilizes faujasite in fluid catalytic cracking, which increases catalyst material lifetimes. Despite its extensive use for decades, many unknowns remain on how lanthanum increases faujasite hydrothermal stability. The hydrothermal stability of zeolites decreases by dealumination, which occurs via reactions that involve water molecules. Brønsted acid sites give desired catalytic properties to zeolites, however they reduce hydrothermal stability by attracting water molecules to aluminum tetrahedra and promoting dealumination. The energetically favored binding sites of water molecules and lanthanum ions in hydrogen exchanged faujasite were determined with density functional theory (DFT) calculations and ab initio molecular dynamics (AIMD) simulations. Potentials of mean force of proton transfer were obtained with constrained AIMD simulations to quantify the Brønsted acidity of faujasite, with and without lanthanum ions. Results indicate that water and lanthanum ions are energetically favored to bind in the same sites in faujasite, and that lanthanum ions increase the acidity of Brønsted sites adjacent to, and not adjacent to, bound lanthanum ions in faujasite. Implications toward hydrothermal stability are discussed, since lanthanum ions change the protonation state, and therefore hydrophilicity, of aluminum tetrahedra in faujasite.

629492 Charge Transport Mechanism in Sodium Selenide (Na_2Se) for Sodium-Selenium Batteries: A First Principles Study

Abstract:

Sodium–selenium (Na–Se) batteries are recognized as promising energy storage systems because of their high theoretical volumetric capacity (~ 3250 mAh/cm^3), which is of particular importance for portable devices and electric vehicles (EVs) with the limited battery packing space. Sodium selenide (Na_2Se) is the final discharge product for the Na–Se batteries and is expected to limit the overall battery conductivity because of the insulating nature. To effectively utilize the active materials, it is necessary to have a better understanding of the charge transport mechanism in Na_2Se. Based on first-principles simulations, we first evaluate the contribution of potential charge carriers (ionic and electronic defects) to the conductivity in Na_2Se by calculating the formation energies and diffusion barriers. Negatively charged sodium vacancies (V_{Na}^-) and positive sodium interstitials (Na_i^+) are found to be the defects with the lowest formation energy (0.80 eV) in Na_2Se, followed by hole polarons (h_p^+) with a formation energy of 1.95 eV. Our simulations demonstrate that while the V_{Na}^- has a higher diffusion barrier than the h_p^+ (0.25 eV *vs.* 0.11 eV), its contribution to the conductivity is the most dominant, on the order of 10^{-15} S/cm, due to the high concentration. On the other hand, the electronic conductivity mediated by the h_p^+ is predicted to be negligible because of its low concentration. Once the vacancy–polaron ($V_{Na}^- - h_p^+$) complex forms, the energy cost to break the pair is predicted to be 0.78 eV, implying that the charge transport could be hindered due to its charge neutrality. Our calculations also illustrate that compressive strain (which is potentially induced by surrounding host materials) could reduce both ionic and electrical conductivity of Na_2Se. Lastly, we will discuss a possible strategy to modify the ionic and electronic conductivities in Na_2Se by controlling the strain.

624018 Nitrogen-Rich Free-Standing Carbon Cathode for Improved Lithium-Sulfur Battery Performance

Abstract:

A study on the production of free-standing carbon cathode is an attractive method to improve the low active material ratio, which is a problem with existing Li-S batteries. This work introduces a free-standing carbon material with superior porosity properties for enhanced Li-S battery performance. The free-standing carbon cathode synthesized in this study shows outstanding physical stability by showing superior elastic properties based on a porous structure. Additionally, the free-standing cathode made through electrospinning of $NaBH_4$ and polyacrylonitrile (PAN) has a high nitrogen atom equivalent to about 20%. This in turn inhibits the shuttle phenomenon of polar lithium polysulfides, which is highly soluble in electrolyte, based on strong chemical adsorption ability between, and at the same times improves conductivity to deliver high specific capacity at high current densities. As a result, the free-standing electrode with 2.54 mg cm^{-2} sulfur loading exhibited a capacity of about 650 mAh g^{-1} at 5.0 C. By maximizing the advantages of the free-standing electrode, the areal loading of the active material has been increased to 6.11 mg cm^{-2}, which delivers an areal capacity of 3.1 mAh cm^{-2} at 500th.

630901 Technoeconomic Analysis of a Microalgae-Based Biorefinery Network for the Production of Advanced Biofuels

Abstract:

The establishment of a circular bioeconomy that satisfies the increased energy demands of a growing population has become a key focus of current governmental policies and initiatives. In Europe, for example, the recast directive 208/2001/EU mandates Members States to reach a minimum 20% share of renewable fuels in the transport sector by the year 2030. The latest statistics, however, indicate that up to 92% of the transport energy consumed in Europe (EU-28) is still satisfied by oil and petroleum fuels (e.g. gasoline and diesel, with only 5.11% deriving from renewable sources [1]. Therefore, the search for more sustainable biomass-derived fuel alternatives and efficient production routes continues to be a challenging but critical bioeconomy target.

Microalgae biomass is a promising production platform for advanced biofuels, although the single fuel conversion strategy has been deemed unfit for commercial applications. Due to their rich chemical composition, microalgal biorefineries are instead regarded as a more economically viable approach for the co-production of fuels and value-added chemicals through the full exploitation of biomass conversion and waste valorisation scenarios [2,3]. Therefore, it becomes necessary to identify those optimal conversion routes leading to reduced processing costs and increased productivity. For example, whilst the feasibility of producing microalgal biodiesel (via transesterification) has been widely evaluated, the co-production of biodiesel along with biobutanol (via fermentation) from microalgae can provide a more favourable outlook for biofuels production through the combined exploitation of microalgal sugars and lipids [4]. Evaluating the economic potential of such a biorefinery network can focalise research developments on critical processing stages, and identify areas of opportunity leading to improved process performance [5].

This work presents a technoconomic analysis of an integrated biorefinery network for the co-production of microalgal biodiesel (via transesterification of lipids) and biobutanol (via fermentation of sugars) along with additional products acetone, ethanol, and glycerol. The biorefinery is comprised of three major areas: i) feedstock pre-treatment, involving acid hydrolysis (AH) and solvent extraction (SX), ii) biodiesel production and purification, and iii) biobutanol production and purification. Two biorefinery configurations were evaluated (Configuration 1: AH followed by SX, Configuration 2: SX followed by AH) and compared against the single-fuel biodiesel or biobutanol conversion scenarios. In addition, a sensitivity analysis was carried out in terms of the Minimum Fuel Selling Price (MFSP) to identify those critical parameters with the greatest effect on fuels price, thus generating insights on the potential effect of process variables (e.g. solvent usage, fuel yield) and market uncertainties (e.g. feedstock costs, biofuel demand, fuel quality standards).

References

[1] European Commission, Energy Balance | 2020 edition, Luxembourg, 2020.

[2] G.M. Figueroa-Torres, J.K. Pittman, C. Theodoropoulos, Optimisation of microalgal cultivation via nutrient-enhanced strategies: the biorefinery paradigm, Biotechnol. Biofuels. 14 (2021) 64.

[3] J. Rajes Banu, Preethi, S. Kavitha, M. Gunasekaran, G. Kumar, Microalgae based biorefinery promoting circular bioeconomy-techno economic and life-cycle analysis, Bioresour. Technol. 302

(2020) 122822.

[4] G.M. Figueroa-Torres, W.M.A. Wan Mahmood, J.K. Pittman, C. Theodoropoulos, Microalgal biomass as a biorefinery platform for biobutanol and biodiesel production, Biochem. Eng. J. 153 (2020) 107396.

[5] C. Sun, C. Theodoropoulos, N.S. Scrutton, Techno-economic assessment of microbial limonene production, Bioresour. Technol. 300 (2020) 122666.

626826 Simulation and Life Cycle Assessment of Biodiesel Production from Micro Algae

Abstract:

The existence of climate change and the connection to green house gas emissions has been studied and debated by scientists, industrialist, and policy makers for roughly 200 years. Today it is widely accepted that warming of the planet is occurring, and that human activity is the primary driver. Fossil fuels have become the source for more than 80% of the world's energy consumption, powering the transportation, industry, electricity production, and commercial/residential sectors' energy demand. The desire for alternative fuels stems from the many problems we face today. Effort to produce renewable cleaner-burning fuels has not yet produced any single method which has the capability of meeting global fuel demand alone, and many of the biofuel sources come with difficulties of their own. Biofuel from oil-yielding biomass sources is a vital part of the solution for climate change. Biodiesel, most commonly made from waste animal fats or vegetable oils, is a renewable fuel with many advantages. The Argonne National Laboratory found that the use of biodiesel blending reduced total carbon dioxide emission by 74%. Lastly, the growth and development of biofuel production serves to lessen the political influence on global energy supply, as these alternative fuel processes are not strictly bound to the occurrence of natural resources in any specific place. However, biodiesel production at mass-scale faces several difficulties regarding its traditional feedstock. Soybeans and other vegetable oil-producing crops require an immense amount of fresh water and land, and commercial-scale biodiesel farms would be in competition with agricultural food production. Microalgae can combat these issues, as it requires much less land to farm and can be manipulated by diet to create a high lipid density crop. For these reasons, algae-based biodiesel is the most promising source for achieving mass-scale production.

In this work, an Aspen Plus® simulation for biodiesel production from microalgae was developed. Modelling parameters were referred from relevant literature reviews and through trial and error during the development of the simulation. Sizing and cost estimates were conducted based on a minimal production capacity of 10,000,000 gallons of biodiesel per year. Equipment Sizing, materials, and their respective prices were taken from literature and online estimators.

The economic considerations include costs such as land, equipment, installation, utilities, and waste treatment, but do not consider labor, permits, or insurance. Finally, the required selling prices of algae biodiesel and DHA, and the relationship between their selling prices, were investigated. For a capital recovery time of five years, and at a moderate interest rate of 10%, economically competitive prices were obtained for both biodiesel and DHA, of $2.04 per gallon of biodiesel and $12.70 per kilogram of DHA. This indicates that the process is economically feasible, though it is heavily dependent on the revenue from DHA.

626982 Techno-Economic and Sensitivity Analysis of Aromatic and Alkene Production from Lignin Via Fast Pyrolysis and Catalytic Upgrading Coupled with Ethanol Production.

Abstract:

Lignin is one of the most abundant existing biopolymers, often underutilized due to its complex structure and chemical composition. In biorefineries, lignin is commonly burned to supply heat and power to the system due to lower costs. While this is currently sufficient, it may be more cost challenging in the future as more advanced biofuel production processes are being developed. Therefore, it is necessary to investigate methods to increase the value of lignin to improve the feasibility of biofuel production. This study looks at the robustness of fast pyrolysis combined with catalytic hydrodeoxygenation process using molybdenum trioxide to convert lignin components into three primary streams: aromatics, alkenes, and alkanes which reduces the need and cost for separation. This process upgrades lignin into aromatics, alkenes, and alkanes to be sold as by-products to increase the value of lignin while improving the cost of ethanol at a biorefinery. Several lignin types (organosolv corn stover, organosolv poplar, Dupont, and Renmatix lignins) were investigated in this study, and the results were compared to a base case for which lignin is simply used as a heat and power source.

The techno-economic analysis (TEA) combines the biochemical process of ethanol production with the thermochemical process of upgrading lignin to chemicals. Based on the combination of various economic parameters, historical market prices, plant life of 30 years, and a 2000 MTPD input biorefinery, capital costs for a 61 million gallon per year (MGPY) cellulosic ethanol plant is estimated at $401 million (MM) while equipment costs were estimated to be $214 MM. The pyrolysis and catalytic upgrading process account for 22% of the total capital costs. TEA shows that the upgrading of lignin into aromatics, alkenes, and alkanes and selling them as by-products can reduce the selling price of cellulosic ethanol by $1.67-2.02/gallon as compared to when lignin is used as power ($2.20/gallon). Sensitivity analysis shows that the minimum selling price of ethanol is highly sensitive to the price of aromatics, operating hours, and capital costs as compared to other operating parameters such as fuel and char output.

631856 Machine-Learning Reduced Order Model for Cost and Emission Assessment of a Pyrolysis

Abstract:

Biomass pyrolysis is a promising approach for producing economical and environmentally-friendly fuels and bioproducts [1]. Biomass pyrolysis experiments show that feedstock properties significantly impact product yields and composition[2], [3]. Scientists are developing detailed chemical reaction mechanisms to capture the relationships between biomass composition and pyrolysis yields [4]. These mechanisms can be computationally intensive[5]. A few studies have investigated the use of machine learning (ML) algorithms in the estimation of pyrolysis products to circumvent the computationally expensive chemical reaction mechanisms and computational fluid dynamics (CFD) models[6], [7]. However, to the best of the author's knowledge, none has integrated ML models into process modeling tools for real-time estimation of pyrolysis products to assess the cost and emissions of a pyrolysis refinery.

In this study, we investigate the use of a machine learning reduced order model (ROM) for assessing the pyrolysis products used in estimating costs and emissions of a pyrolysis biorefinery in real-time. We developed a Kriging-based ROM to predict pyrolysis yields of 314 feedstock samples based on the results of a detailed chemical kinetic pyrolysis mechanism. The ROM is integrated into a chemical process model for calculating mass and energy yields in a commercial-scale (2000 tonne/day) biorefinery. All Kriging models achieved excellent accuracy with an average coefficient of determination score of 99% when the predicted results were compared to the chemical kinetics results. The ROM estimated biofuel yields of 65 to 140 gallons per ton of dry biomass. This results in biofuel minimum fuel-selling prices of \$2.62 to \$5.43 per gallon and emissions of -13.62 to 145 kg of CO_2 per MJ of biofuel. The ROM achieved an average mean square error of 1.8e-20 and a mean absolute error of 0.53%. These results suggest that ROMs can facilitate rapid feedstock screening for biorefinery systems.

Figure 1: Machine learning (Kriging), lifecycle cost (TEA), and emissions (LCA) prediction framework for chemical process simulations with machine learning reduced-order models implemented using the CAPE-OPEN standard. Rounded rectangles represent data, filled squares represent software and sub-processes, and the diamond represents a decision step. Process outcomes include minimum fuel-selling price (MFSP) and greenhouse gas (GHG) emissions.

FUNDING

The authors would like to acknowledge the support of the U.S. Department of Energy through the grant DE-EE0008326.

REFERENCES

[1] D. O. Hall and J. I. House, "Biomass: a modem and environmentally acceptable fuel," *Sol. Energy Mater. Sol. Cells*, vol. 38, no. 1–4, pp. 521–542, Jan. 1995, doi: 10.1016/0927-0248(94)00242-8.

[2] W. Li, Q. Dang, R. C. Brown, D. Laird, and M. M. Wright, "The impacts of biomass properties on pyrolysis yields, economic and environmental performance of the pyrolysis-bioenergy-biochar platform to carbon negative energy," *Bioresour. Technol.*, vol. 241, pp. 959–968, 2017, doi: 10.1016/j.biortech.2017.06.049.

[3] P. A. Meyer, L. J. Snowden-Swan, S. B. Jones, K. G. Rappé, and D. S. Hartley, "The effect of feedstock composition on fast pyrolysis and upgrading to transportation fuels: Techno-economic analysis and greenhouse gas life cycle analysis," *Fuel*, vol. 259, no. September 2019, p. 116218, 2020, doi: 10.1016/j.fuel.2019.116218.

[4] E. Ranzi, P. E. A. Debiagi, and A. Frassoldati, "Mathematical Modeling of Fast Biomass Pyrolysis and

Bio-Oil Formation. Note I: Kinetic Mechanism of Biomass Pyrolysis," *ACS Sustain. Chem. Eng.*, vol. 5, no. 4, pp. 2867–2881, 2017, doi: 10.1021/acssuschemeng.6b03096.

[5] P. Mellin, E. Kantarelis, and W. Yang, "Computational fluid dynamics modeling of biomass fast pyrolysis in a fluidized bed reactor, using a comprehensive chemistry scheme," *Fuel*, vol. 117, no. PART A, pp. 704–715, 2014, doi: 10.1016/j.fuel.2013.09.009.

[6] H. Merdun and I. V Sezgin, "Modelling of pyrolysis product yields by artificial neural networks," *Int. J. Renew. Energy Res.*, vol. 8, no. 2, pp. 1178–1188, 2018, [Online]. Available: https://www.scopus.com/inward/record.uri?eid=2-s2.0-85049086039&partnerID=40&md5=229f1199f048ff3c11db6511fc4fca1c.

[7] Q. Tang *et al.*, "Prediction of Bio-oil Yield and Hydrogen Contents based on Machine Learning Method: Effect of Biomass Compositions and Pyrolysis Conditions," *Energy & Fuels*, Aug. 2020, doi: 10.1021/acs.energyfuels.0c01893.

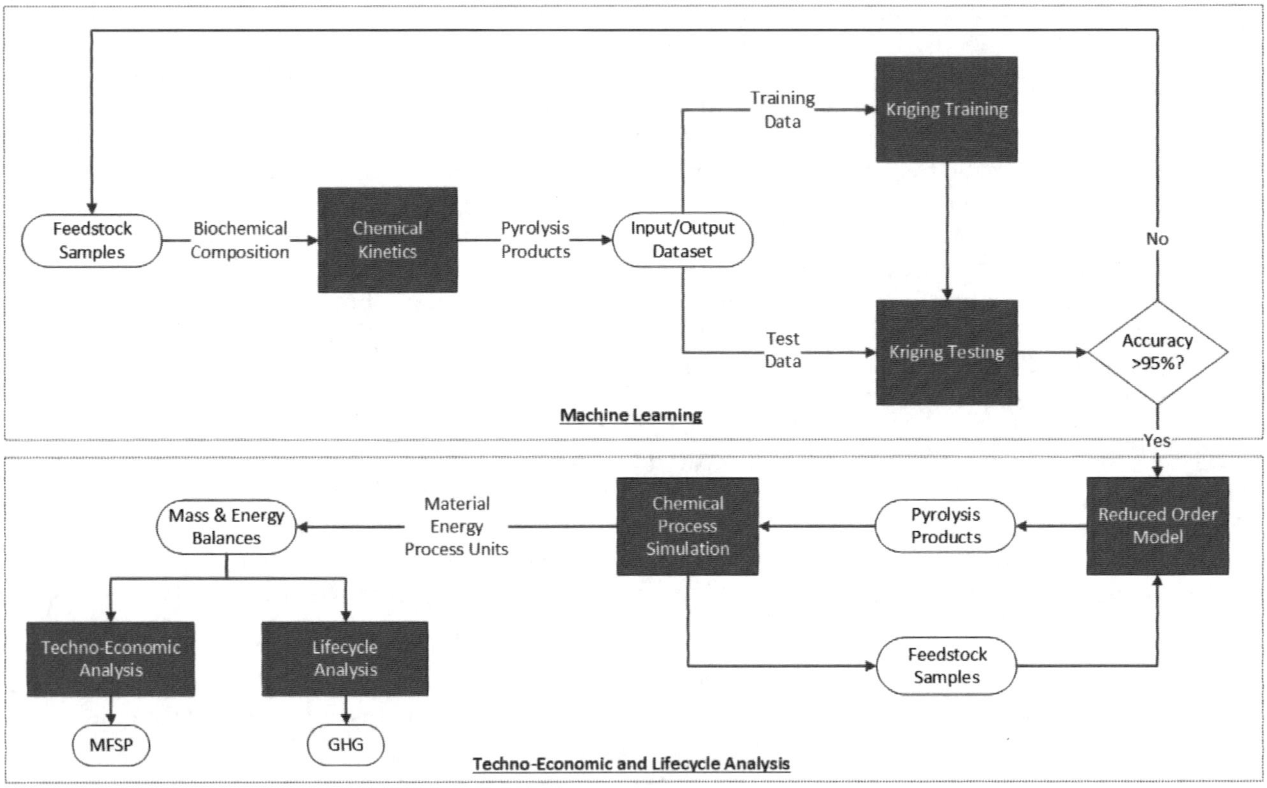

625049 Evaluation of the Economic Viability and Sustainability of Autothermal Pyrolysis Sugars through Computation and Machine Learning

Abstract:

Autothermal pyrolysis (ATP) is a novel approach to converting biomass into chemicals, fuels, and bioproducts. ATP employs a small amount of oxygen to partially combust biomass and some pyrolysis products inside the reactor. This avoids heat losses associated with indirect heating and enables process intensification of pyrolysis reactors. Recent studies demonstrated the production of pyrolytic sugars, bioasphalt, and biochar from red oak and corn stover ATP. Experimental results indicate that ATP product yields are comparable to conventional pyrolysis while reducing the need for heat generation equipment. Feedstock properties could have a significant impact on the yield and quality of ATP products. This study employs computation and machine learning to investigate the impact of feedstock properties on ATP products and their economic and environmental sustainability.

Previous studies indicate that the costs and emissions of pyrolysis products can vary widely depending on the feedstock properties. For example, the minimum selling-fuel price (MFSP) of pyrolysis biofuels could range between $2 and over $5 per gallon. Several methods have been proposed to improve the prediction ability of pyrolysis computational models and reduce commercialization risk. However, few studies have employed machine learning models and techniques. The objective of this paper is to investigate the ability of machine learning models to predict the minimum sugar-selling price (MSSP) and greenhouse gas (GHG) emissions of producing sugars, biochar, and bioasphalt ATP.

This paper integrates machine learning alongside a chemical process model, techno-economic (TEA) and life cycle analysis (LCA) of the ATP system. Commercial biochemical feedstock composition was provided by Idaho National Laboratory (INL). Approximately 3000 cellulose, hemicellulose and lignin data samples were generated using a Generative Adversarial Network (GAN) machine learning model. A Kriging-based machine learning model predicted ATP product yields based on the GAN samples. Product yields were introduced into the chemical process model to generate mass and energy balance for TEA and LCA calculations. Preliminary results estimate an average MSSP of around $400/ tonne of sugars and net GHG emissions of -0.16 kg CO2eqv/MJ of sugar. These results demonstrate that machine learning models could help reduce the commercialization risk of biorefinery techonologies.

622485 A Comparative Modeling Analysis of Electromicrobial Production Process Strategies

Abstract:

Electromicrobial production strategies, in which electrochemically-generated mediator molecules provide energy for microbes to fix CO_2 and generate value-added products, have received significant attention for the sustainable production of commodity chemicals. Here, we describe the development and application of comprehensive process models for three proposed electromicrobial production systems: a Knallgas bacteria-based system, a formatotroph-based system, and an acetate-mediated system in which acetate is first produced by the anaerobic conversion of H_2 and CO_2 via an acetogenic organism and is subsequently consumed for the generation of target products via heterotrophic growth. We consider the formation of several products including biomass, industrial enzymes, and lactic acid.

The lactic acid productivity is highest for the formatotroph system, ~30% higher than the Knallgas bacteria system and ~200% higher than the acetate-mediated system; this is due primarily to the high solubility of formate, which avoids coupling growth substrate availability to the gas-liquid mass transfer rate of sparingly soluble gases such as H_2. Notably, restricting the Knallgas bacteria system to nonflammable gas mixtures reduces the lactic acid productivity to equal that of the acetate-mediated system. However, the Knallgas bacteria system achieves the highest overall efficiency, ~4x greater than the formatotroph system and ~5x greater the acetate-mediated system. This difference is due, in the former case, to the substantial energy penalty associated with concentrating the formic acid effluent from the CO_2 electrolyzer to achieve industrially-relevant lactic acid titers (100 g/L) and the relative inefficiency of CO_2 electrolysis to formate. In the latter case, the lactic acid titer is limited by salt build-up, so the low system efficiency is caused by the energy penalty associated with concentrating the lactic acid effluent. Regardless of the system, the single-pass conversion efficiency of CO_2 is low (~10% in all cases), so gas recycle would be necessary to avoid high energy costs in upstream CO_2 capture devices in a fully integrated process. In sum, our analysis reveals trade-offs between productivity and efficiency in major electromicrobial production schemes, indicates the need for an integrated ecodesign framework relying on life cycle impacts that can evaluate these trade-offs, and demonstrates the utility of process models towards this ecodesign paradigm.

624486 Middle Distillates from Ethanol-Technoeconomic and Life CYCLE Assessment

Abstract:

Ethanol has been established as the most commonly used biofuel. In the U.S. alone, its production capacity has increased from 2 to 16 billion gallons in the last twenty years. However, ethanol has an important limitation, it cannot be incorporated in significant amounts in middle distillates. This limitation becomes more relevant if we consider fuel consumption projections for the upcoming decades [1]. While gasoline demand, where ethanol is typically blended, is expected to decrease, middle distillates demand, especially that of diesel, is expected to increase. An additional consideration is that the electrification of aviation and freight transportation sectors, where middle distillates are used, is not expected in the near term. This scenario leaves us with two problems: a large infrastructure built for the production of a biofuel whose demand may decrease (i.e. ethanol) and the lack of sustainable alternatives for the fossil fuels whose demand is expected to increase (i.e. aviation fuel and diesel). One strategy that can address these two challenges simultaneously relies on the upgrading of ethanol. A large number of different pathways for ethanol upgrading may be used. One particular alternative based on Guerbet coupling and etherification seems promising [2]. In this alternative ethanol is first transformed into higher alcohols using Guerbet coupling, and then these alcohols are used to produce ethers in an etherification reaction. The main advantage of this approach is the possibility of producing diesel blends of high quality with a high content of ethers, known for their high cetane number.

In this work, we present the results of a techno-economic and life cycle assessment for the upgrading of ethanol to diesel using a biorefinery based on Guerbet coupling and etherification. This work is the result of an iterative and collaborative effort between experimentalists and process engineers. This synergistic collaboration has allowed us to explore the economic and environmental impacts of variables such as conversion and feed composition, using data obtained from experiments designed based on process modeling needs. In particular, we have established the effects that changes in conversion and selectivity have on the minimum selling price and the separation train associated with the designed biorefineries. The results presented constitute the first analysis of this kind for this system and reveal the main economic drivers of the process and its environmental impacts. In particular, the feedstock cost appears as the most important cost component, pointing toward the need of minimizing the production of low molecular weight by-products not suitable to be blended in diesel fuel. Notably, using lignocellulosic ethanol as a feedstock can lead to a 50% reduction in greenhouse gas emissions. The properties of the resulting fuel (cetane number, density, viscosity, low heating value, flash point, and distillation curve) are evaluated and satisfy the ASTM requirements for diesel.

1. Dagle, R. A., Winkelman, A. D., Ramasamy, K. K., Lebarbier Dagle, V. & Weber, R. S. Ethanol as a Renewable Building Block for Fuels and Chemicals. *Ind. Eng. Chem. Res.* **59**, 4843–4853 (2020).
2. Eagan, N. M. *et al.* Catalytic Synthesis of Distillate-Range Ethers and Olefins from Ethanol through Guerbet Coupling and Etherification. *Green Chem.* (2019).

630871 Investigation of Suitable Technologies & Solvents for Bio-Ethanol Dehydration with the Help of Aspen.

Abstract:

In the current scenario shortage of fossil fuels has become a major concern for humankind. Such conventional sources of energy have a major effect on the environment in terms of pollution. Due to this, there is an urgent need of exploring alternative sustainable sources of energy, with an ability to replace prevalent fossil fuels fulfilling the augmenting global energy demand.

In this regard, Bio-ethanol derived from sustainable sources of energy has many promising characteristics having its application as a gasoline additive. Notably, the attention of researchers, industrialists, and governments globally is shifting towards Bioethanol and its derivatives. However, the commercial exploitation of Bioethanol production is still infeasible. As per the government mandate globally, 99.6 ± 0.2 % concentration of ethanol is acceptable. Nevertheless, the formation of well-known binary Azeotrope of ethanol-water at 95.63 % ethanol achieving this purity becomes even more challenging. Therefore, a wide variety of technologies are being used globally for the purification of Bio-ethanol such as Extractive Distillation, Azeotropic Distillation, and Dehydration of ethanol using molecular sieves, etc. However, the use of molecular sieves is still improvident as compared to the other two mentioned technologies. Although the countries like Brazil and the USA, the largest producers of ethanol in the world, use Extractive Distillation with solvent Ethylene Glycol and Azeotropic Distillation with solvent Cyclohexane. Whereas these technologies at commercial scale are economically infeasible making the overall cost of Bio-ethanol way higher than non-conventional sources of energy.

Therefore, in this work, Bio-ethanol dehydration using Extractive and Azeotropic Distillation with solvents like Propylene glycol, Pentane, etc. are investigated. The study was conducted using Aspen Plus with the typical industrial considerations of feed having 5 wt.% ethanol & 95 wt.% water. The result obtained using the simulation were optimized based on operational energy and design requirements and are discussed.

628173 Life-Cycle Analysis of Low-Carbon Hydrogen Supply Pathways for the Provision of Heat in Buildings

Abstract:

Decarbonization of the buildings sector in the global economy presents an inherent challenge due to the efforts required to shift from existing fuels and infrastructure to lower-carbon alternatives at both the consumer and system-levels. In this context, recent evidence from the United Kingdom indicates that hydrogen may have a complementary role to electricity in a cost-effective hybrid-heating system [1]. Studies have shown that a significant portion of this hydrogen is likely to be sourced from natural gas reforming with carbon capture and storage (CCS) in the near-term to reduce fuel costs [2][3]. More recently, advanced natural gas reforming processes with higher CO_2 capture rates have been proposed as an economical interim solution to scale-up the use of hydrogen before alternative technologies such as renewable-based water electrolysis are matured over time. However, the overall environmental implications of these processes are poorly understood as previous studies have not accounted for the associated infrastructure requirements. In general, further analysis is required to understand the lifecycle impacts of the fuel supply, H_2 and CO_2 storage infrastructure, and operating practices across different regions for informed decision-making. To this end, this paper presents a comparative lifecycle assessment (LCA) of five different hydrogen supply pathways for meeting the heating demands in the North of England.

The LCA is based on attributional modelling and focuses on cradle-to-gate analysis for the following pathways: a) steam methane reforming (SMR) with CCS, b) autothermal reforming (ATR) with CCS, c) water electrolysis using offshore wind power, d) water electrolysis using grid power, and e) water electrolysis using offshore wind and grid power in a hybrid system. The scope of the H21 North of England project, which is an industry feasibility study, informs the modelling of technical parameters for the supply infrastructure in the inventory dataset. We combine published literature with process inventory modelling to represent various lifecycle stages in the LCA software, SimaPro. Version 3 of the ecoinvent database was used to model the inventory data for background processes and ReCiPe 2016 was used to characterize the impacts using midpoint indicators. The supply pathways were compared on key performance metrics such as global warming potential, fine particulate matter formation, freshwater ecotoxicity, mineral resource scarcity, etc., to estimate and benchmark their overall performance. Preliminary findings suggest that water electrolysis using offshore wind power has the lowest global warming potential as reported in previous literature [4]. However, both SMR and ATR with CCS may result in a CO_2 footprint, which is 2 – 3 times that of water electrolysis using offshore wind. Water electrolysis with grid power has the potential to be low-carbon only when the electricity grid is near zero-carbon. Furthermore, the results stress the need to lower other environmental impacts linked to the offshore wind supply to minimize its impact on freshwater ecosystems and mineral resources. More generally, greenhouse gas emissions associated with the natural gas supply chain and the electricity grid have the potential to increase the global warming potential of both SMR and ATR processes by up to a factor of five depending on the country of operation. Therefore, efficient management of the fuel supply chains and accelerated progress in their decarbonization contributes to the gradual transition of the buildings sector towards net-zero emissions.

[1] Climate Change Committee, "The Sixth Carbon Budget - The UK's path to Net Zero," 2020.

[2] N. Sunny, N. Mac Dowell, and N. Shah, "What is needed to deliver carbon-neutral heat using hydrogen and CCS?," *Energy Environ. Sci.*, vol. 13, no. 11, p. 4204, Nov. 2020, doi: 10.1039/d0ee02016h.

[3] M. Thompson *et al.*, "Hydrogen in a low-carbon economy," 2018.

[4] E. Cetinkaya, I. Dincer, and G. F. Naterer, "Life cycle assessment of various hydrogen production methods," *Int. J. Hydrogen Energy*, vol. 37, no. 3, pp. 2071–2080, Feb. 2012, doi: 10.1016/j.ijhydene.2011.10.064.

624467 Decarbonizing Industrial Heat: The Potential of Low Carbon Technologies in Chemicals Manufacturing

Abstract:

Decarbonizing the industrial sector remains a key challenge for reaching global emissions targets in the next few decades. Despite ongoing efforts of electrification and renewable energy integration in the transportation and power sectors, industry has been slow to adopt low carbon technologies because of its diverse set of energy-consuming processes and competitive international markets for its products. However, one cross-cutting opportunity stands to significantly eliminate sectoral emissions – the decarbonization of heat.

Industrial heat accounts for two-thirds of industrial energy demand and nearly one-fifth of global energy consumption. In the U.S., industrial process heat in manufacturing industries amounts to 8% of total primary energy use and relies predominantly the combustion of fossil fuels, mainly natural gas, coal, and oil. To achieve emissions goals in the next ten to fifty years, massive shifts in the use and deployment of industrial heating technologies are necessary. With the emergence of cost-competitive renewable power and revived interest in climate change mitigation in the U.S., there is significant potential for industry to transition to low carbon process heating through the use of bioenergy, CCUS, electrification, hydrogen, and solar thermal. Our work provides analyses that evaluate these low carbon technologies to determine their technical and economic potential in manufacturing industries.

This national analysis focuses on industrial heating in the U.S. chemicals subsector, which is the single largest energy consumer among all types of industry, comprising almost a third of industrial energy. The analysis specifically assesses:

- facility-level unit processes requiring heat in the production of high energy-consuming chemicals
- thermal energy demand at a spatial-temporal scale, modeled with a forward-looking time horizon
- the characterization of bioenergy, CCUS, electrification, hydrogen, and solar thermal technologies applied to industrial heating, based on resource availability and technical parameters, such as temperature, thermal efficiency, technology readiness level, and equipment lifespan

The presentation will provide a critical synthesis of our findings, with an emphasis on the technical potential analysis, where process-level thermodynamic models are integrated with national manufacturing energy use and technical equipment data to estimate achievable fossil fuel and carbon dioxide emissions savings. To this end, we have developed a model for thermal energy demand in select chemical industries, identified feasible low carbon technologies that meet process requirements and integrate into existing process systems, and quantified their decarbonization potential.

We will also discuss identified barriers, both technical and economic, that must be overcome to promote future adoption and will suggest a research agenda for addressing those barriers moving forward.

631145 Green Hydrogen from PEM Electrolysis: Uncovering Its Worldwide True Cost and Production Boundaries When Accounting for Intermittency

Abstract:

Even when the concept of electrolytic hydrogen has its origins in the 18th century, it was not until the beginning of the last century that hydrogen started to gain attention from an industrial point of view. From 1920 to the present, several hydrogen hype cycles have come and gone, particularly due to the possibility of using free power coming from renewable sources. The production of hydrogen from renewable and non-renewable sources has been highly regarded due to its potential as an energy carrier and its utilisation in several fields, including heat, power and mobility services. In this regard, electrolytic, or green, hydrogen is expected to play a crucial role in any net-zero transition, with polymer electrolyte membrane (PEM) electrolysis considered one of the most promising technologies in this space. However, the costs of electrolytic hydrogen are still high compared to the BAU production routes (grey hydrogen), even when carbon capture and storage is considered (blue hydrogen). This notwithstanding, there is an abundance of literature presenting bullish estimates of significantly reduced future cost. Given that much of this literature rests on so-called expert opinion, or the potential for cost reduction of individual elements of a PEM process, disregarding the entire system, there is a danger of over-estimating the real scope of cost reduction. Furthermore, the intermittent nature of renewables, which is strongly affected by location, is often disregarded in this kind of analysis, although energy storage has been proven to have a high share of the production costs.

Hence, we present a bottom-up analysis for the large-scale implementation of PEM electrolysis, decomposing the whole system into distinct technology components, and evaluate the potential for cost reduction on this basis following a learning rate approach. We conclude that a likely minimum cost for a mature technology cost is in the range of 980 ± 380 USD/MW, relative to current installed costs of 1700 ± 350 USD/MW.

As the cost of electrolytic hydrogen is greatly affected by the intermittency of renewable electricity, we followed a previous framework aimed at the optimal production of green fuels from air, water and electricity (Ganzer and Mac Dowell, 2020). This model accounts for the fluctuating nature of solar energy throughout utilising location-based capacity factors (Pfenninger and Staffell, 2016; Staffell and Pfenninger, 2016) to estimate the electricity produced in any location. For this study, we divided the Earth's surface into 1140 points based on the Universal Transverse Mercator (UTM) coordinate system. We then modified the existing model to allow the use of hybrid systems, including solar and wind power and worked with 20-year hourly-based capacity factors for each location (see Fig. 1). To calculate the production cost of hydrogen in each location, the installed costs for solar and wind systems, electrolysers and energy storage media, and the economic parameters, were chosen according to each region in the map. This MILP model was implemented and solved using GAMS 32.1.0 with CPLEX 12.10.0 as the solver.

We estimated the current (2019) and the future (2050) production cost of green hydrogen at each map point with this information. The obtained costs leased between 3 and 40 USD/kg H_2, depending on the location, proving the considerable influence intermittency has over those production costs.

These values compare relatively poorly with the costs of blue hydrogen, reported in the range of 2-3.5 USD/kg H_2. However, green hydrogen from PEM electrolysis powered with wind electricity could reduce CO_2 emissions to 1.0 kg $CO_{2\text{-eq}}$. This is not the case for hydrogen from solar power owing to the panel manufacturing processes, whose emissions reach values of around 5.0 kg $CO_{2\text{-eq}}$. These emissions surpass those of blue hydrogen from autothermal reforming, *i.e.*, 2.6 kg $CO_{2\text{-eq}}$, and from steam methane reforming, *i.e.*, 4.0 kg $CO_{2\text{-eq}}$. The environmental results also exhibit burden-shifting, particularly concerning human health and ecosystem diversity toward resource availability. A crucial factor identified for the large-scale deployment of PEM electrolysis is that of iridium, vital to the production of PEM electrolysis units. While the total amount of iridium available on the Earth would satisfy the scenarios envisioned for the hydrogen economy, this would appear to exhaust proven reserves of this critical raw material. Thus, identifying a viable substitute for iridium, a balance with different technologies, significant technological improvements and regulatory policies to manage this finite resource will be critical as this agenda progresses.

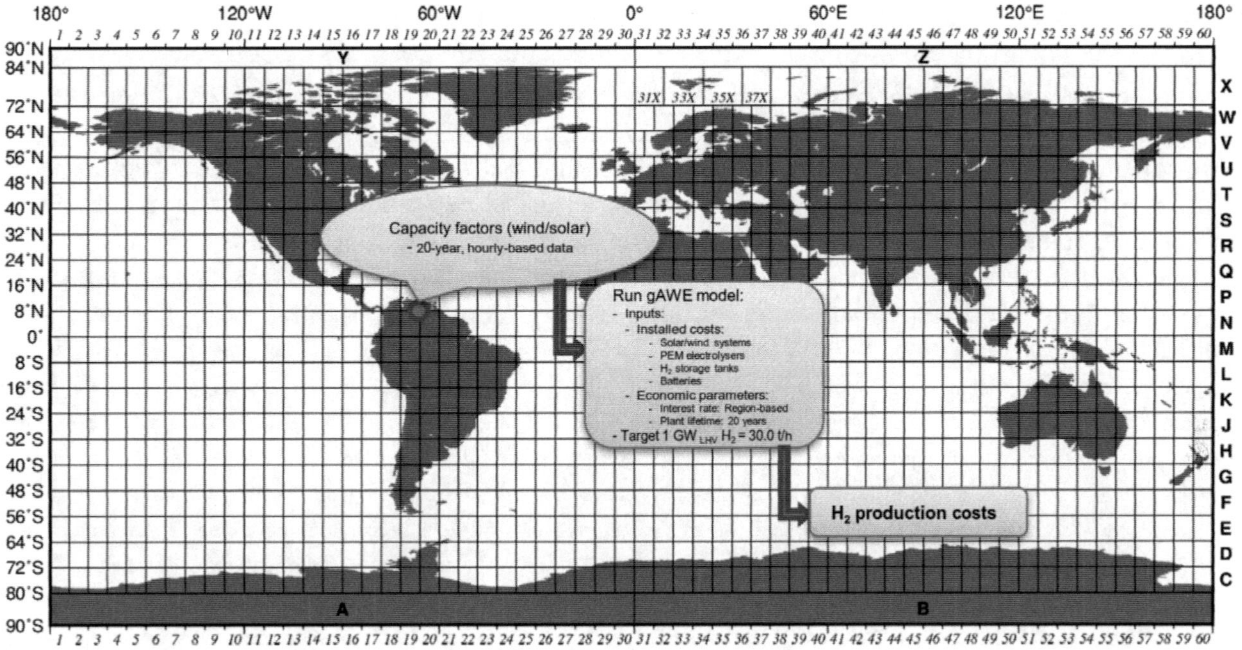

Fig. 1. Summary of the methodology followed in this study to estimate the production cost of hydrogen in each location.

631397 Comprehensive Process and Environmental Impact Analysis of Integrated DBD Plasma Based Fuel Synthesis

Abstract:

Electrification of the existing industrial chemical production processes is the key in progressing towards reducing carbon dioxide (CO_2) emissions and production of near-carbon free fuels. In particular, replacing energy from the burning of carbon-based fuels with that supplied by green, renewable electricity sources is viewed as an emerging chemical engineering concept towards green fuel production. This electrical energy can be utilized in electrochemical or plasma (or combined) processes for chemical transformations, routinely at temperatures much lower than conventional reactions. Plasma technology, in particular, provides unprecedented opportunities in process versatility as well as low investment and even operating costs since plasma reactors can be utilized as *drop-in* technology due to their structure – tubular or parallel plate configurations – and linear scalability of the production capacity. However, with very few exceptions, integration of green plasma related fuel generation processes into existing large scale processes has been lacking. The work presented will focus on rigorous process design calculations combined with life cycle analysis to integrate cold dielectric discharge barrier (DBD) plasma reactors into the existing upstream and downstream chemical processes. Many pitfalls, including those based on the plasma reactors intrinsically operating at low pressures while for fuel production and separation high pressures necessary, will be highlighted and discussed. In particular, the presented work will focus on two key hydrogen carrier molecules, e.g. H2 and NH3, synthesis and integration into the large scale industrial processes using DBD plasma from low quality biomethane. These will be described and compared against the conventional steam methane reforming (SMR), dry methane reforming (DMR) and NH3 synthesis processes. The calculated environmental impacts from life cycle analysis based on green WECC U.S. grid electricity produced in wind 1-3 MW onshore turbines will be considered.

626532 Low-Pressure Microwave Assisted Ammonia Synthesis – Evaluation of Novel Separation Technologies, Plant-Wide Process Design, and Technoeconomic Analysis

Abstract:

Ammonia is one of the key chemicals for producing food and delivering energy[1]. It has been estimated that 40% of the world population rely on ammonia to produce food [2]. Due to its high energy intensity, ammonia is used as a fuel in combustion engines and fuel cells[3]. Ammonia is produced industrially via the Haber-Bosch process. This process involves an exothermic reaction between nitrogen and hydrogen at a temperature between 400-500°C and at a pressure of 150-300 bar[4]. As a result of the extreme conditions, the Haber-Bosch process suffers from high capital cost and high energy consumption mainly due to compression [5] accounting for 1-2% of energy consumption, about 5% of natural gas consumption, and 1.6% of CO_2 emissions worldwide[6]. An alternative to this energy-intensive process with a large carbon footprint is to produce ammonia at considerably lower pressure and temperature using microwave (MW)-assisted catalytic technology.

Recently, several lab-scale experimental studies have reported the promise of ammonia synthesis in MW-assisted reactor at ambient pressure and moderate temperature [7], [8]. However, to the best of our knowledge, no plant-wide modeling and techno-economic analysis has been presented in the open literature for the MW-assisted NH_3 synthesis yet. One of the key challenges in low pressure ammonia synthesis is the separation of unreacted N_2 and H_2 from NH_3 in a cost-effective manner. While the high-pressure Haber-Bosch process facilitates this separation readily using simple flash separators at moderate temperature, the low-pressure synthesis would require much lower temperature for flash separation thus leading to high penalty due to cryogenic conditions. Thus, one of the focus of this work is also selection of a suitable technology by evaluating a number of traditional and novel technologies.

A model of the MW reactor is developed where the reaction rate parameters are estimated by using the in-house experimental data from a laboratory-scale MW-assisted reactor. In this work, a plant-wide model is developed and several separation technologies are evaluated, namely: (i) temperature swing absorption; (ii) cryogenic distillation; (iii) cryogenic flash separation; (iv) absorption using water; (v) absorption using ionic liquid. Each of these separation technologies are optimized along with consideration of optimal heat integration. A detailed techno-economic optimization was carried out considering the tradeoff between the capital costs and the operating costs. As some of the technologies evaluated here such as the temperature swing absorption and ionic liquid absorption are novel, an uncertainly analysis is undertaken. Results shows that the novel separation technologies hold high promise for facilitating commercialization of low-pressure MW-assisted NH3 synthesis technology. Our work also identifies where more research will be desired in reducing the uncertainties in techno-economic analysis.

References

[1] E. R. Morgan, J. F. Manwell, and J. G. McGowan, "Sustainable Ammonia Production from U.S. Offshore Wind Farms: A Techno-Economic Review," *ACS Sustain. Chem. Eng.*, vol. 5, no. 11, pp. 9554–9567, 2017, doi: 10.1021/acssuschemeng.7b02070.

[2] J. W. Erisman, M. A. Sutton, J. Galloway, Z. Klimont, and W. Winiwarter, "How a century of ammonia synthesis changed the world," *Nat. Geosci.*, vol. 1, no. 10, pp. 636–639, 2008, doi: 10.1038/ngeo325.

[3] C. Zamfirescu and I. Dincer, "Ammonia as a green fuel and hydrogen source for vehicular applications," *Fuel Processing Technology*, vol. 90, no. 5. pp. 729–737, 2009, doi: 10.1016/j.fuproc.2009.02.004.

[4] A. Skodra and M. Stoukides, "Electrocatalytic synthesis of ammonia from steam and nitrogen at atmospheric pressure," *Solid State Ionics*, vol. 180, no. 23–25. pp. 1332–1336, 2009, doi: 10.1016/j.ssi.2009.08.001.

[5] X. Cui, C. Tang, and Q. Zhang, "A Review of Electrocatalytic Reduction of Dinitrogen to Ammonia under Ambient Conditions," *Adv. Energy Mater.*, vol. 8, no. 22, 2018, doi: 10.1002/aenm.201800369.

[6] K. H. R. Rouwenhorst, P. M. Krzywda, N. E. Benes, G. Mul, and L. Lefferts, "Ammonia, 4. Green Ammonia Production," *Ullmann's Encycl. Ind. Chem.*, pp. 1–20, 2020, doi: 10.1002/14356007.w02_w02.

[7] C. Wildfire, V. Abdelsayed, D. Shekhawat, and M. J. Spencer, "Ambient pressure synthesis of ammonia using a microwave reactor," *Catal. Commun.*, vol. 115, no. May, pp. 64–67, 2018, doi: 10.1016/j.catcom.2018.07.010.

[8] C. Wildfire, V. Abdelsayed, D. Shekhawat, R. A. Dagle, S. D. Davidson, and J. Hu, "Microwave-assisted ammonia synthesis over Ru/MgO catalysts at ambient pressure," *Catal. Today*, no. January 2020, 2020, doi: 10.1016/j.cattod.2020.06.013.

629758 National Priorities in the Net-Zero Transition of Electricity Systems: Policy and Technology Dimensions

Abstract:

The 20th century electricity system's challenges include the rapidly increasing demand and fossil fuel price volatility in the last quarter of the century. Accordingly, technology improvements in that period were focused on reducing capital cost and increasing efficiency to not only lower the cost but also to preserve fuel resources. In the 21st century, however, challenges also include the effort to reduce greenhouse gas emissions to mitigate climate change. Although it has been recognized that various contributions/roles of existing technologies are critical to delivering the net-zero target, the first 2 decades are characterized by different technologies competing to become a technology capable to play all roles in the system with the lowest cost. Notwithstanding this, owing to the unique characteristics of different systems, there is no one-size-fits-all technology to cost-effectively deliver the target.

In addition to the technological approach, policy design to drive the system to decarbonize is also vital. However, the lack of recognition of the unique role each technology may offer to the system may also inspire our current approach by simply focusing on the carbon tax. Whilst this approach can effectively reduce system carbon intensity, the absence of some roles that cannot be incentivized simply by increasing carbon tax may hinder the transition towards net-zero. Accordingly, the objective of this research is to identify priorities to robustly drive the decarbonization of electricity systems in different national contexts.

In the present study, we employed Electricity System Optimization (ESO) framework to understand how the electricity system might transition from 2020 to 2050. ESO is an MILP unit commitment model for electricity systems capacity expansion planning to minimize the total system cost considering system-wide and technology-wise constraints. To capture the impact of future uncertainty, numerous combinations of technology cost, thermal efficiency, and fuel price are implemented in ESO. Moreover, we also implemented different combinations of the carbon tax and CO_2 removal (CDR) credit to drive the system to decarbonize. Similarly, to investigate the impact of different characteristics a system may have, we include the UK, Poland, and ERCOT and PacifiCorp East (PACE) systems in the USA as the systems' archetypes.

Our results show that the impact of a carbon tax on the total system cost is driven by region-specific factors. For the UK, the impact tends to be marginal as the country has good sources of renewables, moderate price of natural gas, and coal ban. For regions with access to cheap coal and gas, such as ERCOT and PACE, the transition from BAU to low carbon system, which can be delivered by imposing a low carbon tax (£133/t_{CO2}), can be slightly costly. This is because the regions must switch from cheap coal (4.8-5.9 £/MWh) to a slightly more expensive gas (7.4-19.8 £/MWh). As the result, the total system cost is increased to around 130% (£ 80 billion) and 150% (£ 20 billion) of ERCOT and PACE BAU cost, respectively.

In contrast to the other systems, Poland possesses cheap and domestically sourced coal, but the price of imported natural gas is three times more expensive. Moreover, the load profile is less variable, and therefore the least-cost system has been historically achieved by operating coal for baseload generation. Furthermore, the country also anticipates banning further deployment onshore wind.

Accordingly, the system's decarbonization requires £ 50 billion (30% of BAU) of the additional cost compared to that of BAU.

Whilst the results show that carbon tax alone can consistently significantly reduces 2050's carbon intensity from 100-800 kg/MWh to 5-20 kg/MWh in all cases, we observe that the incremental reduction of carbon intensity can be achieved drops with the declining system carbon intensity due to the increasing value of low carbon dispatchable technologies, such as CCS-equipped plants. Accordingly, although carbon tax alone can effectively decarbonize the system to around 50 kg/MWh, it is exceptionally unlikely to deliver the net-zero target. The reason is that achieving this target consistently requires negative emissions technologies, such as BECCS, to offset residual emissions which cannot simply be incentivized by increasing the carbon tax. Accordingly, a combination of both the carbon tax and CO_2 removal (CDR) credit is consistently required; in our cases, central carbon tax (£230/t_{CO2}) and central CDR credit (£125/t_{CO2}) prove to be the optimal combination that can deliver the net-zero target with the highest probability and the lowest cost of transitions.

The results in the central carbon tax and central CDR credit combination reveal that each technology is expected to offer a unique role in the system. Thus, the focus of technology improvements for each technology is unique and country-specific. For instance, a combination of high fuel price and peaky load pattern in the UK requires thermal plants to provide flexible capacity in the future and drive the system to be more diverse. Hence, focusing on thermal plants cost reduction only reduces the range of total system cost by 2% of the initial cost range. Similarly, as the role of thermal plants is mainly to provide flexible capacity, the value of efficiency improvement is marginal. In contrast, Texas' access to cheap natural gas allows the system to maintain significant domination of thermal plants capacity in the system. Accordingly, cost reduction of CCS in the state shows more value than that in the UK, *i.e.,* total system cost reduction can be achieved is 5% of the initial range. Moreover, because the system is also benefited by a prospective iRES potential, iRES cost reduction should remain the focus of technology improvement as the effort can reduce the total system cost by 10% of the initial cost.

In contrast to the UK and ERCOT, PACE demand profile tends to be flat as the demand is mainly dominated by industrial sectors. Hence, the deployment of low carbon technologies which can continuously operate is critical. Accordingly, the role of CCS in the system is expected to be significant, and therefore, cost reduction of the technology yields a 5% total system cost reduction. However, although the share of CCGT-CCS generation in ERCOT and PACE is substantial, the value of thermal efficiency improvement is negligible owing to the low cost of natural gas.

Poland has the same load profile as PACE but the country does not have access to cheap natural gas and further development of onshore wind is banned. Hence, the country is expected to heavily rely on thermal plants and much less of that on iRES. This leads to a substantial benefit of CCS cost reduction whilst the value of iRES cost reduction is negligible. Interestingly, the value of efficiency improvement in Poland is the highest among the other cases owing to expensive natural gas prices and higher residual emissions of coal compared to that of gas. In Poland's case, our results show that increasing thermal efficiency can reduce the total system cost by 11% of the initial cost.

Finally, our study concludes that priorities in the net-zero transition of the electricity system are case-specific. While carbon tax alone can decarbonize the system, the combination of the central carbon tax and CDR credit is necessary and adequate to deliver the net-zero target cost-effectively. Moreover, each system needs to focus on key technology improvements that are more likely to substantially reduce the total system cost depending on the system's characteristics.

References:

1. Heuberger, C.F. et al. Impact of myopic decision-making and disruptive events in power systems planning. Nature Energy 3 (2018) 634–640

2. Heuberger, C.F. et al. Power capacity expansion planning considering endogenous technology cost learning. Applied Energy 204 (2017) 831–845

628259 Could Benzyl Hydroxyl Shielding Promote the Radical Induced Pyrolysis of Lignin?

Abstract:

Introduction

Due to the inherent complexity and changing chemical structure of lignin, a strategy, which has been gradually accepted in liquid depolymerization, is to shield the benzyl hydroxyl to inhibit the self-condensation reaction of lignin. This information on whether benzyl hydroxyl shielded in lignin structure affected the pyrolysis behaviors of lignin, especially explained by the radical theory, however, is scarce in the literature. Accordingly, the influence of the shielded benzyl hydroxyl in lignin structure on thermal behavior and radical evolution was investigated in this study. The benzyl hydroxyl in the lignin structure was first stabilized by propionaldehyde and prepoxidized by 2,3-dichloro-5,6-dicyano-1,4-benzoquinone (DDQ) respectively, compared with the raw lignin as a control (ball-milling wood lignin, MWL) for subsequent investigation. Then, the raw lignin and pretreated lignin were pyrolyzed at different temperatures, and the corresponding products were determined to find the best condition for obtaining organic liquids. In addition, through the electron paramagnetic resonance (EPR) spectroscopy, the evolution of pyrolytic radicals had been comprehensively studied to reveal the connection of the products formation with the radical induced reaction by different method of pretreatment.

Materials and Methods

The maple wood was first prepared for subsequently extraction of different types of lignin. Milled wood lignin (MWL) were obtained from the maple wood via the ball-milling method. Propionaldehyde stabilized lignin (PAL) was obtained by extraction in dioxane: water solutions with HCl and the propionaldehyde. To prepare the DDQ oxidized lignin (DOL), the maple wood was firstly oxidized by the DDQ oxidation with O_2(g) as the co-oxidants. Then the DOL was obtained by the ball-milling method extracted from the oxidized wood. TGA, 2D HSQC NMR and Ultimate analysis of raw and pretreated lignin samples were conducted to investigate the characterization and thermal stability.

Analytical pyrolysis of MWL, DOL and PAL was conducted with pyrolysis gas chromatography/mass spectrometry (Py-GC/MS) at 350-650 °C. Lab-scale pyrolysis was performed on a fixed bed to harvest chars from lignin. The radical concentration of sample was detected on a Bruker EMX-10/12 spectrometer.

Results and Discussion

The results showed that the process of inhibiting the coking of lignin cannot be achieved by shielding the α-OH. The maximum total yields (11.10×10^9 area/mg) of MWL lignin was obtained at 450 °C compared with the oxidized lignin (3.71×10^9 area/mg of DOL, 5.02×10^9 area/mg of PAL), which indicated that the yields of bio-oil was inhibited by oxidized α-OH of the lignin.

To understand the pyrolysis radical induction, the radical evolution of MWL was first detected by EPR from 350 to 650 °C. The pyrolysis radical concentration increased rapidly from 26.82×10^{17} spin/g to 245.04×10^{17} spin/g at 550 °C, which means the lignin undergo an explosive fission reaction. For PAL radical detected at 550 °C, the lower radical concentration (169.32×10^{17} spin/g) and higher char yields (37.10%)was obtained compared to the MWL. As for DOL radical, the liquid yield of the DOL was lowered than the MWL with the higher spin concentration (259.32×10^{17} spin/g for DOL). Therefore, the possible pathway of radical evolution leading to various products distribution from raw and pretreated lignin was given a new sight to understand the lignin pyrolysis reaction.

Conclusions

The shielded α-OH in lignin structure showed significant effect on the thermal stability and yield distribution during fast pyrolysis. It was proved that the radical reactions play an important role and process of inhibiting the coking of lignin cannot be achieved by shielding the benzyl hydroxyl. The bio-oil yield was decreased after shielding the benzyl hydroxyl, while the char yield had been increased. Facts have proved that in the pyrolysis process, due to the violent radical-induced reaction, the caped hydroxyl groups might be quickly converted into oxygen radicals, which further increases the probability of formation of benzoquinone and caused carbonization. Thus, the effect of capping pretreatment on the distribution of lignin pyrolysis products is clarified, and the failure of benzyl hydroxyl shielding pretreatment and the possible inducement of promoting the condensation of lignin to carbon were proved from the perspective of radical induction.

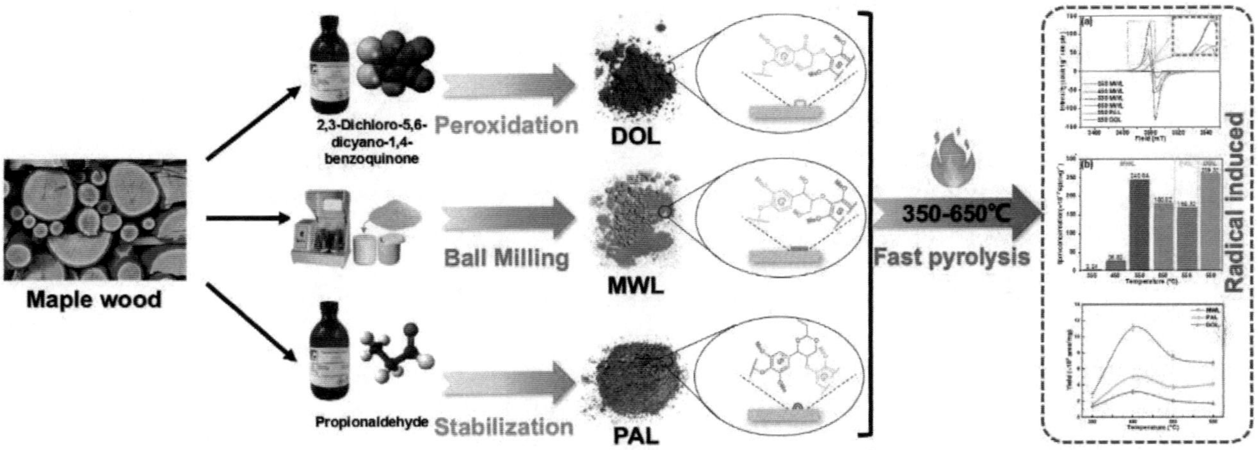

622248 Decarbonising the UK Heating Sector: A Whole-Energy System Analysis

Abstract:

The decarbonisation of heating is crucial in reaching the ambitious emission targets adopted by many countries, including the UK, which aims to reach net-zero emissions by 2050. While the electricity sector has received much attention from policymakers, private companies and researchers and considerable progress has been made on the decarbonisation of power generation, the heating sector has seen only little change. The majority of heat in the UK is still generated from conventional, unsustainable sources, with more than 85 % of households using natural gas boilers as main heating technology [1]. On top of that, 57 % of the commercial heat demand and 51 % of the industrial heat demand are provided from natural gas; with electricity, oil, solid fuel (coal) and biomass & waste also being significant contributors [2]. Consequently, emissions associated with heating are high. Residential combustion of fuel, mainly for heating purposes, alone accounted for 18 % of the UK's carbon emissions in 2019, and commercial and industrial heating add to that total [3]. It was estimated that in 2009, 38 % of the UK's carbon emissions could be attributed to heating [4].

Decarbonisation of heating is challenging due to the systems typically being distributed. Millions of devices in people's homes, commercial buildings and industrial facilities need to be replaced by low-emission heating technologies. These technologies typically rely on carbon-neutral fuels since carbon-capture at the end-user is impractical. Heat pumps and resistive electric heaters can utilise low-carbon electricity for heating, and hydrogen boilers can combust low-carbon hydrogen generated from e.g., methane-reforming with carbon-capture and storage (CCS) or water electrolysis without carbon emissions at the end-user. Solar-thermal collectors directly use a renewable energy source. Another potentially attractive option, especially for urban areas, are low-emission district heating networks that can benefit from centralised low-carbon generation and economies of scale. Additionally, improved building insulation and other energy conservation measures can be valuable as they decrease the demand and therefore reduce the required generation capacities. Various studies presented in literature show that no single technology is clearly superior to others, and most likely a combination of different technologies is most suitable for heating decarbonisation.

We present a newly developed heating sector optimisation model to identify optimal decarbonisation pathways. The model is a deterministic generation-expansion and unit-commitment optimisation model that simultaneously optimises technology portfolios and their operation. It is implemented as mixed-integer linear-programming (MILP) model in GAMS, using Python for data pre- and post-processing. A graphical user interface is also available to make the model accessible and easy to use. The model uses a similar formulation as the Electricity Systems Optimisation (ESO) model [5,6]. Starting from today's technology portfolio, the model optimises the evolution of the heating sector, considering investment decisions every five years. Build-rate constraints limit the deployment of technologies to technically feasible rates. For each of the decision years, the operation schedule of the technologies is optimised. Regarding emissions, the model can consider end-point constraints (e.g., net-zero in 2050), interior-path constraints and cumulative carbon budgets. So far, the model considers residential heating demand in form of representative households, commercial and industrial demand is to be added later. Modelled technologies are vapour-compression heat pumps, electric heaters, gas and hydrogen boilers and solar-thermal systems, and it is planned to include additional technologies such as thermal heat pumps, co-generation plants and district heating in future.

The emissions and fuel costs of electric and hydrogen heating technologies depend strongly on the respective generation mixes. A representation of the hydrogen industry is included in the heating sector model. Considered are generation from natural gas using methane reforming with or without CCS and from electricity via water electrolysis. Large-scale hydrogen storage is also included in the model. To determine the costs and emissions associated with electricity use, it is planned to connect the heating sector model with the existing ESO model. The result will be a comprehensive model of the UK electricity and heating energy systems that allows a simultaneous optimisation of technology capacities and dispatch in both sectors. The integrated approach allows to reduce the model externalities and necessary assumptions. The model can be a powerful tool to e.g., determine the optimal degree of heating electrification, prioritise the use of hydrogen for heating or electricity generation and to consider potential synergies such as co-generation or electricity load-shifting with heat pumps and thermal storage tanks. It can also be used to investigate the effectiveness of different policy measures. Guidelines on how to best support and facilitate the transition towards low-carbon heating and electricity provision can be developed. We present an analysis which robustly quantifies and qualifies the role and value of a range of technologies for providing low carbon heat.

References

[1] Committee on Climate Change, "Heat in UK buildings today" 2016. [Online]. Available: https://www.theccc.org.uk/wp-content/uploads/2017/01/Annex-2-Heat-in-UK-Buildings-Today-Committee-on-Climate-Change-October-2016.pdf.

[2] Department for Business Energy & Industrial Strategy, "Energy consumption in the UK (ECUK) - End use tables." 2019, [Online]. Available: https://www.gov.uk/government/statistics/energy-consumption-in-the-uk.

[3] Department for Business Energy & Industrial Strategy, "Final UK greenhouse gas emissions national statistics: 1990 to 2019." 2021, [Online]. Available: https://www.gov.uk/government/statistics/final-uk-greenhouse-gas-emissions-national-statistics-1990-to-2019.

[4] Department of Energy & Climate Change, "Overview of UK's greenhouse gas emissions from heat related activities" 2012. [Online]. Available: https://www.gov.uk/government/statistics/uk-emissions-from-heat.

[5] C. F. Heuberger, I. Staffell, N. Shah, and N. Mac Dowell, "A systems approach to quantifying the value of power generation and energy storage technologies in future electricity networks" Comput. Chem. Eng., vol. 107, pp. 247–256, Dec. 2017, doi: 10.1016/j.compchemeng.2017.05.012.

[6] C. F. Heuberger, E. S. Rubin, I. Staffell, N. Shah, and N. Mac Dowell, "Power capacity expansion planning considering endogenous technology cost learning" Appl. Energy, vol. 204, no. August, pp. 831–845, 2017, doi: 10.1016/j.apenergy.2017.07.075.

632205 Techno-Economic Assessment of Three Potential Pathways for Biomass Liquefaction

Abstract:

Biomass liquefaction is of interest within the biorefinery as it is a potential answer to the bottleneck formed from front-end materials handling. If the biomass could be transformed into a flowable slurry at the process entry point, then the material could more readily flow into the biorefinery pretreatment. Increased flowability will reduce downtime from any materials plugs that may form in the current design. Currently, the heterogeneous mixture of pelleted biomass and water can impede machine functionality and operability, which results in costly maintenance shutdowns. Liquefaction of biomass can potentially mitigate process bottlenecks. If a liquefaction process were to be installed at the front-end of an existing biorefinery, it would likely reduce or eliminate bottlenecks due to solids agglomeration. However, installation cannot be more costly than downtime losses. Through Techno-economic assessment (TEA), it is possible to quantify the reduction in losses via implementation of a liquefaction process. Three liquefaction pathways were considered: enzyme, maleic acid, and a combination of maleic acid+enzyme. The trade-offs in viscosity and equipment requirements were assessed using TEA to effectively compare the pathways to one another. All three pathways were simulated in Aspen Plus® Software.

623756 Altering Physico-Chemical Properties of Pine-Derived Carbon Quantum Dots By Changing Hydrothermal Treatment Conditions

Abstract:

Carbon quantum dots (CQD) are a relatively new and exciting material that are finding their place in a vast range of industries. These dots are aptly named as they exist within a diameter range of 0-10 nanometers. They are being utilized in everything from material coatings that self-clean or self-repair, to medicines that deliver cancer treatments to targeted areas in the body using UV light, to use in home entertainment systems to produce high quality screen visuals. Efficient methods and sources for producing these quantum dots are still being studied. It is found that the properties of the dots largely depend upon the surface functional groups and the size of the dots. Typically depending upon the method used to develop the dots (top-down or bottom-up) and the feed source the products are derived from will determine the surface groups and sizing.

A large amount of attention has recently been given to producing carbon quantum dots in the environmentally friendly manner of hydrothermal treatment. Many feedstocks have been studied and almost everyone has produced a variant dot. The goal of this research is to determine the effects of varying the hydrothermal treatment temperature and residence time on the characteristics of CQDs. Loblolly pine was chosen is chosen for this study due to abundancy and low cost. The experimental procedure focuses on three different reaction temperature conditions of 180, 220, and 260 °C. The time also varies in increments of 30 min up to 3 hours. After hydrothermal treatment the process fluid will undergo vacuum filtration followed by dialysis in order to separate the dots for characterization. Next the dots will be freeze dried for characterization that requires dry samples. The process fluid and the hydrochar itself will be analyzed for carbon quantum dots and these will be characterized using Fourier Transform Infra-Red Spectroscopy, Ultraviolet-Visible Spectroscopy, Fluorometry, Tunneling Electron Microscopy, and X-Ray Diffraction to determine trends. Altering the treatment time and temperature/pressure the quantity, surface functionalization, and the size can be controlled due to altering the formation of the passivation layer and condensation mechanism that forms the dots. As a result the quantum yield, absorbance, emission, and applicability of the dots are altered and can be fine tuned per application in industry.

623863 Understanding the Effects of Mixed Municipal Solid Waste Streams on HTL Product Characteristics

Abstract:

In 2018, each person in the United States was responsible for the production of nearly five pounds of municipal solid waste (MSW) per day.[1] Co-HTL of food and green waste components of MSW can help reduce waste and achieve economies of scale and the interactions between these feeds must be understood.

This work aims to understand the effect of mixing food and green waste on oil yield and product composition. Real food waste obtained from a veteran's affairs hospital resulted in 35.9% oil yield non-catalytically, in contrast to 12.13% for green waste. Current studies have revealed sustained oil yield (34.1%) with addition of up to 25% lignocellulosic waste (**Figure 1**). Additionally, results show increasing char yields (21 – 33.5%) with increased green waste content, due in part to its high (up to 13%) ash content.

Despite a drastic difference in carbon content in the feeds, 52% for food waste and 19% for green waste, FTIR indicates minimal differences between oil phase functional groups and vibrations. The primary difference occurs at 3400 cm^{-1} where the addition of green waste increases the OH stretching band. Solid phase FTIR mirrors the observed similarities, however, pure green waste solid phase which contains a significant OH stretch band. The similarities observed via FTIR are not sustained with GC/MS, as food waste oil contains large fatty acid peaks that decrease with the addition of green waste, only to be replaced by the presence of ketones and aromatics.

Qualitative observations also show improved homogeneity of food waste oil, noting that it is less sticky. Green waste oil, on the other hand appears to contain two distinct phases, a free-flowing liquid, and a sticky residue. This works provides useful insight into the feasibility of mixed MSW feeds for use in hydrothermal liquefaction.

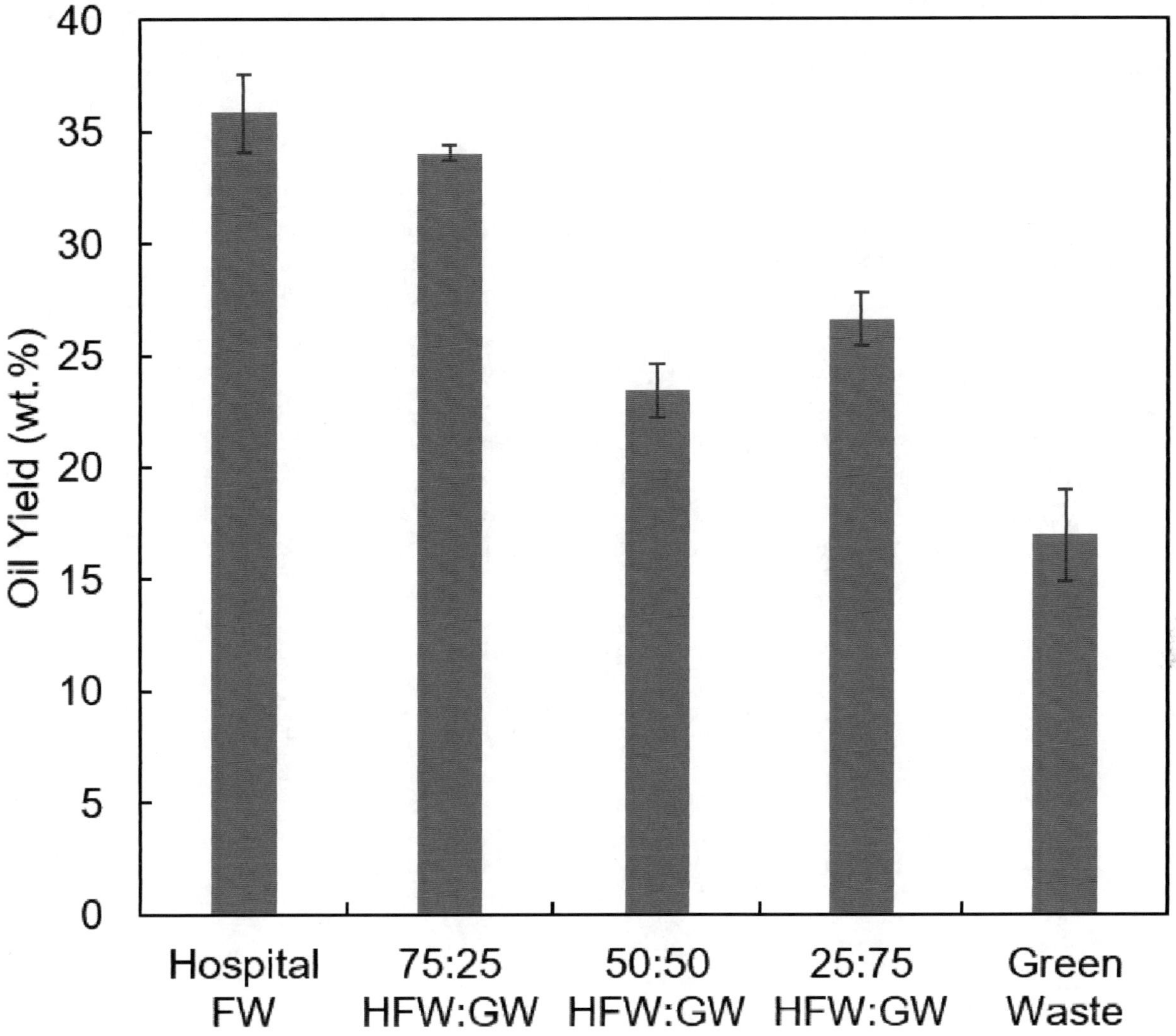

Oil Yield (wt.%)
0
5
10
15
20
25
30
35
40
Hospital FW
75:25 HFW:GW
50:50 HFW:GW
25:75 HFW:GW
Green Waste

632957 Adsorption Kinetics Studies of Catalytic HTL Derived Biochar

Abstract:

Ultra-porous carbonaceous powdered material derived from lignin rich biomass can be useful for water purification, energy storage, soil amendment, cement additive, wastewater treatment, gas purification etc.. In this study, we used corn stover agricultural residue as a cheaper feedstock and treated with hydrothermal liquefaction (HTL) at 250-300°C and 800-1600 psi to derive a solid residue, which was thermally and chemically post-treated to obtain mesoporous carbon. This carbon material was thoroughly characterized using the SEM/EDX, TEM and FTIR. Nitrogen adsorption isotherms such as Langmuir, Freundlich and Brunauer-Emmett-Teller (BET) were analyzed to determine the kinetic parameters. Barrett, Joyner, and Halenda (BJH) method was used to determine the pore-size distribution. Results obtained with specific surface area, pore size distribution, adsorption rate constant, monolayer/multilayer adsorption characteristics and mesoporosity will be presented.

633005 Magnesium Oxychloride Formation Kinetics and Enhanced Water Stability for Sustainable Building Materials Applications

Abstract:

Magnesium oxychloride (MOC) is a ceramic material with significant fire resistant properties and growing potential as a light-weight, structurally strong alternative building material. Despite many advantages, MOC has limitations as a construction material, particularly water stability. While researchers have explored solutions to these shortcomings, fundamental understanding of this material is still urgently needed. Thus we have investigated the fundamentals of the formation cure reaction and mechanisms of water stability enhancement reactions. The formation cure reaction kinetics for magnesium oxychloride 5-phase was monitored from 35 to 55°C using time-resolved quantitative x-ray diffraction and differential scanning calorimetry (DSC). The reaction was characterized as a two-step process: dissolution of magnesium oxide into a gel state followed by crystallization of magnesium oxychloride. Assuming first-order kinetics for both MgO dissolution and MOC crystallization, a kinetic model predicts 42.4 kJ/mol and 26.1 kJ/mol for dissolution and crystallization activation energies respectively. Alternatively, the Avrami nucleation and growth model was fit to DSC measurements predicting diffusion controlled, one-dimensional growth with an activation energy of 72.4 kJ/mol, accounting for both dissolution and crystallization.

For the water stability enhancement, two methods of water stability enhancement have been explored: chlorartinite formation from CO_2 exposure and phosphoric acid addition. The conversion of magnesium oxychloride to chlorartinite by CO_2 forms a protective, semi-insoluble chlorartinite layer on the surface of the magnesium oxychloride crystals, which improves water stability. Phosphoric acid (2.5 to 10 wt. %) was added to the MOC slurry before the cure reaction. Additions of 2.5 wt. % and above had positive impacts on the water stability, preserving ~50 wt. % crystalline MOC after the water stability test, but has a significant impact on the MOC formation kinetics. This is achied via. the formation of an amorphous phase on the MOC crystal surface that contains structural motifs related to insoluble MgHPO4·3H2O (newberyte) and Mg2P2O7·3.5H2O (magnesium pyrophosphate) phases. Results from this work are significantly impactful for the use of MOC in more sustainable alternatives for conventional materials in residential and commercial building applications.

624619 Optimization of TEG Dehydration Process in Natural Gas Processing Under Metamodel Uncertainty

Abstract:

Optimization of TEG Dehydration Process in Natural Gas Processing under Metamodel Uncertainty

Rajib Mukherjee[1,2], Urmila M Diwekar[2*]

[1]*Department of Chemical Engineering, The University of Texas Permian Basin, Odessa, TX 79762, USA*

[2]*Vishwamitra Research Institute, Crystal Lake, Illinois, 60012, USA*

Abstract

Natural gas processing involves removal of acidic gases followed by dehydration that is mainly carried out through absorption in tri-ethylene glycol (TEG). The dehydration process is accompanied by the emission of volatile organic compounds, including BTEX. There are several process parameters associated with major equipment that may impact BTEX emission. In our previous work, a multi-objective optimization is performed for optimal operating conditions identification related to the process parameters to mitigate BTEX emission using data driven metamodeling using machine learning followed by optimization with metaheuristic technique [1]. The metamodel is generated with data obtained from process simulation carried out in ProMax® process simulator. The metamodels are created using limited samples which introduces metamodeling uncertainty [2]. Previous metamodel-based robust designs treated a metamodel as the real model and ignore the influence of metamodeling uncertainty. Thus, the performance of the resulting optimized process parameters may have shortcomings due to the lack of adequately accounting for the uncertainty introduced by the metamodel. In the present work, the bias of metamodel uncertainty has been addressed for parameter optimization. An algorithmic framework is developed to address the challenge of parameter optimization under metamodel uncertainties. We used the better optimization of nonlinear uncertain systems (BONUS) algorithm to solve the problem. BTEX mitigation is used as the objective of the optimization [3]. Our algorithm allows the determination of the optimal process parameters for BTEX emission mitigation from TEG dehydration process under metamodel uncertainty.

Keywords: Natural gas processing, TEG dehydration, BTEX mitigation, metamodeling uncertainty, support vector regression (SVR), BONUS algorithm

[1] Mukherjee, R., & Diwekar, U. M. (2021). Multi-objective Optimization of the TEG Dehydration Process for BTEX Emission Mitigation Using Machine-Learning and Metaheuristic Algorithms. ACS Sustainable Chemistry & Engineering, 9(3), 1213-1228.

[2] Zhang, S., Zhu, P., Chen, W., & Arendt, P. (2013). Concurrent treatment of parametric uncertainty and metamodeling uncertainty in robust design. *Structural and multidisciplinary optimization*, *47*(1), 63-76.

[3] Sahin KH, Diwekar U, M. (2004) Better Optimization of Nonlinear Uncertain Systems (BONUS): a new algorithm for stochastic programming using reweighting through Kernel density estimation. Ann Oper Res 132:47–68

633725 Mapping Environmental and Economic Analysis of Decentralized Cogeneration Energy Management Centers

Abstract:

In the midst of a global conversation on how to reduce GHG emissions, Ontario's electrical grid stands out due to its extensive use of low carbon-intense processes such as nuclear and hydropower that acts as baseload energy sources. However, for peak electricity demands that require a quick ramp up of production, natural gas (NG) turbines are deployed [1], which causes the carbon intensity of Ontario's electrical grid to vary hourly [2]. Another intensive and permanent Canadian energy requirement is heating. Currently, it is provided mostly by NG -fired boilers [3]. The pursuit for better NG utilization is the motivation for this study, which proposes a design of combined heat and power energy management center (CHP-EMC).

CHP-EMC is composed of an NG-powered internal combustion engine connected to an electrical generator (NG-CHP), a hot water tank storage, an NG boiler, and a cooling tower to keep the water temperature within acceptable bounds. EMC is connected to a water loop similar to the ones used in district heating (DH) by a heat exchanger and responds to a community's heating demand using an hourly lumped-demand approach. The simulations are developed in TRNSYS, which is a capable heat transfer and fluid dynamic simulator [4]. The storage tank uses a built-in dynamic model whilst the rest of the components operate in pseudo steady-state. These components are connected by pipe models that account for time delay and flowrate.

The equipment presence and its sizes are decided by minimizing the natural gas consumption while simultaneously maximizing thermal and electrical generation. The optimization strategy used is a black-box approach using modeFrontier software [5]. This optimization and data visualization software can be coupled with any other software that uses input/output files -such as TRNSYS- and it executes several simulations guided by the optimization strategy and its constraints. For this work, classic Multi-Objective Genetic Algorithm (MOGA) is chosen for design variables (including the presence of specific components) and the pareto points are analyzed in more detail. TRNSYS simulation has internal simple controllers that take care of schedule and operational constraints. The result of the optimization aims to create an EMC flexible enough to endure deviations from expected demand profiles.

The emissions of proposed EMC designs are compared to the status quo (electricity supplied by electrical grid and thermal demand supplied by NG boilers) for a year of operation. Total emissions are calculated for each possible value of gCO_{2eq}/kWh to better represent possible scenarios of carbon intensity. The results are summarized in an easy-to-read emissions vs technology map that defines the conditions under which the EMC is less pollutant than status quo (SQ). The optimal design and selection of the EMC (vs. SQ) system is strongly dependent on the carbon intensity of the electric grid during peak-load times, the electric demand profiles, and the cooling demand profiles over the course of its operation. These are best characterized by the combination of greenhouse gas emissions per kWh during peaking operation, the ranges of heating and electricity demand over the course of the year (the maximum minus the minimum), and the kurtosis of the demands.

Following this analysis, an economic evaluation of both systems is carried out, demonstrating the cost of carbon avoided (CCA). CCA is a metric that compares CO_2 and cost of two technological strategies and calculates the cost per ton of GHG emissions potentially reduced [6]. These results are compared to possible carbon tax values, in a similar way to carbon intensity. A second map, cost vs technology, is then developed for easy visualization of the impact of carbon taxes on EMC and SQ operation costs.

The goal of this work is to apply the benefits of decentralized EMCs to reduce GHG emissions from both heating and electricity generation. The results are generalizable and can be applied to any location due to its methodology in representing carbon intensity and carbon tax.

References

[1] IESO, "Power Data," 30 04 2021. [Online]. Available: https://www.ieso.ca/power-data. [Accessed 30 04 2021].

[2] L. Pereira and I. D. Posen, "Lifecycle greenhouse gas emissions from electricity in the province of Ontario at different temporal resolutions," *Journal of Cleaner Production,* vol. 270, no. 122514, 2020.

[3] S. Canada, "Primary heating systems and type of energy," 03 04 2021. [Online]. Available: https://www150.statcan.gc.ca/t1/tbl1/en/tv.action?pid=3810028601. [Accessed 28 04 2021].

[4] TRNSYS, 2019. [Online]. Available: http://www.trnsys.com/index.html. [Accessed 30 04 2021].

[5] ESTECO, "modeFrontier," 2021. [Online]. Available: https://engineering.esteco.com/modefrontier/. [Accessed 30 04 2021].

[6] K. Gillingham and J. H. Stock, "The Cost of Reducing Greenhouse Gas Emissions," *Journal of Economic Perspectives,* 2018.

627430 Development of an Interactive Software Tool for Designing Industrial Solvent Recovery Processes

Abstract:

Abstract

In 2017 the chemical industry was reported to be the world's second-largest manufacturing sector (United Nations Environment Programme 2019). Solvents are used in most operations in chemical and pharmaceutical industries as a reaction medium, selective dissolution and extraction medium, and dilution agents. Thus, there is a sizable amount of solvent waste generated due to process inefficiencies such as inefficient mixing, and inappropriate process equipment being utilized. There are several options for solvent waste disposal, however, the three most common are on-site disposal, off-site disposal, and incineration (J. Raymond, Stewart Slater, and J. Savelski 2010). Even though these methods of solvent waste disposal carry considerable negative environmental impacts they are still widely used since they are readily available and provide an easy method to dispose of large quantities of waste. Solvent recovery is typically not used because of the potential difficulties in achieving a required purity to meet the stringent guidelines imposed on the products in pharmaceuticals and specialty chemical industries as well as the additional infrastructure and investments needed. To this end, this problem needs to be studied carefully by involving all aspects in regard to capital needs, environmental benefits, comparison to traditional disposal methods, while achieving the required chemical purity.

Thus, designing a solvent recovery method is a multi-objective problem focused on minimizing capital investments of the new separation technologies while maximizing the amount of solvent recovery matching the desired purity requirements. This task is further complicated because there is a lack of a guidance system to assist professional engineers in the modeling and design of solvent recovery processes. Such a guidance system or tool that could predict the value of solvent recovery and reuse for industry users, can generate an environmentally friendly trend to shift away from using incineration, on-site disposal, and off-site disposal as primary solvent waste handling methods. As seen in Figure 1A, not only will this tool provide a greener alternative, but also aid in solvent recycling into the chemical plant, which is economically favorable. To this end, if we can develop a user-friendly tool that allows engineers to easily access solvent recovery options, we can provide an economical and environmentally favorable strategy.

To account for all possible waste streams a robust framework is needed. This consists of a maximal process flow diagram that encompassed multiple stages of separations and multiple technologies within those stages. Together this process flow diagram develops the superstructure that provides multiple technology pathway options for any solvent waste stream. We established the different stages in the superstructure as Solid Removal, Recovery, Purification, and Refinement by reviewing the literature on existing materials recovery and separation processes (Yenkie, Wu, and Maravelias 2017; Biegler 1997; Geankoplis 2003). Common technologies are placed in different stages depending on the component they can separate in terms of their physical and chemical properties. A comprehensive chemical database is created using sources, such as the Design Institute for Physical Properties (DIPPR), to store all relevant chemical and physical properties ("DIPPR" 2012). Using General Algebraic Modeling Systems (GAMS) to model the various separation units and stream flows as a multi-objective optimization problem, the amount of solvent recovered can be optimized while minimizing the environmental impacts and total cost of operation (Chea et al. 2020; Yenkie et al. 2016). With the GAMS code as the backbone, a Graphical User Interface (GUI) is created to provide a user-friendly tool to the chemical industry. Figure 1B shows how the GUI connects the superstructure modeled in GAMS model, the chemical database, and user inputs to allow for anyone in the industry to find an optimal technology pathway for their solvent waste.

The GUI will allow for a user to input chemical species and parameters that are specific to the process that is being observed. By connecting MATLAB to the GAMS code, it is possible to model a solvent recovery process solely in the GUI with these input parameters. This tool reports economic, chemical, and environmental factors for the user to consider. The economic factors include an annualized cost breakdown, utility costs, consumable costs, labor costs, and overhead costs. These costs can be seen for the whole selected recovery pathway with details of the individual technologies. The chemical factors report the amount of solvent the predicted pathway can recover and its purity level. The environmental factors show the estimated emissions from the technology pathway and compare these emissions against those for incinerating the waste stream. Thus, this tool is a powerful option for those in the industry who want to look for a cost-effective way of recycling their solvents from the waste streams.

Keywords: Solvent Recovery, Recycling, Graphical User Interface

Acknowledgments

This research was supported by the United States Environmental Protection Agency's Pollution Prevention (P2) Program (NP96259218). The authors thank the Department of Chemical Engineering at Rowan University for their continued assistance with acquiring the tools required for this research.

References:

Biegler, Lorenz. 1997. *Systematic Methods of Chemical Process Design*. 1st ed. Prentice Hall PTR.

Chea, John D., Austin L. Lehr, Jake P. Stengel, Mariano J. Savelski, C. Stewart Slater, and Kirti M. Yenkie. 2020. "Evaluation of Solvent Recovery Options for Economic Feasibility through a Superstructure-Based Optimization Framework." *Industrial & Engineering Chemistry Research* 59 (13): 5931–44. https://doi.org/10.1021/acs.iecr.9b06725.

"DIPPR." 2012. The American Institute of Chemical Engineers. March 29, 2012. aiche.org/dippr.

Geankoplis, Christie J. 2003. *Transport Processes and Unit Operations*. 4th ed. Englewood Cliffs, NJ: P T R Prentice-Hall, Inc.

2010. Raymond, Michael, C. Stewart Slater, and Mariano J. Savelski. 2010. "LCA Approach to the Analysis of Solvent Waste Issues in the Pharmaceutical Industry." *Green Chemistry* 12 (10): 1826–34. https://doi.org/10.1039/C003666H.

United Nations Environment Programme. 2019. "Global Chemicals Outlook II - From Legacies to Innovative Solutions - Synthesis Report (2019)." Global Chemicals Outlook II. United Nations Environment Programme. https://wedocs.unep.org/bitstream/handle/20.500.11822/27651/GCOII_synth.pdf?sequence=1&isAllowed=y.

US EPA, OEI. 2013. "TRI Data and Tools." Data and Tools. US EPA. March 3, 2013. https://www.epa.gov/toxics-release-inventory-tri-program/tri-data-and-tools.

Yenkie, Kirti M., WenZhao Wu, Ryan L. Clark, Brian F. Pfleger, Thatcher W. Root, and Christos T. Maravelias. 2016. "A Roadmap for the Synthesis of Separation Networks for the Recovery of Bio-Based Chemicals: Matching Biological and Process Feasibility." *Biotechnology Advances* 34 (8): 1362–83. https://doi.org/10.1016/j.biotechadv.2016.10.003.

Yenkie, Kirti M., Wenzhao Wu, and Christos T. Maravelias. 2017. "Synthesis and Analysis of Separation Networks for the Recovery of Intracellular Chemicals Generated from Microbial-Based Conversions." *Biotechnology for Biofuels* 10 (1). https://doi.org/10.1186/s13068-017-0804-2.

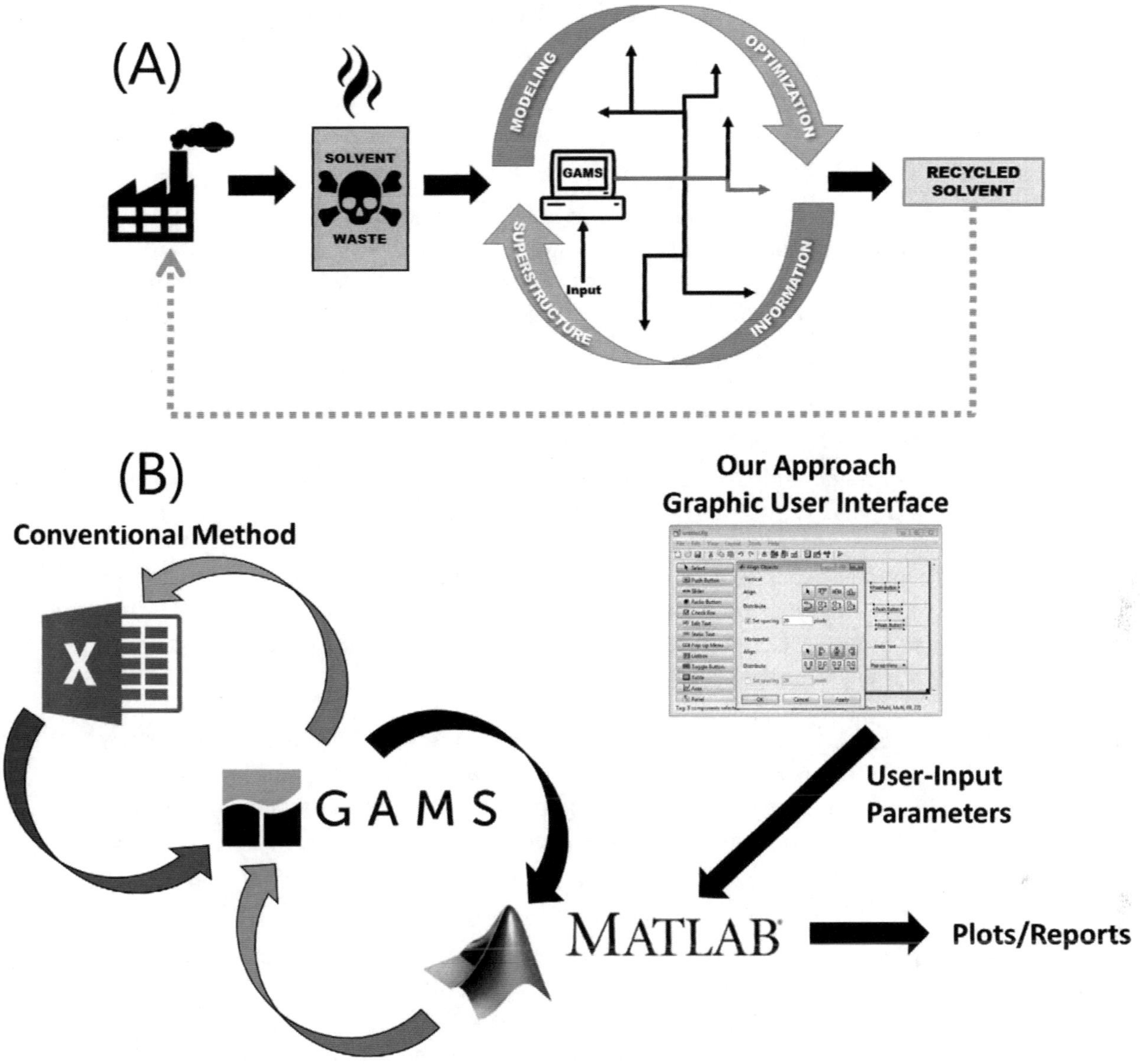

Figure 1(A) Proposed Process for Modeling Solvent Recovery for use in an Industrial setting.
(B) Flow chart of Information Processing of the Proposed Computational Software Tool

625002 High Flux CO2 Selective Membranes for Renewable Natural Gas and CO2 Capture

Abstract:

Reducing carbon emissions is key to preventing further climate change and in correcting the damage that has already been done. Compact Membrane Systems is developing a carbon dioxide selective membrane designed to operate at low pressure with minimal pretreatment. Results from two applications will be discussed: removal of carbon dioxide from flue gas and upgrading of biogas to renewable natural gas.

According to the World Biogas Association, anerobic digestors for biogas production can reduce greenhouse gas emissions by 10% in 2030. Separating carbon dioxide from methane is a key challenge for the renewable gas industry. Biogas is produced largely on farms or small waste centers, not at large centralized plants. While there are existing technologies to separate methane from carbon dioxide, they require multiple stages and high pressures, and are sensitive to process poisons. An efficient and cost-effective solution is needed to bring down barriers that previously prevented small scale operations from transforming their biogas into usable fuel. CMS has developed a membrane aimed at allowing users to capture, separate and upgrade existing biogas streams that would otherwise be flared or released into the atmosphere.

Results from the first field demonstration of Optiperm™ *biogas* will be shared. Working with Energy Projects Manager Matt Steiman at the Dickinson College Farm, a small spiral wound module system has been installed. The system takes raw biogas feed directly from the digester with no pretreatment. The pilot is intended to demonstrate successful operation under field conditions for an extended period of time. Optiperm™ *biogas* technology will be tested for a period of at least 2 months to demonstrate stability in the field and validate performance measured in the CMS laboratory. Optiperm™ *biogas* upgrading streams from 60% to 90% methane in a single stage.

Finally, initial results for utilizing the CMS membranes to capture carbon dioxide from flue gas, including the effect of operating conditions and initial system modeling, will be shared. This application is earlier in development, but results thus far have shown a very high flux membrane that could result in substantially lower capture costs than the current state of the art.

AICHE Fall 2021

Optimizing Energy Efficiency of Ammonia Production via Electrochemical Reaction and Haber-Bosch process.

Gbemisola Ojo and Kyle Camarda
[1]The University of Kansas, Department of Chemical & Petroleum Engineering
4132 Learned Hall, 1530 W 15th Street, Lawrence, KS 66045

Ammonia (NH_3) synthesis is crucial to the production of fertilizers, and accounts for nearly 2% of total global energy consumption. Therefore, finding methods to produce ammonia employing only renewable energy and sustainable feedstocks is paramount. This project uses dynamic simulation to design a 100% renewable ammonia synthesis plant, using only air and water as raw materials. Energy is supplied from a solar source, which is supplemented with industrial batteries. The intermittent nature of solar energy production requires the development of a dynamic production profile for the plant.

In this work, an electrolytic cell powered by a time-varying solar source is employed for hydrogen production. Air is separated via a membrane to provide a nitrogen source, and then the Haber-Bosch reaction is used to produce the product. The degrees of freedom in this problem are the number of batteries purchased, as well as the dynamic production and charging schedules. To determine the optimal production profile, a mixed integer non-linear programing model is formulated and solved. Results demonstrate how varying levels of production and energy storage usage correspond to different levels of energy availability, and show the feasibility of completely sustainable ammonia production, using no raw materials other than water and nitrogen, and creating zero emissions.

630714 Electrochemical Reduction of Nitrates to Ammonia on Oxide Derived Cobalt

Abstract:

NH_3 is manufactured in an industrial scale by Haber-Bosch process which is highly energy intensive as it requires high temperatures (500 to 600 °C) and high pressures (150 to 200 atm), and it leaves a massive carbon footprint. Electrochemical route is an alternate process to synthesize NH_3 by using renewable electricity at ambient conditions. Direct electrochemical reduction of dinitrogen would be ideal, but the NH_3 Faradaic efficiency and yield is low in this case due to high mass transfer resistances and the competing hydrogen evolution reaction (HER). Electrochemical reduction of NO_3^- to NH_3 is less challenging when compared to direct N_2 reduction to NH_3 due to lower dissociation energy of $N = O$ bond (204 kJ/mol) than $N \equiv N$ (941 kJ/mol). A catalyst that can selectively reduce nitrates to ammonia is required. DFT calculations were performed on late transition metals to identify the most active metal for the electrochemical nitrate reduction to NH_3. Ni and Co were found to be most active among the late transition metals. Experimental investigations corroborated with the theoretical findings and Co was found to give the highest NH_3 Faradaic efficiency. Oxide-derived Cobalt was used for the electrochemical nitrate reduction as it had low onset potential compared to other forms of Co. We obtained a maximum NH_3 Faradaic efficiency of 68.13 % at -0.6 V vs RHE (pH = 14), and a high NH_3 current density of 245.34 mA/cm^2 at -0.8 V vs RHE (pH = 14). The oxide derived Co was found to be stable during the test period of 24 h. NMR Control studies were performed using N-15 nitrates to confirm that the source of NH_3 is nitrates.

623825 Discovery of Catalysts for the Application of Nitrate Reduction and Water Purification Using Machine Learning Techniques

Abstract:

Nitrate (NO_3) runoff has become a leading pollutant to nearby water supplies in areas with intense agricultural and industrial processes. Redox reactions facilitated by ion exchange is currently the dominant means of converting NO_3 into more benign components such as N_2. However, such processes require the continuous replenishment of ions and produces brine as a by-product that must be subsequently removed. Electrocatalysis can provide an alternative means of reducing NO_3 to N_2 without requiring the continuous regeneration of the material. Furthermore, the absence of ions and other chemicals provides a brine-free product. However, the majority of known electrocatalysts that can select N_2 as a by-product is limited to costly precious metals and the overpotentials required to facilitate these reactions are often very high. The discovery of more cost- and energy-efficient electrocatalysts is essential to making NO_3 reduction with electrocatalysts viable. In this abstract, we screened for potential cost efficient electrocatalysts by querying the Materials Project and Aflow databases for bimetallic alloys. We predicted the oxygen and nitrogen adsorption energies for these alloys using a Graph Neural Network machine learning model previously developed by the Open Catalyst Project (OC20). Using BEP scaling relationships and microkinetic models developed by Singh and Goldsmith group, we estimated the turnover frequency for NO_3 consumption with the predicted adsorption energies. We then performed DFT calculations facilitated by a high-throughput framework on alloy surfaces with an estimated turnover frequency above 0.1/s to accurately assess the selectivity, overpotential and turnover frequencies of each candidate alloy in order to propose a novel set of materials for NO_3 reduction reaction.

623921 Controlling Electrocatalytic Nitrate Reduction to Ammonia through Adsorption Energies

Abstract:

Nitrate pollution from agricultural and industrial waste is causing an imbalance in the nitrogen cycle and poses an immediate threat to environmental and human health. Although there are numerous methods to treat this nitrate, the cost of treatment is often prohibitive, which could potentially be addressed by converting the waste nitrate to a valuable product such as ammonia. In this work we discuss the use of electrocatalysis to convert nitrate to ammonia, where renewable electricity can be used to drive the reaction, minimizing the issue of emissions of carbon dioxide. A major challenge with electrocatalytic nitrate reduction is decreasing the overvoltage of the reaction and increasing the reaction rate without losing faradaic efficiency towards the desired product. Through the use of kinetic studies, in situ spectroscopy, microkinetic modeling, and density functional theory calculations, we show how the adsorption energetics of nitrate and hydrogen can be used to understand the rates of nitrate reduction on metals and metal sulfide surfaces. By tuning these adsorption energies with alloying, as shown below by alloying Pt with Ru, we can increase the rates of nitrate reduction. Our measurements of the product distribution show that this alloying does not significantly change the high faradaic efficiency towards ammonia production (>90%). By comparing to thermal nitrate catalytic reduction (i.e., using molecular H2 rather than an applied potential to reduce nitrate) we identify numerous similarities between electrocatalysis and thermal catalysis, but also specific factors unique to electrocatalysis that impact the rate of reaction.

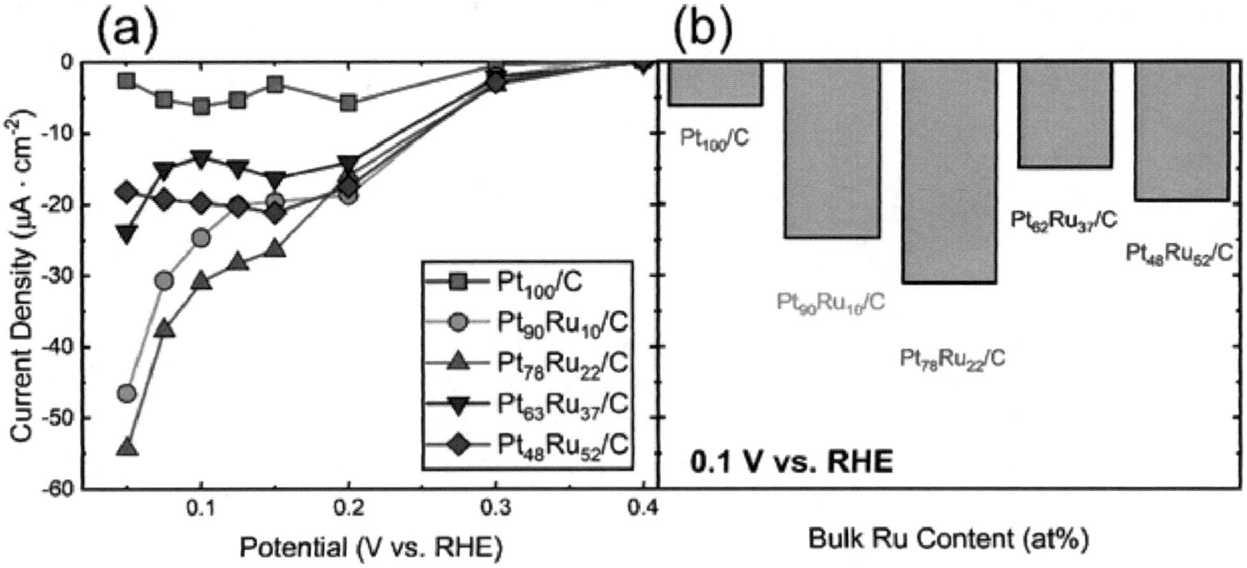

629137 Advanced Reactor Design for Ammonia Production from Electrochemical Nitrogen and Nitrate Reduction

Abstract:

Motivated by the increasing demand for flexible and sustainable routes of ammonia (NH_3) production, the electrochemical nitrogen (N_2) and nitrate reduction reaction (NRR and NO3RR) have attracted intense research interest in the past few years. NRR allows for distributed and on-site generation of NH_3 with water as the proton source, thereby reducing the transportation and storage costs of NH_3 and H_2. Besides N_2 which is highly abundant in the atmosphere, nitrate-N exists widely in agricultural and industrial wastewaters. NO3RR offers a promising strategy of simultaneously removing the harmful nitrate-N and generating NH_3 as a useful product from those waste streams.

While research activities on both NRR and NO3RR are blooming with substantial progress, some major challenges remain unresolved so far. Due to the wide existence of reactive N-containing species, the source of NH_3 in NRR measurements is sometimes elusive and requires rigorous examination by control experiments with costly $^{15}N_2$. On the other hand, while the electro-reduction of nitrate is much more facile, additional costs arising from the enrichment and purification of nitrate in contaminated waste resources have challenged the practical feasibility of NO3RR both technically and economically.

In this talk, we will present our latest research progress as part of the solutions to these challenges in state-of-the-art NRR and NO3RR studies, from the perspective of reactor design. An improved $^{15}N_2$ circulation system, which allows for more accurate and economized quantitative assessment of NH_3 origins, was suggested herein. For NO3RR, we developed a compact reactor system for rapid and efficient NH_3 production from real nitrate-containing waste sources, accompanied by the concurrent separation and enrichment of the produced NH_3 to yield pure ammonium compounds. Our work highlights the importance of advanced reactor design in N-related electrochemistry research, which will facilitate the transformation of the current N-centric chemical industries towards a sustainable future.

631601 Harnessing Mass Transport to Optimize Ammonia Production during Electrochemical Nitrate Reduction

Abstract:

Nitrate is a prevalent waterborne pollutant that threatens the health of humans and aquatic systems. By selectively producing ammonia, electrochemical nitrate reduction reaction (NO_3RR) can directly transform nitrate pollutants into a widely used commodity chemical and fertilizer, thus balancing the nitrogen cycle while reducing energy consumption from the traditional Haber-Bosch process. Efforts to date have primarily focused on improving NO_3RR at the electrode surface. In parallel, mass transport throughout the system dictates the supply of reactants and supporting ions while defining the local reaction environment (i.e., electric potential, pH, ion concentration). Therefore, mass transport plays an important, and sometimes determining, role in NO_3RR activity and selectivity.

We used a membrane-separated flow cell and titanium foil electrode to study the influence of mass transport (Fig. 1a). Considering the two serial interfaces in the working electrode chamber (electrode/electrolyte and electrolyte/membrane), different strategies are employed to enhance mass transport across them. Nitrate removal, ammonia N-selectivity and Faradaic efficiency are chosen as key performance metrics.

Cross-membrane ion transport resistance can significantly limit the total current density of electrochemical systems. We have found that pairing membrane type (proton vs. cation exchange membrane) with the primary charge carrier species in bulk electrolyte can improve total current density and nitrate removal (Fig. 1b). Meanwhile, the structure of the diffusion layer at the electrode/electrolyte interface is strongly impacted by bulk electrolyte flow rate, which can be utilized to tune nitrate removal and ammonia production. Interestingly, the dependence of nitrate selectivity on flow rate shows opposite trends in different co-existing cation concentrations (Fig. 1c). Continuum models will be applied to solve interfacial potential and concentration profiles to illustrate the underlying mechanisms.

As an underexplored topic, mass transport processes in electrochemical NO_3RR are being systematically studied and will be engineered to optimize ammonia production from polluted waters.

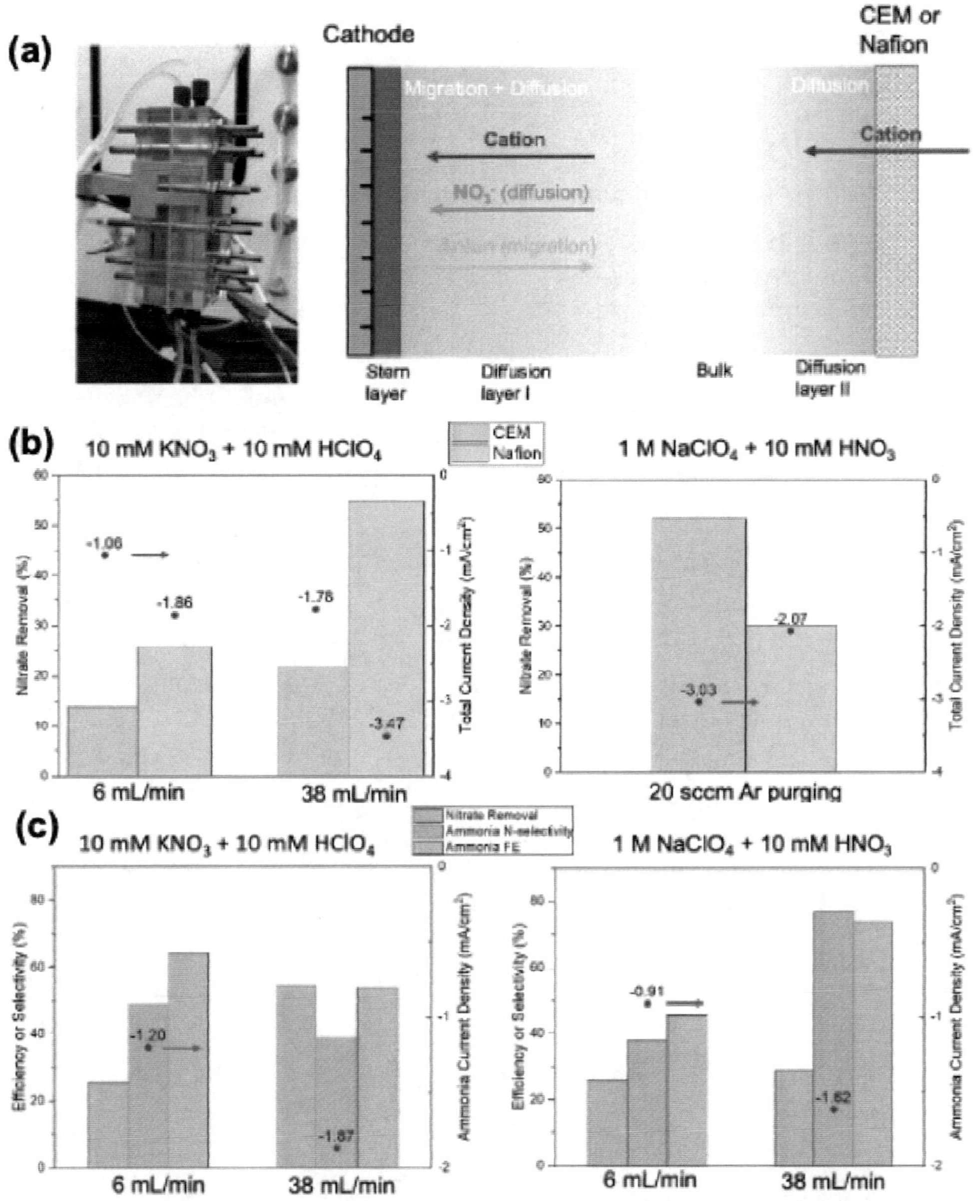

Fig. 1 (a) Picture of the membrane-separated electrochemical cell used in this study and diagram illustrating the two serial mass transfer interfaces in the working electrode chamber. (b) Effects of membrane choice on total current density and nitrate removal. Electrolyte I: 10 mM KNO_3 + 10 mM $HClO_4$. Electrolyte II: 1 M $NaClO_4$ + 10 mM HNO_4. In Electrolyte I, system with proton exchange membrane (Nafion 117) shows better performance across two electrolyte flow rates. In Electrolyte II, system with cation exchange membrane (CEM, CMI-7000) shows better performance. (c) Effects of electrolyte flow rate on nitrate removal, ammonia N-selectivity, Faradaic efficiency (FE) and partial current density. Electrolyte I is paired with Nafion, and Electrolyte II is paired with CEM. In both systems, higher flow rate leads to higher nitrate removal. Ammonia N-selectivity and FE are improved at low flow rate in Electrolyte I yet impeded in Electrolyte II.

632184 Enabling Nitrate Valorization to Ammonia through Electrocatalyst Design

Abstract:

Nitrate (NO_3^-) is a widespread surface and groundwater contaminant which if not treated cause adverse affects on human health such as methemoglobinemia ("blue baby syndrome") and cancer. Electrocatalytic routes toward nitrate transformation have centered on remediation strategies. Electrocatalytic remediation occurs through a two-step process. First, nitrate reduction (NO_3RR) transforms nitrate to nitrite, and then nitrite reduction (NO_2RR) transforms nitrite to nitrogen gas. This approach produce no waste as nitrogen can be released into the atmosphere. Others routes toward nitrate transformation have centered on valorization strategies [2,3]. In this approach, electrocatalytic NO_3RR and NO2RR are performed on a catalyst, however, ammonia is the product of NO_2RR instead of nitrogen. Independent of the transformation path, choice of catalytic material to perform NO_3RR and NO_2RR reduction is imperative to achieve desirable activity and selectivity. NO_3RR electrocatalytic activity has been previously shown to be high on Cu, Sn, and In, which precious metals such as Pd and Pt show poor activity. Conversely, most electrocatalytic NO_2RR reaction investigations have found Pd and Pt to demonstrate ideal activity. Thus, there is a strong need to design multi-functional catalytic materials which aim to accelerate both NO_3RR and NO_2RR. Here, we aim to begin to the elucidate the impact of catalyst structure has on NO3RR and NO2RR for valorization (Nitrate to Ammonia). To accomplish this we develop Pd shape-controlled nanoparticles (e.g. nanocubes, cuboctahedrons, octahedrons, and concave nanocubes).

629230 Computational Prediction of Polymer Solubilities and Its Application in Multilayer Plastic Film Recycling

Abstract:

Billions of pounds of multilayer plastic films are produced annually as packaging materials. These films are made of distinct polymer layers, which makes it a challenging task to recycle them. Recently, we developed a strategy called solvent-targeted recovery and precipitation (STRAP) that can deconstruct and recycle such films through a series of solvent washes. It utilizes the solubility difference of the polymers to selectively dissolve them in carefully chosen solvent systems. One key step in designing a STRAP process for a target film is to identify selective solvent systems and operating temperatures, which could be prohibitively time-consuming using experimental screening alone. Here, we report a computational scheme for polymer solubility prediction and solvent selection that employs Hansen solubility parameters, molecular dynamics simulations, and the COnductor-like Screening Model for Realistic Solvents. These tools take molecular interactions and conformational information into consideration to yield quantitative, temperature-dependent solubility predictions of polymers in pure solvents and solvent mixtures. We demonstrate that this computational scheme can efficiently screen appropriate solvents to guide the design of solvent-mediated processes for multilayer plastic film recycling.

631196 Calibration of a CFD-DEM Fluidised Bed Simulation for Plastic Recycling Using Positron Emission Particle Tracking

Abstract:

Industrial-scale plastic recycling using pyrolysis in fluidised beds is an emerging technology aiming to address the ever growing plastic waste crisis. Whilst we developed a basic understanding of fluidised beds with a single solid phase, such as the fluid catalytic cracking in modern oil-refineries which have been used for decades, binary systems which introduce a non-identical second solid species often lead to difficulties in reactor design. Uneven particle distributions, segregation and hot-spots lead to not only product issues but also safety issues which need to be understood and removed.

The Discrete Element Method (DEM) does, when calibrated properly, reproduce the behaviour of moving granular material very well. Coupled with the well developed computational fluid dynamics (CFD) it provides a powerful tool in predicting the behaviour of a fluidised granular system such as fluidised beds.
These systems however need to be calibrated and even though the conventional techniques of calibration for DEM models such as shear cells or angle of repose measurements can lead to quantitative granular properties, the characterisation and calibration of the fluid phase is currently much less precise.

We use positron emission particle tracking, a Lagrangian particle tracking technique, to directly compare coupled DEM-CFD simulations of fluidised beds with real world fluidised beds. The rich, high-resolution data yielded by PEPT allows us to fine tune drag models and particle parameters in a significantly more precise manner than is possible using conventional measurements such as pressure drop alone.

631881 Use of Waste Plastic in Asphalt Binder: Theoretically and Experimentally Justified Identification of Compatible Blends

Abstract:

Previous attempts to add waste plastic to asphalts were not met with great success primarily due to the incompatibility of common plastic and asphalt cement, which resulted in poor performance. If the compatibility between waste plastic and asphalt binder can be systematically understood and improved, then large amounts (millions of tons annually) of previously land-filled waste plastic can be repurposed in asphalt binder. Here, we fundamentally investigate the compatibility of waste plastic and asphalt using a combination of atomistic simulation, coarse-grained simulation, and experimental validation to provide guidelines for predicting the compatibility between asphalt and waste plastic. Waste plastics such as HDPE, LDPE, PVC, PET, and PS are considered, along with some approaches to improve compatibilitzation such as crosslinking with ground tire rubber, aminolysis of PET, and wax residue addition. These materials and approaches are considered in terms of fundamental polymer science concepts.

630282 A Model-Based Approach for the Shift from Batch to Continuous Production of Succinic Acid from Glycerol

Abstract:

Developing a bioprocess can be a very cumbersome task, with a multitude of experiments required to affirm its feasibility. Usually, such a bioprocess is initially developed in batch mode which is sufficient most of the times.

Nevertheless, there are scenarios, where a batch bio process is not fit for purpose and continuous operation mode should be considered, e.g. to efficiently integrate the bioprocess with continuous downstream processing or within an integrated biorefinery.

The production of succinic acid form glycerol [1] fits well such a scenario.

Glycerol is the main by-product of biodiesel production, hence the biological production of succinic acid serves two purposes: a. the production of a significant added value product (succinic acid) and b. the valorisation of a biofuel byproduct, which can help to make the biofuel production more profitable [1].

In our bioprocess, we are using and adapted A. succinogenes strain, which naturally overproduces succinic acid, that it is able to grow by consuming glycerol [3]. Initially the production of succinic acid from glycerol with A. succinogenes was developed as a batch process. For this process we constructed a robust kinetic model which was used for bioprocess design and optimisation [3, 4].

Nevertheless, to improve commercialisation potential, and to be better integrated into a biodiesel biorefinery, the bioprocess should have higher productivity, than that of the batch and should be able to handle continuously and efficiently all of the glycerol substrate. To achieve this, the process should move from a batch to continuous mode of operation. This can be efficiently achieved, through the use of a predictive bioprocess model, which can compute the behaviour of the process under different conditions. As with the batch model the development of a continuous model requires experimental data, in order to calibrate and validate it. Here we used the intrinsic kinetics of an existing batch process model [4], to develop our continuous process model.

In this work, two continuous bioprocesses were developed: a. through the design of a continuous stirred tank reactor (CSTR) with cell recycling and b. through the design of a packed bed bioreactor (PBBR) It should be noted that it is the first time succinic acid is successfully produced in a continuous manner using a packed-bed bioreactor. A. succinogenes was entrapped in sodium alginate beads, which then were used as the packing material for the bioreactor. The biomass concentration inside the beads remained stable and the beads were used for prolonged fermentation time, without showing any substantial reduction in process yield. A multiscale diffusion-advection-reaction model was developed, using the batch process parameters. The model is able to compute concentration profiles of the substrate and the products inside the catalyst beads and in the bulk phase of the bioreactor.

The developed models for both processes are able to successfully simulate both the dynamic and the state behaviour of the process. The model predictions are validated with experimental data. The computed error between the predicted concentration profiles and the actual experimental results varies from 0 % to 6 % for the CSTR and 2 % to 4% for the packed bed reactor.

Both continuous processes outperform the batch process in terms of succinic acid productivity, with the PBBR being the best between them. The PBBR process is 4 times more productive than the CSTR and 6.6 time more productive than the batch process. Furthermore, the PBBR is able to achieve complete conversion of the substrate in contrast to the CSTR case.

This in conjunction with the fact that the bacterial culture is immobilised, results in an effluent consisting mostly of succinic acid, leading to reduced separation costs.

Continuous processing can make this succinic acid bioproduction easily integrable with a biodiesel biorefinery, helping to significantly improve both its economic potential and its environmental impact.

References

[1] A. Vlysidis, M. Binns, C. Webb, and C. Theodoropoulos, \A techno-economic analysis of biodiesel biore_neries: Assessment of integrated designs for the co-production of fuels and chemicals," Energy, vol. 36, 4671-4683, 2011.

[2] H. W. Tan, A. R. A. Aziz, and M. K. Aroua, Glycerol production and its applications as a raw material: A review, Renewable and Sustainable Energy Reviews, vol. 27, pp. 118-127, 2013.

[3] A. Vlysidis, M. Binns, C. Webb, and C. Theodoropoulos, Glycerol utilisation for the production of chemicals: Conversion to succinic acid, a combined experimental and computational study," Biochemical Engineering Journal, vol. 58-59, pp. 1-11, 2011.

[4] A. Rigaki, C. Webb, and C. Theodoropoulos, Double Substrate Limitation Model for the Bio-Based Production of Succinic acid from Glycerol, Biochemical Engineering Journal, p. 107391, 2019.

625596 Low Temperature Reactors for Biomass Processing and Carbon Modification Using Atmospheric Pressure Plasmas

Abstract:

This presentation will cover advances from the Center for Plasma-Material Interactions in the design and operation of atmospheric pressure plasma devices for the purposes of chemical refining. Atmospheric plasmas driven by gliding arcs or dielectric barrier discharges and A/C currents have been under investigation as a mechanism for decomposing complex biomass, as well as the production of specialty chemicals. These reactors often suffer from poor selectivity, as well as high capital costs. More recently, scalable microwave driven atmospheric pressure plasma devices have produced reactors capable of operating further from chemical equilibrium in a highly non-thermal state. This enables reactive species densities to remain high with fewer concerns for thermal decomposition and gasification of feedstocks or products. Results will be shared on microwave atmospheric plasma device diagnostics related to reactor design as well as results discussing the decomposition of anisole and other lignin model molecules into feedstock chemicals for other processes. The production of carbon nanostructures with high thermal and electrically conductivity, and other value added materials, using these reactors will also be presented.

625782 Effect of Particle Size on Supercritical Water Gasification of Biomass

Abstract:

Introduction
Supercritical water gasification of biomass is exactly in the scope of reactor engineering of biomass feedstocks. Specific to biomass is the very complex reaction compared to conventional fossil fuel conversion in chemical industry. Biomass is more reactive than coal and oil, and thus undergoes various reactions at supercritical temperature (600 °C) which is much lower than conventional fossil fuel conversion temperature (1000 °C). Biomass is fed as solid, but it is known that most of biomass species can get dissolved in supercritical water, and then furthermore decomposed in the molecular scale. This dissolution-decomposition model has been long mentioned when supercritical water gasification of biomass is discussed. However, apparently, complete dissolution should be difficult when particle size is sufficiently large. Chemical engineering textbooks tell us dissolution is usually mass transfer limited phenomena, and when diffusion or mass transfer is slow, complete dissolution takes time. When temperature inside the particles gets high before dissolution, pyrolysis inside the particle should take place. Then, the question arises: what is the diameter needed for complete dissolution for usual biomass gasification in supercritical water? The purpose of our study was to answer this question for effective commercialization of supercritical water gasification technology.

Experimental
A laboratory scale continuous flow reactor for supercritical water gasification was employed. Reaction temperature and pressure were 600 °C and 25 MPa, respectively. Feedstock was slurry of particles of biomass or cellulose. The size of these particles was set to several to sub- micrometer using special apparatus developed in Nara Machinery Co., Ltd. So far no one has worked on these fine particles for supercritical water gasification.

Results and discussion
For EFB particle, the effect of particle size is observed. If complete dissolution is achieved, this should not happen. Thus, it is found that 3.5 μm is not sufficiently small to achieve complete dissolution.
Cellulose particle size in the range of much smaller size was prepared. To achieve this sub-micrometer size, we had to employ flow separation of particles. Interestingly, effect of particle size disappeared for small particles. This implies complete dissolution.

Conclusion
We successfully found the criteria of complete dissolution of particles in a laboratory scale continuous supercritical water gasification reactor. This result shows that when actual biomass is used in the continuous reactor, dissolution is not complete, and behavior of solid particles should be considered. This is the new and important finding in terms of supercritical water gasification of biomass.

CATALYTIC DEOXYGENATION OF RAPESSED OIL INTO DROP-IN FUEL UNDER HYDROTHERMAL CONDITION CAUSING IN-SITU HYDROGEN PRODUCTION

Masato Kouzu, Minato Kojima, Katsuyuki Mori, Haruka Ishida, Jun Suzuki,
Tokyo City University, Tokyo, JP

Introduction

One of the global goals for the sustainable development is to achieve carbon-neutral in near future, for limiting the temperature elevation to 2 degrees C. In the recent COP conference, reduction of the elevating limitation seems to be studied. In the carbon-neutral world, the transportation sector will require drop-in fuel to fly aircrafts and drive heavy-duty vehicles. They are fueled with drop-in fuel without any modification of their engine since drop-in fuel satisfy perfectly the quality standard of the petroleum transportation fuel [1-2]. As you know, drop-in fuel is produced by decompose the ester bond of the bio-synthesized triglycerides with the catalytic process into which the petroleum hydrotreatment technology is applied. Neste Oil's NExBTL and Eni/UOP's Ecofining™ are the representative catalytic process for the current production of drop-in fuel [3].

The reaction routes to decompose the triglycerides ester bond is mainly classified into the decarbonylation emitting CO and the direct hydrogenation by-producing much H_2O. Since drop-in fuel is produced through the latter route in the current catalytic process, a large amount of hydrogen is consumed wastefully. For the former route, the emitted CO would probably be converted into hydrogen according to water-gas shift reaction by consuming the by-produced H_2O. Moreover, the unsaturated fatty acid group in triglyceride is hydrogenated prior to the ester bond decomposition. It was reported that the hydrogen consumption of Ecofining™ process ranged from 1.5-3.8 wt% based on the triglyceride feedstock [4]. Also, the collocated facility to produce the required hydrogen is one of the reasons that drop-in fuel is inferior to petroleum transportation fuel on economic competitiveness. To make matters worse, source of the required hydrogen seems to be fossil fuel causing greenhouse gas emission [5]. This damages renewability of drop-in fuel.

For resolving the external addition of the fossil-derived hydrogen, our attention is focused on the hydrothermal processing to split triglyceride into fatty acid. The by-produced glycerol could fill the role of the hydrogen source if the hydrothermal condition is operated in the presence of the proper catalyst that is active in aqueous phase reforming. Aqueous phase reforming (APR) is the catalytic process to convert carbohydrates into the value-added chemicals [6]. Types of the major product varies according to chemical composition of the employed catalyst, and platinum is one of the metals suitable for hydrogen production [7]. There are some previous works verifying production of hydrogen by APR of glycerol [8-9]. Our recent work elucidated that rapeseed oil was converted into C_{17} hydrocarbons mainly under hydrothermal condition in the presence of platinum supported on alumina catalyst [10]. Also, we proved that hydrogen was produced from glycerol under the same operating condition. These results indicated that triglyceride could be deoxygenated with the renewable hydrogen without its external addition and production. On the other hand, the branched alkanes required for satisfying the quality standard of the transportation fuel was not detected in the crude product oil for the same work.

In the present research work, the protonated ZSM5 (H-ZSM5) zeolite was selected as the catalyst support, with expectation that its strong acidity accelerates the isomerization to produce the branched alkanes. The prepared catalyst, Pt supported on H-ZSM5, was employed for the test reaction to deoxygenate rapeseed oil, and the experimental data collected was in line with our expectation. To the best of our knowledge, the present work is the first successful trial to produce the branched alkanes from triglyceride by its hydrothermal deoxygenation, even though there are some research groups focusing on the same conversion pathways shown in Fig.1 [11-13].

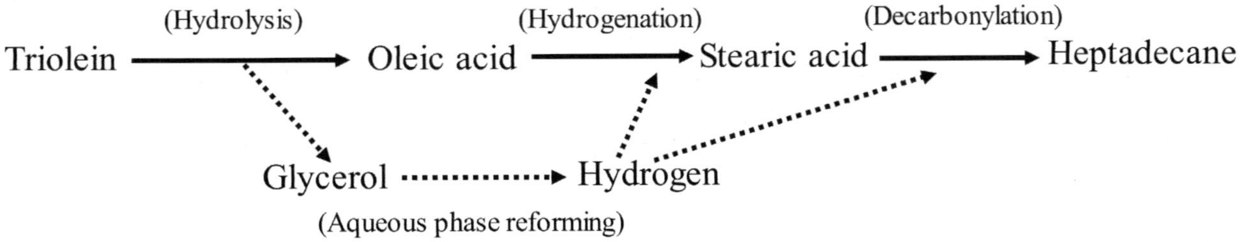

Figure 1. Simplified reaction pathways for the catalytic hydrothermal deoxygenation of triglyceride into hydrocarbon, exampled with triolein.

Experimental

Catalyst Preparation

H-ZSM5 was purchased from TOSOH (Japan), and was employed as the catalyst support after vacuum drying at 353 K. Properties of the purchased H-ZSM5 are opened by its manufacture, 40 of the Si/Al molar ratio, 330 m^2/g of the surface area and 1.3 mmol/g of the acidic site density measured by NH_3-TPD. Additionally, our present work elucidated that the acid site strength measured by Hammett indicator reagents (Ho) was more than -8.2. As the referential support to compare with H-ZSM5, alumina of gamma phase was prepared by calcining Boehmite at 873 K in air.

Assuming that the *in-situ* production of the hydrogen required for the triglyceride deoxygenation is crucial to its reaction rate, Pt was selected as the catalytically active metal. Primarily, hydrogen hexachloroplatinate in the aqueous solution was impregnated over the catalyst support. The impregnated amount corresponded to 4 wt% in metallic form. Then, the obtained catalyst precursor was calcined at 673 K in air for finishing the catalyst preparation. The calcination temperature was based on the research paper about the Pt-catalyzed hydrodeoxygenation of plant oil [14]. For the prepared catalyst, dispersion of the supported Pt particles was observed by using TEM.

Test Reaction

Rapeseed oil, purchased from a neighboring supermarket, was employed for the test reaction to investigate the hydrothermal triglyceride deoxygenation involving the *in-situ* production of the hydrogen required. The fatty acid composition of the rapeseed oil is elucidated in our previous research paper [15].

After 8ml of rapeseed oil, 4ml of distillated water and 0.5 g of the catalyst were put in the reaction vessel of autoclave batch reactor, headspace in the sealed reaction vessel was purged away with nitrogen gas. Then it pressurized the purged headspace to 4 MPa., and the electromagnetically induced stirring of the oil-water-catalyst mixture started with heating it up to the reaction temperature. The test reaction was carried out at 593 K for 1.5-4.5 h. Here, it should be noted that a volume of the autoclave reaction vessel was 100mL. The liquid product mixed with the used catalyst was withdrawn in the propanol solution from the reaction vessel, following the post emission of the headspace gas. The used catalyst was filtered out, and the filtrate formed into the single liquid phase was analyzed on GC-FID and GC-MS, for examining the products distribution.

Additionally, the APR reaction of the reagent glycerol was tested at 593 K for 1.5 h in order to estimate the *in-situ* production of the hydrogen required for the rapeseed oil deoxygenation. For the additional test reaction, 8 ml of 1-tetradecene was added into 4 ml of 10 wt% glycerol aqueous solution mixed with 0.5 g of the catalyst in the reaction vessel, assuming that all the hydrogen produced by the APR reaction is chemically accommodated into the 1-tetraddecene additive due to its double bond saturation. Based on the assumption, a yield of the hydrogen was calculated from the hydrogenation ratio of 1-tetradecene, by the theoretical production according to a reaction stoichiometry of the glycerol APR; $C_3H_5(OH)_3 + 3H_2O \rightarrow 7H_2 + 3CO_2$. For determining the hydrogenation ratio, ^{1}H-NMR spectrum was measured for the organic fraction collected after the APR reaction.

Results and Discussion

Hydrothermal deoxygenation catalyzed by Pt/H-ZSM5

Table 1 lists distribution of the products for the main test reaction to deoxygenate rapeseed oil under the hydrothermal condition. Since hydrolysis of rapeseed oil was perfected independent of the reaction time, the triglyceride conversion ratio was not listed in Table 1. The hydrolysis perfected at 593 K was consistent with the previous work by Sugami et al [16]. They reported that rapeseed oil was hydrolyzed perfectly at 593 K for 1 h. Here, it should be noted that the main test reaction operated for 3.0 h was repeated three times, and that the reproducible deviation was within 5 %. Also, the time dependency of the rapeseed oil deoxygenation is appreciably positive independent of type of the catalyst because the extended reaction time brought about a decrease in the fatty acids and an increase in the hydrocarbon. The small deviation and the positive time dependency indicated that the experimental result of the present work was worthy of the academic discussion.

Table 1. Products distribution for the catalytic hydrothermal deoxygenation of rapeseed oil.

Catalyst	Time [h]	Hydrocarbons [%] < C16	C17	C18	Total	Intermediates [%]	Fatty acids [%]
Pt/H-ZSM5	1.5	4.3	7.0	0.5	11.8	6.7	81.4
Pt/H-ZSM5	3.0	6.3	11.8	1.8	19.9	8.2	71.9
Pt/H-ZSM5	4.5	10.9	13.4	2.2	26.5	12.0	61.5
Pt/Al_2O_3	1.5	2.2	9.5	0.4	12.1	1.9	86.0
Pt/Al_2O_3	3.0	3.0	15.4	0.4	18.8	3.9	77.3
Pt/Al_2O_3	4.5	3.8	17.6	0.6	22.0	5.3	72.7

Pt/H-ZSM5 resulted in the lower yield of fatty acids than Pt/Al_2O_3. The similar result was reported in the research paper describing hydrodeoxygenation of methyl laurate over the beta-zeolite supported Ni-Mo catalyst [17]: As compared to the pure-silica zeolite containing none of the acid site, the protonated beta zeolite brought about the fast conversion of methyl laurate. The research paper concluded that the acid site functioned as the adsorption one for carbonyl group in the molecule of methyl laurate. Focusing on the hydrocarbon yield, there was none of the appreciable difference between Pt/H-ZSM5 and Pt/Al_2O_3 in the present research work since a yield of the deoxygenation intermediates such as stearyl aldehyde was larger for Pt/H-ZSM5. It was possible that dispersion of the Pt particles over the catalyst support affected the deoxygenation rate. As shown in Figure 2, the supported Pt particles was more dispersive over Al_2O_3 than over H-Y zeolite (550m^2/g) even though Pt/Al_2O_3 was approximately equivalent to Pt/H-Y in the size of the supported Pt particles. Assuming that there is no appreciable difference in the dispersion of the Pt particles between the zeolitic materials, the Pt particles were less dispersive than over H-ZSM5. Here, it should be noted that we are now preparing TEM observation of Pt/H-ZSM5.

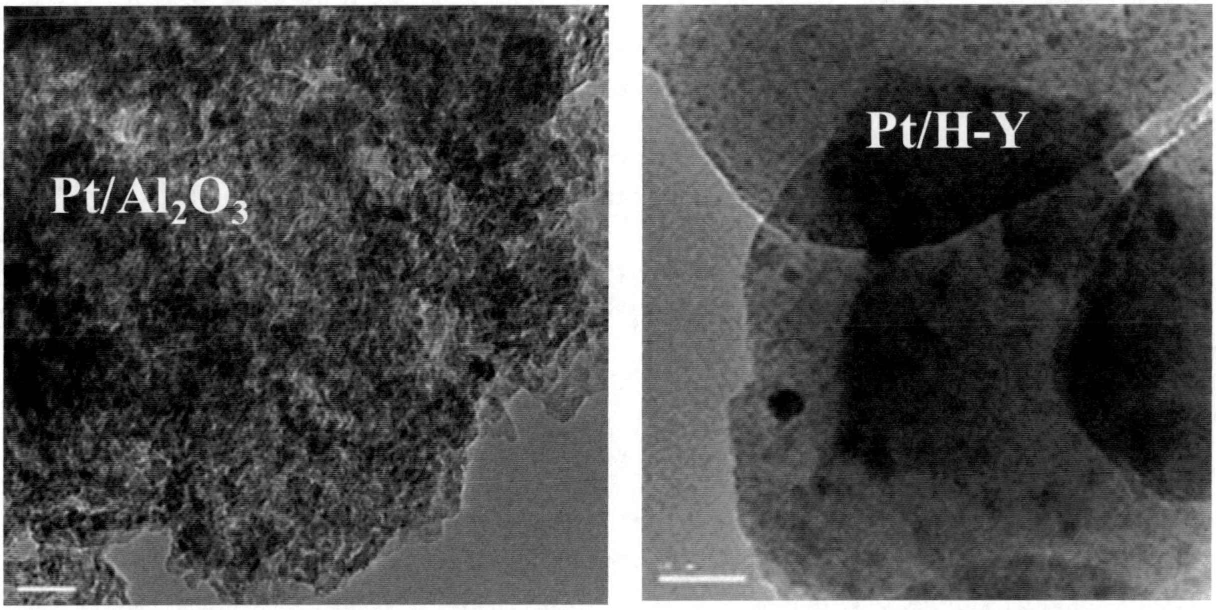

Figure 2. TEM images of Pt/Al_2O_3, as compared to Pt/H-Y.

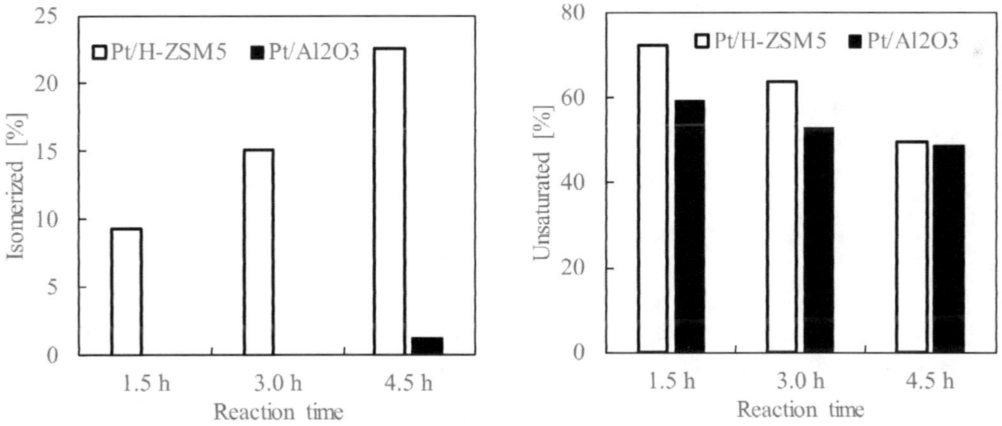

Figure 3. Selectivity to saturated isomers and unsaturated for the C17 hydrocarbons produced.

Classifying the produced hydrocarbons based on their carbon number, the C17 was yielded much more than the C18. It was certain that the major deoxygenation route was the decarbonylation, independent of type of the catalyst, taking the fatty acid composition of the triglycerides employed for the present research work into consideration. On the other hand, it was interesting that the selectivity to C18 was larger for Pt/H-ZSM5 than Pt/Al_2O_3. This seemed to be one of the effects of the acidic support on the catalytic deoxygenation of triglyceride. The same result was described in the research paper mentioned above [17]. Another effect of the H-ZSM5 support, as compared to the Al_2O_3, was the enhanced production of < C16 fraction corresponding the hydrocarbons whose carbon number were less than 16. This fraction included the cracked fragments like C 11 or C12 hydrocarbons.

For the C17 hydrocarbons of the major fraction, the selectivity to the isomer and the unsaturated are drawn in Figures 3. The isomer selectivity increased from 9.3 % to 22.6 % by extending the reaction time for Pt/H-ZSM5, whereas it was hard for Pt/Al_2O_3 to isomerize the produced hydrocarbon. As a result, the liquid product collected after the test reaction catalyzed by Pt/H-ZSM5 was coagulated at the lower temperature. The appreciable isomer selectivity, as well as the hydrocarbon post-cracking described above, was due to the strong acidity of the H-ZSM5 support. Some researchers elucidated that C-C bond scission of the multi-branched alkane occurs under the alkane isomerization condition [18]. Also, the selectivity to the unsaturated seemed to be higher for Pt/H-ZSM5 than for Pt/Al_2O_3. It was deduced that the hydrocarbon post-cracking withdrew atomic hydrogen from the uncracked hydrocarbons. Moreover, there was a possibility that the higher selectivity to the unsaturated was due to the inhibition of the *in-situ* hydrogen production by APR of the by-produced glycerol. It was reported that the acidic catalyst support causes reduction of the selectivity to hydrogen production by the APR reaction due to the enhanced production of alkanes [6-7].

Glycerol APR tested separably

While 1-tetradecene added as the hydrogen scavenger was converted into only tetradecane for the separable APR reaction catalyzed Pt/Al_2O_3, the uncountable hydrocarbons were produced from 1-tetradecene for the Pt/H-ZSM5 catalyzed reaction. As a result, the hydrogenation ratio of 1-tetradecene was determined from ^{1}H-NMR spectrum of the organic phase collected after the reaction. The relative intensity of the chemical shift deriving from the double bond hydrogen was reduced by the additional test reaction independent of the type of the catalyst. Thus, it was evident that the double bond in 1-tetradecene was saturated with the hydrogen produced from the glycerol APR catalyzed not only by Pt/Al_2O_3 by but also Pt/H-ZSM5. The hydrogenation yield estimated from the variation in the ^{1}H-NMR spectrum was 98 % and 91 % for Pt/Al_2O_3 and Pt/H-ZSM5, respectively. However, the estimated hydrogenation yields seemed to include some errors because performance of the employed NMR apparatus was quite low.

From the experimental results and the concerning discussion, we would like to modify the catalyst by highly dispersing the Pt fine particles over the acidic support for not only enhancing the total hydrocarbon yield but also reducing formation of the unsaturated hydrocarbon, in our future work.

Conclusions

For studying economical and eco-friendly production of drop-in fuel, deoxygenation of rapeseed oil with the renewable hydrogen produced *in-situ* was tested under the hydrothermal condition in the presence of the supported Pt catalyst. For the test reaction catalyzed by Pt/ZSM-5 at 593 K, the hydrocarbon yield was raised from 11.8 % to 26.5 % by extending the reaction time from 1.5 h to 4.5 h. Although there was no appreciable difference in the hydrocarbon yield and the major deoxygenation route between Pt/ZSM-5 and Pt/Al_2O_3, it was evident that the former catalyst resulted in the higher selectivity to the isomer and the cracked hydrocarbon fragment. This was probably due to the strong acidity of H-ZSM5 support. Moreover, Pt/ZSM-5 resulted in the higher selectivity to the unsaturated hydrocarbon. These results indicated the further modification of the catalyst is required for our research work.

Acknowledgment

We would like to note that TEM observation for the prepared catalyst was performed in Interdisciplinary Research Center for Nano Science and Technology of Tokyo City University.

References

1 A. Galadima, O. Muruza "Catalytic upgrading of vegetable oil into jet fuels range hydrocarbons using heterogeneous catalysts: a review", *J. Ind. Eng. Chem*. 29, 12–23, 2015.
2 I.H. Choi, K.R. Hwang, J.S. Han, K.H. Lee, J.S. Yun, J.S. Lee, "The direct production of jet-fuel from non-edible oil in a single-step process", *Fuel*, 158, 98-104, 2015.
3 C. Perego, D. Bianchi, "Biomass upgrading through acid-base catalysis", *Chem. Eng. J.*, 161, 314-322, 2010.
4 T.N. Kalnes, K.P. Koers, T. Marker, D.R. Shonnard, "A technoeconomic and environmental life cycle comparison of green diesel to biodiesel and syndiesel", *Environ. Prog. Sustain. Energy*, 28 111-120, 2009.
5 R. D. D. Putra, H. L. Trajano, S. Liu, H. Lee, K. Smith, C. s. Kim, "In-situ glycerol aqueous phase reforming and phenol hydrogenation over Raney Ni", *Chem. Eng. J.*, 350, 181-191, 2018.
6 G.W. Huber, J.A. Dumesic, "An overview of aqueous-phase catalytic process for production of hydrogen and alkanes in a biorefinery", *Catal. Today*, 111, 119-132, 2006.
7 R. R. Davda, J. W. Shabaker, G. W. Huber, R. D. Cortright, J. A. Dumesic,"Aqueous phase reforming of ethylene glycol on silica-supported metal catalysts", *Appl. Catal. B: Environ.*, 43, 13-26, 2003.
8 D.O. Ozgur, B.Z. Uysal, "Hydrogen production by aqueous phase catalytic reforming of glycerin", *Biomass Bioenergy*, 35, 822-826, 2011.
9 A.V.H. Soares, G. Perez, F.B. Passos, "Alumina supported bimetallic Pt-Fe catalysts applied to glycerol hydrogenolysis and aqueous phase reforming", *Appl. Catal. B Environ.*, 185, 77-87, 2016.
10 M. Kouzu, M. Kojima, K. Mori, S. Yamanaka, "Catalytic deoxygenation of triglyceride into drop-in fuel under hydrothermal condition with the help of in-situ hydrogen production by APR of glycerol by-produced", *Fuel Process. Technol.*, 217, 106831, 2021.
11 S.A.W. Hollak, M.A. Ariens, K.P. de Jong, D.S. van Es, "Hydrothermal deoxygenation of triglycerides over Pd/C aided by in situ hydrogen production from glycerol reforming", *ChemSusChem*, 7, 1057-1060, 2014.
12 D.R. Vardon, B.K. Sharma, H. Jaramillo, D. Kim, J.K. Choe, P.N. Ciesielski, T. J. Strathmann, "Hydrothermal catalytic processing of saturated and unsaturated fatty acids to hydrocarbons with glycerol for in-situ hydrogen production", *Green Chem.*, 16, 1507-1519, 2014.
13 V. Dominguez-Barroso, C. Herrera, M.A. Larrubia, L.J. Alemany, "Coupling of glycerol-APR and in-situ hydrodeoxygenation of fatty acid to produce hydrocarbons", *Fuel Process. Technol.*, 190, 21-28, 2019.
14 A.T. Madson, E.H. Ahmed, C.V. Christensen, R. Fehrmann, A. Riisager, "Hydrodeoxygenation of waste fat for biodiesel production: Study on model feed with Pt/alumina catalyst", *Fuel*, 90, 3433-3438, 2011.
15 M. Kouzu, A. Kajita, A. Fujimori, "Catalytic activity of calcined scallop shell for rapeseed oil transesterification to produce biodiesel", *Fuel*, 182, 220-226, 2016.

16 Y. Sugami, E. Minami, S. Saka, "Renewable diesel production from rapeseed oil with hydrothermal hydrogenation and subsequent deoxygenation", *Fuel*, 166, 376-381, 2016.
17 H. Imai, M. Abe, K. Terasaka, H. Yamazaki, R. Osuga, J.N. Kondo, T. Yokoi, "Hydroconversion of methyl laurate over beta-zeolite-supported NiMo-catalysts: effect of acid and base treatments of beta zeolite", *Fuel Process. Technol.*, 197, 106182, 2020.
18 M. Guisenet, F. Alvarez, G. Giannetto, G. Perot, "Hydroisomerization and hydrocracking of n-heptane on PtH zeolites. Effect of the porosity and of the distribution of metallic acid sites", *Catal. Today*, 1, 415-533, 1987.

624715 Ash Adhesion on Inconel Mesh during Bamboo Powder Combustion

Abstract:

In this study, mass transfer of ash during bamboo powder combustion on a flat flame is investigated experimentally and numerically. The bamboo powder combustion consisted of volatiles emission, volatiles combustion, char formation, char combustion, and finally ash formation. The mass of ash formed was about 3 wt% that of raw bamboo powder. The mass of ash decreased with the equivalence ratio *F* for flat flame combustion. Part of the ash fused and then adhered to the Inconel mesh for wrapping of bamboo powder as bottom ash, flowing through the mesh. The mass adhered to the Inconel mesh depended strongly on the equivalence ratio *F*. The fusion tendency of the atmospheric oxidization ash agreed qualitatively with Liu's standard bamboo ash fusion test. On the other hand, as ash fusion started locally in this work, it suggests that the concentration of metal oxides of the atmospheric oxidization ash was not uniform. For the atmospheric oxidization ash at 600°C, $K_2Si_2O_5$, $KAlSiO_4$, $Na_2CaP_2O_7$, Mg_2SiO_4, K_2SO_4, and Fe_2O_3 were predicted to be the major eutectic materials. At 1120°C, only Fe_2O_3 remained and the liquid components accounted for about 95%. On the other hand, the major eutectic materials for the combustion ashes of $\Phi = 0.85$ and $\Phi = 1.0$ at 1260°C were MgO and Fe_2O_3, respectively. Therefore, the predicted eutectic materials of the combustion ash were changed appropriately in comparison with those for the atmospheric oxidization ash because of the fusion and disappearance of some metal oxide components during combustion.

623495 Delineating and Controlling the Electrode-Electrolyte Interfacial Reactions in High Energy Density Batteries

Abstract:

Rapid increase in global energy use and growing environmental concerns have prompted the development of clean, sustainable, alternative energy technologies. Renewable energy sources like solar and wind are a promising solution, but electrical energy storage (EES) is critical to efficiently utilize electricity produced from renewable sources as they are intermittent. EES is also the only viable near-term option for electrification of transportation sector. Rechargeable batteries are prime candidates for EES, but their widespread adoption for electric vehicles and grid electricity storage requires optimization of cost, cycle life, safety, energy density, power density, and environmental impact, all of which are directly linked to severe component materials challenges and the interfacial reactions between them. This presentation will focus on delineating and controlling the electrode-electrolyte interfacial reactions and consequent crossover of chemical species in high energy density batteries.

With an aim to develop sustainable battery technologies at an affordable cost with high energy density and free from supply chain issues, the presentation will focus on interfacial reactions in two battery systems: (i) lithium-ion batteries with low- or cobalt-free, high-nickel cathodes and graphite or lithium-metal anodes and (ii) lithium-sulfur batteries with sulfur cathode and lithium-metal anode or anode-free cells with Li_2S cathode. The high-nickel cathodes are synthesized with a coprecipitation of transition-metal hydroxide precursors followed by calcining with LiOH under controlled temperature and oxygen flow in order to obtain commercially-relevant particle size and morphologies. Practically-relevant pouch lithium-ion cells or lithium-sulfur cells fabricated with the cathodes and anodes are disassembled and characterized after hundreds or thousands of cycles with advanced characterization techniques, such as X-ray photoelectron spectroscopy, time-of-flight secondary ion mass spectrometry, and high resolution transmission electron microscopy, to track the anode-electrolyte interfacial reactions, cathode-electrolyte interfacial reactions, as well as the crossover of chemical species between the two electrodes through the electrolyte/separator during cycling. Based on the data, the impact of interfacial reactions and crossover of chemical species between the two electrodes on cell performance is established. The understanding gained is utilized to condition the electrode surfaces or modify electrolyte compositions in order to minimize undesirable interfacial reactions and enhance the cycle life of lithium-ion and lithium-sulfur cells.

623899 Investigating Li-Ion Behavior in ALD Coated NMC Cathode Materials Via Molecular Dynamics

Abstract:

High nickel content $LiNi_{1-y-z}Mn_yCo_zO_2$ (NMC) has emerged as a cathode capable of delivering high capacities in Li-ion batteries. In particular, high-nickel NMCs are being used commercially in electric vehicles (EVs) and are projected to comprise over a third of EV cathode chemistries by 2026.[1] However, high-nickel cathodes experience structural instability with the charge/discharge cycling of the battery, which leads to large capacity fading, voltage decay, and low rate capability of the overall battery. Atomic layer deposited (ALD) lithium-containing nano-coatings of NMC cathodes have been shown to help stabilize the transition metal sites while still allowing the movement of Li-ions across the coating, leading to greater capacity retention of the Li-ion battery.[2]

Our prior experimental work has demonstrated that alumina particle ALD produces a less than two nanometer, mixed $Al_2O_3 - LiAlO_2$ film that preferentially covers the transition metals of the cathode powder while leaving lithium relatively uncovered and presumably available for transport. Expanding upon this work, molecular dynamics simulations were used to investigate the behavior of surface lithium of uncoated cathodes and ALD coated cathodes. The Interface Force Field in the 12-6 Lennard Jones form was used to model the pairwise interactions.[3] Both amorphous and crystalline films were simulated at 393 K, with the crystalline film being used as a comparison to the amorphous, ALD-like one. The thickness and uniformity of the ALD coating affects the surface lithium movement. For example, lithium is able to completely travel through one simulated Al_2O_3 layer, while it cannot for two or more simulated Al_2O_3 layers. Further, the energy of the interface between the cathode and coating continues to decrease (i.e., become more favorable) as the simulated coating thickness increases until the energy plateaus off and is about the same for all later coating thicknesses. At this energy plateau, the representative energy of deposition for Al_2O_3 on $LiCoO_2$ is -1760 mJ/m^2 for the amorphous film and -2180 mJ/m^2 for the crystalline film.

By 2030, electric vehicles are projected to comprise almost 30% of new car sales globally. A protective nano-coating on the cathode that preserves battery capacity should help hasten the global transition from combustion vehicles to EVs.

References:

(1) Frith, James. Battery Prices and Manufactured Costs (Highlights from Bloomberg NEF EVO Report 2020), 2021.

(2) Hoskins, A. L.; McNeary, W. W.; Millican, S. L.; Gossett, T. A.; Lai, A.; Gao, Y.; Liang, X.; Musgrave, C. B.; Weimer, A. W. Nonuniform Growth of Sub-2 Nanometer Atomic Layer Deposited Alumina Films on Lithium Nickel Manganese Cobalt Oxide Cathode Battery Materials. *ACS Appl. Nano Mater.* **2019**, *2* (11), 6989–6997. https://doi.org/10.1021/acsanm.9b01490.

(3) Dhakane, A.; Varshney, V.; Liu, J.; Heinz, H.; Jain, A. Molecular Dynamics Simulations of Separator-Cathode Interfacial Thermal Transport in a Li-Ion Cell. *Surf. Interfaces* **2020**, *21*, 100674. https://doi.org/10.1016/j.surfin.2020.100674.

632863 Reserve Lithium-Ion Batteries for Lithium-Ion Free Cathodes

Abstract:

Lithium-ion Batteries' (LIBs) influence on every aspect of human lives has been unimaginable. One sector, which has received a tremendous boost because of LIBs is transportation. Electric vehicle (EV) sales have increased almost 20 folds from 2011 until now. This has caused a huge increase in the demand for LIBs across the world. Presently, there is heavy dependence on cathodes like $LiFePO_4$, $LiNiO_2$, $LiCoO_2$, and $LiNi_xMn_yCo_zO_2$ to meet energy expectations. Compounds without lithium in their lattice rarely have been used as cathodes either due to inability for reverse reactions or require lithium electrode, which causes lithium dendrites on continuous cycling. This has resulted in fewer efforts towards the development of such lithium-free cathodes. Some of these materials are ZnO, TiO_2, Cr_3O_4, MnO_2, V_2O_5, etc. Although these materials possess high theoretical capacities, many of them have a low potential for faradic reactions and have the risk of reducing to metal states during the electrochemical reactions. Hence, for such materials to commercialize, few measures have been taken namely, pre-lithiation of the cathode, film-forming additives, lithium additives, blended cathodes, etc. Some strategies like pre-lithiation are not practical and feasible, even though they help in overcoming solid electrolyte layer interface (SEI). Other strategies alter the cell voltage characteristics due to the presence of voids in the electrode.

We present a pioneering configuration called reserve lithium-ion batteries for in situ lithiation of the lithium-free electrodes with the use of lithium-electrode as a reservoir. For the initial formation of the SEI and lithiation of the MCMB electrode, a reservoir lithium electrode was brought into the cell circuitry. V_2O_5 cathode with meso carbon microbeads (MCMB) was used to fabricate high specific capacity and energy-dense reserved lithium-ion batteries (RLIBs). The RLIB system delivered a high charge specific capacity of 245 mAh g^{-1} at 0.2C rate. The tri-electrode system provided an initial high energy density of >880 Wh kg^{-1}. A parallel comparison of electrochemical performance between V_2O_5 half-cell, V_2O_5-MCMB full-cell, and V_2O_5-MCMB RLIB was conducted and observed. There is a distinct resemblance in the performance of RLIB performance with half-cell profile and in terms of capacity versus cycle number. The pouch-cell configurations of this work yielded a high energy-dense system, which cycled stably over more than 200 cycles at 0.5C rate. Using this innovative concept of reservoir charging would be of great importance for applications in various sectors of the society viz., space exploration, UPS, healthcare, EVs, grid storage, etc.

629713 Toward Rapid Chemical Process-Based Life Cycle Inventory Data Generation: From Modeling Frameworks to Simulation to Machine Learning

Abstract:

Commercial LCI databases account for less than 1% of the more than 80,000 chemicals available in the US market alone. This large gap signals the need to estimate or directly collect LCI data for the large majority of chemicals that are currently not represented. This abstract covers findings from multiple projects with the overall aim of providing LCA practitioners with data sets and tools that will enable accurate estimation of chemical LCI data based on chemical engineering design principles. The work includes detailed comparison of existing estimation methods; application of those methods to generate a large, harmonized LCI data set for organic chemicals; and application of statistical and machine learning methods to build a model for LCI data generation.

In the absence of actual data from production plants or industrial process information, material input for the LCI is usually estimated based on balanced stoichiometric reactions. These also provide information on the output from the processes in form of co-products and emissions. Additional process related details such as temperature of reaction, separations operations, and properties of chemicals involved can and should be used for estimating the process energy usage. Chemical process data can often be found in form of laboratory scale experiments in patent literature. Scale-up considerations when using such data have been explored by Piccinno et al., while Smith et al. presented a methodology combining process simulation and synthetic organic chemical manufacturing industry (SOCMI) emission factors for generating detailed chemical LCIs. Others have proposed methodologies for specific sectors such as bio-based chemicals, fine and specialty chemicals, and pharmaceuticals to address the scarcity of LCIs in these sectors. Other methods including proxy LCIs, similarity-based approaches, and molecular structure-based models (MSM) capitalize on the existing chemical LCIs in different databases to estimate missing LCI data for other chemicals. While proxy method assumes that similarity in processes would imply similar LCI input based on expert judgment, similarity-based approach presented by Hou et al. applied computational data analysis to accurately estimate missing LCI data when less than 5% data is missing in a process. Molecular structure-based models developed by Wernet et al. and Song et al. estimate life-cycle impact assessment results, rather than inventory data.

All such methodologies for building chemical LCIs are based on varying data availability, and were explored in terms of their accuracy against actual plant data. The global warming results for "shortcut" methods such as stoichiometry and proxy LCIs were 35-50% lower than that of LCI based on actual plant data. Whereas the process-based methods performed much better with the difference of 1-10% compared to plant LCI in results for the greenhouse gas impact category. The energy data obtained from process simulation was closest to the actual plant data, which emphasizes the need for mechanistic models that capture the complexities of actual chemical plant operation.

Measuring energy consumption for each process in a chemical plant is not always possible due to complex plant geometries, with return flows, batch sequencing, split process streams and utility consumptions are available only at plant level. Jiménez-González et al., in their seminal work use traditional chemical process design principles and rules of thumb in developing a methodology for gate-to-gate LCI data for chemicals. This methodology is later used by Kim et al. to estimate gate-to-gate manufacturing energy data for 43 organic and 43 inorganic chemicals. Some of these results are still used in the ecoinvent database. Other disaggregated LCI datasets in the ecoinvent database use estimations based on design calculations or energy for similar processes, other literature sources, and in rare cases data based on industry sources. Several other disaggregated chemical datasets use an average value for heat and electricity based on energy consumption for manufacturing chemical products at the GENDORF Chemical Park. For a multi-product batch plant, a bottom-up approach is more appropriate, wherein process energy is calculated based on a model for each unit operation and aggregated to determine total energy.

In our work we demonstrate a streamlined process simulation-based methodology to estimate energy consumption in chemical manufacturing when the data availability for the process is limited. The methodology is applied to 151 different chemical processes using Aspen Plus to estimate their gate-to-gate process energy use, representing the largest such simulation-based LCI dataset to date. Further, pinch analysis is used for process heat integration instead of rules of thumb. Results for different utility types for each of the chemical processes further enhances the information provided by the LCI data.

The total heating requirement for chemicals assessed ranges from 0.1-24 MJ/kg with an average of 3.1 MJ/kg product, while the average cooling requirement before heat integration is 4.5 MJ/kg. More than half of the total energy requirement comes from the separation section. Steam is the most used hot utility in the chemical industry which reflects in the simulation results, accounting for 70% of the heating requirements, while air and cooling water accounts for 40% of the cold utilities used. The results are closely correlated with the subset ecoinvent and other existing chemicals LCI data that is based on process modeling, but do not match with existing data that have used simpler estimation methods. This data set represents the largest, harmonized set of chemicals process energy data. In the final project, we utilized this data set to build predictive models.

Use of statistical and machine learning methods for prediction of life-cycle inventory data has so far been limited due to unavailability of chemical production process data required for calibration/training but is beginning to be explored. Classification and regression tree (CART), and Random Forest (RF) methods have been used to estimate emissions from (rather than inputs to) chemical processes. The symmetric mean average percentage error for the CART and RF models estimating process emissions on out of sample data were 80% and 92%, respectively. Pereira et al. used classification and regression trees to predict the range of steam consumption in batch chemical processes using predictor variables such as reaction type, reaction time, and presence of distillation operation in the process. The study also presents models using probability density functions (PDFs) fitted on 250 data points obtained from batch chemical processes, classified based on their reaction types. The accuracy of these PDF models evaluated on an additional industrial dataset with 17 data points was 40% while for classification trees it ranged from 35% to 80% for five different models with varying process inputs as classification rules.

These prior studies address specific LCI data such as emissions and energy use and are limited by its application to certain unit operations such as distillation, and chemical reaction. There is therefore a compelling need for computational models that use process parameters and chemical properties that have physical significance in the manufacturing process, rather using molecular descriptors of only the target molecule to predict the total energy use in chemical manufacturing, particularly as 20 of the top 26 chemicals in energy consumption and GHG emissions are organic chemicals manufactured with continuous processes.

Using our simulation-based dataset, we selected a mix of predictor variables including process parameters, thermodynamic properties of reactants and products, and reaction types, to develop multiple linear regression (MLR) and artificial neural network (ANN) models for the predictive analysis for both heating and cooling. Examining model performance over ten different runs, the mean R^2 for *T_heat* and *T_cool* was 0.65 and 0.6, respectively whereas the mean absolute percent error (MAPE) was 0.39 and 0.56, demonstrating that the selected models perform equally well when trained and tested on different subsets of the data. Some models performed with R^2 values up to 0.82. Further investigation of test data from cross-validation runs showed that one-third of the predicted test data for *T_heat* had a percentage error of less than 25%. Processes with high heating requirements such as furan (23.7 MJ/kg) and methyl methacrylate (12.7 MJ/kg) due to higher energy in the separation section were underestimated by 10 to 30%. Different chemical reaction types were also investigated and incorporated into the predictive modeling.

Estimating total heating and cooling requirements based on physically relevant predictors with an error of less than 50% for more than 70% of the predictions is a significant improvement for current LCIs that are using single value for all organic processes based on proxy or historical plant data. A major challenge remains to test the accuracy of these simulation-based models against industrial data on a wide range of chemicals to train these models to be truly representative of real-world conditions.

We are at a turning point in LCA where advanced data science and machine learning approaches are being introduced, allowing us to model complex processes and products with high resolution. These new approaches will greatly benefit from consistent, abundant LCI data that accurately represent chemical manufacturing practices. Our efforts resulted in a model that is an improvement on existing ones, but did not reveal a truly well-performing set of predictive parameters or modeling approach. As a community, we should discuss opportunities for new data streams, data fusion techniques, and the relative benefits of mechanistic versus statistical approaches.

632241 Planetary Boundaries, Ecosystem Services, and Life Cycle Assessment – Novel Insight By a Case Study and Computational Framework

Abstract:

Recent extensions of life cycle assessment are including the role of ecosystems and nature's carrying capacity. Such extensions are needed to ensure that decisions toward sustainability respect nature's limits and do not cause unintended harm by ecosystem degradation. Such approaches are essential to meet the basic requirement of sustainability, which is to keep human activities within the earth's carrying capacity.

In this work, we compare two such extensions: planetary-boundary-based LCA (PB-LCA) [1,2] and techno-ecological synergy in LCA (TES-LCA) [3,4]. PB-LCA is based on the planetary boundary framework developed by Rockström and colleagues. PB framework defined nine key Earth System processes and identified quantitative 'Planetary boundaries' which delimit a "safe operating space" for humanity to stay within [5]. Approaches have been developed by including the PB-framework in LCA [1]. The original PB-LCA method downscales 9 planetary boundaries to smaller scales and allows to find out whether the environmental impact of an activity is sufficiently low for it to be considered as environmentally sustainable. However, according to the first planetary boundary paper, downscaling boundaries is not recommended [6]. TES-LCA expands the steps in conventional LCA to incorporate the demand and supply of ecosystem goods and services at multiple spatial scales. This enables the calculation of local and absolute environmental sustainability metrics and helps identify improvement opportunities by protecting and restoring ecological systems.

In comparing the PB-LCA and TES-LCA frameworks, we find that both frameworks can evaluate how the assessed activity contributes to occupying the Earth's carrying capacity which in line with the idea of absolute sustainability [2]. These two methods identify potential hotspots in the life cycle, quantify spatial specific supply, demand and characterization factor which break the limitation of geographical resolution. Relationships between 9 planetary boundaries and 24 ecosystem services are studied in this work. Multiple ecosystem services could contribute to one planetary boundary. In general, there are 3 main differences between PB and ES framework: scope, categories included in the framework and the way to define the threshold. The scope of PB framework is planetary while for ES framework it's multiscale such as local and serviceshed scale. For the categories, each of the 9 planetary boundaries is related to an ecosystem service such as climate change boundary responses to the carbon sequestration service. However, not every ecosystem service has a corresponding planetary boundary. For example, pollination, air and water quality regulation, and cultural services do not have a direct boundary. The last point is how these two frameworks define the limit. In PB-LCA, the PB is directly downscaled to process-level, but in TES-LCA, process-level supply consists of local supply and allocation of serviceshed supply. Recently, works have been done on localized boundaries [7] which is similar to the local supply in TES-LCA. Comprehensive comparisons of such concepts, terminologies, and governing characterization factors have been done for these two frameworks.

Aiming at accounting for absolute sustainability and intuitive comparison, these two methods are demonstrated based on comparing the life cycles of products from biomass versus conventional sources. It involves multiple planetary boundaries like climate change and global freshwater use boundaries; multiple ecosystem services including carbon sequestration, water provisioning, etc. Different allocation principles like demand-based, GDP-based are used and compared for downscaling boundary and ecosystem services allocation. Compared with PB-LCA, TES-LCA not only downscales ES of the serviceshed which the process located in, but also takes ES's ownership and finer local-level ES into account. To enable practical use of these methods, computer implementation is developed and applied with an emphasis on climate change, water provisioning, etc. This computational framework accounts for spatial variation, aggregates geographically resolved environmental impact from process-level to larger scales like state-level, and global-level and it could be used for both PB-LCA and TES-LCA. A general computer framework is coded in python for further development into a user-friendly software tool.

Based on this work, we find that there are three main differences between planetary boundary and ecosystem service framework, however despite these differences, PB-LCA and TES-LCA show high similarity and can converge to one general framework under specific conditions. We show how this insight can be sued to develop inventory datasets and tools for practical application of these approaches.

1. Ryberg, Morten W., et al. "Development of a life-cycle impact assessment methodology linked to the Planetary Boundaries framework." Ecological Indicators 88 (2018): 250-262.

2. Bjørn, Anders, et al. "A proposal to measure absolute environmental sustainability in life cycle assessment." Ecological Indicators 63 (2016): 1-13.

3. Liu, Xinyu, and Bhavik R. Bakshi. "Ecosystem services in life cycle assessment while encouraging techno-ecological synergies." Journal of Industrial Ecology 23.2 (2019): 347-360.

4. Bakshi, Bhavik R., Guy Ziv, and Michael D. Lepech. "Techno-ecological synergy: A framework for sustainable engineering." Environmental science & technology 49.3 (2015): 1752-1760.

5. Rockström, Johan, et al. "A safe operating space for humanity." nature 461.7263 (2009): 472-475.

6. Rockström, Johan, et al. "Planetary boundaries: exploring the safe operating space for humanity." Ecology and society 14.2 (2009).

7. Bjørn, Anders, et al. "Life cycle assessment applying planetary and regional boundaries to the process level: a model case study." The International Journal of Life Cycle Assessment 25.11 (2020): 2241-2254.

628587 Estimating the Greenhouse Gas Emissions Payback Period of Planned Offshore Wind Energy Using Multiregional, Environmentally Extended Input-Output Analysis

Abstract:

There is a substantial projected increase in energy generation from offshore wind farms in the US over the next three decades due to an increase in legislative commitments and funding from federal and state governments. Achieving ambitious offshore wind targets will be dependent on the expansion of domestic manufacturing and significant transfers of knowledge from world leaders in this technology. Offshore wind energy will provide GHG emissions-free energy during the use phase of the farm, but project construction is associated with significant environmental impacts that are less commonly acknowledged, which offset the touted benefits to a degree. Furthermore, offshore wind energy will be a competing technology for advanced energy materials. Developing offshore wind as a reliable domestic energy source demands a multiregional impact assessment of the economic and environmental spill-over effects of constructing offshore wind farms in major lakefronts and coastal regions. Despite the current lack of commercially operating offshore wind farms in the US, seven states have announced cumulative capacity commitments of over 28 GW by 2035. In this study, the spatial economic impacts of planned projects are estimated by combining the NREL Jobs and Economic Development Impact (JEDI) model for offshore wind energy with a multiregional input-output (MRIO) model of the United States built with the Virtual Industrial Ecology Laboratory (IE Lab). We also use a newly developed multiregional GHG emissions dataset for the US IE Lab to estimate the supply chain emissions of constructing and installing the offshore wind projects. These emissions estimates provide a more complete understanding of the time it will take for proposed cleaner energy projects to offset the environmental impacts of their manufacture and installation compared to those from energy generated from fossil fuel combustion. The multiregional framework will also be used to identify which states may see the largest spill-over effects in terms of emissions generation and economic activity required to support offshore wind projects.

630587 A Hierarchical Techno-Ecological Decision Procedure for Sustainable Design

Abstract:

In 1985, J. M. Douglas published "A Hierarchical Decision Procedure for Process Synthesis". This paper outlined tiers for decision-making in chemical manufacturing processes with a conceptual design approach aimed to reduce backtracking for project decisions. In practice, the idea is to generate and filter design alternatives at appropriate stages, adding details to the process flow sheet with each phase. This approach is widely adopted in chemical engineering curriculum and design software. With rapid growth of sustainable engineering and new approaches to process synthesis, there is a need for an updated hierarchical approach to incorporate more current sustainable design research. Such a hierarchy can be applied in many ways including determining the proper sustainability assessment methods to conduct, ensuring innovative solutions outside of the technological system boundary are considered, and assisting in the navigation of when to consider certain details or higher resolutions of variables.

Previous work on techno-ecological design has led to innovative approaches for manufacturing sites, supply chains, academic campuses, and others. These approaches include ecosystems and their services within design frameworks, leading to innovative solutions that can reduce environmental and health impacts and, in some cases, costs. Other parties, such as the Intergovernmental Panel on Climate Change, have also identified nature as a significant role in mitigating climate impacts and increasing community resiliency based on the many co-benefits provided by healthy ecosystems. In the pursuit of sustainability, this new hierarchical approach considers techno-ecological systems. Further, the approach is guided by six previously identified necessary (but not sufficient) considerations for claiming sustainability: impacts across multiple spatial scales, temporal changes, cross-disciplinary effects, interactions between multiple stocks and flows, ecosystem demands, and the capacity for ecosystems to supply demanded inputs. This approach is constructed based on previous experience, literature reviews, and within techno-ecological sustainable design.

This work provides both a new conceptual approach and an example toy problem that highlights potential benefits and insights for application to sustainable design projects. The approach presents hierarchical levels of analysis for both technological process design and the surrounding socio-ecological landscape. Each level considers different scales and resolutions of time and space along with decisions around choosing available technological and ecological solutions, or unit operations. One of the main benefits of utilizing this approach is to simplify quantitative analysis and limit decision variables and equations within computable limits. One challenge in designing techno-ecological systems is the inclusion of spatial variables and boundaries that landscapes inherently have. Further, the temporal dynamics of technological and ecological systems are drastically different where ecosystems take years to develop and mature while technological systems aim to control and adjust systems based on much finer temporal resolutions. This hierarchical techno-ecological decision procedure reveals a process for generating new techno-ecological design solutions and assists in limiting mathematical complexities and program size for optimization, or other quantitative design methods, based on its multi-phase approach. The goal of this concept is to bridge academic sustainable design work with industrial practice and curriculum in the classroom. Further, we aim to continue to highlight the opportunities of synergizing technological design with the functions of nature in the search for sustainable solutions.

635419 Reinventing the Chemical/Materials Company: Transitioning to a Sustainable Circular Enterprise

Abstract:

In this paper we advance the proposition that decarbonizing the Chemicals and Mate-rials Industry (CMI) to meet the Paris Agreement goals is impossible without the transformation of the industry into a Sustainable Circular Chemicals and Materials In-dustry (CCMI). The implications of such transformation are extensive and multifacet-ed and would lead to a re-invention of the chemical company. CMI is defined as the set of industrial activities, which cover the transformation of raw material deposits in earth to consumer products, and include cement, iron and steel, non-ferrous materials (aluminum, magnesium, nickel, copper, rare earths, others), chemicals (organic and inorganic) and polymeric materials. Decarbonization of CMI is technically feasible, even though technical and economical hurdles exist. However, given the prevailing constraints from existing industrial infrastructure, limitations in the growth of zero-carbon electricity, limitations in the sustainably available biomass, and the competition for renewable energy by other sectors of the economy, notably buildings and transpor-tation, industry cannot achieve the Paris Agreement solely on the basis of technological solutions. Furthermore, extensive research on the decoupling of GDP from resource utilization indicates that it is impossible to meet the Paris Agreement goals with ob-served historical rates of decoupling. Decoupling needs to be complemented by suffi-ciency-oriented strategies and strict enforcement of absolute reduction targets.
Individual companies cannot transform the CMI to CCMI on their own. They need to reinvent themselves in strategic alliances with information, financial and management companies in the circular supply chain. In this paper we will discuss the primary incentives leading this transformation, such as: (a) Improved economic returns for the chemical/materials companies. (b) Significant expansion of global economic activity. (c) Broad expansion of economic benefits to underserved social groups. The paper will sketch the transition path, as well as the fundamental questions that the academic community must answer, and which underpin the sustainability of the transition.

636154 A Multi-scale Systems Engineering approachtowards Sustainable Energy Transition Strategies

Abstract:

Multi-scale Systems Engineering provides a methodological, generic framework to arrive at realistic integrated solutions to complex energy problems by adopting a holistic, system-based approach. This framework address the complex energy and environmental problems existing in design, control and operation of energy systems and their supply chains in an integrated manner, by producing sustainable design and operational plans for systems ranging from nanoscale, micro-scale, mesoscale to mega-scale levels over horizons that range from milliseconds to months or years.

Methodologies in multi-scale energy systems engineering include superstructure optimization with high-fidelity models & data-driven modeling, mixed integer optimization strategies, integrated design, control & operations under uncertainty, and life-cycle assessment, with circular economy considerations. Such an approach is particularly suitable and powerful to analyze future energy scenarios towards establishing sustainable and business-sensible energy transition strategies. We will illustrate the concepts and methods of such a holistic strategy by its application to renewable resources utilization focusing on the production & delivery of hydrogen-based Dense Energy Carriers, where issues related to intermittency and geographical variability, storage options and transportation will be addressed.

627850 System-Level Approaches for Intensifying the CO_2 Electrolysis Proccss

Abstract:

Electroreduction of CO_2 (CO_2RR) is a potentially carbon-neutral method to utilize a fraction of the excess CO_2 emissions. CO_2RR can convert CO_2 into various carbon-based chemicals such as methanol, formic acid, ethylene, and carbon monoxide. Significant research efforts by academia and industry have resulted in various energy-efficient cell configurations and active/selective catalysts for CO_2RR, however, significant electrochemical performance improvements are still needed for economic feasibility at scale.

This talk will cover multiple system-level approaches that can be further explored to intensify the electrochemical performance of a flow electrolyzer for CO_2RR.

First, we focus on the role of electrolyte composition in the mechanism of CO_2RR to CO on Ag nanoparticles (NPs). Insights obtained from polarization curves, Tafel slopes, Pourbaix diagrams, onset potentials, and electrochemical impedance spectroscopy will be used to explain the effects of electrolyte identity, concentration, and pH on the rate determining step.

Next, we cover the systematic process optimization of CO_2RR to CO on Ag NPs resulting in state-of-the-art electrochemical performance: CO partial current density (j_{CO}) exceeding 850 mA/cm^2 with a CO Faradaic efficiency (FE_{CO}) of 98% at a cell potential (V_{cell}) of -3 V corresponding to a full cell energy efficiency for CO production (EE_{CO}) exceeding 40% for a conversion per pass of CO_2 to CO exceeding 35%.

Finally, we discuss establishing quantitative/qualitative functional property relationships between system parameters (e.g.: pH, electrolyte flow rate) and system performance (e.g.: cathode overpotentials, j_{CO}) to help in rational system design for CO_2RR to CO on Ag NPs. We also briefly discuss the effect of electrolyte composition on performance stability over time for CO_2RR to CO on Ag NPs.

636155 Sustainable synthesis of NH_3 via electrochemical reduction of N_2 on an earth abundant Cu gas diffusion electrode at ambient conditions

Abstract:

The electrochemical reduction of N_2 to produce NH_3 at ambient conditions is an effective and sustainable route to store H_2, balance the N_2 cycle, and provide means to produce on-demand fertilizers. The efficient electrosynthesis of NH_3 is challenging because of the high energy of $N \equiv N$ triple bond (941 kJ/mol), dominant undesired H_2 evolution reaction (HER), and lower solubility of N_2 in aqueous solution. Here, we propose theory-guided activity descriptors to identify an efficient N_2 reduction reaction (NRR) catalyst, followed by its implementation in a flow through gas diffusion electrode (GDE) to quantify the effects of pH, cation identity, H_2O saturation, and N_2 concentration on the kinetics of the NRR. The identified Cu catalyst with dominant (111) facets electrodeposited on a carbon paper provides optimal active sites to obtain maximum NH_3 Faradaic efficiency (FE) of 18 ± 3 % at -0.3 V vs RHE and the maximum NH_3 current density of 0.25 ± 0.03 mA/cm^2 at -0.5 V vs RHE in an alkaline medium. The electrolyte pH, counter cations and the H_2O saturation on the catalyst are optimized to suppress the HER. The fixed potential DFT calculations reveal an associative distal mechanism for the NRR over Cu where the hydrogenation of $*N_2$ is the rate limiting step.

633557 Thiol-functionalized Hyper-Cross-Linked Milk Protein polymers for Mercury Removal

Abstract:

Hyper-cross-linked organic polymers (HCPs) have recently attracted a lot of attention as their large surface area and strong interactions with other molecules making them highly appropriate for different applications. The production of this type of microporous polymers is usually accomplished by employing pricey reagents. As an inexpensive way of synthesizing HCPs scaffold, we create one step Friedel–Crafts alkylation reaction with milk caseins, existing in high quantity and exhibiting unique polymer properties. To optimize the surface area of the polymer, the ratio of the catalyst ($AlCl_3$) to the cross-linker (2,4,6-Trichloro-1,3,5-triazine) was studied. Then, the polymer was decorated with sulfur functional groups to become a suitable platform for mercury separation from the water and gas phases. Our approach together with characterization techniques (FTIR, ICP, XPS, BET) validated that the integration of thiol-groups into HCP leads to remarkable de-mercuration performance. Thiol-functionalized casein polymer scaffold demonstrates excellent capacity and selectivity for trapping heavy metal ions. The high capabilities of developed sustainable adsorbents from waste streams make them comparable to current state-of-the-art nanoporous materials.

623747 Effect of Selective Impurities on Carbon Capture from Biogas Using Deep Eutectic Solvents

Abstract:

Anaerobic digestion (AD) of food wastes is one of the most promising pathways to convert wastes to value added fuels (e.g., biogas) and products (e.g., fertilizer). As food wastes are heterogeneous and their composition changes from household-to-household, the resulting biogas composition varies significantly. Biogas from food wastes often contain 40-60 % methane (CH_4), 40-60 % carbon dioxide (CO_2), and 1-5 % impurities. Although the actual concentration may vary from the source of biomass and AD operating conditions, most common impurities found in the biogas are hydrogen sulfide (H_2S), ammonia (NH_3), octamethyltrisiloxane, and decamethylpentasiloxane. carbon capture and removal of the impurities are required prior to use or injection into a natural gas pipeline. CO_2 from the gaseous stream can be separated by physical (ionic liquid) or chemical solvents (amine-based solvents). Amine-based chemical solvents have higher adsorption capacity but are often expensive to regenerate and toxic for the environment. Physisorption solvents are generally expensive and their adsorption capacity is lower, meaning that significantly higher amounts of physisorption solvents are required. In the context of carbon capture, the impurities are likely to play a significant role in adsorptive capacity of the solvents due to an influence on the molecular environment.

In this study, deep eutectic solvents (DES) are evaluated to upgrade biogas. The benefit of DES is that they are made from the physical mixing of a hydrogen bond donor (HBD) and a hydrogen bond acceptor (HBA), generally made from a quaternary salt and acid/alcohols. In this study, a conductor like screening model (COSMO) is adopted to screen suitable HBA's and HBD's for biogas upgrading based on hydrogen bond properties using a triple-zeta valence polarization (TZVP) basis set. This model is evaluated compared to literature experimental studies to verify integrity of the results. Of the five DES studied, it was found that quaternary-octyl ammonium bromide with decanoic acid (N_{8888}Br: DA) showed the highest CO_2 adsorptive capability. However, the choline-chloride (ChCl) based solvents had a much higher selectivity for CO_2 over CH_4. Through an activity coefficient analysis, it was determined that choline chloride-based DES are much more affected by the presence of contaminants than the ammonium salts with regards to adsorptive capabilities. It has been observed that the HBD is the main determinant of contaminant sensitivity rather than alkyl chain length. Due to these several trade-off factors apparent from the results, it is determined that the concentration of impurities should be accounted for when determining an appropriate DES for carbon capture from biogas.

629073 Design, Simulation, and Validation of 3-D Printed Hydrocyclones for Microbial Cell Concentration

Abstract:

Concentration of microbial cells is a key step in bioprocessing and laboratory operations. At the laboratory scale, concentration is typically conducted in centrifuges, however, this requires energy and consumables, and is not compatible with autonomous, continuous operation. Hydrocyclones are promising alternatives because the only energy input is to operate a pump, there are no consumables, the hydrocyclone requires no labor after setup, and it can operate continuously. This study provides a parametric open source design of a 3-D printable hydrocyclone with interchangeable parts for ease of testing different hydrocyclone dimensions during experimental optimization. The hydrocyclone was simulated using COMSOL Multiphysics® software using the k-ε turbulence physics to simulate rotational fluid flow and coupled with the particle tracing module to track individual cells or flocs. The simulation was able to replicate the rotational vortex and central air core and could track individual cells to the hydrocyclone overflow and underflow. The simulation was used to estimate the concentration efficiency for different hydrocyclone geometries. Simulations were experimentally validated for *E. coli* cultures mixed with varying concentrations of food-grade flocculant to increase the size of particles in solution and improve cell concentration efficiency.

629770 Switching Bioprocesses into Zero-Waste Virtual Power Plants By Flexible Operation of Electrochemical pH-Swing Extraction Processes

Abstract:

Carboxylic acids are important platform molecules that enable a sustainable chemistry based on renewable resources. [1] Although the biochemistry of the production is well established the successful industrial application is still challenging. As an example the US-Canadian company Bio-Amber, once considered a rising star for their ambition to establish a fermentation based production route for succinic acid had to file for bankruptcy. [2] This demonstrates that up to today advantages in environmental metrics cannot sufficiently compensate an economic disadvantage of the bio-based production. In order to make bio-based chemistry successful a holistic process development is required and must strive to achieve high yields in the fermentation steps, a low-cost and low-waste separation process and exploit all opportunities to harvest every added value feasible in the process.

We developed an electrochemically driven pH-swing extraction process for the recovery of succinic acid from aqueous solutions. [3] [4] This concept overcomes the issue of salt waste production otherwise encountered during the separation of bio-based succinic acid. [5] Instead of waste generating acid and base this technology uses electricity to drive the pH-swing extraction. The calculated electricity costs required to operate the electrochemical pH-swing extraction with the current prototype equipment are comparable to the costs for acid, base and waste disposal in state of the art process designs. A more detailed techno-economic assessment of the operational expenses (OPEX) of our electrochemical separation concept and four other process designs proposed in the literature was conducted with ASPENplus. The study revealed a cost advantage of the electrochemical processing route compared to other downstream options. Beyond the promising OPEX values determined for the electrochemical pH-swing process the technology enables utilization of electricity instead of fossil fuel derived heat for the separation of succinic acid.

Preferably the electricity used for the separation should be harvested from renewable sources like wind and solar power. These energy sources are inherently fluctuating and an extensive use of these sources will impose a significant stress on the stability of the power grid in the future. This talk demonstrates how a flexible operation of the electrochemical pH-swing extraction process can turn a bio-based succinic acid production into a virtual power plant. We implemented an algorithm in MATLAB that given the price profile of the electricity market calculates the cost-optimal load profile for the electrochemical unit operations. Modulating the power consumption according to the electricity prices provides a 7% saving on the electricity costs. A dynamic model of the electrochemical pH-swing extraction is used to calculate the electric current density, pump-speed and other operational parameters from the dynamic load profile. With the determined operational parameters the load flexible operation of a 100cm^2 electrolysis cell was tested experimentally. Throughout the experiment a constant degree of protonation close to 100% was measured, indicating that despite the load flexible operation a constant separation efficiency is viable with the operational conditions suggested with the model.

The developed method enables a load flexible and cost optimal operation of the electrochemical pH-swing extraction of succinic acid. In addition to the cost advantage harvested from times of abundant renewable electricity, hydrogen and oxygen are concurrently produced in the electrochemical pH-swing step. The cost advantage together with the generation of valuable co-products constitute an added value that can improve the overall economic competitiveness of the bio-based succinic acid production. Beyond that, load flexible operation and demand side management of large consumers aids in stabilizing the power grid and will likely become more important in the renewable centered electricity supply of the future.

Acknowledgements

The authors gratefully acknowledge the financial support of the Kopernikus project SynErgie (03SFK3L1-2) by the Federal Ministry of Education and Research (BMBF) and the project supervision by the project management organization Projektträger Jülich (PtJ).

Literature Cited

1. Becker J, Lange A, Fabarius J, Wittmann C. Top value platform chemicals: Bio-based production of organic acids. *Current Opinion in Biotechnology.* 2015;36:168–175.
2. Michael McCoy. Succinic acid maker BioAmber is bankrupt. *Chemical & Engineering News.* 2018.
3. AiChE Proceedings, editor. *An Electric Drive for Biorefineries: Electrochemical pH-Shift Liquid-Liquid Extraction of Succinic Acid.*, 2019.
4. Gausmann M, Kocks C, Doeker M, Eggert A, Maßmann T, Jupke A. Recovery of succinic acid by integrated multi-phase electrochemical pH-shift extraction and crystallization. *Separation and Purification Technology.* 2020;240(240):116489.
5. Kiss AA, Lange J-P, Schuur B, Brilman DWF, van der Ham AGJ, Kersten SRA. Separation technology–Making a difference in biorefineries. *Biomass and Bioenergy.* 2016;95:296–309.

632720 Electrokinetic Modeling of Salt Transport and Rejection in Graphene Oxide Nanofiltration Membranes

Abstract:

Graphene oxide (GO)-based membranes have gained tremendous interest in nanofiltration (NF) and reverse osmosis (RO) applications. A major challenge in GO membrane science is to improve their salt rejection in NF and RO processes. Previous researchers have proved that the interlayer spacing between the GO layers as well as the charge density of the GO layers can strongly influence salt rejections from ideal/dilute solutions. However, many practical desalination and dewatering processes operate at much higher concentrations. In this work, our goal is to systematically understand and predict the effect of salts concentration on GO membrane properties by electrokinetic modeling supported by experimental measurements. The correlations between feed concentration, interlayer spacing, and charge density of these GO membranes were determined by single-component salt permeation experiments and ion adsorption models. The modified Donnan steric pore model was then used to model, fit, and predict the salts rejections at different conditions. We then use the transport model investigate in detail the contributions of convection, diffusion, and electro-migration as a function of feed conditions. Finally, the combined effects of initial charge density and interlayer distances on salts rejection are evaluated through the plotting of 2D heat maps. Our study indicates that for neutral pH, under low salt concentrations (~ 0.01 M), modifying the charge density of GO will be more helpful for increasing salt rejections. On the other hand, under high concentrations (above 0.1 M), decreasing the interlayer distance of the GO membranes (i.e., reduction of effective pore size) would be a better strategy. Finally, we show the accurate prediction of binary/multicomponent salt mixture permeation behavior using the model. Overall, this study provides mechanistic insight into salt transport and rejection in GO membranes, and useful guidance for modification of GO membranes in a number of desalination processes.

622690 Process Design and Scale-up Study for the Production of Polyol-Based Biopolymers from Sawdust

Abstract:

Polymers are materials with exceptional properties that contributed to the development of modern societies (Thompson et al., 2009; Yarsley and Couzens, 1945). However, multiple environmental threats associated with their extensive use have been revealed. The difficult and slow degradability of these materials, which leads the generation of micro-plastics polluting soils and fresh and marine waterbodies, is an environmental issue of major concern. This requires the development of new environmentally and economically sustainable materials with adequate properties capable to replace traditional petroleum-based polymers. Different wastes are emerging as interesting raw materials for this purpose, which use to produce new polymers also contribute to reduce the environmental footprint of modern societies. Particularly, sawdust is a by-product of the forestry industry which can be used as raw material for the production of added-value bio-based chemicals, leading to the development of a circular economy based on this organic resource. One of the most promising paths for the reuse of sawdust is the production of polyols, which in turn can be used as raw material for obtaining biopolymers capable of replacing crude derived polymers, changing the paradigm of the polymer industry towards a more sustainable model.

A conceptual design and techno-economic evaluation of a process for the production of polyol-based biopolymers from sawdust at full-scale have been performed in this work. The process has been designed following the premise of being as environmentally sustainable as possible, but using simple equipment feasible to be deployed and operated by the timber facilities. It consists of preparing the raw material, the catalysis and glycerol for the solvolysis reaction followed by the purification of the polyol and finally the production of the bio polymer adding starch. A fraction of the sawdust processed is combusted to cover the thermal energy requirements of the process rather than using external fossil fuel supplies. The performance of the process stages is assessed based on experimental data (CIPA, 2017).

The results show that, as a result of the low capital expenses required, the deployment of the process proposed has a low economic entry barrier. However, the intensive use of raw materials results in most of the production costs are due to raw materials cost. In order to evaluate the effect of the scale in the economy of the process, a scale-up study has been performed. It is revealed that production costs are largely affected by the economies of scale, allowing a significant reduction of the investment per unit of waste processed as the facilities increase their size. For the typical range of market prices of glycerol and starch, the production cost ranges from 0.10 to 0.66 USD per kilogram of biopolymer produced.

References

CIPA, 2017. Biodegradable polymer based materials using Wood residues of SME from wood industry in Chile's Biobío region. R15F10009. Technical Report. Centro de Investigación de Polímeros Avanzados.

Thompson, R.C., Moore, C.J., Vom Saal, F.S., Swan, S.H., 2009. Plastics, the environment and human health: current consensus and future trends. Philosophical Transactions of the Royal Society B: Biological Sciences 364, 2153-2166.

Yarsley, V., Couzens, E., 1945. Plastics Middlesex. Penguin Books Limited.

624814 Phasins Employed By *Rhodopseudomonas Palustris* CGA009 for Bioplastic Production from Lignocellulosic Biomass

Abstract:

Lignocellulosic biomass is considered to be the most economic carbon source in the world, and is thus an ideal candidate for cheaper production of bioplastics. *Rhodopseudomonas palustris* CGA009 is a very metabolically diverse microbe that can utilize lignin breakdown products to produce bioplastics. Our recent efforts suggest cytoplasmic space is a limiting factor of maximum production of the bioplastic polyhydroxybutyrate (PHB) by *R. palustris.* Phasins are the dominant proteins surrounding PHB granules that control the size, shape, and abundance of granules. In this study, the Phap1 phasin from *Cupriavidus necator* H16 was expressed in *R. palustris* with the aim of overproducing PHB from lignocellulosic sources. Phap1 leads to smaller and more abundant PHB granules in *C. necator*, and it was hypothesized that expression of this phasin would increase overall titers from *R. palustris.* Transmission electron microscopy revealed that indeed Phap1 expression yielded smaller and more abundant granules in both aerobic and anaerobic conditions compared to the wild type strain. Expression of Phap1 yielded PHB production from *R. palustris* aerobically (0.7 g/L), which does not occur in the wild type strain. Although the Phap1 expression modified the granules in anaerobic conditions, the PHB titer was not as high compared to the wild type strain. This suggests that phasin performance may be optimal under certain conditions since *C. necator* is typically an aerobic microbe. Our ongoing efforts are aimed at quantifying changes in gene expression to develop a more holistic analysis of *R. palustris'* PHB production framework, including Phap1 as well as native phasins employed by *R. palustris* that have never been reported.

628379 Engineering Functional Materials from Cellulose Nanocrystals By Exploring Their Structure and Property Relationships

Abstract:

Cellulose nanocrystal (CNC) films possess chiral nematic structure by virtue of its chiral nematic self-assembly and therefore exhibits iridescence or structural color. Although, the structural color has been exploited for various application, it also restricts CNC's applications in packaging or coating industry where transparency is a prime factor.

The goal of this study is to evaluate the combined effect of different electrolyte addition and mechanical force alignment on CNC structure and properties. We found that particle order parameter of CNC increased from 0.32 to as high as 0.90 under the combined effect of electrolyte addition and mechanical force at a very low speed of 50 mm/sec. The scope of this study also includes comprehensive analysis of CNC structure and properties such as crystallinity, mechanical, optical, electrical, rheological, and barrier properties.

629832 Discovery and Development of New Sustainable Polyesters from Biomass

Abstract:

One of the greatest challenges facing the world today is the accumulation of plastic waste in the environment. Researchers have long pursued the development of biodegradable plastics, and some are already widely used, for example the popular poly(lactic acid) PLA or poly(hydroxybutyrate) PHB. However, the poor thermo-mechanical properties of such polymers limit their use for many applications. Therefore, it has become essential to discover and develop a new biodegradable polymer platform with robust mechanical properties, which can be derived from biosources to meet societal needs. In this work, we have explored the development of such new polyesters based on hydroxyacid monomers that can be produced from biomass. After searching the literature for a chemically diverse set of target molecules, this class of monomers was carefully evaluated for its polymerizability to narrow down to the most promising candidates. Initial efforts to synthesize high molecular weight homopolymer was challenging due to high crystallinity of short alkyl chain hydroxyacid compounds. However, copolymer made out of 3-hydroxypivalic acid and 6-hydroxyhexanoic acid exhibited high molecular weight as well as good thermo-mechanical properties that allowed the thermal processing window to be systematically widened by tuning copolymer composition. Lastly, a high throughput degradation was introduced and performed to assess the biodegradation of such polymers. The results from biological discovery, chemical synthesis and environmental metabolic engineering indicate that polyesters made from biosourced hydroxyacids have potentials to simultaneously meet the need for polymers with improved performance and improved sustainability.

Chemical upcycling of waste polyolefinic plastics to low-carbon synthetic naphtha for closing the plastic use loop

Leilei Dai*, Nan Zhou, Paul Chen, Kirk Cobb, Roger Ruan*

Center for Biorefining and Department of Bioproducts and Biosystems Engineering
University of Minnesota, 1390 Eckles Avenue, St. Paul, MN 55108, USA

dai00087@umn.edu (L. Dai); ruanx001@umn.edu (R. Ruan)

ABSTRACT

Plastic materials are extremely popular all over the world and used in the different walks of life due to the inherent properties of being strong, lightweight, and easily shaped. However, the vast majority of waste plastics ever produced enters into landfills or our ecosystems, creating a plastic waste crisis. Polyolefins, such as high density polyethylene (HDPE), low density polyethylene (LDPE), and propylene (PP), represents the largest amount of discarded waste plastics, accounting for over half of the waste plastic stream.

Currently, the most common solution to recycle the plastic waste is mechanical method that is limited by the corresponding challenges. The recycled plastics after melting and remolding generally have poorer properties than those of the virgin plastics, enabling the recycled plastics to be only applicable to lower quality products. This downcycling process makes the waste plastic recovery not attractive in the industry. Developing an effective pathway to remove waste plastics from landfill and incineration plants and create a circular economy requires a more appropriate technology beyond the conventional mechanical recycling by melting and re-molding. Pyrolysis has shown the great potential of plastic recycling for energy, fuels, chemicals, and materials production, enabling the plastic wastes to stay in the economy and out of the environment. Recently, we have screened different catalysts to improve the yield and quality of the liquid hydrocarbons from plastic waste pyrolysis, with the aim of maximizing naphtha fractions for new plastic production. Notably, relay catalysis of Al_2O_3 followed by ZSM-5 achieved up to 100% conversion into monoaromatics and C5-C12 alkanes/olefins at a catalyst to plastic ratio of 4:1. In order to further mitigate the aromatic formation, Al_2O_3 pillared clay catalyst was developed and tested, resulting over 50% C5-C12 alkanes and less than 10% monoaromatics. The liquid yield can reach over 70 wt.% with 52% of C5-C12 alkanes, 17% of C5-C12 olefins, 14% of C5-C12 aromatics, and 17% of C13+ hydrocarbons from catalytic pyrolysis of polyethylene. More importantly, this process is also suitable for waste plastic mixtures, e.g., LDPE 41%, HDPE 24%, and PP 35%, which represents one of the typical plastic waste streams. Furthermore, we proposed a novel approach (**Figure 1**) of combining two catalytic reforming zones, where the first catalytic reforming zone was intended to improve the cracking of polyolefins into short chain olefins, and the second, lower-temperature catalytic reforming zone was intended for the hydrogenation process to convert these olefins to C5-C12 alkenes. The tests were very successful and very promising results (60-75% C5-C12 alkanes, 3-5% C5-C12 olefins, 5-15% mono-aromatics) were obtained as hypothesized.

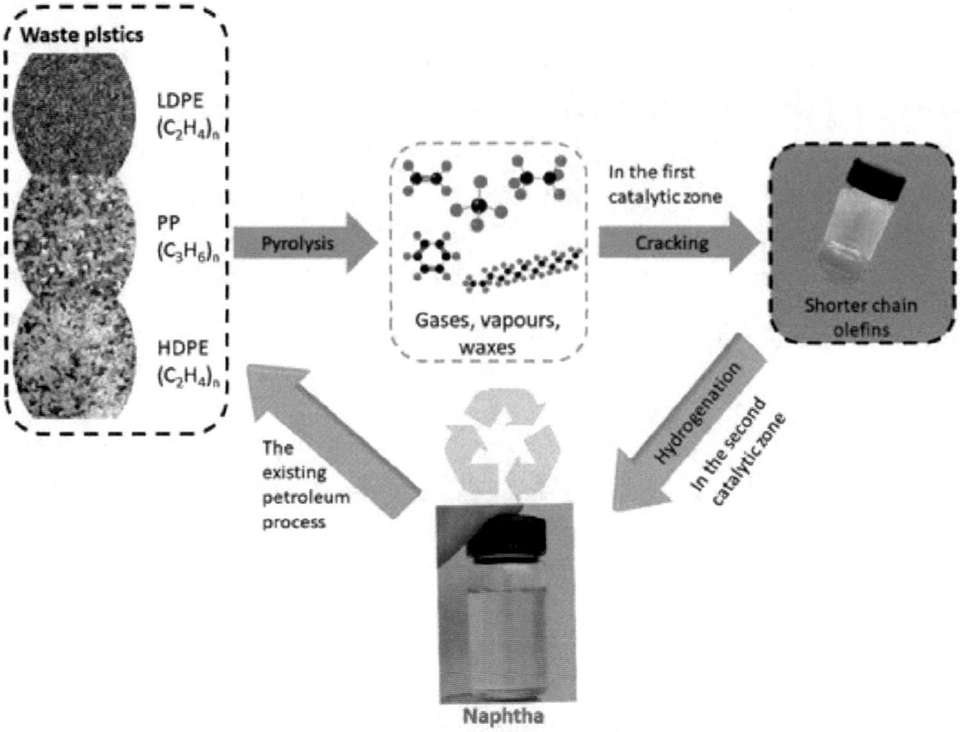

Figure 1. Converting polyolefinic plastics to high quality naphtha via pyrolysis coupled with a tandem catalysis process.

Table 1 shows the mass balance of the plastic-to-naphtha technology. When just using one catalytic zone with Zn/SBA-15 as a catalyst, 51.67 wt.% of liquid oil can be obtained, which is higher than that of the two catalytic zone (36.65 wt.%). It may be because the second catalytic zone was operated at lower temperature and some unconverted long chain hydrocarbons with high boiling temperature were trapped in this zone. This also highlights the importance of wax cracking in the first catalytic zone. It can be noted that after adding the second catalytic zone, hydrogen content in the gas product experiences a decline. That supports our hypothesis that the olefins produced in the first catalytic zone will be converted into paraffins by consuming the spontaneous hydrogen or hydrogen transfer reaction. The liquid oil composition verifies that the tests were very successful and very promising results were obtained as hypothesized.

Table 1. Mass balance for the catalytic reforming of HDPE pyrolysis volatiles.

	One catalytic zone	Two catalytic zone
Product yields (wt.%)		
Gas yield	34.99	30.60
Liquid oil yield	51.67	36.65
Coke on the first catalytic zone	13.34	15.05
Coke on the second catalytic zone		17.70
Chemical composition of products		
Dry gas (vol.%)		
Hydrogen	25.97	17.62
Methane	11.72	21.94
Ethylene	9.72	23.44
Ethane	15.04	12.27
Liquefied petroleum gas (vol.%)		
Propylene	9.07	13.43

Propane	9.36	6.80
Butylene	4.61	12.21
Butane	1.68	5.12
Liquid oil (%)		
C5-C12 n-paraffins	5.42	17.61
C5-C12 iso-paraffins	2.41	41.89
C5-C12 olefins	22.49	4.78
C5-C12 naphthene		7.83
C5-C12 aromatics	4.66	10.56
C13-C22 hydrocarbons	63.29	17.33
C23-C30 hydrocarbons	1.73	

Reaction conditions: Pyrolysis temperature = 500 °C, weight of HDPE, Zn/SBA-15, and Al2O3 pillared Montmorillonite clay = 6, 3, and 15 g, first catalytic zone temperature = 500 °C, second catalytic zone temperature = 250 °C, flow rate of nitrogen = 50 ml/min.

The promising results of the two catalytic zone is mainly attributed to the cracking and hydrogenation in the first and second catalytic zone, respectively (**Figure 2**). The catalytic activity of Al_2O_3 pillared M-clay may be due to the special features of this material, including the large pore size and Al_2O_3-induced active sites (more Brønsted acid sites). On one hand, the large pore size alleviates the steric and diffusional hindrances for entering the catalyst channels because the bulky nature of the polyolefins or long chain hydrocarbon intermediates in the pyrolysis reaction; on the other hand, Brønsted acid sites can be responsible for carbonium ion formation which is cracked into alkanes further. Based on the results above, it is supposed that this catalyst increases the share of C5-C12 alkanes by primarily facilitating the following reactions.

$$\underset{\text{Olefins}}{3C_nH_{2n}} + \underset{\text{Cyclo-alkanes (Olefins)}}{C_mH_{2m}} \longrightarrow \underset{\text{Alkanes}}{3C_nH_{2n+2}} + \underset{\text{Aromatics}}{C_mH_{2m-6}} \qquad \text{Hydrogen transfer reaction}$$

$$\underset{\text{Olefins}}{C_nH_{2n}} \xrightarrow{\text{H addition}} \underset{\text{Alkanes}}{C_nH_{2n+2}} \qquad \text{Hydrogenation}$$

In the HDPE degradation, the long polymer chain is initially cracked into shorter chain olefins, which would then be converted to alkanes and aromatics via the hydrogen transfer reaction over Al2O3 pillared M-clay catalyst.

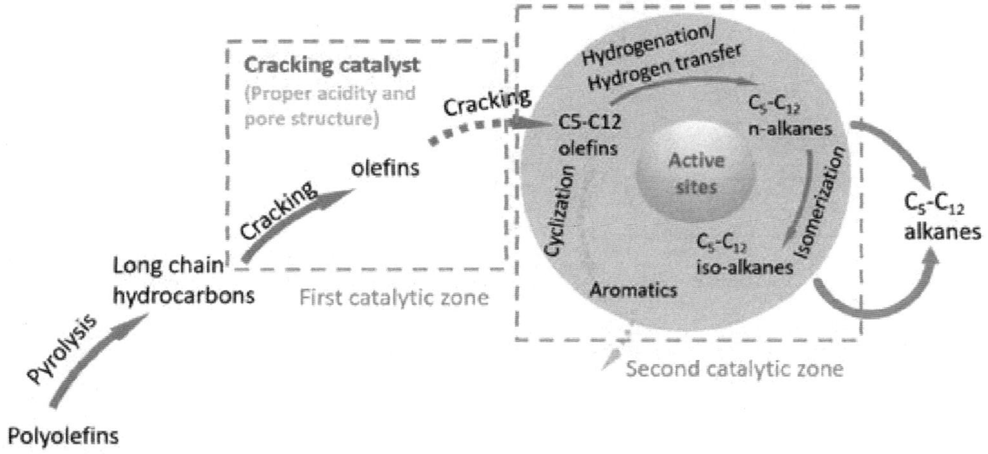

Figure 2. Reaction scheme for the two catalytic zones.

To advance this technology towards industrial applications, a continuous catalytic microwave-assisted pyrolysis (cMAP) system with a processing capacity of 1 ton/day was developed. The system features a mixing bed of silicon carbide balls enabling fast, uniform and energy-efficient heating for the process. Through this system, 57 wt.% C5-C22 liquid hydrocarbons was obtained from polyethylene, allowing over 32% energy saving compared to production of similar products from virgin materials.

KEYWORDS: Catalysis; Environmental sustainability; Chemical Reaction Engineering

631656 Deconstruction of Waste Polyethylene into Liquid Hydrocarbon Fuels and Lubricants over Ru Catalyst

Abstract:

Polyethylene (PE) is the most popular plastic globally, and the widespread use of plastics has created severe environmental issues. High energy consumption in the current process makes its recycling a challenging problem. In our report, the depolymerization of high-density polyethylene (HDPE) was conducted in various liquid-phase solvents with the Ru/C catalyst under relatively mild conditions. The maximum yields of the jet-fuel-range and lubricant hydrocarbons were 60.8 wt% and 31.6 wt%, respectively. After optimizing the reaction conditions (220 °C and 60 bar of H_2), the total yield of liquid hydrocarbon products reached approximately 90 wt% within only one hour. The product distribution could be tuned by the H_2 partial pressure, the active metal particle size, and the solvents. The solvation of PE in the different solvents determined the depolymerization reaction kinetics, which was confirmed by the molecular dynamics (MD) simulation results.

621594 Over Half a Century of Protecting the Environment through Various Policies, Administrations, Successes, & Failures; An Overview of 50+ Years of EPA Regulations and Standards

Abstract:

Since the establishment of the U.S. Environmental Protection Agency (EPA) on December 2nd, 1970, the Agency has implemented a vast number of regulations to benefit the American people and the environment. Over the last 51 years, the EPA and its dedicated workers have endeavored to fulfil its mission to protect human health and the environment; the Agency has sustained its critical function through 51 years and 10 different administrations (6 Republican and 4 Democratic presidents). The change of administration leads to change and priorities, and at times new or rolled back regulations. One such case is the Bush-era fuel standard increasing the fuel standards (established in 1975) to a corporate average of 35 miles per gallon by 2020, further increased by the Obama-era fuel efficiency standard of 54.5 mpg by 2025, then replaced by the Trump-era standard requiring a small 1.5% increase in average fuel economy compared to the pre-Obama standards, and then upon the inauguration of the Biden-era, auto makers themselves began to support higher fuel efficiency standards as the Administration foreshadowed increasing the standards akin to the Obama-era. This presentation aims to provide an overview of current U.S. environmental regulations, where we currently are, where we have come from, and where we are likely to go in the future.

628185 Decarbonising Residential Heating: Cost-Benefit Analysis to Support Policy Design

Abstract:

The UK's legally binding targets to achieve "net zero" greenhouse gas emissions by the year 2050 may require full decarbonisation of residential heating. This will involve deploying low-carbon heating technologies (such as heat pumps, electric storage heaters, and/or various hydrogen heating technologies) across the country, including in the 24 million homes that currently use natural-gas-based central heating systems. A key challenge of this transition is that the capital and operating costs of low-carbon heating systems are often greater than current natural-gas-based systems. In addition, the UK has a wide variety of building stock, which defies a one-size-fits-all solution.

The aim of this study is to assess the suitability of various policy instruments to promote the uptake of low-carbon heating technologies. A cost-benefit model is being developed based on housing stock data from the English House Survey, performance data from field trials, and cost estimates from the literature. The model will be used to calculate the lifetime cost and levelised cost of heat for a wide range of households and the results analysed to consider how the costs of different low-carbon heating options vary with characteristics such as building type, number of occupants, thermal efficiency, and geographical location. In addition, an uncertainty analysis will used to evaluate the impact of different fuel prices, outdoor temperatures, and personal preferences and to identify the household types likely to be impacted most by these uncertainties.

The results of this study will be presented and discussed in relation to current policies designed to promote low-carbon heating. This study will consider what level of government support will be sufficient to promote low carbon heating over the current natural-gas-based heating systems and will identify if a targeted approach based on geographic location, building type, tenure type, or other characteristics would be more appropriate.

625673 Improving the Economics of Industrial Battery Storage: A Proactive Policy and Management Approach

Abstract:

Historically, implementation of battery energy storage has been proportionally higher in commercial and residential systems than industrial systems [1]. Economic and operational constraints of industrial facilities make battery installation close to infeasible. Industrial energy usage accounts for about 25% of total energy usage in the US. As the deployment of battery storage increases, lack of implementation in the industrial sector creates a gap, bridged by adapting control and policy strategies to constraints of industrial users. Projections indicate that battery prices will continue to fall, making it even more important to find ways to implement batteries in the industrial sector [2].

This research aims to show that the utilization of electric battery storage to decrease the energy costs and peak power usage of industrial energy users can be made attainable by identifying and closing the policy gap between affordability and utility of industrial battery installations. Similar studies have been done to improve economics and management of storage in residential installations [3]. Other studies have looked at regulatory design of energy storage coupled with renewable sources such as solar and wind [4]. Though these efforts have furthered the penetration of battery storage in the residential and commercial sectors, this same penetration has not been seen in the industrial sector from lack of analysis and specificity to the needs of industry. Policies specific to large electrical users in the industrial sector have also not been explored.

Load data from actual industrial facilities is used spanning a full year of typical manufacturing operations in which the daily electrical peak is about 2200 to 2500 kW. Two models are used, a battery model utilizing the facility load data to look at battery dynamics and effects of system load, and a financial model to analyze overall cost and effect of policy on project economics. The study identifies policies that are currently in place in parts of the US or similar policies that would allow batteries to be financially viable for industrial facilities.

[1] EIA, "Battery Storage in the United States : An Update on Market Trends," Washington, DC, 2020.

[2] W. Cole and A. W. Frazier, "Cost Projections for Utility-Scale Battery Storage," 2019.

[3] M. N. Sheha and K. M. Powell, "An economic and policy case for proactive home energy management systems with photovoltaics and batteries," *Electr. J.*, vol. 32, no. 1, pp. 6–12, 2019, doi: 10.1016/j.tej.2019.01.009.

[4] O. Zinaman, T. Bowen, and A. Aznar, "An overview of behind-the-meter solar-plus-storage regulatory design: Approaches and case studies to inform international applications (Report no. NREL/TP-7A40-75283)," *NREL Rep.*, no. March, pp. 1–76, 2020.

630728 Enabling Low Carbon Hydrogen Production Using Resource Integration Approach

Abstract:

Under the Paris emissions reduction protocol, the global energy sector is expected to shift from carbon-heavy industries to achieve a climate-safe target of 1.5 ^{o}C. To divest from fossil fuel, however, turns out to be more challenging to achieve. Concerns about environmental degradation and dwindling fossil fuel supplies have spurred efforts into adopting renewable energy resources. Hydrogen has been recognized as a key material in the decarbonization of the energy sector and in tackling the global warming challenge. Nevertheless, hydrogen can only exist in combined with other elements in the chemical-bound form. Thus, it has to be produced by means of known energy-intensive processes and thus generates significant amounts of CO_2. One promising alternative in making hydrogen production greener is to incorporate renewable energy in hydrogen production. However, the green hydrogen pathway seems to be more optimistic to see the light in near terms since it suffers from the high cost and the scalability of the hydrogen produced. Therefore, the need for an effective and efficient solution to allow simultaneous production of hydrogen, assess renewable energy use, and incorporate CO_2 management, which is possible through carbon capture utilization and storage. This could serve as the missing link towards a profound energy transformation. This work presents a framework for hydrogen infrastructure design and CO_2 utilization strategies in an industrial park while identifying optimum decisions related to production, storage, and transportation technologies. A techno-economic optimization model based on a superstructure approach has been developed to identify the best possible hydrogen network configurations from a number of technologies, which is achieved by a combination of many possible hydrogen networks (this includes storage and transportation routes) and CO_2 utilization scenarios. A case study that resembles Qatar's industrial sector is set out to investigate the application of the proposed approach.

628182 The Effect of Geological Properties of Underground Storage on Its Suitability for Inter-Seasonal Storage of Hydrogen and Its Role in Achieving Net Zero

Abstract:

Sourcing a sustainable low-carbon energy supply is one of the key requirements towards attaining Net Zero by 2050. Although the use of renewable energies is on the rise, there remain some barriers to their full roll-out, including their intermittency. Underground storage of green hydrogen, generated by electrolysis using renewable energy, across a variety of geological formations and underground features, such as salt domes, rock cavities and aquifers, presents a solution to a year-round national-level supply of renewable energy. Although there are current debates for decentralising the energy grid, the use of inter-seasonal storage provides a good argument for the use of the existing centralised infrastructure.

This study aims to evaluate the role played by underground inter-seasonal storage in decarbonising energy in the UK. It will investigate key factors such as the potential of the UK for inter-seasonal storage, gather data on the properties of the inter-seasonal storage, evaluate what sector is most important for this application, and assess the potential environmental impacts. The effect of properties, such as permeability of surrounding layers, integrity of formation and storage pressure, on the amount of hydrogen that can be stored, how long it can be stored, the rate of loss during storage, and the rate at which hydrogen can be injected into and withdrawn from storage will be investigated. These findings will then be used as inputs to the Value Web Model, a mixed-integer linear programming model for the optimal design and operation of integrated energy value chains. Different optimal scenarios for the use of inter-seasonal storage in the UK will be developed, based on a range of probable costs, efficiencies, and capacities.

The results will be presented in best scenario terms, controlled by factors such as available geologies, demand and other local considerations. These scenarios will be discussed by the role they play for the UK and evaluated on their feasibility. The ultimate goal for the utilisation of inter-seasonal storage is to provide a solution to the current use of non-renewable fuels with the potential in aiding the UK through the transition to renewable energies by 2050.

633237 Stone-Wales Defect-Rich Carbon-Supported Dual-Metal Single Atom Sites for Zn-Air Batteries

Abstract:

This work aims to obtain a fundamental understanding of active sites near stone-wales defects rich nitrogen-doped graphene (DG) with specific coordination of carbon atom rings. It reveals that the stone-wales rich defects (*e.g.,* pentagon (5), pentagon—octagon—pentagon (i.e. 585), or pentagon-heptagon-heptagon-pentagon (5775) rings, appears correspondingly with carbon rings that brought active sites during catalytic reactions. Moreover, we anchored dual isolated metallic atoms (Ni/Fe) on DG support via linkers (O/N) called NiFe-DG. X-ray absorption spectroscopy indicates Ni/Fe metal single atoms are embedded via $Fe\text{-}N_4$ and $Ni\text{-}N_4$ coordination on DG surfaces. It exhibits high catalytic activity for oxygen reduction reaction (ORR) with an onset potential of 0.97 V, a half-wave potential of 0.86 V, and diffusion current density of 5.7 mA/cm^2, which is at par with commercial Pt/C. The catalyst shows superior stability, retained 82% of the initial current density even after 12 hours under an applied potential of 0.86 V. Similarly, the oxygen evolution reaction (OER) overpotential of 358 mV was achieved at 10 mA/cm^2 with a lower Tafel slope value (76 mV/dec) than commercial Pt/C. It maintains 85% stability for 12 h at a constant potential of 1.588 V, shows better stability than commercial Pt/C.

629975 Determination of Safety Parameters in Lithium-Ion Batteries

Abstract:

Lithium-ion batteries (LIBs) have raised increasing interest due to high energy density and efficient energy storage. LIBs are used not only in portable electronics, such as cell phones and computers, but also for energy storage systems (UPS and Grid applications.), and electric vehicles. LIB's offer; excellent performance, high energy density, high specific capacity & voltage, no memory loss, very little self-discharge, and a wide operating temperature range. This has made LIBs increasingly popular. However, safety is a major concern, and a large number of LIBs have been recalled due to explosion and fire accidents. Consequently, increasing attention is being paid to LIB safety issues and a large number of safety strategies are being developed.

The safety of a battery is determined by the battery chemistry, its operating environment, and its abuse tolerance. The internal failure of a LIB is caused by electrochemical system instability. Thus, understanding the material properties, electrochemical reactions, and side reactions occurring in LIBs is fundamental in assessing battery safety. Temperature and voltage are the two controlling factors of the battery reactions. Safety accidents are accompanied by continuous heat and gas generation, which causes battery rupture and ignition of the combustible materials. Cells are equipped with safety features such as separator shut down to stop the flow of current and opening of the vent to help prevent the cell from bursting and/or catching fire in addition to others.

In this study, a testing method was developed using accelerating rate calorimeter (ARC®) to determine thermal behavior of a cell under thermal abuse conditions. Onset temperature of self-heating of the cell, separator shutdown temperature, leak/vent temperature, and ignition temperature of the cell were all determined. The two types of cells that were tested were cylindrical 18650 cells and polymer pouch cells. All tests were run with 100% state of charge. Attached below is an ARC test result from a polymer pouch cell (4.2V, 1800 mAh). It was found that the separator shutdown temperature was 140°C, the primary self-heating onset temperature was 134°C, and ignition temperature was 160°C. Cell leak temperature was found to be in the range of 141°C to 147°C.

Silica Gel/ $MgSO_4$ Hybrids for Thermal Energy Storage

Suboohi Shervani, Curtis Strong, and F. Handan Tezel*

Department of Chemical and Biological Engineering, University of Ottawa, 161 Louis-Pasteur, Ottawa, Ontario K1N 6N5, Canada.

* Handan.Tezel@uottawa.ca (author for correspondence)

Abstract

In the past few years, fossil fuels usage have been increased significantly, which produce large amount of carbon-dioxide, a greenhouse gas, and causes climate change. Therefore, a clean technology is needed to fulfill the global energy demand. Renewable energy sources, such as solar thermal, can be used as potential alternatives to the fossil fuels. However, thermal energy storage is required to fully utilize the solar energy.

Thermo-chemical energy storage technology is a promising technology used in space heating and cooling applications. Salt hydrates show high energy storage density and appeared as encouraging candidate for thermal energy storage. Nevertheless, lack of stability/ repeatability are the main issue. The current study focuses on $MgSO_4$-based hybrids for thermal energy storage. The structural properties of the hybrid is measured by SEM, and the adsorption/regeneration characteristics are measured by a lab scale thermal energy storage apparatus. The hybrid shows high energy storage density ~ 150 kWh/m^{3}at 50% relative humidity and 120°C regeneration temperature.

Keywords: Salt Hydrates, Adsorption, Thermal Energy Storage, Energy Storage Density, Hybrids

632187 Thermochemical Heat Recuperation for Compressed Air Energy Storage

Abstract:

Compressed Air Energy Storage (CAES) suffers from low energy and exergy conversion efficiencies (ca. 50% or less) inherent in compression, heat loss during storage, and the commonly employed natural gas-fired reheat prior to expansion. Previously, isothermal, and adiabatic (or 'advanced' adiabatic) compressed air energy storage have been proposed to enhance the efficiency of the process. Among the adiabatic schemes, thermal storage can be used to store heat from compression process. Prior to expansion, the air can recuperate the stored thermal energy, which could reduce or even eliminate the need for natural gas-fired reheat. Although sensible heat storage (SHS) and latent heat storage (LHS) have been previously investigated for this purpose, there are very few published studies looking at the use of thermochemical energy storage (TCES) for the CAES application. Here, a direct heat transfer scheme with combined thermochemical energy storage and sensible energy storage is proposed. The present study develops a mathematical model to address the conceptual design of the proposed idea. A detailed model and analysis are performed for TCES-rock-filled PBTES (packed bed filled with both barium oxides and rocks) and rock-filled PBTES. Results showed that similar round-trip efficiency is achieved between rock-filled and TCES-rock-filled PBTES due to the relative low heat capacity and heat of reaction for the barium oxides materials used in the current study. However, with TCES material placed on top of the packed bed, more stable turbine air inlet temperature (less than 200 K temperature variation) is obtained compared with all rock-filled packed bed (more than 300 K temperature difference) if idle time after charge is less than 20 h. In addition, with future advanced materials, the proposed idea could potentially improve the roundtrip efficiency of the process, have longer duration of storage and stable turbine air inlet temperature under suitable operating conditions. The concept could also be applied to Liquid Air Energy Storage.

623606 Exploring the Role and Value of Grid-Scale Energy Storage in Deep Decarbonisation

Abstract:

As the power system in the UK undergoes decarbonisation, the transformation of the capacity mix remains a topic of discussion. Not only novel generation technologies, such as power plants with carbon capture and storage (CCS), but also storage technologies are to be considered. The increasing share of intermittent renewables in the energy mix creates the need for higher storage capacity and/or more flexible plants. Furthermore, the additional requirements of other sectors of the energy system, specifically the added demand from electric vehicles (EVs) and from heat pumps (HP), may impact the power demand profile and not only raise overall demand but also increase peak demand. This raises the question which storage technologies and how much storage capacity is required to satisfy the power demand at all times.

In the UK context, pumped hydro storage (PH) is already a part of the system capacity. Interconnection with neighbouring countries is also used to balance demand peaks and renewables fluctuations. These technologies, however, are mainly used for short-term storage, as in, storage within a day or between days. It is of interest if long-term storage, such as power-to-gas storage technologies, could also have a place in the capacity mix. In this work, we therefore examine under which circumstances power-to-gas storage has value in the future UK energy system under a net zero carbon target by 2050. Specifically, we model a route of power to synthetic methane to salt cavern storage to CCGT (P2M) with recycling of CO_2, which has been determined to be a promising power-to-gas storage route based on techno-economic analysis by Yao *et al.*[1]. We use the electricity systems optimisation model with capacity expansion (ESO-X) developed by Heuberger *et al.*[2]. The model constraints include build rates, ancillary services requirements, plant flexibility constraints, etc. We use one year of full-hourly renewables availability (solar, onshore wind and offshore wind) and demand profiles. This allows the analysis of seasonal storage. The demand is considered to be comprised of three parts, power demand, added demand from electric vehicles, and added demand from heat pumps as shown in Figure 1. Owing to the high amount of uncertainty regarding the level of electrification for both transport and heat, we consider four scenarios for both with varying levels of electrification ranging from minimal to total. The same applies to the generating technologies allowed in the system. We analyse scenarios with and without gas-fired power plants with CCS (CCGT-CCS) as a form of stable, low-carbon generation, and bioenergy with CCS (BECCS), as negative emissions technology, since the deployment of these technologies is uncertain. Furthermore, we analyse the value of key technology parameters of P2M such as input-to-output power ratio and round-trip efficiency and perform sensitivity analysis regarding the CAPEX of the technologies.

Excerpts of the dispatch schedules for scenarios with and without CCS and P2M are shown in figure 2. Our results suggest that in a scenario where all technologies are available for deployment with historical build rate limits, the increasing seasonality of the demand curve and the fluctuation of the increased share of renewables would optimally be absorbed by a combination of all archetypes of technologies. Renewables contribute a large share of power on days with high availability. CCGT-CCS and CCGT provide load-following and flexibility on peak days, respectively. The continuous operation of BECCS offsets emissions from gas-fired power plants. Using our central numbers for technology features and costs, P2M would have value in a cost-optimal technology mix by absorbing excess renewable energy and releasing it on peak days and hours (around 6 GW with 3.6 TWh storage capacity in 2050).

When CCS is removed from the capacity mix, P2M becomes a key technology. In our scenario without CCS or P2M, meeting demand with renewables and short-term storage (PH and battery storage) requires large amounts of curtailment, incurs lost load, and hence doubles the total system cost. When P2M as seasonal storage is added (26 GW and 13 TWh installed in 2050), it is used in the system to absorb power from renewables instead of curtailing it and providing flexible generation on low-availability days. This reduces the necessary amount of renewables capacity substantially. Notably, in scenarios with P2M available, it is deployed in greater amounts than battery storage despite the higher CAPEX and lower round-trip efficiency, indicating the increased need for seasonal as opposed to short-term storage.

Our sensitivity analysis shows a link between the deployment of onshore wind and power-to-gas storage. Lower seasonal storage CAPEX enables the deployment of both in conjunction, raising the renewables capacity as shown in Figure 3. Furthermore, our model suggests that renewables in combination with seasonal storage could be cost-competitive with CCGT-CCS in case of continuously high renewables build rates. We further identified the ratio of charging power to discharging power of the power-to-gas technology as a parameter with impact on its deployment potential and found that cost reductions of the technology may provide more value to the system than technology improvements.

In summary, grid-level power-to-gas storage may have a place in the future energy system in the UK. Whether or not it has value depends heavily on the build rates of renewables and on the relative cost of power-to-gas storage and low-carbon dispatchable power such as CCGT-CCS. Our results underscore the value of a diverse energy system with a mix of multiple technologies for generation and storage.

References

1. Yao JG, Bui M, Mac Dowell N. Grid-scale energy storage with net-zero emissions: comparing the options. *Sustainable Energy & Fuels* 3.11 (2019): 3147-3162.
2. Heuberger CF, Rubin ES, Staffell I, Shah N, Mac Dowell N. Power capacity expansion planning considering endogenous technology cost learning. *Appl Energy*. 2017;204:831-845.

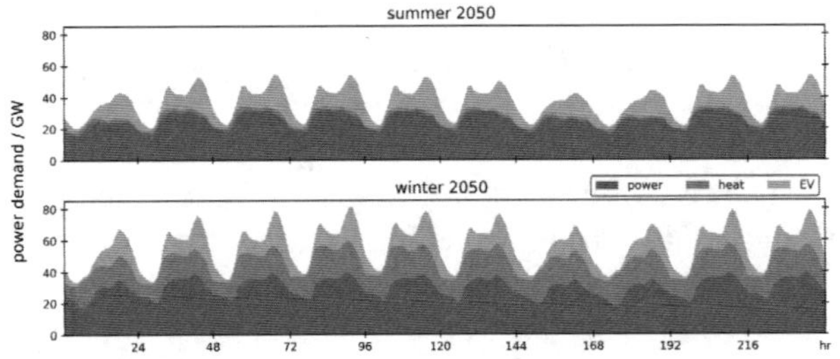

Figure 1: Example of the demand profile used in the model. Hourly power demand, added demand from heat pumps and electric vehicles, for ten days in summer and winter.

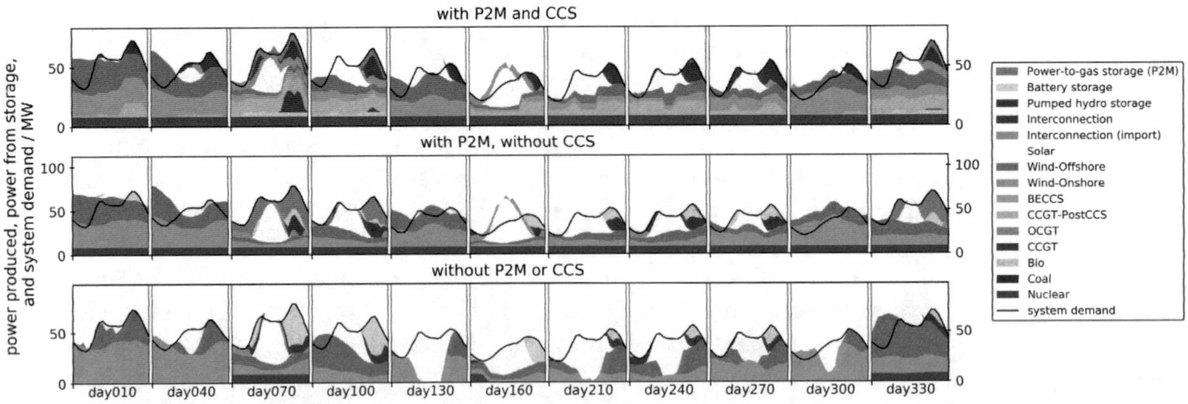

Figure 2: Excerpts of hourly dispatch schedules throughout the year (power produced and power from storage) for central electrification scenario. Stack higher than demand indicates power is fed to storage.

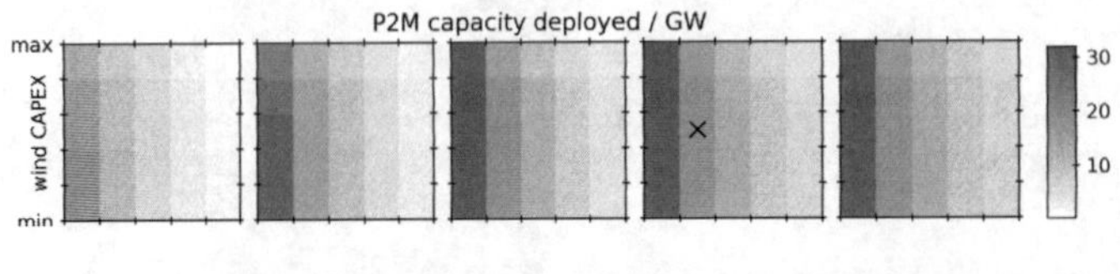

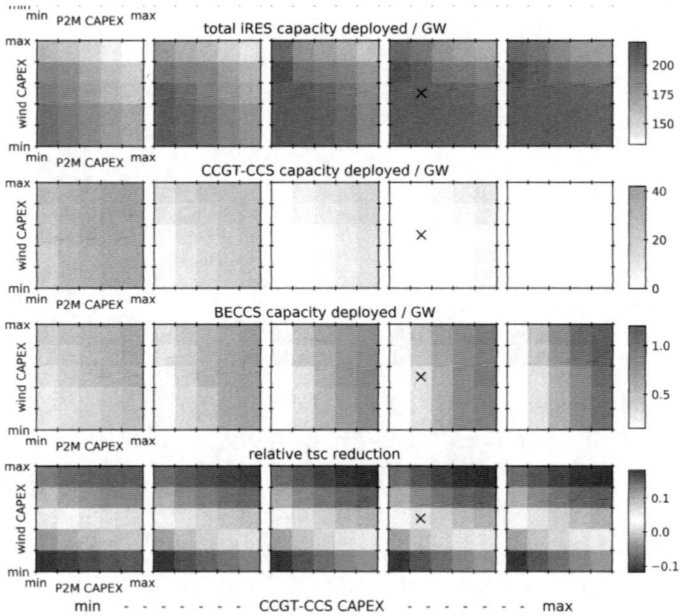

Figure 3: Deployment of power-to-gas storage (P2M), intermittent renewables (iRES), CCGT-CCS, and BECCS, and relative reduction of total system cost (tsc) depending on technology CAPEX. "min" values are estimated lower bounds, "max" values are 1.5 of the current CAPEX. Build rate limits set to 2x historical values. Cases marked with (x) indicate model did not converge within allocated time; values are interpolated based on neighboring cases.

629997 Data-Driven Incipient Fault Management for Proton Exchange Membrane Fuel Cell

Abstract:

In the efforts of establishing cleaner energy systems, hydrogen is receiving increasing attention for transportation and energy storage applications. Among various hydrogen systems of interest, fuel cell vehicles (FCVs) are well studied as the next-gen energy vehicles. Specifically, proton exchange membrane fuel cell (PEMFC) is considered the leading technology for handling hydrogen as fuel. It converts chemical energy from hydrogen to electrical energy via an electrochemical reaction and leaves water as the only byproduct. Therefore, PEMFC is attracting the interest of both the fuel cell community and the automobile industry. However, there are challenges associated with large-scale commercialization of PEMFC for electric vehicles such as operational safety, reliability, performance, and cost [1]. Though several prognostics and health management (PHM) [2, 3] and fault diagnosis [4, 5] studies have been performed to overcome these challenges, the problem of incipient fault diagnosis in PEMFC has not been fully examined. Incipient faults are usually masked by disturbances or other system uncertainties due to their small magnitudes and thus, are unnoticeable and difficult to detect [6, 7]. If no corrective action is taken, incipient faults can evolve into functional faults and lead to catastrophic failures. Therefore, early detection and isolation of incipient faults is highly valuable for the on-board safety of automobiles.

Motivated by the aforementioned considerations, we propose an incipient fault management framework that uses an adaptive data-driven modeling method, OASIS [8] (to detect faults), along with contribution analysis (to isolate faults) through a moving window strategy [9]. First, we obtain a data-driven process model using OASIS for the current operating condition. The model is used to predict the real-time dynamics. Based on this prediction, we continuously evaluate and monitor process risk associated with the fuel cell operation. Here, we incorporate moving window approach for risk assessment to improve the detection accuracy of incipient faults. Also, the adaptive model obtained through OASIS can cope up with any changes in the system, which further helps with decoupling incipient faults from process uncertainties. Next, after the detection step, we initiate contribution analysis to identify the location of the detected incipient fault. Finally, we isolate the process variable with the maximum contribution to the process risk from the remaining components of PEMFC. In this work, we investigate a scenario of hydrogen leakage with a small flowrate. As hydrogen is highly inflammable, it is important to diagnose any small traces of hydrogen leakage to facilitate safe operation of the PEMFC system. To conclude, the proposed framework has the ability to 1) predict the fuel cell operation under varying operating conditions, 2) detect and isolate an incipient fault before it propagates as a functional fault, and 3) assess the process risk to ensure safe operation of the PEMFC system.

Literature cited

[1] Zhang, X., Zhang, T., Chen, H. and Cao, Y., 2021. A review of online electrochemical diagnostic methods of on-board proton exchange membrane fuel cells. *Applied Energy*, *286*, 116481.

[2] Jouin, M., Gouriveau, R., Hissel, D., Péra, M.C. and Zerhouni, N., 2013. Prognostics and Health Management of PEMFC–State of the art and remaining challenges. *International Journal of Hydrogen Energy*, *38*(35), 15307-15317.

[3] Meraghni, S., Terrissa, L.S., Yue, M., Ma, J., Jemei, S. and Zerhouni, N., 2021. A data-driven digital-twin prognostics method for proton exchange membrane fuel cell remaining useful life prediction. *International Journal of Hydrogen Energy*, *46*(2), 2555-2564.

[4] Tian, Y., Zou, Q. and Lin, Z., 2020. Hydrogen Leakage Diagnosis for Proton Exchange Membrane Fuel Cell Systems: Methods and Suggestions on its Application in Fuel Cell Vehicles. *IEEE Access*.

[5] Li, Z., Outbib, R., Giurgea, S. and Hissel, D., 2018. Fault diagnosis for PEMFC systems in consideration of dynamic behaviors and spatial inhomogeneity. *IEEE Transactions on Energy Conversion*, *34*(1), 3-11.

[6] Safaeipour, H., Forouzanfar, M. and Casavola, A., 2021. A survey and classification of incipient fault diagnosis approaches. *Journal of Process Control*, *97*, 1-16.

[7] Chen, H., Jiang, B., Zhang, T. and Lu, N., 2020. Data-driven and deep learning-based detection and diagnosis of incipient faults with application to electrical traction systems. *Neurocomputing*, *396*, 429-437.

[8] Bhadriraju, B., Bangi, M.S.F., Narasingam, A. and Kwon, J.S.I., 2020. Operable adaptive sparse identification of systems: Application to chemical processes. *AIChE Journal*, *66*(11), p.e16980.

[9] Ji, H., He, X., Shang, J. and Zhou, D., 2016. Incipient sensor fault diagnosis using moving window reconstruction-based contribution. *Industrial & Engineering Chemistry Research*, *55*(10), 2746-2759.

630373 Experimental and Theoretical Studies of an Iron-Based Oxygen Carrier with a Core-Shell Structure of Fe_2O_3@CeO_2 for Chemical Looping Biomass Gasification

Abstract:

Biomass that is originated from plants or animal waste is one of abundant renewable resources. Chemical looping gasification (CLG) that is considered as one of crucial biomass utilization strategies to produce syngas, is the thermochemical conversion technique that benefits from the use of lattice oxygen in an oxygen carrier such as metal oxides as a gasifying agent to avoid direct contact between fuel and air. The promotion of iron oxide with ceria will create a synergetic effect between iron oxide and cerium oxide as a bimetallic oxygen carrier with high oxygen capacity, reactivity and selectivity for syngas production in CLG. In this research, a comparative theoretical study to investigate the role of oxygen vacancy formation energy of Fe_2O_3 and CeO_2 oxygen carriers and adsorption pathways of pyrolyzed gaseous compounds of CO, H_2, CO_2 and CH_4 has been conducted. The Density Functional Theory (DFT) studies of Fe_2O_3 and CeO_2 surfaces showed the lower energy for the formation of oxygen vacancy in CeO2 (1 1 0) than Fe2O3 (0 0 1), which implies the better performance of CeO_2 in gasification reaction. In addition, CeO_2 has lower activation energy and reaction energy in the conversion of CO_2 into CO, which is beneficial in gasification reaction. Consistent with theoretical results, thermogravimetric analysis of the oxidization of biomass with the oxygen carriers showed the increase of CeO_2 loading of the Fe_2O_3@CeO_2 oxygen carrier from 50% to 60% increased the ability of the carrier to release oxygen.

631150 One-Step Catalytic Fast Hydropyrolysis and Upgrading of Biomass to Biofuels Using MCM-41/ZSM-5 Composites

Abstract:

Biomass is a widely available and renewable resource that is considered as a promising alternative to fossil fuels. The direct production of high quality bio-oil from biomass can be realized via catalytic fast pyrolysis (CFP). Nevertheless, low bio-oil yields and rapid coke formation on catalysts are still the main challenges to the commercialization of the CFP technology [1]. Catalytic fast hydropyrolysis (CFHP) is an alternative to CFP that leads to deeper deoxygenated fuels and less coke. Nevertheless, the bio-oil produced via one-stage CFHP has a low atomic H:C ratio that needs to be increased via hydrotreatment to obtain gasoline and diesel range fuels. Our previous studies have shown that a second stage hydrotreatment process following the first CFHP unit can shift the product distribution from aromatics to alkanes, which increases the atomic H:C ratio [2].

Our objective in this study is to use Ni impregnated MCM-41/ZSM-5 composites with unique structure as catalysts to facilitate the tandem CFHP and hydrotreatment reactions in one step, therefore, minimizing the number of required reactors. The composites have been prepared using the aerosol method [3] and were fully characterized by N_2 sorption-desorption, scanning electron microscope, transmission electron microscopy, X-ray diffraction and inductively coupled plasma atomic emission spectroscopy. The CFHP experiments were performed using a PyGC-MS system and the biomass feedstock was miscanthus × giganteus. The initial results indicate that the composite was able to integrate the two-step CFHP and hydrotreating steps in one, which enhanced the yield of alkanes without sacrificing the yield to monoaromatic hydrocarbons.

[1] D.P. Gamliel, H.J. Cho, W. Fan, J.A. Valla, Appl. Catal. A Gen. 522 (2016) 109–119.

[2] D.P. Gamliel, G.M. Bollas, J.A. Valla, Fuel 216 (2018) 160–170.

[3] L. Yu, A. Farinmade, O. Ajumobi, Y. Su, V.T. John, J.A. Valla, Appl. Catal. A Gen. 602 (2020) 117727.

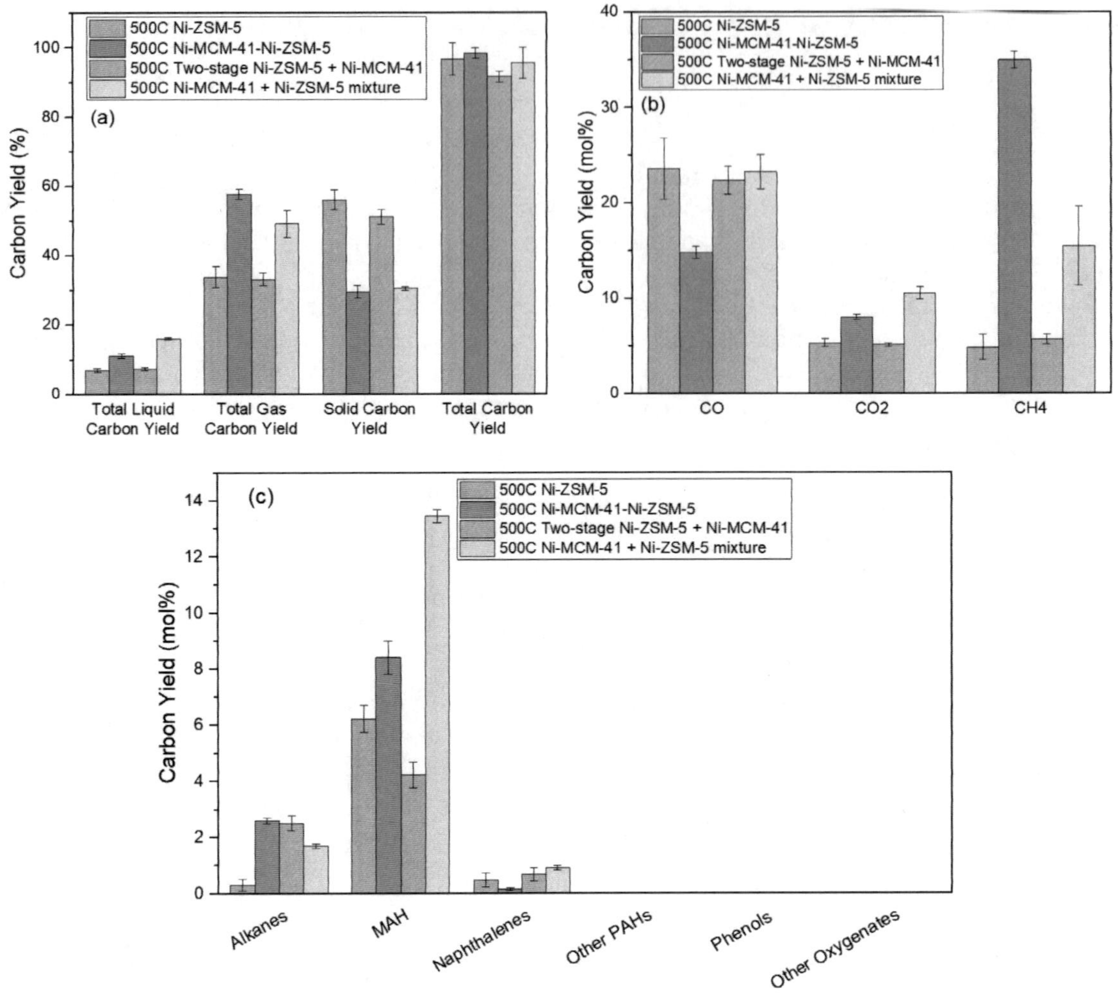
500C Ni-ZSM-5
500C Ni-MCM-41-Ni-ZSM-5
500C Two-stage Ni-ZSM-5 + Ni-MCM-41
500C Ni-MCM-41 + Ni-ZSM-5 mixture
(a)
Carbon Yield (%)
0
20
40
60
80
100
Total Liquid Carbon Yield
Total Gas Carbon Yield
Solid Carbon Yield
Total Carbon Yield
500C Ni-ZSM-5
500C Ni-MCM-41-Ni-ZSM-5
500C Two-stage Ni-ZSM-5 + Ni-MCM-41
500C Ni-MCM-41 + Ni-ZSM-5 mixture
(b)
Carbon Yield (mol%)
0
10
20
30
40
CO
CO2
CH4
(c)
500C Ni-ZSM-5
500C Ni-MCM-41-Ni-ZSM-5
500C Two-stage Ni-ZSM-5 + Ni-MCM-41
500C Ni-MCM-41 + Ni-ZSM-5 mixture
Carbon Yield (mol%)
0
2
4
6
8
10
12
14
Alkanes
MAH
Naphthalenes
Other PAHs
Phenols
Other Oxygenates

630063 CFD Model Investigation of Syngas Production Process through Chemical Looping Gasification of Biomass in Fluidized Bed Reactor

Abstract:

Chemical looping gasification as an integrated technology in energy conversion of biomass fuel is a promising route for producing syntheses gasses (H_2 and CO) from carbonaceous feedstocks with a very low energy penalty and cost. In this work, Eulerian multicomponent multiphase flow simulation is carried out to investigate the performance of CLG system in bubbling fluidized bed (BFB) reactor using iron-based oxygen carrier as partial combustion oxygen provider and biomass as dispersed solid fuel. The kinetic theory of granular particle is applied to evaluate constitutive properties of the solid phase of granular flow. Chemical kinetics of biomass pyrolysis, char gasification, and metal oxide reduction along with drag coefficients model are employed to consider the effects of homogenous and heterogeneous reactions, heat, and momentum exchanges between gases and solid particles. The predicted time averaged outlet concentrations of gas components agree well with the experimental data from the literature. The impact of operation temperature, biomass particle size and gasifying agent flow on gas products during the CLG process are evaluated.

623347 Upcycling Plastic Waste into Plastic Derived Cooking Fuel in Developing Countries

Abstract:

Plastic waste is a critical challenge in developing countries, however, implementing circular economy solutions to address this challenge are problematic due to the lack of the necessary infrastructure to collect, sort and remanufacture plastic. Waste management solutions for developing communities must be based on the principles of appropriate technology to be successful. One potential solution is the conversion of post consumer polyolefin plastic waste into fuel oil by the process of slow pyrolysis. The process for converting polyolefin plastic in to fuel oil is well known, but the impacts of using this fuel in a developing world setting have not been properly explored. We will present the results of research to measure the use of cooking fuel produced from plastic waste as a clean cooking fuel. The results of laboratory and field experiments to measure the air emissions from the use of plastic derived fuel oil as a cooking fuel will be presented. The results of PM2.5, PM10, CO, NOx and SOx measurements using protocol established by the Clean Cooking Alliance will be included. This research is important because it addresses to global challenges in the developing world: 1) the accumulation of mismanaged post-consumer plastic waste, and 2) the issue of lung disease, especially in women, caused by open fire indoor cooking.

623589 A Comparison of Solar Methanol and Ammonia Production Explicitly Accounting for Intermittency

Abstract:

The combustion of fossil fuels in the transport sector contributes to anthropogenic CO_2 emissions and climate change. Sustainable fuel concepts are needed to replace established fuels and move toward a sustainable transport sector. Ways to decarbonise transport are electric vehicles (EVs), hydrogen fuel cell vehicles (HFCVs), biofuels, and carbon-neutral synthetic fuels (CNSF). While EVs are making strong advances in short-distance and personal transport, CNSFs may be needed for decarbonising long-distance and heavy goods vehicles. Among CNSFs, carbon-based fuels such as methanol have received considerable attention. However, in a post-fossil economy the prevalence of carbon is not a given. Nitrogen-based fuels like ammonia should be carefully examined. Not only is CO_2 more dilute in air, its reduction is more energetically expensive compared to that of N_2. Furthermore, sustainable fuels are required to be produced using renewable energy. When the energy source of the process is intermittent, as is the case with wind, solar, and tidal energy, intermittency needs to be taken into account in the design of the process. When the analysis of sustainable fuels only includes the synthesis plant assuming fixed cost and availability for the building blocks, the cost of intermittency is neglected and importance of flexibility in processes is ignored.

We developed a framework for the comparison of sustainable fuel production from air, water, and renewable energy. We perform simultaneous design and scheduling of sustainable fuel production including renewable energy source, air separation, electrolysers, and synthesis. All power requirements have to be met by the renewable energy source; heat requirements can be met via heat integration, residual heat requirements are met using hydrogen. In our work, we compare solar methanol and ammonia in London and Dubai with regard to process design and operation as well as cost.

Our results show that the optimal route depends on the location. In London, where the solar pattern exhibits high seasonality, the route requiring the least amount of gas storage is optimal. In Dubai, with lower seasonality, the most energy efficient route, requiring the least amount of PV and electrolyser capacity, is the optimal choice. Figure 2 shows the optimal routes for methanol.

We further determined that nitrogen-based fuels may be competitive with carbon-based fuels. When taking into account the energy needed to separate CO_2 or N_2 from air, it becomes apparent that the production of nitrogen-based fuels requires less energy and hence may be less costly. Our results support that carbon's dominance in fuels may not prevail throughout the decarbonisation of transport.

We quantify the cost of intermittency to be as high as two thirds of the total cost. Figure 3 shows the cost of intermittency in London and Dubai for methanol and ammonia. This implicates that for CNSF, specifically solar fuels, the intermittency needs to be taken into account when making cost predictions. The value of flexibility of processes is accordingly high. Operating the synthesis plant flexibly reduces the seasonal storage needed and with it the overall cost. This implies that novel fuel production processes, such as one-step synthesis like photocatalytic ammonia production, or more flexible processes, may have advantages over established production routes even if they have higher individual CAPEX.

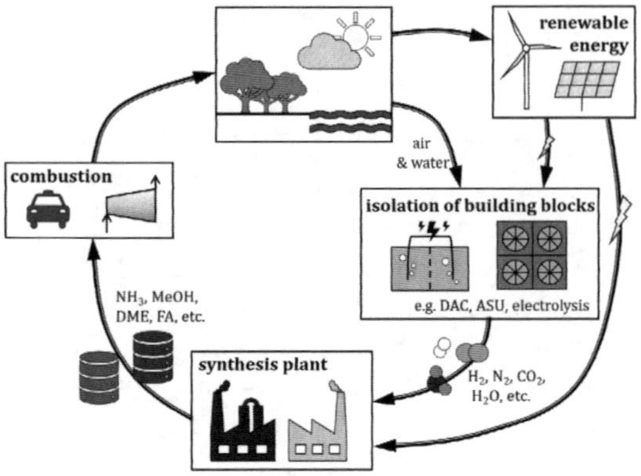

Figure 1: Circular economy for fuels.

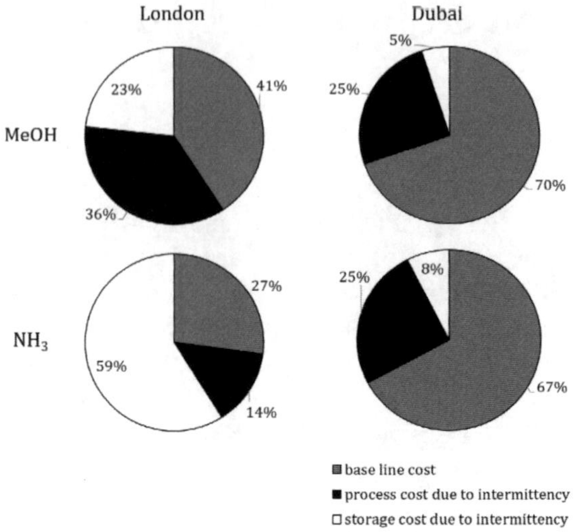

Figure 3: Cost of intermittency.

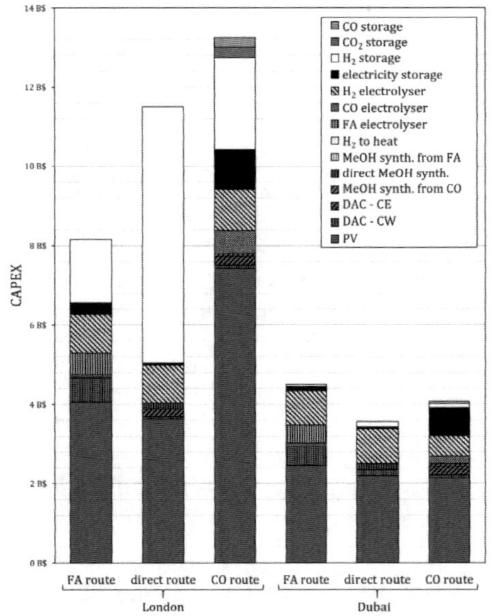

Figure 2: Comparison of methanol production routes via formic acid (FA) and carbon monoxide (CO) as well as direct route in London and Dubai.

*** Formatting Issue - Best Available Paper/Graphic ***

623630 Decarbonisation of Power and Industry in the UK

Abstract:

Power and industry are the largest contributors to point source CO_2 emissions in the UK. Their decarbonisation is critical for the UK to reach carbon targets and minimise climate change and the associated social and economic loss. The two sectors vary greatly with regard to their pathways to net-zero. While decarbonisation of the power sector is well-analysed, industrial emissions are considered notoriously difficult to abate. Many options for low- and zero-carbon power (renewables, carbon capture and storage (CCS), storage) exist, and the provision of negative emissions through bioenergy with CCS (BECCS) can lead to carbon-negative electricity. In industry, no clear pathways forward exist, many technologies rely on CCS, and substantial residual emissions are to be expected. In addition to residual and negative emissions, power and industry are connected through the additional power demand arising from abatement in industry. Much of the important work on industrial decarbonisation is currently focused on individual technologies. However, there is interest in analysing the whole of industry and how it might reach net-zero together with the power sector, which is where this study is aimed at.

We build our model on ESO[1], a power systems model, extend it to include industry, and apply it to the UK. The cement, steel and refinery sectors and options for abatement in these sectors are modelled explicitly. Remaining industrial sectors, such as chemical industry, paper, natural gas processing, are included as fixed emissions requiring offsets. Abatement for these small point sources is difficult to achieve and likely uneconomical unless incentives are particularly high. The additional power demand caused by abatement in industry is added to the baseline power demand. We model import and export of industrial products explicitly and in the first part of the results analyse four scenarios with regard to abatement and trade. In the "BAU & offset" scenario industry is restricted to conventional technologies, and BECCS is required to provide offsets for the entire sector. For "abate & offset", abatement is made available. In the "import & offshore" scenario, all industrial products are imported and therefore the emissions offshored. Lastly, the "abate & export" scenario assumes an increase in UK production of low-carbon products and subsequent export. A net-zero carbon target on the total emissions with a linear trajectory is enforced for all scenarios. For the second part of the results, we investigate how trajectories to net-zero might be incentivised using a carbon price (CP) and negative emissions credit (NEC).

Figure 1 shows the results for industrial production for the four scenarios. When abatement is available, oxy-combustion is the optimal choice for cement, production with biochar and CCS is determined preferable for steelmaking, and a high amount of post-combustion capture (PCC) is optimal for refineries. Notably, all optimal abatement technologies require CCS. Abatement is deployed in the abate & offset scenario – indicating it is preferred over BECCS offsets. Examination of the emissions trajectories and emissions for transport and storage in figure 2 reveals that residual and negative emissions are highest (54 Mt-CO_2/yr) when all industrial emissions are offset (BAU & offset). CCS in industry reduces the emissions by 90%, resulting in only 3 Mt-CO_2/yr in residual emissions from cement, steel and petrochemicals. When industrial emissions from the three sectors are offshored, the negative emissions requirement is only marginally lower. This is one indicator that the economic damage from offshoring industrial emissions might not be worth the reduction in system cost. The impact of an abate & export scenario on the power sector is minor as residual emissions from cement, steel and refining are small. Consequently, strategies for exporting low-carbon British products can be pursued without major adjustments in power. The CO_2 emissions requiring transport and storage range between 65 and 130 Mt-CO_2/yr, with the abate & offset scenario capturing 85 Mt-CO_2/yr. With only carbon budgets to drive decarbonisation, CCS is deployed first in the industrial sector in 2035, then CCGT-CCS and BECCS are added in the power sector.

The decarbonisation of power and industry can be incentivised using a carbon price (CP), *i.e.*, a penalty paid by all emitters in the system, and a negative emissions credit (NEC), *i.e.*, a reward per t-CO_2 removed

***** Formatting Issue - Best Available Paper/Graphic *****

by negative emissions technologies (in this case, BECCS). For this study, ranges of 0-300 £/t-CO_2 for the CP and 0-200 £/t-CO_2 for the NEC are simulated. The results are detailed in figure 3. Many combinations of CP and NEC are able to deliver net-zero in our model (thin solid contour line). Higher combinations of both lead to cumulative net-zero (thick solid contour line). NEC values between 90-135 £/t-CO2 across the range of CP values achieve net-zero in 2050. A NEC alone allows to reach net-zero at the minimum cost to the private sector. In this scenario, no abatement in industry is deployed and the burden of decarbonisation is placed entirely on BECCS. Adding the CP reduces the cost to the public sector. It also decreases cumulative emissions (not shown). Indicated in the graph is the break-even line for the public sector. For every NEC granted to the negative emissions technologies, there is a minimum CP necessary to offset the cost to the public. An even higher CP allows the public sector to use the penalties on emitters to support other decarbonisation efforts, such as funding R&D. The highest possible CP and the lowest required NEC constitute the public sector optimum, combining the highest income through penalising emissions and lowest expenditure towards BECCS. In our model, this public sector optimum is located at a NEC of 90 £/t-CO_2 and a CP of 300 £/t-CO_2. This optimum is preferable to the private sector optimum from a burden-sharing perspective, as it leads to CCS in industry, renewables expansion, and therefore a lower demand for BECCS. Arguably, a CP this high may place an undue burden on emitters. A more sensible optimum might therefore be located at the CP beyond which no significant decreases in cumulative emissions are gained.

In this work, we have modelled power and industry together and evaluated trajectories to net-zero. We found that emissions in cement, steel and refining can be reduced by 90% with technologies heavily relying on CCS. Under carbon budgets, CCS is deployed first in industry and later in power. We quantified the required amount of carbon offsets for varying scenarios as well as the volume of captured CO_2. We identified combinations of CP and NEC that incentivise decarbonisation to net-zero and evaluated optimal pathways from both the private and public sector perspective. Future work may involve including more industrial sectors explicitly and modelling industrial demand side response.

References

1. Heuberger CF, Rubin ES, Staffell I, Shah N, Mac Dowell N. Power capacity expansion planning considering endogenous technology cost learning. *Appl Energy*. 2017;204:831-845.

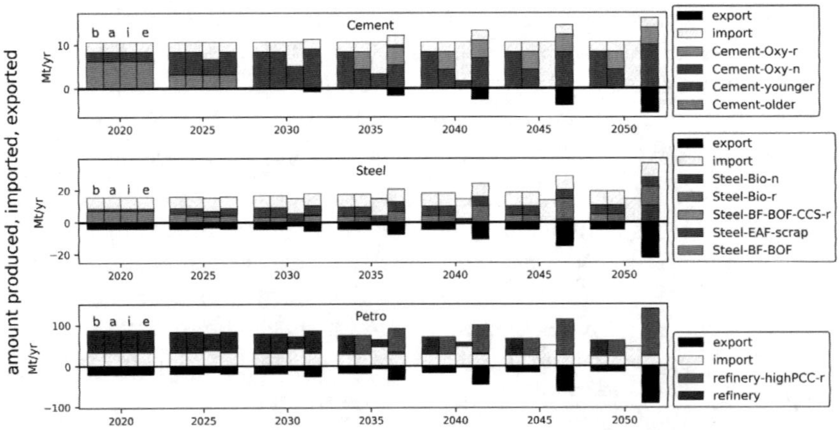

Figure 1: Industrial sector results for BAU & offset (b), abate & offset (a), import & offshore (i) and abate & export (e) scenarios.

*** Formatting Issue - Best Available Paper/Graphic ***

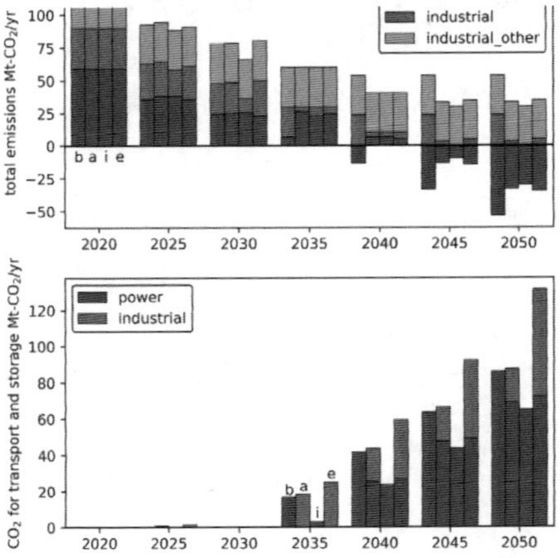

Figure 2: System emissions and emissions for transport & storage for BAU & offset (b), abate & offset (a), import & offshore (i) and abate & export (e) scenarios.

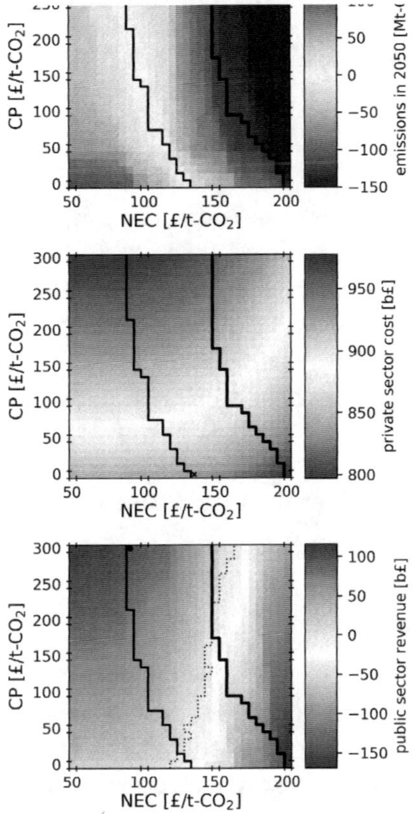

Figure 3: Emissions in 2050, private sector cost and public sector revenue for varying carbon price (CP) and negative emissions credit (NEC). Thin and thick solid contour lines represent boundaries for net-zero in 2050 and cumulative net-zero, respectively. ··· shows the break-even for the public sector. × denotes private sector optimum. • denotes public sector optimum.

630583 Remediating Corn Ethanol Waste Stillage with *Chlorella vulgaris*

Abstract:

Corn Ethanol industries require energy and cost intensive operations for the treatment of corn ethanol generated stillage before safe disposal to the environment. The purpose of this study is to observe the viability of introducing the microalgae *Chlorella Vulgaris* to a pond simulated environment and characterize its growth on nutrients derived from the stillage and Bold's Basal Medium (BBM) which contains required nutrients such as NPK. The Corn Ethanol stillage contains significant amounts of Nitrogen in various forms along with considerable organic matter presence indicated by COD tests, both of which are good sources of nutrition for *Chlorella Vulgaris*.

Growth conditions for the cultivations of *C. Vulgaris* were first screened in mini-photo bioreactors with working volumes of 35 mL and average light intensity of 100 μmol m^{-2} s^{-1} incident to the light source facing lateral surface area of the reactors. Air flow to reactors was averaged over seven days to 1.49 min^{-1} Vessel volumes per min (VVM). Change in Nitrate (mmol L^{-1}) and COD concentrations (mg L^{-1}) were observed and growth was characterized by spectrophotometric absorbance and biomass density (mg L^{-1}). Based on the results, we scaled up experiments to working volumes of 650 mL in a Phenometrics PBR. The culture to media volume ratios were kept the same (1:9), the vvm was kept at 1.27 min^{-1} to match the 35 ml PBRs. The incident light intensity on the surface of the reactor volume was set to 1500 μmol m^{-2} s^{-1} and this was kept constant by maintaining the initial reactor volume via addition of 25 ml of fresh media daily to account for sampling volumes (25ml) extracted. Experimental variables were the growth in illuminated environment with light intensity of 1500 μmol m^{-2} s^{-1} and dark environment, culture media composition was tweaked to either include or exclude the presence of nitrogen to observe the nitrogen consumption, solely on the stillage, and Additional control experiments were performed without algae to determine the effects of aeration and light. The Change in nitrate concentration (mmol L^{-1}), the change in absorbance (at 680nm), The growth of biomass density (g L^{-1}), the decrease in COD (mg L^{-1})and the decrease in total Nitrogen (mg L^{-1}) were sampled daily over a seven day period. Using these measures growth was characterized on the basis of biomass productivity (*P*, g L^{-1} hr^{-1}), specific growth rate (μ, hr^{-1}), Removal Efficiency, R.E. (%) and elimination capacity (mg L^{-1} hr^{-1}).

We observed the highest Removal Efficiency of 70.6 % for N_2, 87.2 % for COD for *C. Vulgaris* in the culture environment with light and increased nitrogen concentration. The removal efficiencies of COD was similar for both light and dark environments which suggests COD assimilation is isn't dependent on light intensity. The highest overall elimination capacities for N_2 (0.65, 0.74 mg L^{-1} $hr^{-}1$) and COD (7.38, 7.57 mg L^{-1} hr^{-1}) were achieved by the *C. Vulgaris* in the culture environments with increased nitrogen concentration (dark, light). The specific growth rate over seven days was observed to be highest for *C. Vulgaris* in the culture environment with light and increased nitrogen concentration (0.0136 hr^{-1}) and was almost 70 % higher than the next highest values. Plotting the biomass productivity of *C. Vulgaris* in different cultivation environments indicated a more sustained productivity for the *C. Vulgaris* in the culture environment with light and increased nitrogen concentration.

630782 A Community-Level Energy Planning Model for the Dynamic Optimization of Waste-to-Gas and Waste-to-Power Anaerobic Digestion Supply Chains

Abstract:

Society has growing concerns over the long-term environmental impacts of climate change and the accumulation of organic waste in local communities. Community energy planning is increasingly focused on the use of novel technologies to provide sustainable and cost-effective heat and power. Furthermore, changes in energy demand patterns support the development of dynamic optimization models of energy dispatch. This study focuses on the use of anaerobic digestion (AD) technology for converting human, agricultural, and industrial waste into renewable natural gas (RNG) and power. We develop a dynamic mixed integer linear programming (MILP) in Calliope to optimize the production and dispatch of RNG and power.

Anaerobic Digestion (AD) is a microbiological process that is used in processing organic, biodegradable wastes and side streams into methane-rich biogas and nutrient-rich digestate. Previous techno-economic (TEA) and lifecycle (LCA) analyses indicate that AD can be a commercially competitive and sustainable technology for communities to generate clean energy. In this study, we evaluate the use of AD in Ames, Iowa to convert human and animal waste, biomass, and wastewater into RNG and power. RNG and power are consumed locally to provide industrial, commercial, and residential heat and power demand. The AD process generates by-product digestate that is returned to Iowa farms. The model covers simulates 1 year of energy demand in hourly and monthly increments.

The MILP identified 13 locations in the vicinity of Ames, Iowa capable of producing power or RNG. The electricity price ranged between \$0.065 - \$0.18/kWh and the RNG price is estimated at \$0.026/kWh (\$7.62/mmbtu). Electricity production varied between 30 and 50 million kWh per month. Gas production varied between 15 and 40 million kWh per month. All the electricity demand in Ames, Iowa could be supplied from local waste resources. Summer gas demand can be supplied locally. However, the gas supply will need to be supplemented with gas from the external grid to meet winter demand. Technological improvements and public incentives could increase the adoption of AD in community energy planning systems. Further research will evaluate the impacts of variabilities in energy supply and demand to address grid disruptions.

631619 CO_2 Capture Using Amine-Functionalized Fumed Silica Via Molecular Layer Deposition

Abstract:

CO_2 capture from power generation sources provides a concentrated CO_2 stream and the possibility of incorporating the captured CO_2 into useful industrial processes. However, in order to implement these technologies on the scales we require, we must develop economically feasible methods and materials. Traditional CO_2 capture technology relied on liquid amines that requires substantial energy to strip the CO_2 that leads to substantial costs for the process.[1] Addressing the limitations of the liquid process, solid adsorbent materials have become a research focus over the past few decades. Popular materials have included zeolites, activated carbons, metal organic frameworks and supported amines. Supported amines are an attractive material as they take advantage of the traditional basic chemistry, are tolerant of humidity, and require moderate regeneration conditions. As such, they have received much attention in literature as researchers vary the functionalization method, support materials, and amine-containing groups to increase the adsorption capacity. Current processes of loading the amines through impregnation face poor regeneration stability and transport limitations, while grafted sorbents tend to have low adsorption capacities.[2] Mesoporous silicas such as MCM-41 and SBA-15 are very popular support materials due to high surface areas above 800 m^2/g. However, these materials are not produced on commercial scales and are expensive to make which would inhibit the implementation of this technology.

This research focuses on development of a regenerable, high capacity, and inexpensive supported amine adsorbent materials. Here, we introduce particle molecular layer deposition (MLD) as a novel functionalization method to improve upon the limitations of current methods. MLD is closely related to atomic layer deposition (ALD). Both processes involve stepwise reactions between gaseous reactants on a solid surface to create thin films.[3] Here we use MLD to deposit a covalently bonded aminopropylsiloxane network on fumed silica nanoparticles (380 m^2/g). Two MLD chemistries have been confirmed: (3-aminopropyl)triethoxysilane (mono-amine) and N1-(3-trimethoxysilylpropyl)diethylene triamine (tri-amine). Both precursors were deposited at 150°C. Adsorption capacity and regeneration stability were investigated using a thermogravimetric analyzer (TGA). Adsorption capacity of the amine functional groups increased to ~2mmol/g or 0.005 $mmol/m^2$ as the number of MLD cycles increased. On the basis of surface area, this capacity is competitive with other surface functionalization methods and well performing adsorbents in literature. Low regeneration temperatures were demonstrated by both the mono-amine and tri-amine. The MLD sorbents with 10 cycles of MLD underwent 25 regeneration cycles with adsorption at 30°C and desorption at 80°C. Both materials appear stable and robust over these cycling conditions. These materials will be stable for adsorption/desorption cycling temperatures below the deposition temperature of 150°C. Preliminary cost approximations estimate the cost of this sorbent is less than $10/kg. This work demonstrates a new functionalization method and capture material to the field of solid CO_2 capture research that could aid in widespread CO_2 capture implementation.

(1) Wang, J.; Huang, L.; Yang, R.; Zhang, Z.; Wu, J.; Gao, Y.; Wang, Q.; O'Hare, D.; Zhong, Z. Recent advances in solid sorbents for CO2capture and new development trends. *Energy Environ. Sci.* **2014,** *7* (11), 3478.

(2) Ünveren, E. E.; Monkul, B. Ö.; Sarıoğlan, Ş.; Karademir, N.; Alper, E. Solid amine sorbents for CO 2 capture by chemical adsorption: A review. *Petroleum* **2017,** *3* (1), 37.

(3) George, S. M. Atomic Layer Deposition: An Overview. *Chemical Reviews* **2010,** *110* (1), 111.

632292 Lignin-First Strategy for Upgrading Lignin into Value-Added Products: Maximizing Feedstock Utilization

Abstract:

Lignocellulosic biomass - a bountiful source of renewable carbon for the sustainable production of fuels and chemicals - is composed of 70-85 wt.% polysaccharides (cellulose and hemicellulose) and 15-30 wt.% lignin, a heteropolymer of aromatic compounds. For a successful bio-refinery, it is imperative to maximize the value obtained from all lignocellulosic biomass fractions. Polysaccharides have been extensively studied and various processes have been developed to produce valuable liquid fuels and commodity chemicals from the polysaccharide fraction of lignocellulosic biomass. Lignin, being difficult to fractionate and process, is generally burnt for its calorific value. However, lignin is the largest source of renewable aromatics and, as such, strategies to valorize lignin are required for a bio-refinery to economically compete with fossil-fuel refinery.

"Lignin-first" strategies have been employed to valorize lignin before upgrading polysaccharide fractions [1]. This strategy liberates lignin in "native-like" form from the plant cell wall and further prevents degradation through either catalytic processing [2] or by protection-group chemistry [3].

In our previous work, we developed a lignin-to-bioproduct processing chain for the production of 2-pyrone-4,6-dicarboxylic acid (PDC) through microbial funneling of the phenolic monomers obtained from catalytic depolymerization of lignin [4]. We used chemical and biological upgrading in tandem to extract greater value from the lignin fraction by converting a fraction of lignin to high value PDC (Figure 1A). In this previous work, we first isolated lignin from lignocellulosic biomass under mild reaction conditions using g-valerolactone (GVL) and water as the solvent system and dilute sulfuric acid as a catalyst. In this scheme, lignin degradation was minimized during biomass fractionation step due to the protection by the solvent system and low process temperatures (<120 °C). However, the techno-economic analysis of the processing chain showed the cost of lignin isolation to have the largest impact on the minimal-selling-price (MSP) of PDC.

In this work, we modified the lignin-to-bioproduct processing chain by eliminating the costly biomass fractionation step (Figure 1B). In the new scheme, we combine biomass fractionation and catalytic depolymerization of the lignin into a single step, while simultaneously preserving the structure of the polysaccharide fraction of the lignocellulosic biomass. We demonstrate that the lignin fraction of the whole-cell-wall poplar can be successfully depolymerized by hydrogenolysis over a Pd/C catalyst into a mixture of monomeric and oligomeric phenolic compounds. We show that, using this strategy we were able to increase the monomer/oligomer product yield on a per kg of biomass basis. We further show that using an engineered strain of *Novosphingobium aromaticivorans* DSM12444, this complex mixture of aromatic compounds containing syringyl, guaiacyl, and *p*-hydroxyphenyl can be upgraded to PDC. Moreover, we show that PDC can be extracted from the culture broth with a simple separation and purification step (*e.g.*, precipitation with sodium chloride). Furthermore, the sugar stream was subjected to enzymatic and microbial digestion to liquid fuels to produce value added products from both the lignin (phenolics) and polysaccharide fractions.

Combining these improvements over the previous process we demonstrate that the economics of PDC production from lignocellulosic biomass can be further improved. Most importantly, we show that tandem processes utilizing both chemical and biological upgrading can significantly improve the upgrading of a complex feedstock like lignocellulosic biomass.

References:

[1] Abu-Omar M.M., Barta K., Beckham G.T., Luterbacher J.S., Ralph J., Rinaldi R., Roman-Leshkov Y., Samec J.S.M., Sels B.F., Wang F., Energy Environ. Sci., 2021, 14, 161-292.

[2] Luterbacher J.S., Azarpira A., Motagamwala A.H., Lu F., Ralph J., Dumesic J.A., Energy Environ. Sci., 2015, 8, 2657-2663.

[3] Shuai L., Amiri M.T., Questell-Santiago Y.M., Heroguel F., Li Y., Kim H., Meilan R., Chapple C., Ralph J., Luterbacher J.S., Science, 2016, 354(6310), 329-333.

[4] Perez J.M., Umana G.E., Sener C., Misra S., Coplien J., Haak D., Li Y., Maravelias C., Karlen S.D., Ralph J., Donohue T.J., Noguera D.R., "Integrating lignin depolymerization and microbial funneling processes using agronomically relevant feedstocks", in preparation.

This material is based upon work supported by the Great Lakes Bioenergy Research Center, U.S. Department of Energy, Office of Science, Office of Biological and Environmental Research under Award Number DE-SC0018409.

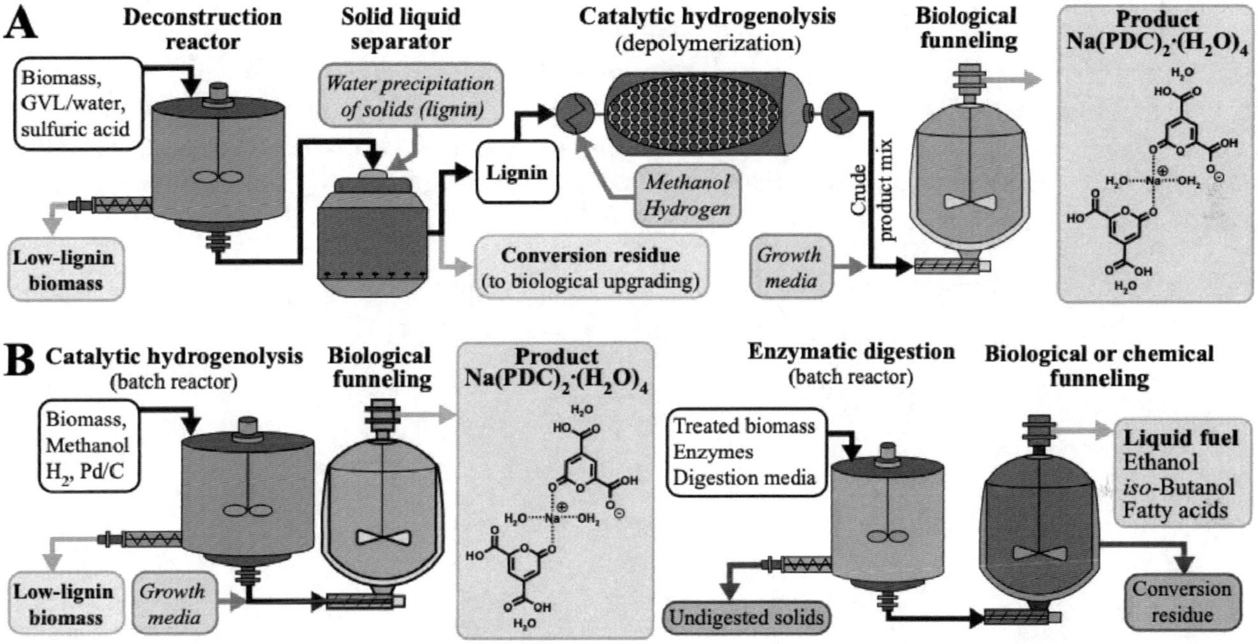

Figure 1. Lignin-to-bioproduct processing chain (A) with biomass fractionation (B) without biomass fractionation.

632848 Water Retention Value As a Characterization Approach for Predictive Modeling of Corn Stover Deconstruction

Abstract:

Chemical and physical heterogeneity in herbaceous biomass feedstocks due to substantial differences in plant tissue types can contribute significant challenges to handling, preprocessing, and conversion in biorefining processes. A second, related challenge is quantifying this heterogeneity within feedstocks as it relates to processing in a biorefinery. Water sorption properties vary significantly with plant tissue and particle properties. In this work, we will apply several techniques for quantifying water-biomass interactions in combination with other characterization approaches for developing predictive models. These models will predict relative abundance of anatomical fractions and response to processing by pretreatment and enzymatic hydrolysis of air-classified corn stover. One technique, water retention value (WRV) assay, is employed as a tool for screening biomass response to pretreatment and enzymatic hydrolysis in combination with other characterization approaches. We have further developed this assay to increase response sensitivity and identify the operating limits, while the impact of particle size, centrifugation speed and filter loading on WRV were explored.

624850 Pretreatment of Used Cooking Oils and Recovery of Free Fatty Acids Via Liquid-Liquid Extraction in a Falling Film Contactor

Abstract:

Exploitation of waste lipids in the production of valuable green chemicals and biofuels is becoming an extended practice worldwide. Currently these wastes are mainly used in the manufacture of biodiesel and green diesel (i.e. hydrotreated vegetable oils), but they can only be harnessed after suitable pretreatment to reduce problematic impurities that affect the valorization processes. Some of the most challenging and abundant waste lipids are the used cooking oils (UCOs) and trap greases. These residues are globally generated, and when mismanaged, they generate a cascade of problems mainly in urban centers. Alternatively, UCOs and waste greases could be reclaimed and harnessed as raw material in the production of value-added chemicals.

Despite the economic potential and the environmental benefits, harnessing of such waste lipids via chemical and biochemical transformations requires the reduction of impurities in the waste material. Due to the processing and storage conditions, and compared with traditional oleochemical feedstocks (i.e. refined oils and fats), waste lipids have higher acidity, larger humidity, and a high content of degradation products (i.e. polar compounds, polymers, gums, etc.) and contaminants (e.g. water, food residues, packing residues, etc.). In order to reduce such impurities, waste lipids are normally processed via filtration, degumming to reduce water soluble compounds, neutralization to reduce acidity, and vacuum distillation to reduce volatiles and water content. Despite the process is very effective, it is highly energy- and materials-intensive, producing large losses due to acid neutralization, and generating large amount of residues (e.g. spent filtration earths, soapstocks, distillation toppings). In this regard, alternative pretreatment processes are required to reduce waste exploitation costs, to reduce losses and to mitigate emissions.

As an alternative approach, solvent extraction can be used as a pretreatment process. The solvents can be selected in such a way that can enable a further valorization of the impurities. In this regard, this work assessed the pretreatment of used cooking oils by using liquid-liquid extraction with alcohols. While the acylglycerides in UCOs are barely soluble in alcohols, the free fatty acids can be readily dissolved, thus obtaining a reactive mixture for further esterification. In this case acidity and other polar compounds are removed from the waste lipids, while the impurities can be recovered in the alcoholic phase and further valorized.

Initially, preliminary tests were carried out at the lab scale to identify the best operating conditions regarding temperature, UCO-to-solvent mass ratio, and water content in the solvent. Takins into account liquid-liquid equilibria and the subsequent valorization routes, acidity removal was assessed using ethanol and ethanol as solvents. The extraction process was carried out under isothermal condition at 40, 50, 60 ºC in a stirred jacketed reactor. The UCO-Solvent mass ratio was assessed in-between 1:1 to 1:2, and the water content in the solvent was varied in the range of 0.5 to 5% wt. The effectiveness of the process was monitored by tracing acid value, humidity, color, total polar content, iodine value, peroxide value, and saponification value of the pretreated UCO.

Once the best operating conditions regarding impurities removal were obtained in the lab scale, a subsequent evaluation was carried out in a pilot-scale falling liquid film contactor. In this case, a counter-current multi-filament packed column was used as contactor under continuous operation at constant temperature. According to results, the liquid-liquid extraction process effectively reduced the acid content of UCO to a value less than or equal to 0.05%, with a simultaneous reduction of polar compounds (25%) and other impurities. The extraction process helped to enhance the color of the UCO (30%) which is positive for further valorization. Additionally, the operation under laminar flow conditions enabled a fast phase splitting in the decanters reducing liquid-liquid separation and enhancing process performance. Finally, and to validate the process concept, the extract was subjected to further esterification to verify the potential harnessing of the extracted free fatty acids obtaining high conversion to fatty esters.

621699 The Hybrid Kinetic-Optimization Modeling and Machine-Learning (ANN and SVM) of Biomass (Maple Leaf) Pyrolysis

Abstract:

Biomass can be utilized to generate heat and valuable chemicals through pyrolysis. Hundreds of chemicals can be produced during the process. Kinetic models involving more reaction steps were generally developed to provide detailed information on the process. However, finding proper kinetic parameters for each reaction step is very time-consuming. In this work, a hybrid kinetic-optimization scheme was proposed to address this issue. A 24-parallel reactions mechanism was developed in the hybrid scheme to describe the thermal decomposition of maple-leaf. A kinetic model using the 24-parallel reaction mechanism was established and coupled with an optimization model to find optimal kinetic parameters of each reaction. The hybrid kinetic-optimization model predicted well-matched results at various conditions. In addition to the kinetic model, the machine learning on biomass pyrolysis was also studied. The artificial neural network (ANN) and support vector machine (SVM) methods were applied to describe pyrolysis of maple-leaf. The impact of important factors for the machine learning methods were investigated. The backpropagation algorithms such as the Levenberg-Marquardt, Bayesian Regularization, Scaled Conjugate Gradient, Adam Stochastic Gradient were examined to ensure that the machine learning models can predict accurate and "unbiased" results. Finally, the performances of the kinetic models and machine learning models were compared and evaluated to explore better modeling options for biomass pyrolysis.

627576 Optimization Framework for the Electroreduction of CO_2 into Chemicals

Abstract:

The past few years have seen rapid advances in the experimental design of the electroreduction of carbon dioxide (ECO2R) for the synthesis of chemicals and fuels, leading to its foreseeable industrial implementation. A transition to carbon-free chemicals could have a considerable impact on the future of the chemical industry but requires holistic methods to assess the technological, economic, and environmental performance of ECO2R processes and optimize process network designs. In this work, we propose a methodology to assess the large-scale implementation of ECO2R based on the analysis of market data, traditional process synthesis, techno-economic assessment, and supply chain optimization.

We build upon previous studies that have dealt with the modeling and assessment of ECO2R processes (Jouny et al., 2018; Orella et al., 2019; Roh et al., 2020), hybrid fossil- and CO_2-based routes (Ioannou et al., 2020) and supply chain optimization of carbon utilization processes (Leonzio et al., 2020; Zhang et al., 2020). Our method consists of a multi-level approach to network synthesis with detailed models of each one of the technologies involved. The network is modeled as a superstructure that includes different options of carbon capture, electrolysis, and separation.

To illustrate the capabilities of the method, we define a case study on the substitution of fossil-based ethylene, ethanol, and formic acid by their CO_2-sourced alternatives. Different scenarios based on technology adoption and efficiency, electricity prices, and CO_2 source are considered to target the break-even conditions necessary to adopt the technology. The results obtained show that the model is a useful tool to identify the most promising routes to convert CO_2 into valuable chemicals. The methodology is valuable for the multiple actors involved in the waste revaluation sector, from private companies to policymakers.

References

Ioannou, I., D'Angelo, S.C., Martín, A.J., Pérez-Ramírez, J., Guillén-Gosálbez, G., 2020. Hybridization of Fossil- and CO 2 -Based Routes for Ethylene Production using Renewable Energy. ChemSusChem cssc.202001312. https://doi.org/10.1002/cssc.202001312

Jouny, M., Luc, W., Jiao, F., 2018. General Techno-Economic Analysis of CO 2 Electrolysis Systems. Ind. Eng. Chem. Res. 57, 2165–2177. https://doi.org/10.1021/acs.iecr.7b03514

Leonzio, G., Foscolo, P.U., Zondervan, E., 2020. Optimization of CCUS Supply Chains for Some European Countries under the Uncertainty. Processes 8, 960. https://doi.org/10.3390/pr8080960

Orella, M.J., Brown, S.M., Leonard, M.E., Román-Leshkov, Y., Brushett, F.R., 2019. A General Technoeconomic Model for Evaluating Emerging Electrolytic Processes. Energy Technol. 1900994. https://doi.org/10.1002/ente.201900994

Roh, K., Bardow, A., Bongartz, D., Burre, J., Chung, W., Deutz, S., Han, D., Heßelmann, M., Kohlhaas, Y., König, A., Lee, J.S., Meys, R., Völker, S., Wessling, M., Lee, J.H., Mitsos, A., 2020. Early-stage evaluation of emerging CO 2 utilization technologies at low technology readiness levels.

Green Chem. 22, 3842–3859. https://doi.org/10.1039/C9GC04440J

Somoza-Tornos, A., Giraldo-Carvajal, V., Espuña, A., Graells, M., 2019. Targeting material exchanges in industrial symbiosis networks, Computer Aided Chemical Engineering. https://doi.org/10.1016/B978-0-12-818634-3.50255-1

Zhang, S., Zhuang, Y., Liu, L., Zhang, L., Du, J., 2020. Optimization-based approach for CO2 utilization in carbon capture, utilization and storage supply chain. Comput. Chem. Eng. 139, 106885. https://doi.org/10.1016/j.compchemeng.2020.106885

627651 Material Property Targets for Emerging Adsorptive Water Treatment and Resource Recovery Systems

Abstract:

There is a critical need to develop sustainable separations technologies to meet current and emerging challenges, for example, i) the removal of trace metals such as lead from water to ensure acceptable contaminant levels[1] and ii) the recovery of lithium from brines to keep up with the raw material demand due to vehicle electrification[2]. Yet, the synthesis of numerous advanced materials in the past three decades have failed to produce a corresponding revolution in separation applications[3]. In this talk, we present a mathematical modeling framework for material property targets in general adsorption separation systems; using the framework, we benchmark existing materials and identify theoretical limits for the lead removal and lithium recovery applications [4].

In the paradigm of process targeting[5], feasible high-level goals such as solute recovery targets are fixed before detailed process optimization. Once targets are fixed, fundamental thermodynamics and material and energy balances are used to find alternate, often non-trivial pathways to achieve the targets. Process targeting is commonly applied to flowsheet optimization wherein the individual unit operations are optimized after setting a feasible overall production (or utility consumption) target[6]. Traditionally, design is carried out using sequential techniques in which unit operations are optimized in a logical order which is most often the sequence based on the process flowsheet. Sequential design techniques often place a huge demand on the downstream processes which makes them very difficult to design efficiently. Process targeting overcomes the limitations of sequential design by explicitly accounting for inter-unit interactions early in the design process.

We extend process targeting to study adsorptive separation systems from a multiscale perspective: we set process targets at the device-scale to study the feasibility of using existing adsorbent materials as the separating medium. We use material balances and isotherms to set process targets. At the materials scale, we use structure-property relationships to define the theoretical limits of adsorptive membranes and packed-bed adsorbers. We demonstrate that while there is still scope for improvement, the sorbents for next-generations separations such as the lead removal from water and lithium extraction from brine already exist and are limited by engineering at the device-scales or higher. In other words, our framework helps illuminate if specific applications are process- or materials-limited. We highlight our belief that adsorption-based devices will play a key role in realizing the fit-for-purpose paradigm of water treatment which places emphasis on treating water to meet the varied specifications of its end users[7].

References

[1] Hanna-Attisha, M., LaChance, J., Sadler, R., & Schnepp, A. (2016). Elevated Blood Lead Levels in Children Associated With the Flint Drinking Water Crisis: A Spatial Analysis of Risk and Public Health Response. *American Journal of Public Health*, 106(2), 283–290. https://doi.org/10.2105/AJPH.2015.303003

[2] Li, H., Eksteen, J., & Kuang, G. (2019). Recovery of lithium from mineral resources: State-of-the-art and perspectives – A review. *Hydrometallurgy*, 189. https://doi.org/10.1016/j.hydromet.2019.105129

[3] Luo, J., & Crittenden, J. (2019). Nanomaterial Adsorbent Design: From Bench Scale Tests to Engineering Design. *Environmental Science & Technology,* 53(18), 10537–10538. https://doi.org/10.1021/acs.est.9b04371

[4] Eugene, E. A., Phillip, W. A., Dowling, A. W., (2020). Material Property Targets for Emerging Nanomaterials to Enable Point-of-Use and Point-of-Entry Water Treatment Systems. *Preprint submitted to ACS ES & T Engineering*
https://chemrxiv.org/articles/preprint/Material_Property_Targets_for_Emerging_Nanomaterials_to_Enable_Point-of-Use_and_Point-of-Entry_Water_Treatment_Systems/12526190/1

[5] Fox, J. A., & Stacey, N. T. (2019). Process targeting: An energy based comparison of waste plastic processing technologies. *Energy (Oxford)*, 170, 273–283. https://doi.org/10.1016/j.energy.2018.12.160

[6] Hildebrandt, D., Glasser, D., Patel, B., Sempuga, B. C., & Fox, J. A. (2015). Making processes work. *Computers & Chemical Engineering*, 81, 22–31. https://doi.org/10.1016/j.compchemeng.2015.03.022

[7] Zodrow, K., Li, Q., Buono, R., Chen, W., Daigger, G., Dueñas-Osorio, L., Elimelech, M., Huang, X., Jiang, G., Kim, J.-H., Logan, B., Sedlak, D., Westerhoff, P., & Alvarez, P. (2017). Advanced Materials, Technologies, and Complex Systems Analyses: Emerging Opportunities to Enhance Urban Water Security. *Environmental Science & Technology,* 51(18), 10274–10281. https://doi.org/10.1021/acs.est.7b01679

628097 Integrated Waste-to-Energy and Energy Management Platform: A Demonstration in an Eco-Industrial Park

Abstract:

The rapid expansion of urban populations leads to the increases in the generation of municipal solid waste (MSW) and energy consumption, causing great pressure on the waste treatment and energy systems. Waste-to Energy (WtE) technologies, which processes waste into electricity, heat, and biogas, are popular waste treatment alternatives to landfill that address the problem of rising energy consumption simultaneously. Nowadays, both the conventional centralized MSW and energy management systems are evolving toward a more decentralized form with smaller capacity. This shift would requires an adapted decision-support method to assist decision-makers in analyzing and optimizing the WtE systems. In this work, an integrated waste-to-energy and energy management platform was developed and it is comprised of two parts: 1) waste-to-energy decision-support system (WTEDSS) and 2) energy management decision support system (EMDSS).

The WTEDSS addresses the process of electricity generation from waste. Firstly, the historical amount of waste generation is analysed to projects future waste generation for different waste streams. Then, mixed-integer linear programming (MILP) optimization is used to determine the optimum capacities of waste treatment facilities (centralized vs decentralized, anaerobic digestion (AD) vs incineration) and the allocation of waste intakes from different waste generation sites for different facilities for each year by maximizing the net present value (NPV) of the waste management system.

The EMDSS deals with the management of electricity generated by the WtE facilities using a game-theory-based demand-side management (DSM) incorporating energy storage components. The DSM is used to reduce the Peak-to-Average ratio smoothen the dips in load profile caused by supply constraints. A novel data-driven optimization framework is incorporated to select and size energy storage systems at suitable locations of the electricity network. The emerging blockchain technologies are also introduced to guarantee the seamless and secure implementation of the proposed energy management scheme, illustrating how a decentralized DSM approach could be implemented in practice.

As a demonstration, the integrated platform is applied to Jurong Island, an eco-industrial park in Singapore. The platform help decision-makers with examining the design and operation of the waste-to-energy and energy management systems under different scenarios. A case study showed that decentralized waste-to-energy systems were preferred if the current capital cost of the decentralized AD system could be lowered. The incorporation of DSM and energy storage systems were able to maintain the supply and demand balance, reduce utility bills of consumers, and reduce stress on grid in daily operation.

630716 Design and Integration of Thermal Energy Storage Systems for Power Plants

Abstract:

Integration of intermittently available renewable energy with electricity grids leads to increased variability and uncertainty in power generation, thereby pushing many conventional power plants to speedily ramp their loads to maintain grid reliability. Considerable loss of power plant energy is incurred due to rapid cycling and time delay during energy transmission. Frequent cycling can also induce low overall efficiency and high failure rates of high-temperature units in the long run [1]. Energy storage can resolve these challenges. Among the several existing energy storage alternatives, thermal energy storage is particularly promising because of cost-effective design, operational flexibility, high efficiency, stability and long storage lifetime. Thermal storage has been widely employed in the cooling and heating system of buildings, chemical processes as well as solar plants in the past [2-5]. However, the optimal design and integration of thermal energy storage with large-scale fossil-fueled power plants have not been investigated rigorously. There are two types of thermal energy storage, namely high-temperature thermal storage (HTTS) and cryogenic energy storage (CES). HTTS employs molten salt or phase change materials (PCM) as storage medium. The operating temperature of molten salt ranges from 70 to 800 °C. Conversely, the operating temperatures of PCM are between 130-580 °C. Depending on the storage medium, process configuration and operation, the costs range from \$0.5–60 per kg of the material used. CES provides a highly efficient mechanism to store excess energy by liquefying air to obtain cryogen [6-8]. The stored cryogen is further used to generate energy during high energy demand scenarios. To increase efficiency of the CES process, cryogenic turbines and internal heat transfer loops with intermediary thermic fluids are used for energy retention. There also exists a large range of options for thermal fluids, storage pressures, trains for compressing and cooling, and expansion cycles for energy generation, which affect the capital and operational cost of a CES system.

We propose a model-based process design and synthesis approach for thermal energy storage systems that store energy from high-/medium-/low-pressure steam extracted from the power plant and discharge energy on demand by regenerating steam from stored energy which is sent back to the power plant. The designs are complicated as the heat transfer involves both sensible and latent heat, different types of storage media, various techno-economic and energetic trade-offs, and numerous possible configurations and operational stages for heat integration. To this end, we have developed a superstructure configuration of the storage systems, thereby allowing the various alternative configurations to adapt different charging, discharging, and integration strategies effectively. The superstructure optimization problem is formulated as a mixed-integer nonlinear program (MINLP), which is then solved to identify the optimal process configurations, number of cycles, charging/discharging strategies, and storage medium. The objective function and the constraints consider various trade-offs between ramp rates, load balances, lifespan, and capital costs, and leading to optimal energy storage operations. Our results indicate that the performance and the cost of energy storage depend strongly on charging steam conditions and the qualities of both charging and discharging steam. Using realistic power plant conditions and charging/discharging demands, we illustrate how the MINLP-based synthesis of storage configurations provides the target process configurations, capacities, and operational decisions with an efficiency of 85%-90%, thereby reducing the overall cost of energy storage for a conventional power plant. The insights obtained from the process synthesis study is further incorporated in developing surrogate models for dynamic integration of storage technologies for distributed energy storage for different scenarios. The development of these technology models will pave way for solving an overall storage downselection problem for reducing the cycling of conventional power plants in face of intermittent renewable energy integration.

References:

[1]Kumar, N., Besuner, P., Lefton, S., Agan, D., & Hilleman, D. (2012). Power Plant Cycling Costs. doi: 10.2172/1046269

[2]Edwards, J., Bindra, H., & Sabharwall, P. (2016). Exergy analysis of thermal energy storage options with nuclear power plants. *Annals of Nuclear Energy*, *96*, 104–111. doi: 10.1016/j.anucene.2016.06.005

[3]Richter, M., Möllenbruck, F., Starinsk, A., Oeljeklaus, G., & Görner, K. (2015). Flexibilization of Coal-fired Power Plants by Dynamic Simulation. *Proceedings of the 11th International Modelica Conference, Versailles, France, September 21-23, 2015*. doi: 10.3384/ecp15118715

[4]Miró, L., Gasia, J., & Cabeza, L. F. (2016). Thermal energy storage (TES) for industrial waste heat (IWH) recovery: A review. *Applied Energy*, *179*, 284–301. doi: 10.1016/j.apenergy.2016.06.14

[5]Zhu, Y., Zhai, R., Peng, H., & Yang, Y. (2016). Exergy destruction analysis of solar tower aided coal-fired power generation system using exergy and advanced exergetic methods. Applied Thermal Engineering, 108, 339–346. doi: 10.1016/j.applthermaleng.2016.07.116

[6]Zhang, Q., Grossmann, I. E., Heuberger, C. F., Sundaramoorthy, A., & Pinto, J. M. (2015) Air separation with cryogenic energy storage: optimal scheduling considering electric energy and reserve markets. AIChE Journal. doi: 10.1002/aic.14730

[7]Li, Y., Cao, H., Wang, S., Jin, Y., Li, D., Wang, X., & Ding, Y. (2014) Load shifting of nuclear power plants using cryogenic energy storage technology. Applied energy. doi: 10.1016/j.apenergy.2013.08.077

[8] Abdo, R. F., Pedro, H. T., Koury, R. N., Machado, L., Coimbra, C. F., & Porto, M. P. (2015). Performance evaluation of various cryogenic energy storage systems. Energy. doi: 10.1016/j.energy.2015.08.008

632270 A Multiscale Electro-Chemical Model for Simulating Dendrite Formation in Lithium-Ion Batteries

Abstract:

The growing market demand for Lithium-ion batteries (LIBs) with a high energy density has given an impetus to a plethora of studies investigating the utilization of Lithium anode, which is considered as the 'Holy-Grail' anode due to its high energy density (~ 3860 mAh/g) as compared to traditional graphite anodes (~ 372 mAh/g) [1]. However, commercial application of Li metal anode has not been possible because it is plagued by irregular Li deposition, which causes dendrite formation and results in poor cycling efficiency, capacity fade, and in extreme cases, short-circuiting of batteries [2]. Previous studies show that dendrite formation is majorly affected by microscopic battery parameters [1] (i.e., mechano-chemical properties of the SEI species), and macroscopic operating conditions (i.e., cell voltage, current density, and Li-ion concentration) [3]. However, the complex interaction between these two types of factors (i.e., macroscopic and microscopic) has not been adequately addressed in the literature.

Inspired by this, we have constructed a multiscale model that combines a high-fidelity kMC simulation with an electrochemical continuum model, which in this case is a modified form of the single-particle model (SPM) [4]. Specifically, the kMC model employs a series of microscopic events like the desolvation of Li-ions from the electrolyte, their diffusion in the SEI,[1][6] and finally the electrodeposition of Li-ions on the anode to simulate the microscopic dendrite formation phenomenon [5],[6]. The electrochemical model considers Fickian diffusion of Li-ions in the electrolyte region to trace the spatio-temporal evolution of , and then couples this with various equations to determine cell voltage and the state of charge (SOC). The proposed multiscale model is simulated for varying current densities at different values of initial Li-ion concentration to investigate its effect on microscopic dendrite formation (quantified using roughness), and the temporal evolution of various macroscopic variables. Furthermore, the simulation results for dendrite growth and SOC (which is a measure of the charging status of LIBs) are successfully validated with experimental results from the literature.

Overall, the proposed multiscale electrochemical model can (a) generate the evolution of macroscopic variables such as current density, Li-ion concentration, cell voltage, and SOC, (b) explain their effect on the microscopic spatio-temporal evolution of dendrites (kMC model), and (c) be extended to simulate multiple charge-discharge cycles to help devise effective dendrite mitigation strategies in future studies.

Literature Cited:

1. Qin X, Shao M, Balbuena PB. Elucidating mechanisms of Li plating on Li anodes of lithium-based batteries. Electrochemica Acta. 2018 Sep 10; 284:485-94.
2. Wen J, Yu Y, Chen C. A review on lithium-ion batteries safety issues: existing problems and possible solutions. Materials Express. 2012 Sep 1;2(3):197-212.
3. Hao F, Verma A, Mukherjee PP. Mechanistic insight into dendrite–SEI interactions for lithium metal electrodes. Journal of Materials Chemistry A. 2018;6(40):19664-71.
4. Zhang D, Popov BN, White RE. Modeling lithium intercalation of a single spinel particle under potentiodynamic control. Journal of the Electrochemical Society. 2000 Mar 1;147(3):831.
5. Sitapure N, Lee H, Ospina-Acevedo F, Balbuena PB, Hwang S, Kwon JS. A computational approach to characterize formation of a passivation layer in lithium metal anodes. AIChE

Journal. 2021 Jan; 67(1): e17073.

6. Angarita-Gomez S, Balbuena PB. Insights into lithium ion deposition on lithium metal surfaces. Physical Chemistry Chemical Physics. 2020;22(37):21369-82.

624138 Computationally Efficient Distillation Energy Targeting Model for Superstructure-Based Process Synthesis

Abstract:

Separation processes account for a significant fraction of the total energy consumption in chemicals and fuels facilities[1]. Thus, to design energy efficient systems, especially during the early design phase, it is important to develop methods that allow us to easily estimate separations-related energy requirements. Accordingly, given the importance of distillation in the process industries, the goal of this work is to develop an improved method for the calculation of energy demand of complex distillation networks. Shortcut methods, which do not rely on rigorous mass-energy balances, are particularly useful for this purpose[2,3]. In particular, the Underwood equations[2] have been adopted in many works because of their simplicity, that is, the computationally cheap calculation of minimum vapor flow rates to achieve the separation task and from the flow the calculation of the minimum energy requirement. Notably, to solve the Underwood equations, components that are present in the stream to be separated need to be known a priori so that all relevant roots of the Underwood equations can be calculated.

However, when synthesizing an integrated reactor-separation system, component flow rates of the stream to be separated can vary, and some components can have even zero flow rates, depending on decisions in the upstream reactor subsystem(s). Thus, components in the stream to be separated may not be known a priori. Such streams are referred to as *undetermined*[4]. The presence of undetermined streams complicates the use of the Underwood equations; for instance, existing approaches disaggregate the stream so that each disaggregated stream can have a known set of components. Then, each disaggregated stream is connected to a separate distillation network/column. This approach can easily lead to a large optimization problem and therefore prevent the use of Underwood equations (and other shortcut models) when multiple undetermined streams are present.

To address this limitation, we propose a novel reformulation of the Underwood equations, which allows the automatic calculation of all relevant roots efficiently even for undetermined streams. We introduce binary variables to detect components in the stream to be separated, and these binary variables are then used to constrain the Underwood roots accordingly. Also, we introduce strong valid constraints which significantly enhance computational efficiency. Then, utilizing this reformulation, we propose a distillation energy targeting model inspired by the Fully-Thermally-Coupled (FTC) configuration[5], which has the minimum energy demand among all configurations for the distillation of a zeotropic mixture. The proposed model can be used to calculate the energy target to separate a stream, potentially undetermined, into pure components without finding detailed distillation network configurations. Thus, it is particularly useful for preliminary process synthesis with multiple candidate reactions, where many undetermined streams can naturally appear. Also, we introduce some extensions to consider cases where some components do not need to be separated from each other. For example, for fuel production, some components do not need to be separated from each other if all property constraints can be met.

We present a number of case studies. First, we present the preliminary synthesis of a multi-stage itaconic acid production process. Then, we present the preliminary synthesis of ethanol upgrading bio-refinery, which has more complexity. These case studies illustrate how the proposed energy targeting model can be used to formulate and solve interesting and challenging process synthesis problems.

References

1. Sholl DS, Lively RP. Seven chemical separations to change the world. *Nature News*. 2016;532(7600):435
2. Bausa J, Watzdorf RV, Marquardt W. Shortcut methods for nonideal multicomponent distillation: I. Simple columns. *AIChE journal*. 1988;44(10), 2181-2198.
3. Underwood AJ V. Fractional distillation of multicomponent mixtures. *Ind Eng Chem*. 1949;41(12):2844-2847.
4. Ryu J, Maravelias CT. Computationally efficient optimization models for preliminary distillation column design and separation energy targeting. *Computers & Chemical Engineering*. 2020;143:107072.
5. Halvorsen IJ, Skogestad S. Minimum energy consumption in multicomponent distillation. 3. More than three products and generalized Petlyuk arrangements. *Ind Eng Chem Res*. 2003;42(3):616-629.

624709 Dynamic Simulation of an Ammonia Synthesis Plant Fed By Stranded Natural Gas in Aspen Hysys

Abstract:

The Haber Bosch process is one of the most important industrial processes to modern society. In 2015, 160 million tons of ammonia was produced mainly from the Haber Bosch process, with a global value of $250 billion [1,2]. The process has benefited from over 100 years of improvements and modifications. Today, the Haber Bosch process is being modified to operate on alternative feedstocks in an attempt to preserve the world's natural gas reserves [2]. Modifying the Haber Bosch process to use Stranded Natural Gas (SNG) as a feedstock, can allow for the monetization of an unexploited resource [3]. SNG, is natural gas that is in a location that is uneconomical to recover due to its quantity, composition or both. SNG sources are intermittent and the quality of natural gas from the source can fluctuate dramatically. Previously a Haber Bosch process fed by SNG was studied at steady state [3], which does not accurately capture the transient nature of the feed source. Dynamic simulation of a Haber Bosch process fed by SNG will allow for accurate: operational, economic and environmental assessments of the modification.

This presentation will communicate the development, performance and validation of a dynamic model of a 100 tNH_3/day plant in Aspen Hysys [4]. The process of Fig. 1 uses feed streams of natural gas, water, and air to synthesize ammonia in a three-stage fixed bed reactor. Steam reforming of methane is performed to generate hydrogen from a variable stranded natural gas feed modeled after natural gas wells in the Bakken Basin of North Dakota. A water gas shift reactor is employed to convert CO from the steam reformer to CO_2. The CO_2 is removed ahead of the ammonia synthesis reactor using a Monoethanolamine stripping unit. Ammonia is synthesized at 200 atm and 450°C, and is then refrigerated to -20°C before being stored. The model includes heat integration of wastewater streams to reduce water consumption and pressure relief systems to prevent over-pressurization of system components. The model also includes controllers programmed to control system temperature, pressure, and flows. The steady state results of the model are compared and agree with the performance of previous works [4]. The following case studies are presented in this paper: startup under variable natural gas composition and flow, and normal operation under variable natural gas composition and flow. The model is found in good agreement with prior work and, therefore, it can be used as a benchmark virtual testbed for evaluation of intensified schemes of ammonia synthesis such as those presented in [4].

Acknowledgements
This work was partially sponsored by the United Technologies Corporation Institute for Advanced Systems Engineering (UTC-IASE) of the University of Connecticut. Any opinions expressed herein are those of the authors and do not represent those of the sponsor.

References

[1] Fao, *World fertilizer trends and outlook to 2018;* Annual Report 14; Food and Agriculture Organization of the United Nations (FAO): Rome, Italy, 2015; ISBN 978-92-5-108692-6.

[2] L. Wang *et al.*, "Greening Ammonia toward the Solar Ammonia Refinery," *Joule*, vol. 2, no. 6, pp. 1055–1074, 2018.

[3] Z. Gou, Monetization of Stranded Gas through Ammonia and Urea Production, Texas A&M University 2017

[4]. L. Burrows *et al.*, (In Review) "Dynamic Simulation of an Ammonia Synthesis Plant fed by Stranded Natural Gas in Aspen Hysys^{â}," *Chem. Eng. J.*

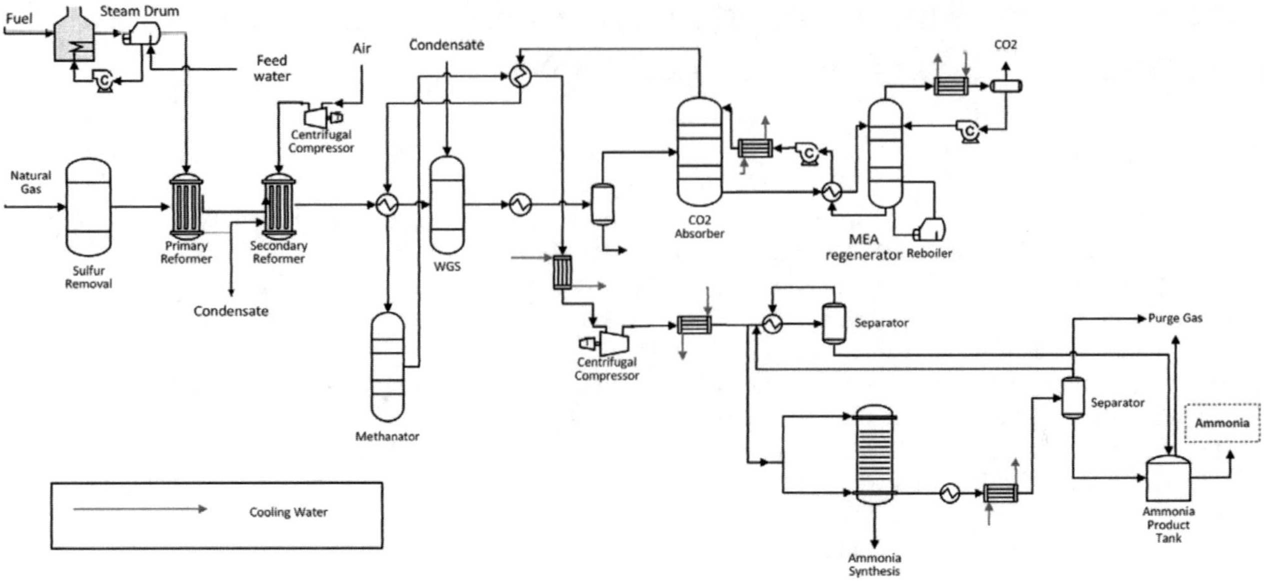

Fuel
Steam Drum
Feed water
Air
Condensate
CO2
Centrifugal Compressor
Natural Gas
Primary Reformer
Secondary Reformer
Sulfur Removal
Condensate
WGS
CO2 Absorber
MEA regenerator
Reboiler
Separator
Purge Gas
Centrifugal Compressor
Separator
Ammonia
Methanator
Cooling Water
Ammonia Product Tank
Ammonia Synthesis

625781 Effects of Microporous Structure on Transient Diffusion and Adsorption of Hydrolytic Enzymes into Biomass

Abstract:

Generation of biofuels from lignocellulosic biomass is an emerging technology which is being investigated all around the world. Enzymatic hydrolysis is a major step in the contemporary process of generation of biofuels. There are many factors which affect the yield of enzymatic hydrolysis such as particle size, porosity, density, maximum adsorption capacity of the biomass about which the data is very limited. Understanding the impact of these parameters on the diffusion and adsorption of enzymes into the biomass is essential for achieving maximum conversion. Guided by pore size distribution measured using NMR cryoporometry, in this study we develop pore-enzyme diffusion models for both adsorbing and non-adsorbing enzymes. By simulating these models in MATLAB software package, we investigate the effects of various biomass particle-related parameters (particle dimensions, porosity) on the characteristic time of enzyme diffusion and adsorption for various biomass types. The model for diffusion of adsorbing and non-adsorbing enzymes is in agreement with the characteristic times from the literature. The model was applied to predict characteristic times for diffusion and adsorption of hydrolytic enzymes into dilute acid-pretreated biomass particles (hybrid poplar).

632608 Characterization of Cellulose Biopolymer Synthesizing Enzymes Reconstituted *in Vitro*

Abstract:

Cell walls are one of the most important constituent of plant cells composed predominantly of complex polysaccharides. One such critical plant cell wall polysaccharide component composed of b-1,4-linked glucose units is cellulose. This abundant biopolymer is processively synthesized by integral membrane proteins called cellulose synthases (CesAs). These CesAs are arranged in a rosette-shaped manner and forms a larger complex containing six trimers of individual CesA units. Together they form a cellulose synthase complex (CSC). These cellulose synthase complexes contain several different cellulose synthase isoforms involved in primary and secondary cell wall formation. To understand how they synthesize cellulose at an enzymatic level, we are heterologously expressing two putative CesAs in *Pichia pastoris* and reconstituting them into lipid carriers. Such a system will aid in understanding the structural and biochemical characteristics of this challenging enzyme involved in the synthesis of this essential biopolymer. The reconstituted enzyme polymerizes UDP-glucose to cellulose by releasing UDP as detected by luminescence assays. The synthesized cellulose will be characterized using microscopic and enzymatic deconstruction studies. Ultimately, this work will help advance our mechanistic and holistic understanding of both *in vitro* and *in vivo* cell wall synthesis that would accelerate the development of improved transgenic crops for bioenergy and bioproducts relevant applications.

631068 Lignin Valorization By Integrating Chemical Depolymerization and Microbial Funneling: New Strategies to Produce 2-Pyrone-4,6-Dicarboxylic Acid

Abstract:

Lignocellulosic biomass - a bountiful source of renewable carbon for the sustainable production of fuels and chemicals - is composed of 70-85 wt.% polysaccharides (cellulose and hemicellulose) and 15-30 wt.% lignin, a heteropolymer of aromatic compounds. For a successful bio-refinery, it is imperative to maximize the value obtained from all lignocellulosic biomass fractions. Polysaccharides have been extensively studied and various processes have been developed to produce valuable liquid fuels and commodity chemicals from the polysaccharide fraction of lignocellulosic biomass. Lignin, being difficult to fractionate and process, is generally burnt for its calorific value. However, lignin is the largest source of renewable aromatics and, as such, strategies to valorize lignin are required for a bio-refinery to economically compete with fossil-fuel refinery.

"Lignin-first" strategies have been employed to valorize lignin before upgrading polysaccharide fractions [1]. This strategy liberates lignin in "native-like" form from the plant cell wall and further prevents degradation through either catalytic processing [2] or by protection-group chemistry [3].

In our previous work, we developed a lignin-to-bioproduct processing chain for the production of 2-pyrone-4,6-dicarboxylic acid (PDC) through microbial funneling of the phenolic monomers obtained from catalytic depolymerization of lignin [4]. We used chemical and biological upgrading in tandem to extract greater value from the lignin fraction by converting a fraction of lignin to high value PDC (Figure 1A). In this previous work, we first isolated lignin from lignocellulosic biomass under mild reaction conditions using g-valerolactone (GVL) and water as the solvent system and dilute sulfuric acid as a catalyst. In this scheme, lignin degradation was minimized during biomass fractionation step due to the protection by the solvent system and low process temperatures (<120 °C). However, the techno-economic analysis of the processing chain showed the cost of lignin isolation to have the largest impact on the minimal-selling-price (MSP) of PDC.

In this work, we modified the lignin-to-bioproduct processing chain by eliminating the costly biomass fractionation step (Figure 1B). In the new scheme, we combine biomass fractionation and catalytic depolymerization of the lignin into a single step, while simultaneously preserving the structure of the polysaccharide fraction of the lignocellulosic biomass. We demonstrate that the lignin fraction of the whole-cell-wall poplar can be successfully depolymerized by hydrogenolysis over a Pd/C catalyst into a mixture of monomeric and oligomeric phenolic compounds. We show that, using this strategy we were able to increase the monomer/oligomer product yield on a per kg of biomass basis. We further show that using an engineered strain of *Novosphingobium aromaticivorans* DSM12444, this complex mixture of aromatic compounds containing syringyl, guaiacyl, and *p*-hydroxyphenyl can be upgraded to PDC. Moreover, we show that PDC can be extracted from the culture broth with a simple separation and purification step (*e.g.*, precipitation with sodium chloride). Furthermore, the sugar stream was subjected to enzymatic and microbial digestion to liquid fuels to produce value added products from both the lignin (phenolics) and polysaccharide fractions.

Combining these improvements over the previous process we demonstrate that the economics of PDC production from lignocellulosic biomass can be further improved. Most importantly, we show that tandem processes utilizing both chemical and biological upgrading can significantly improve the upgrading of a complex feedstock like lignocellulosic biomass.

References:

[1] Abu-Omar M.M., Barta K., Beckham G.T., Luterbacher J.S., Ralph J., Rinaldi R., Roman-Leshkov Y.,

Samec J.S.M., Sels B.F., Wang F., Energy Environ. Sci., 2021, 14, 161-292.

[2] Luterbacher J.S., Azarpira A., Motagamwala A.H., Lu F., Ralph J., Dumesic J.A., Energy Environ. Sci., 2015, 8, 2657-2663.

[3] Shuai L., Amiri M.T., Questell-Santiago Y.M., Heroguel F., Li Y., Kim H., Meilan R., Chapple C., Ralph J., Luterbacher J.S., Science, 2016, 354(6310), 329-333.

[4] Perez J.M., Umana G.E., Sener C., Misra S., Coplien J., Haak D., Li Y., Maravelias C., Karlen S.D., Ralph J., Donohue T.J., Noguera D.R., "Integrating lignin depolymerization and microbial funneling processes using agronomically relevant feedstocks", in preparation.

This material is based upon work supported by the Great Lakes Bioenergy Research Center, U.S. Department of Energy, Office of Science, Office of Biological and Environmental Research under Award Number DE-SC0018409.

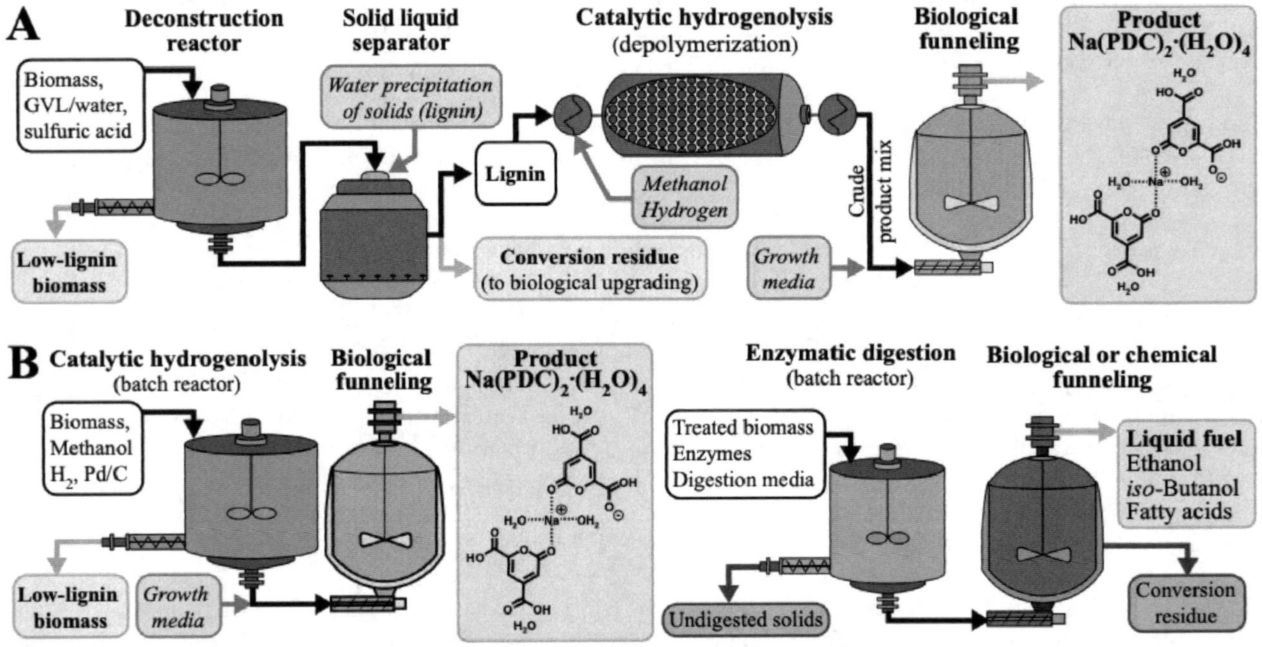

Figure 1. Lignin-to-bioproduct processing chain (A) with biomass fractionation (B) without biomass fractionation.

632470 Understanding the Impact of High Solids Loadings on Deconstruction of Celf Pretreated Poplar By *C. Thermocellum* consolidated Bioprocessing (CBP)

Abstract:

Co-solvent Enhanced Lignocellulosic Fractionation (CELF) pretreatment of Poplar was employed to produce glucan-enriched solids as a highly reactive substrate for application to consolidated bioprocessing (CBP) by wild type *C. thermocellum* DSM 1313. Our previous studies showed that deconstruction of CELF-pretreated poplar by *C. thermocellum* in shake flasks achieved nearly complete solids solubilization in 24-48 hours at glucan loadings of 5-10 g/L. However, because higher product titers would reduce the cost of downstream product recovery, total sugar solubilization was followed over CBP culture time for 50, 75, and 100 g/L of CELF-pretreated poplar solids generated at different pretreatment severities. In a run at 50 g/L solids loadings, 90% carbohydrate solubilization was achieved at lower CELF pretreatment severity while 96% carbohydrate solubilization was reached at higher severities. In order to gain insights into how solids composition, structural properties, and lignin linkages affected CBP performance, pretreated poplar and fermentation residues were both probed with NMR and Stimulated Raman Spectroscopy, and fermentation broths were characterized to determine if metabolite yields changed with solids loadings. This presentation will be report key factors that impacted the rate and extent of solids solubilization from synergistic pairing of CELF pretreatment with *C. thermocellum* CBP.

624938 Comparison of Five Acetogens for Production of C_2 – C_6 Alcohols and Acids from CO_2

Abstract:

The use of fossil fuels, which results in huge CO_2 emissions causing negative environmental impacts, is unsustainable. In addition, there are over 15 billion gallons of corn ethanol produced per year in the United States generating over 43 million tons per year of high purity biogenic CO_2. This biogenic CO_2 can potentially be converted to additional ethanol using autotrophic microorganisms. The conversion of waste carbon streams to valuable products is an emerging research field that addresses the need to reduce greenhouse (GHG) emissions, improve carbon conversion efficiency and generate additional revenues towards a circular carbon economy. Thermochemical and electrocatalysis methods have been developed to convert CO_2 to formic acid, CO, H_2 and CH_4. However, these methods suffer from a low carbon conversion efficiency and catalyst poisoning. Biological CO_2 conversion is a novel process in which the cells directly convert CO_2 at near ambient temperatures and pressures into biofuels and biobased products. This can generate additional revenue from CO_2-containing waste streams for biorefineries and other industries. The objective of this study is to compare five different acetogens (*Clostridium ragsdalei strain P11, C. carboxidivorans* strain P7A and three new clostridial strains A, B and C) for their ability to convert CO_2 to C_2 – C_6 alcohols and acids via the *acetyl-CoA* pathway. In addition, the effects of inoculum preparation on cell activity and product profiles were examined. Experiments were performed with inocula prepared with syngas ($CO:CO_2:H_2$) and $CO_2:H_2$. Inoculum preparations and fermentations were performed in 250 mL bottle assays with 50 mL working volume at 37°C and 125 – 150 rpm. After inoculum preparation, fermentations were performed with a gas mix containing $CO_2:H_2:N_2$ (20:60:20) that was fed daily for 360 h. Results showed that all strains converted CO_2 into various amounts of C_2 – C_6 alcohols and fatty acids depending on the inoculum preparation technique. All five strains prepared with syngas inocula produced alcohol. However, acetic acid was mostly produced by all strains with $CO_2:H_2$ inocula. *S*train P11 was the best CO_2 and H_2 consumer regardless of the inoculum preparation technique, producing 2 g/L ethanol and 6 g/L acetic acid from syngas inoculum. All strains with $CO_2:H_2$ inocula produced mostly acetic acid (~ 12 g/L) from CO_2, with the highest from strains P11, P7A and strain A. Notably, strain C produced ethanol, butanol, hexanol and acetic, butyric and hexanoic acids with either syngas or $CO_2:H_2$ inocula. Additional results from enzyme activities of the five strains from syngas and $CO_2:H_2$ inocula will be presented to explain the variations in C_2 – C_6 alcohols and fatty acids profiles during CO_2 conversion. This research is expected to advance our understanding of biological CO_2 conversion into valuable products with increased carbon utilization towards realizing the circular carbon economy.

624306 System-Level Analysis of Isobutanol Production in Lignocellulosic Biorefineries

Abstract:

Lignocellulosic biomass has received a lot of attention as a promising feedstock for biofuel production. The use of lignocellulosic biomass avoids the food versus fuel competition and is largely available at lower costs. Isobutanol has been considered as an interesting biofuel alternative to bioethanol. Isobutanol has higher energy content and lower vapor pressures compared to ethanol, and it can be blended with gasoline in higher proportions, resulting in blends less corrosive compared to gasoline-ethanol blends. However, the fermentation step to convert sugars to isobutanol is more challenging. Most microorganisms used in isobutanol production have relatively low tolerance to isobutanol, limiting the concentration of isobutanol achievable in the fermentation broth. Several works in the literature have suggested *in situ* product removal strategies to increase sugar conversion and isobutanol yields without achieving toxic levels of isobutanol during fermentation.

Given the advantages of isobutanol and technological feasibility of sugar-to-isobutanol fermentation, the goal of this work is to perform a system-level analysis of an isobutanol biorefinery. Our goal is to provide useful insights into improvements required to make isobutanol production more economically attractive. We use superstructure-based optimization to study different biorefinery configurations, and we evaluate the impact of critical parameters on the energy efficiency and economics of the system.

The biorefinery is represented via blocks described in terms of conversion, energy demand, and cost parameters utilized to calculate outlet component flows, heat and electricity demands, and production cost. Base values for the block parameters are calculated from the literature and offline simulations. Given that sugar conversion to isobutanol is an active area of research, parameter values for blocks associated with the sugar fermentation and isobutanol recovery are calculated over a wide range of sugar conversion using detailed process simulations. The obtained parameter values, expressed as functions of sugar conversion, are utilized in the biorefinery superstructure optimization model.

Due to the low tolerance of microorganisms to isobutanol, two configurations are considered. The first includes direct fermentation and separation of isobutanol from the fermentation broth and is applicable when sugar conversion is low. The second configuration includes a modified fermentation block to simultaneously remove isobutanol from the broth. The broth is recycled, after an isobutanol-rich vapor stream is removed, to the bioreactor diluting the fermentation broth and avoiding toxic concentrations of isobutanol.

The need for additional unit operations in the second configuration result in higher production cost and energy (i.e., heat and electricity) demand in the fermentation block. Furthermore, these values increase as a function of sugar conversion due to higher isobutanol-rich vapor flows removed from the broth. The production cost and energy demand in the separation block remain approximately constant as a function of sugar conversion.

The results of the sugar-to-isobutanol biorefinery optimization model indicate that sugar conversion is the most important driver in terms of economics. The isobutanol production cost is reduced to \$1.2/kg if sugar-to-isobutanol conversion increases to 100% of the theoretical yield. Furthermore, an additional 4–8% reduction in the production cost of isobutanol is attainable if a 50% reduction of the heat demand during isobutanol recovery can be achieved. The optimal biorefinery configuration does not change as a function of sugar conversion. The biorefinery is energetically self-sufficient at most conversion values studied, producing a surplus of electricity to be sold to the grid for extra credit. Finally, we expand our study to analyze a biorefinery co-producing isobutanol and ethanol.

628738 Synthesis of Bio-Sourced 4,4'-Dimethylbiphenyl

Abstract:

4,4'-dimethylbiphenyl (DMBP) is one of the most promising platform chemicals for the production of polymers, plasticizers, and metal-organic frameworks. For instance, 4,4'-biphenyldicarboxylic acid (BDA), prepared via the oxidation of DMBP, can be used in engineering plastics, and has been shown, as a co-monomer, that it greatly enhances mechanical properties—glass transition temperature, strength, modulus and barrier properties—of polyethylene-terephthalate (PET). However, DMBP is difficult to produce from petroleum feedstocks with good selectivity given that there are six possible isomers of similar physical properties. It is desirable to have a biomass-derived substitute that could transform non-edible sugars into advanced materials.

We have developed a two-step process to produce DMBP selectively using biomass-derived 2-methylfuran (MF) and ethylene (Scheme). The first step consists of the oxidative coupling of MF to form 5,5'-dimethyl-2,2'-bifuran (DMBF) through Pd(II)-based homogeneous catalysis, and the second step consists of the addition of ethylene to DMBF via a Diels-Alder cycloaddition-dehydration tandem reaction using heterogeneous acid catalysts. We will discuss an optimized protocol for the oxidative coupling of MF to DMBF with $Pd(OAc)_2$ as catalyst precursor. Site-time-yield (STY) of the reaction (7.0 h^{-1}) was improved by a factor of ~20 in presence of trifluoroacetic acid (TFA), compared with the initially reported STY. High O_2 pressure and high TFA concentration are necessary to improve the DMBF formation rate. Phosphoric acid-on-silica (P-SiO_2) catalyzes the Diels-Alder dehydration tandem reactions of DMBF with ethylene to generate DMBP (83% yield), with a carbon balance of ~93%. The moderate Brønsted acidity of P-SiO_2 leads to the high conversion and selectivity. This alternative provides a completely renewable path to DMBP synthesis with a combined yield of 49% and selectivity of 78%. This result enables multiple opportunities for the production of promising and sustainable polymers and chemical products.

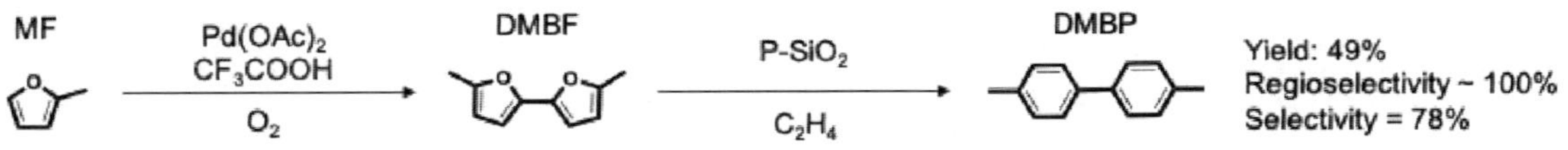

632596 Meso-Scale Studies of Simultaneous-Isomerization-Reactive-Extraction of Glucose to Fructose in a Flow Reactor

Abstract:

Glucose, a lignocellulose-derived sugar, can be converted into liquid fuels - ethanol by fermentation and other hydrocarbons by thermocatalytic conversion. In many biochemical and thermochemical processes, the yield and specificity of the reactions is much higher with fructose, the keto-isomer of glucose. However, the isomerization of glucose to fructose is equilibrium limited and does not favor fructose formation. The inability to fully utilize glucose or isomerize glucose to fructose limits the overall efficiency of the downstream chemistries and the economics of the products. We previously reported a bench-scale process - simultaneous-isomerization-and-reactive-extraction (SIRE), developed to shift the equilibrium isomerization to higher levels of fructose. Herein, we present results from experiments in a 2 L meso-scale flow set-up with mass transfer analysis of the SIRE processes to provide metrics for further scale-up. To characterize the SIRE process in the meso-scale flow set-up, isomerization and reactive-extraction were first decoupled and analyzed as stand-alone processes using a packed bed reactor and hollow fiber membrane contactor (HFMC), respectively. The sugar isomerization and extraction kinetics were compared with those from bench-scale shake-flask experiments. The packed bed reactor and HFMC module were integrated and operated together to implement the simultaneous-isomerization-and-reactive-extraction process for fructose production. A theoretical mass balance/mass transfer model of the reactive-extraction process was developed and predictions were in good agreement with the experimentally measured sugar concentration profiles. Diffusion of through the hydrophobic membrane in the extraction module was found to be the rate-limiting step for the process. Alternative contacting methods for extraction should be evaluated for the next level of scale-up.

633090 Hydrolysis of Bio-Oil Anhydrosugars Derived from Autothermal Pyrolysis of Cellulosic Biomass

Abstract:

Anhydrosugars are the major carbohydrate products of autothermal pyrolysis of lignocellulosic biomass that can be upgraded into biofuels and bio-based chemicals. Our research group has developed a fast pyrolysis process, namely autothermal pyrolysis which can produce high yields of bio-oil by thermal depolymerization of pretreated biomass. A significant fraction of the heavy ends of bio-oil contains sugars, mainly levoglucosan, but also xylose as the major carbohydrate products. While some of the anhydrosugars can be directly fermented, most yeasts prefer glucose, which can produce renewable fuels such as ethanol, butanol, and value-added platform chemicals such as organic acids and 5-hydroxymethylfurfural. The anhydroglucans can be acid-catalyzed to glucose; however, the currently available bio-oil hydrolysis processes could be potentially cost-intensive for scaling up the hydrolysis reactor and integrated biorefinery scenario.

The requirement of pressurized vessel at hydrolysis temperature (~2.7 bar at 135°C), complex handling of high concentrations of sulfuric, and difficulty of handling water-insoluble solids associated with the use of some common neutralization agents (e.g. calcium hydroxide) could increase the overall cost of production of biofuel and biochemicals from biomass. Furthermore, a high sulfuric acid concentration could be detrimental to the downstream fermentation unit if it ends up as a neutralized salt surpassing the admissible toxicity level for the fermenting microorganisms to remain reactive for glucose conversion.

We have a developed a process for hydrolyzing pyrolytic bio-oil at relatively mild conditions, with temperatures as low as 95-105°C and sulfuric acid concentrations of only 50-150 mM. A response surface method was used to optimize the hydrolysis reaction conditions for levoglucosan. Overall, a 95% conversion of levoglucosan could be achieved and a high concentration of glucose (160.8 g/L) could be produced at 115°C using only 150 mM sulfuric acid at 3 h of reaction time. Modeling of bio-oil hydrolysis shows a good overall fit with first order conversion of levoglucosan to glucose with possibility of other C_6 anhydrosugars conversion to excess glucose.

628903 Preparation of Pt/ZrO_2 Catalyst with Different Morphologies for the Selective Hydrogenation of Vegetable Oil

Abstract:

In the field of food industry, a large amount of trans fatty acids are produced during the hydrogenation of vegetable oils. Scientific studies have shown that high intake of trans fatty acids can cause cardiovascular disease. As the requirement of trans fatty acids in food becomes more and more strict, it is imperative to reduce the content of trans fatty acids in hydrogenated vegetable oils. Therefore, reducing the formation of trans fatty acids during hydrogenation has important engineering practical implications. However, trans fatty acids are more easily generated during the reaction. In order to overcome this adverse effect, designing a catalyst that can accelerate the formation of cis fatty acids is an ideal method to improve the reaction selectivity. At present, monatomic catalysts are widely used in hydrogenation because of their better catalytic performance and higher atomic utilization. In this work, the single-atom Pt catalysts supported on ZrO_2 with different morphologies (monoclinic structure, tetragonal structure, cubic structure) were prepared as highly efficient and stable catalysts for selective hydrogenation of vegetable oil. The prepared single-atom Pt/ZrO_2 catalyst (tetragonal structure) had a high cis fatty acid yield (about 50%). According to the reaction results, the catalytic performances of Pt/ZrO_2 catalysts with different morphologies were different. XPS and O_2-TPD showed that there were abundant oxygen vacancies on the surface of ZrO_2, and the number of oxygen vacancies on the surface of ZrO_2 with different morphologies was different. Combined with the results of reaction and in-situ infrared analysis, these oxygen vacancies can regulate the electronic structure on the surface of the catalyst, enhance the electron enrichment, and promote the adsorption and activation of reactants. DFT calculation showed that oxygen vacancies on Pt/ZrO_2 catalyst (tetragonal structure) affected the adsorption form of the C=C bond, thus promoting the formation of cis fatty acids.

628159 Aqueous Phase Reforming of Glycerol for Production of Hydrogen over Non-Noble Metal Nano-Catalysts

Abstract:

During the bio-diesel production process, glycerol is surplus as a by-product. In order to fully utilize oversupply glycerol, a quantity of method has been established to produce value-added chemicals, such as propylene glycol, lactic acid. Because of the increasing environmental issues, a concept of carbon neutral and green hydrogen has been put forward. Therefore, many researchers are devoted to transforming glycerol into green hydrogen through reforming reaction. However, the glycerol reforming reaction is a steam reforming reaction in most cases, and requires the use of a high content of precious metals, which have restricted its large-scale application in industry. In an effort to reduce energy consumption and economic cost, non-noble metal catalysts are developed to catalyze this reaction in aqueous phase. Moreover, we attempt to coupling aqueous phase reforming reaction with the hydrogenation reaction of phthalic acid to 1,2-cyclohexanedicarboxylic acid. The aqueous phase reforming reaction have a lower conversion compared with published paper. Consequently, more efficient hydrogen production catalysts should be exploited.

628253 A Study of Thermal Management in Commercial Face Masks

Abstract:

The indisputable and widespread Coronavirus disease (COVID-19) has caused tremendous social life changes. COVID-19 has expeditiously taken the lives of innocent individuals. Combined with social distancing and quintessential hygiene habits, facemasks can slow the coronavirus spread and protect wearers and the people around them. This project aims to study the thermal management of commercially available face masks using FLIR One thermal camera. The goal is to enhance the user's thermal comfort as heat generated from prolonged use causes fatigue, skin rash, and discomfort to an individual. In this experiment, a healthy young man volunteered to wear the masks and conduct the study on himself. Three commercially used masks, KN95, disposable masks, and reusable cloth masks were utilized in this experiment. The volunteer wore each mask for 15 minutes, and the thermal images were continuously captured using the FLIR One camera five feet apart. The breathing temperature in the masks was recorded as a function of time. The ambient temperature and humidity were also recorded at the time of the experiments. The Heat Flow Rate equation was used to find the heat transfer through each mask's membranes. A gradual increase in temperature on the mask's surface was observed. The thermal images show that the KN95 has the best thermal management overtime. The morphology and chemical structure of the different layers of the mask were investigated using the Dino-lite premier and the Attenuated Total Reflection (ATR) respectively. The rate of heat flow of the Non-surgical disposable masks is higher than its counterparts in comparison. Face masks with better thermal management will be helpful for prolonged use and enhance comfort. The future goal of this work is to study the effects of Stimuli-responsive polymer to advance thermal management of the ideal mask.

Keywords: Thermal management, thermal image, face mask, coronavirus.

633277 Synthetic biology platforms for the sustainable production of superior flavor and fragrance ingredients

Abstract:

In the wake of the looming climate crisis and the rapid transformation in both economy and society caused by the coronavirus pandemic, corporations see an opportunity to redouble their efforts towards a new, more sustainable development model for the years to come. In the past decade, biotechnology and synthetic biology have made great advances in the sustainable biomanufacturing of precision chemicals that are clean, safe, natural, and have a low carbon footprint. The flavors and fragrances market is estimated at over $20 billion and is one of the industries that is experiencing a more rapid transition from chemical to natural. Moreover, the industry's susceptibility to supply chain disruptions has prompted the major F&F companies to invest heavily in biotechnology, as it does allow a very rapid development and commercialization of new ingredients, with delocalized production independent of environmental and geopolitical instability, yielding products of reliable and reproducible quality. In this context, Conagen has developed synthetic biology platforms that leverage both bioconversion and fermentation-based production of ingredients with improved sustainability, cost, and quality to support the F&F industry and others. Conagen's Terpenoid, Lipid & Lactone, and Phenolic & Flavonoid platforms will be introduced. Besides low-calorie, high-intensity sweeteners, sweetener enhancers and bitter blockers, Conagen's product portfolio includes several fruit and vanilla flavors and floral, fruity, citrusy, creamy and woody fragrances. Some of these are existing ingredients with improved sustainability, cost, and/or quality attributes while others are novel ingredients with improved performance or that have been developed to fill a market gap. Overall, Conagen offers one of the largest and most diverse portfolios of SynBio-derived F&F ingredients in the industry.

624866 On the Quantitative Investigation of the Antimicrobial Efficacy of Grape Seed Extracts Against Food-Related Bacterial Pathogens.

Abstract:

INTRODUCTION

Over the last years, the trends in food consumption have changed significantly. Consumers are demanding products that are minimally processed, without the use of chemical preservatives or antibiotics. In addition, there is a growing concern about the role of antimicrobial resistance (AMR) in disease outbreaks. Antimicrobial resistance (AMR) is driven by the extensive-use of antibiotics and preventive measures need to be taken against their excessive use[1]. Moreover, due to climate change, environmental awareness is more present as compared to the past, urging the food industry follow a more sustainable approach on the production design. Traditional processing methods such as pasteurization or sterilization cannot meet all the above demands and novel processing control strategies should, therefore, be developed[2].

The possibility to use natural antimicrobials to replace chemical preservatives and the extensive-use of antibiotics has been one of the major goals of the food industry. Fruit and vegetable by-products are particularly interesting as sources of natural antimicrobials. Valorising waste of a food processing line which otherwise would have been disposed is an added value for researching their antimicrobial activity, as it contributes to the reduction of food waste[3]. Grape by-products are derived from grape processing to produce wine and juice. They represent 20% of the total weight of the fruit of the grape, making their disposal a challenging problem for wineries and for the grape juice industry. Grape by-products are composed by the skins, the seeds and in some cases the stems of the fruit. However, the seeds contain higher concentrations of antioxidant and antimicrobial compounds, such as polyphenols, in comparison with other grape by-products. Most studies conducted to date, are focused on the extraction of polyphenols from locally produced grape pomaces and test whether these extracts have a microbial inhibitory effect, without specific systematic monitoring of the microbial dynamics/kinetics[4]. The limited studies that have quantified the inhibition following treatment with extracts from grape by-products used specific food products, deeming the results relevant only for those specific food products[5].

Overall, in order to use grape by-products effectively as a natural antimicrobial strategy in a larger scale, it is important to study the detailed microbial dynamics of, Gram-negative and Gram-positive, foodborne pathogens robustly controlled systems (2D or 3D food models) instead of food products, as different physicochemical and rheological properties can be better monitored. Such fundamental studies will enable a better understanding of the exact concentration ranges at which grape pomace extracts are bacteriostatic and bactericidal, their mode of action and how they could be utilized, individually or in combination with other novel technologies (ultrasound, cold atmospheric plasma, pulsed electric fields) to attain sustainable food processing. *The aim of our work is therefore to perform fundamental systematic studies of the impact of grape seed extracts on the microbial dynamics of food related pathogens in model systems.*

METHODS

Grape seed extracts (GSE) were purchased from Bulk powders. The inactivation effect of GSE was tested against *L. monocytogenes and E. coli* inoculated in TSBYE supplemented with 1% and 4% GSE. Additionally, two levels of initial microbial population were tested, i.e., 10^5 and 10^8 CFU/ml. The effect of 1% GSE against mid-exponential or early stationary phase *L. monocytogenes* was also studied, to enable an understanding of the impact of the growth phase on the potential antimicrobial resistance. After the inoculation of *L. monocytogenes* in the nutrient broth, microbial dynamics were monitored for 24h. Furthermore, potential morphological changes of the cells treated with GSE, were observed using scanning electron microscopy (SEM). To identify the percentage of sub-lethally injured cells and dead cells, flow cytometry methods were carried out, staining with propidium iodide (PI) and bis-(1,3-dibutylbarbituric acid) trimethaine oxanol (BOX), respectively.

RESULTS

Treatment with GSE showed great antimicrobial activity against Gram-positive *L. monocytogenes* and Gram-negative *E. coli,* especially for lower initial inoculum levels (10^5 CFU/ml) and for mid-exponential cultures phase cells. The microbial inactivation was substantially enhanced for higher GSE concentrations ,i.e., 4%. Analysis of the treated samples using flow cytometry showed significant differences on the ratio of dead and sub-lethally injured cells when the initial inoculum concentration, the GSE concentration and/or growth phase were changed, in alignment with the microbial inactivation kinetics.

CONCLUSIONS

This work shows that GSE have good potential as a natural antimicrobial strategy against food related bacterial pathogens. This insight to the microbial dynamics is essential for the development of novel and robust green/sustainable processing control strategies, i.e., application of GSE individually or coupled with other novel technologies in a hurdle approach.

Acknowledgements:

This work was supported by the Department of Chemical and Process Engineering of the University of Surrey, the Doctoral College of the University of Surrey and R-processing Limited. E.V. is grateful to the Royal Academy of Engineering, for an industrial Fellowship.

References:

1. European Food Safety Authority. The European union summary report on antimicrobial resistance in zoonotic and indicator bacteria from humans, animals and food in 2017. *EFSA J.* 2019;17(2). doi:10.2903/j.efsa.2019.5598
2. Teixeira JA. Grand Challenges in Sustainable Food Processing. *Front Sustain Food Syst.* 2018;2(June):2017-2019. doi:10.3389/fsufs.2018.00019
3. Villalobos-Delgado LH, Nevárez-Moorillon GV, Caro I, Quinto EJ, Mateo J. *Natural Antimicrobial Agents to Improve Foods Shelf Life.* Elsevier Inc.; 2019. doi:10.1016/b978-0-12-817190-5.00004-5
4. Xu Y, Burton S, Kim C, Sismour E. Phenolic compounds, antioxidant, and antibacterial properties of pomace extracts from four virginia-grown grape varieties. *Food Sci Nutr.* 2016;4(1):125-133. doi:10.1002/fsn3.264
5. Sagdic O, Ozturk I, Yilmaz MT, Yetim H. Effect of Grape Pomace Extracts Obtained from Different Grape Varieties on Microbial Quality of Beef Patty. *J Food Sci.* 2011;76(7). doi:10.1111/j.1750-3841.2011.02323.x

627237 Systematic Screening of Materials for Consumer Products Using Semantically Enabled LCA

Abstract:

In the European Union, the manufacture and use of chemical substances as well as produced and imported goods must comply with a comprehensive legislative framework (e.g. REACH, RoHS). The main focus of this legislation is the protection of human health and the environment that can be posed by chemicals. Thus, substances can be restricted or banned if they are characterized as hazardous. The manufacture of consumer products, such as home appliances, is a complicated process involving a wide range of materials and chemicals. Health and environmental impacts of such processes should be evaluated, monitored and adjusted to the lowest achievable level taking into account the legislative framework and the competitiveness of the industry.

The LIFE MATHER Platform, developed in the context of the LIFE MATHER Project, constitutes a new chemical monitoring data platform that integrates corporate data (materials, chemicals, manufacturing process) with health and environmental data in order to evaluate possible alternative materials according to existing and/or upcoming legislation. The principles of LCA (Life Cycle Analysis) are applied, thus building a rich inventory regarding consumer products (home appliances) with detailed data flows: items of the appliance (Bill of Materials, BoM), construction of materials (Full Material Disclosure, FMD) and manufacturing data (energy, electricity etc.). The inventory is enriched with open and commercial LCA databases (e.g. Ecoinvent).

The LIFE MATHER Platform conceptualizes health and environmental aspects using knowledge models, enables reasoning by employing semantics and ontologies, builds properties and its automated technology enables high-throughput screening and analysis. It is prepared to manage large volumes of data streams as may be available from customers or external providers in general. The substitution of chemicals in existing products leads to “new products” (green products) ahead of design and new product features emerge. It is noteworthy that the approach is generalized, thus enabling the incorporation of other classes of appliances, while the challenge lies with the inclusion of processes beyond appliances.

The authors acknowledge financial support by the H2020-EU.1.3.3. Grant RENESENG II (H2020-778332).

625233 Innovating a Bendable Concrete Railroad Tie with Enhanced Fatigue Durability Via CO_2 Utilization

Abstract:

The concrete industry is a significant CO_2 emitter, with one of its primary components, Portland cement, accounting for 8% of anthropogenic CO_2 emissions through its production. As the global construction market is continuously expanding, effective strategies for mitigating concrete CO_2 footprint are much needed. In this study, we propose a novel approach that utilizes and sequesters CO_2 to create a highly durable bendable concrete (technically known as *engineered cementitious composites*, ECC) through precast carbonation curing of ECC rail ties. Our goal was to examine the potential for this production strategy to enhance rail tie service life and reduce overall lifecycle CO_2 emissions in railroad tie applications. To do this, we verified fatigue resistance of the CO_2-sequestered ECC with respect to conventional concrete, as this is a critical criterion for railroad tie service life prediction, and developed a comprehensive lifecycle assessment for the CO_2-cured ECC ties.

Preliminary results suggest that CO_2-sequestered ECC can achieve >30% CO_2 uptake per cement mass after 24-hour carbonation curing. Carbonation curing increased ECC's flexural strength by 20% and promoted crack width control capability with maximum post-fatigue crack width being reduced from 128μm to 79μm. The flexural strength of carbonated ECC reached 12.03 MPa, exceeding the American Railway Engineering and Maintenance-of-Way Association (AREMA) requirement of 5.89 MPa for rail seat positive, negative, and center bending tests. Moreover, carbonation curing significantly improved the fatigue life of non-reinforced ECC to pass the standard-required 3 million cycles under flexural loading, with comparable performance to traditional prestressed railroad ties. This eliminates the need for the costly and time-consuming prestressing process for traditional tie production. Carbonation curing can also increase the fatigue life by 20% (under 0.5 stress level) and 12% (under 0.4 stress level). The enhanced fatigue life through CO_2 utilization is anticipated to improve the railroad safety, resiliency, and sustainability, while mitigating CO_2 emissions both at concrete manufacture and through infrastructure lifetime.

620054 Efficient Deoxygenation of Sludge Vegetable Oil over Bimetallic Doped Activated Carbon Catalyst for the Production of Green-Diesel-like Fuel

Abstract:

Deoxygenation (DOx) of waste sludge palm oil (WSPO) into green fuel using activated carbon-based catalyst supported with tungsten and lanthanum oxides (La-W/AC) is reported in this study. La-W/AC catalyst was synthesized via co-precipitation (CP) of tungsten and lanthanum onto the activated carbon. La-W/AC catalyst was characterized using various characterization techniques and the results showed high surface area and acidity. Furthermore, the effect of catalyst loading, reaction time, and reaction temperature on deoxygenation of WSPO were investigated. The catalyst demonstrated a superior catalytic performance in deoxygenation reaction under optimal reaction conditions (0.5% catalyst loading, 1 hour and 250 ^{o}C reaction temperature), recording 98% conversion of WSPO to green fuel. Reusability test of the catalyst was performed and the results showed that the synthesized catalyst can be reused up to 8 times maintaining green fuel conversion of more than 87%. The catalytic activity of the catalyst was attributed by the presence of the acid and basic sites on the synthesized catalyst. The TOFs of the waste sludge palm oil was also determined in this work.

631776 Bio-Based Amines Production Via Selective Deoxygenation of N-Containing Biomass

Abstract:

Today, industrial amines are synthesized from ammonia, which is produced on a large scale by the highly energy-intensive Haber-Bosch process. In contrast, the production of bio-based amines from the abundant N-containing bio-based feedstock is attractive and can save a lot of energy. Two types of biobased amines sources could be highlighted; the first one is proteins or amino acids, and the second one has functions with nitrogen, such as chitin. Lysine is the second most-produced amino acid in industry, next to glutamic acid. Lysine has two amino groups and one carboxylic group. Hence, to maximize the production yield of diamine from lysine, the oxygen in lysine must be removed, while two amino groups being kept intact. Bio-based diamines, such as pentamethylene diamine (cadaverine), are the key building block monomers to synthesize bio-based polyamides, such as Nylon 5.6, 5.10, etc. While cellulose is the most abundant polysaccharide in nature, chitin is the second most abundant and a major component of the exoskeletons of insects and crustaceans. As one of three main components in those shells, chitin is a polymer of the amino sugar N-acetylglucosamine (NAG). Besides protein or its derivatives, chitin is the most important organic substance having amino groups on their polysaccharide backbones. Utilization of the ocean fixed nitrogen made a greener production of N-containing chemicals since industrial ammonia/amines synthesis is energy-intensive and consumes more than 1% of the total energy generated in the world. Through the selective deoxygenation of chitin monomers/L-lysine over Ru/C in acidic solutions, bio-based amines/diamines were obtained respectively in one-pot reactions, in which the critical step is to protect the amino group against deamination.

632556 The Role of Water in the Hydrodeoxygenation of 5-Hydroxymethylfurfural from Lignocellulosic Feedstocks

Abstract:

Finding fossil fuel replacements as carbon sources is crucial to a carbon-neutral future. 5-hydroxymethylfurfural (HMF) is one of the platform chemicals that can serve as an answer to this demand. The hydrodeoxygenation reaction of HMF over Ru/C yields 2,5-dimethylfuran (DMF), paving way to the broad utilization of HMF. We recently reported effective solvent extraction of HMF from the reactive aqueous phase using water-organic biphasic systems.[1] In these systems, the organic phase is saturated with water, typically around ten weight percent of the organic phase. Separation of water from such organic phase *via* distillation is difficult due to the instability of HMF and the azeotropic systems often formed between water and the organic solvents. Interestingly, we report here that water improves the selectivity of the HMF HDO reaction while decreasing the HMF conversion. Reaction pathway analysis *via* monitoring the species distribution over the course of the reaction sheds light on the water's role in the HMF HDO reaction. To elucidate whether catalyst reoxidation occurs, the Ru/C catalyst is subject to a variety of treatment conditions. Besides, the effects of temperature, the presence of levulinic and formic acids, and the presence of chloride ions from the previous fructose dehydration step, are also investigated. Additionally, we complement our experimental investigation with a technoeconomic analysis (TEA) of a proposed water-organic biphasic system for HMF production from fructose to show the impact of the design on the overall economics of future lignocellulosic biorefineries.

1. Wang, Z.[#]; Bhattacharyya, S.[#]; Vlachos, D. G., Solvent selection for biphasic extraction of 5-hydroxymethylfurfural via multiscale modeling and experiments. *Green Chem.* **2020,** *22* (24), 8699-8712.

633087 Nutrient Recovery and Fuel Precursor Production from Extracted Algae Residues Using Mild Oxidative Treatment

Abstract:

Valorization of algal biomass to fuels and chemicals frequently requires pretreatment to lyse cells and extraction of lipids, producing an aqueous lysate and an extracted solid residue as intermediates. Wet oxidation is a promising route to simultaneously convert nitrogen contained in the lysate and residues to easily-recyclable ammonia and to convert carbon in the same fractions to carboxylic acids that can be converted to biofuels by ketonization, condensation, and hydrodeoxygenation. We show that for a *Scenedesmus* algae under certain oxidation conditions, nearly all of the nitrogen in the residues can be converted to ammonia and recovered by cation exchange, while almost half of the carbon can be converted to carboxylic acids. At the same time, we also show that soluble phosphorus in the form of phosphate can be selectively recovered by anion exchange, leaving a clean aqueous carboxylic acid stream for upgrading to fuels.

630547 Supported Amine Adsorbents for CO_2 Capture Via Particle Molecular Layer Deposition

Abstract:

CO_2 capture from power generation sources provides a concentrated CO_2 stream and the possibility of incorporating the captured CO_2 into useful industrial processes. However, in order to implement these technologies on the scales we require, we must develop economically feasible methods and materials. Traditional CO_2 capture technology relied on liquid amines that requires substantial energy to strip the CO_2 that leads to substantial costs for the process.[1] Addressing the limitations of the liquid process, solid adsorbent materials have become a research focus over the past few decades. Popular materials have included zeolites, activated carbons, metal organic frameworks and supported amines. Supported amines are an attractive material as they take advantage of the traditional basic chemistry, are tolerant of humidity, and require moderate regeneration conditions. As such, they have received much attention in literature as researchers vary the functionalization method, support materials, and amine-containing groups to increase the adsorption capacity. Current processes of loading the amines through impregnation face poor regeneration stability and transport limitations, while grafted sorbents tend to have low adsorption capacities.[2] Mesoporous silicas such as MCM-41 and SBA-15 are very popular support materials due to high surface areas above 800 m^2/g. However, these materials are not produced on commercial scales and are expensive to make which would inhibit the implementation of this technology.

This research focuses on development of a regenerable, high capacity, and inexpensive supported amine adsorbent materials. Here, we introduce particle molecular layer deposition (MLD) as a novel functionalization method to improve upon the limitations of current methods. MLD is closely related to atomic layer deposition (ALD). Both processes involve stepwise reactions between gaseous reactants on a solid surface to create thin films.[3] Here we use MLD to deposit a covalently bonded aminopropylsiloxane network on fumed silica nanoparticles (380 m^2/g). Two MLD chemistries have been confirmed: (3-aminopropyl)triethoxysilane (mono-amine) and N1-(3-trimethoxysilylpropyl)diethylene triamine (tri-amine). Both precursors were deposited at 150°C. Adsorption capacity and regeneration stability were investigated using a thermogravimetric analyzer (TGA). Adsorption capacity of the amine functional groups increased to ~2mmol/g or 0.005 $mmol/m^2$ as the number of MLD cycles increased. On the basis of surface area, this capacity is competitive with other surface functionalization methods and well performing adsorbents in literature. Low regeneration temperatures were demonstrated by both the mono-amine and tri-amine. The MLD sorbents with 10 cycles of MLD underwent 25 regeneration cycles with adsorption at 30°C and desorption at 80°C. Both materials appear stable and robust over these cycling conditions. These materials will be stable for adsorption/desorption cycling temperatures below the deposition temperature of 150°C. Preliminary cost approximations estimate the cost of this sorbent is less than $10/kg. This work demonstrates a new functionalization method and capture material to the field of solid CO_2 capture research that could aid in widespread CO_2 capture implementation.

(1) Wang, J.; Huang, L.; Yang, R.; Zhang, Z.; Wu, J.; Gao, Y.; Wang, Q.; O'Hare, D.; Zhong, Z. Recent advances in solid sorbents for CO2capture and new development trends. *Energy Environ. Sci.* **2014,** *7* (11), 3478.

(2) Ünveren, E. E.; Monkul, B. Ö.; Sarıoğlan, Ş.; Karademir, N.; Alper, E. Solid amine sorbents for CO 2 capture by chemical adsorption: A review. *Petroleum* **2017,** *3* (1), 37.

(3) George, S. M. Atomic Layer Deposition: An Overview. *Chemical Reviews* **2010,** *110* (1), 111.

632565 Migration-Assisted Water Gradient Process for a Continuous CO_2 Capture

Abstract:

Stabilizing atmospheric CO_2 will require a significant reduction in anthropogenic CO_2 emission sources like coal-fired power plants. The current industrial processes such as calcium looping, amine scrubbing, and oxyfuel combustion capture CO_2 from the exhaust flue gases but are extremely energy-intensive. Water-driven CO_2 capture techniques are in their infancy but are particularly attractive due to their low energy penalty. In this work, we develop a continuous CO_2 capture device utilizing the ability of CO_2 to be in a carbonate-bicarbonate equilibrium in the presence of water. The incoming CO_2 is captured in a CO_2 -binding organic liquid. The captured CO_2 is allowed to diffuse through an anion exchange membrane to an aqueous medium where it can either be sequestered or transformed into usable products. To enhance the CO_2 capture rate, the diffusion of the captured CO_2 is accelerated by applying an electric field across the device. The performance of this capture technique is evaluated under various operating parameters such as the composition of the gas, relative humidity, temperature, the effect of concentration of SO_x , and gas flow rate. Additionally, for the device configuration, effects of the operating temperature, type of membranes, percentage of the solvent, solvent circulation flow rate, and applied current are reported.

621723 Instantaneous Nucleation of Carbon Dioxide Hydrates Using Magnesium

Abstract:

Gas hydrates offer technological solutions in areas like CO_2 capture and sequestration, desalination and gas separation. Formation of gas hydrates is severely constrained by long induction (wait) times, which can range from hours to days in the absence of external promotion techniques. Presently, we report that magnesium (in pure and alloyed form) can significantly accelerate the nucleation of CO_2 hydrates. We report magnesium-promoted induction times of just a few minutes (in the absence of surfactants and mechanical stirring), via statistically meaningful measurements of nucleation kinetics in milli-liter volume reactors. Interestingly, in larger reactors (half liter volume) we observe instantaneous nucleation, where hydrates nucleate even before the reactor is completely pressurized with CO_2. This is the fastest reported nucleation time of gas hydrates in quiescent gas-water systems. Visualization of hydrate formation coupled with post experiments-XPS and FTIR analysis reveals important insights underlying nucleation promotion. Overall, this simple, passive, environmental-friendly technique opens up new pathways towards instantaneous nucleation, thereby solving a key challenge in the field of hydrates.

Introduction

Gas hydrates are ice-like crystalline solids consisting of cages of water molecules enclosing a 'guest molecule'. Gas hydrates offer promising solutions in areas like CO_2 sequestration, desalination, gas separation, storage etc. Nucleation of hydrates has been a top challenge in this field; nucleation refers to the formation of the first 'stable' crystal of hydrate that can subsequently grow[1]. Subsequently, hydrates grow on the gas-liquid interface as a film [2]. The induction (wait) time for hydrates to nucleate can range from hours to days in the absence of external promotion techniques. An important recent discovery by the present group involved the use of aluminum surfaces to promote nucleation of CO_2 hydrates[3]. Presently, we report that magnesium (Mg) induces nucleation of CO_2 hydrates within a few minutes; this represents a >10X improvement from our previous result, with the observed reduction in nucleation time being as high as 98%, compared to [3]. Both pure Mg, as well as a common Mg alloy (AZ31) display enhanced nucleation with pure water as well as salt water. X-ray photoelectron spectroscopy (XPS) and Fourier-transform infrared spectroscopy (FT-IR) provided insights into the surface chemical changes controlling the nucleation of CO_2 hydrates. Fig 1 shows the images of carbon dioxide hydrate formation using Magnesium plates.

Methods

Figure 2 shows a schematic of the experimental setup. A custom-built 450 mL, pressure vessel (Parr Inst.) with sapphire windows (for visualization) was used. Experiments were conducted in 3 ml capacity polystyrene spectrophotometric cuvettes; 4 cuvettes could be accommodated in the visualization window per experiment. The pressure vessel was located in an environmental chamber (ESPEC) to cool it. Hydrate nucleation was detected visually using a Nikon DSLR 850 fitted with a Tokina macro lens. Mirror-polished (surface roughness: 137 nm) Mg alloy (AZ31) plates (from Goodfellow), and pure Mg rods (99.9 % purity, from Sigma Aldrich) were used.

Results

Table 1 shows the induction time in small scale polystyrene cuvette-based reactors. Both pure Mg and Mg alloy AZ31 show very effective promotion with average nucleation times around 10 minutes in deionized (DI) water. With saltwater with salinity of 35 g/L, the average nucleation time with Mg alloy is about 24 minutes which is still very favorable, compared to the two hour induction time in the absence of Mg (with saltwater). Without the presence of either Mg alloy or pure Mg, no nucleation of CO_2 hydrates was observed in 15 hours.

Table 2 shows the influence of various configurations of Mg alloy plates on the nucleation of CO_2 hydrates in DI water. The different configurations are depicted in Fig 3. It is clearly observed that the three-phase contact line of Mg-water-CO_2 plays a critical accelerating role in nucleation of CO_2 hydrates. Also, placing the Mg alloy plates close to the cuvette surface accelerates nucleation even further and average nucleation times as low as 3 minutes is observed.

Next, we studied this concept in larger reactors of stainless steel with 350 ml of deionized water. We observed instantaneous nucleation, wherein CO_2 hydrates nucleated during the pressurization stage at pressures ranging between 350-400 psi. Statistical measurements of induction time resulted in a mean of 0.7 mins and standard deviation of 0.1 mins. Without the presence of Mg, CO_2 hydrates nucleated with a mean induction time of 53 minutes and standard deviation of 16 mins.

XPS and FTIR analysis reveal interesting mechanistic insights related to Mg-based promotion of nucleation. Post experiments, there were two distinct regions on the Mg alloy plate, one much darker than the other. XPS characterization showed the presence of minor alloying elements such as Zn, Al in the darker region which was absent in the lighter region. FTIR analysis showed a strong chemisorption of CO_2 on the surfaces of Mg during the hydrate reaction. Fig 4 shows the images of the Mg alloy plates post reaction and the results of XPS and FTIR.

Conclusions

Presently, we describe the significant benefits of pure Mg and Mg alloy AZ31 for promoting the nucleation of CO_2 hydrates. Results from more than 80 experiments clearly demonstrate that Mg-based promotion can trigger hydrate nucleation in a few minutes, with the fastest nucleation obtained in less than a minute. Our experimental and surface characterization efforts suggest that multiple interfacial phenomena could be at play to achieve ultrafast nucleation[4]. This simple and passive technique for nucleation promotion opens up new avenues for faster hydrate formation, which is critical to realizing the benefits of hydrates in practical applications.

Acknowledgments

Research on CO_2 hydrates was supported by ExxonMobil through its membership in The University of Texas at Austin Energy Institute. This acknowledgment should not be considered an endorsement of the results by ExxonMobil. V.B. acknowledges NSF-CBET grant 1653412 for partial support of this work. The authors also acknowledge the support provided by Jamie and Israel from the machine shop in the department of Mechanical Engineering.

References

[1] M. Khurana, Z. Yin, P. Linga, A review of clathrate hydrate nucleation, ACS Sustain. Chem. Eng. 5 (2017) 11176–11203. https://doi.org/10.1021/acssuschemeng.7b03238.

[2] A. Kar, A. Bhati, P.V. Acharya, A. Mhadeshwar, P. Venkataraman, T.A. Barckholtz, V. Bahadur, Diffusion-based modeling of film growth of hydrates on gas-liquid interfaces, Chem. Eng. Sci. 234, 116456 (2021). https://doi.org/10.1016/j.ces.2021.116456.

[3] P. V. Acharya, A. Kar, A. Shahriari, A. Bhati, A. Mhadeshwar, V. Bahadur, Aluminum-Based Promotion of Nucleation of Carbon Dioxide Hydrates, J. Phys. Chem. Lett. (2020) 1477–1482. https://doi.org/10.1021/acs.jpclett.9b03485.

[4] W. Ke, T.M. Svartaas, D. Chen, A review of gas hydrate nucleation theories and growth models, J. Nat. Gas Sci. Eng. 61 (2019) 169–196. https://doi.org/10.1016/j.jngse.2018.10.021.

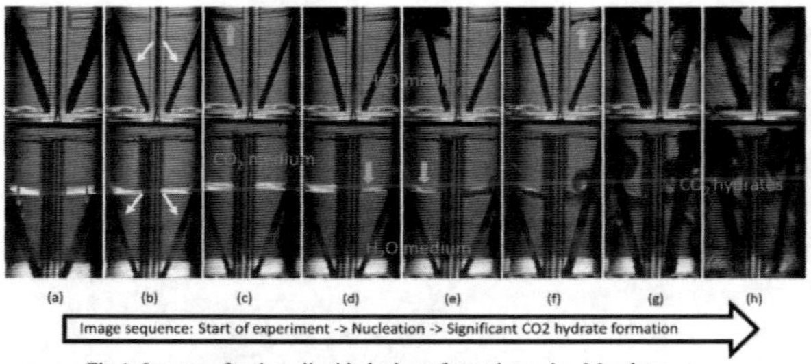

Fig 1: Images of carbon dioxide hydrate formation using Mg plates.

Promoter	Water	Mean induction time (min)	Standard deviation (min)	Minimum induction time (min)	Maximum induction time (min)
Mg alloy plate (AZ31)	Deionized	8.76	0.84	7.3	10.0
	Salt water (35 g/L)	23.52	6.92	11.0	29.5
Pure Mg rod	Deionized	13.26	3.94	7.9	20.8
	Salt water (35 g/L)	120.78	87.72	19.4	239.6
None	Deionized	No nucleation (>15 hours)			
	Salt water (35 g/L)	No nucleation (>15 hours)			

Table 1: Summary of induction time for various experiments. Each data point is a repeat of 8 sets

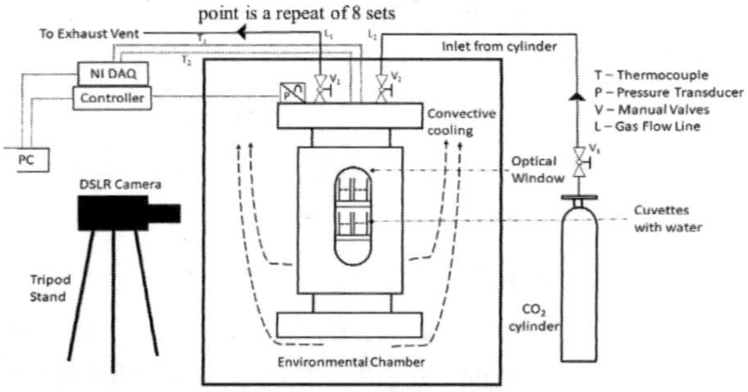

Fig 2: Schematic of experimental setup

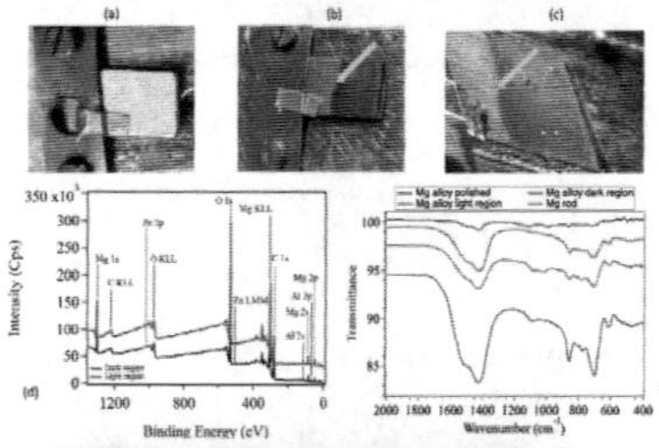

Fig 4: Images of Mg alloy plates post-experiment and XPS and FTIR results.

Table 2: Induction time summary for various Mg alloy plate configurations. Each data is a repeat of 8 sets

Configuration	Mean induction time (min)	Standard deviation (min)	Minimum induction time (min)	Maximum induction time (min)
Inclined (a)	8.76	0.84	7.3	10.0
Immersed – no three-phase line (b)	46.4	3.94	37	50.1
Upright and centered (c)	10.36	1.66	8.8	13.5
Upright and flush with side (d)	2.92	1.83	1.12	6.47

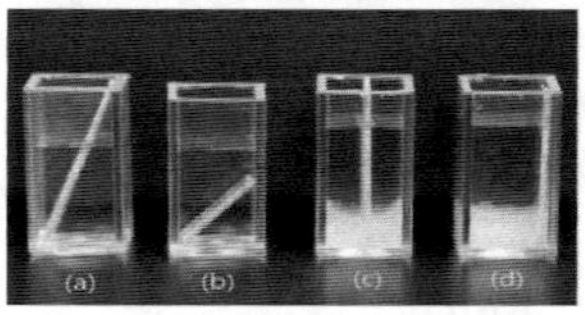

Fig 3: Various configurations of Mg alloy plates in Table 2

624097 Mass Transfer Intensification Using 3D Printing Novel Dynamic Polarity Packing for Post Combustion Carbon Capture

Abstract:

Global warming has drawn world-wide attention as it impacts global sustainable development. Since fossil fuel are still a major part of energy production in the foreseeable future, CO_2 capture plays a significant role in the mission of reduced CO_2 emissions, or transition to a negative CO_2 emission future [1]. The post combustion carbon capture using amine solutions as the CO_2 absorbent is the most mature and scalable technology to target CO_2 in dilute, low pressure flue gas. Despite this technology having been studied for decades, there still has some limitations that remain unsolved [2]. A major concern is the high cost in terms of energy and capital investment [3]. The energy cost is mostly attributed to the regeneration in the stripper, which strongly depends on the absorbent properties and how the stripper is operated. The capital investment is also significant because of the column size requirement to handle large volume of flue gas.

Herein, we describe the use of 3D printing to develop novel structured packings for the absorber in a CO_2 capture system. 3D printing technique allows for fast and low cost printing of prototypes for objects and patterns, which enjoys various benefits like freedom of design, waste minimization and ability to manufacture complex structures [4]. Our motivation is to develop new concept of packings with adjustable surface properties to further intensify mass transfer in the absorber column. By doing this, the absorber size can be shrunk to reduce the capital cost or/and a higher CO_2 loading can be achieved in the rich solvent which benefits absorbent regeneration with higher CO_2/H_2O ratio and less energy consumption.

Novel packings, what we call dynamic polarity packings, are shown in this work with similar geometric structure as conventional Mellapak 250Y steel packing (3-inch diameter). The internal structure of dynamic polarity packings has a unique surface that alternatively combines different segments with different material surfaces in order to manipulate the absorbent flow on the packing surface. Due to the inherent properties of these materials, one segment is more hydrophobic and the other one is more hydrophilic. It is known that the aqueous amine CO_2 absorbents will have better wettability (a smaller contact angle) on the hydrophilic surface while it tends to gather and resist spreading on the hydrophobic surface (a larger contact angle). When the absorbent flows down on the packing surfaces, the regular alternation of surface property generates local turbulence within the solvent which will consequently refresh liquid film surface and improve the gas liquid mixing effect. In addition, the proposed dynamic polarity packings are designed with a surrounding hydrophobic shell to reduce the potential wall effect and channeling issues to maintain a better liquid loading on the packing surface.

Based on these considerations, several versions of dynamic polarity packings were printed with dissimilar pattern designs. The printed packings (denoted as DP-1, DP-2 and DP-3) are shown in Figure 1a, along with Mellapak 250Y steel packing. The internal packing surface of dynamic polarity packing has unique design to study the influence of printing pattern on CO_2 absorption. DP-1 and DP-3 packings have 5% of hydrophobic high impact polystyrene (HIPS) surface embedded in hydrophilic nylon-based packing surface, where the HIPS surface pattern differs from each other. The HIPS surface pattern is staggered rectangles for DP-1 packing and "V" shape for DP-3 packing. DP-2 packings have alternative segments with sole HIPS or nylon materials and the height of each segment is 0.5 inch. All the dynamic polarity packings are surrounded by the more hydrophobic HIPS shell. Primary amine 1-amino-2-propanol (1A2P) solution is used as the CO_2 absorbent for the test. To quantify the wettability on different materials, the contact angle of 1A2P solution was measured by sessile drop method as shown in Figure 1b. Depending on the CO_2 loading of the absorbent, there will be around 13 to 18° difference in contact angle serving as driving force to manipulate absorbent flow. The packing performance is tested in an integrated bench scale CO capture unit which mainly includes an absorber column and a stripper column both with ~2 meters of packing height. In these preliminary tests, only the steel packing at the top of the absorber was replaced with dynamic polarity packings to screen the best design for gas absorption (~18" replaced out of 60" total packing height). First a parametric study was carried out to determine the optimal operating conditions in the current unit with absorber filled with the steel packing to be used as a baseline date set. Next, the different 3D printed packing was placed in the absorber and the parametric conditions were repeated. Preliminary testing results of performance including CO_2 absorption efficiency and the energy demand are summarized in Figure 1c. The distribution of data points depends on the operating conditions for the capture system. In general, replacing the steel packings with the dynamic polarity packings shifts the CO_2 capture efficiency higher and leads to lower energy penalty. Comparing with Mellapak 250Y steel packing under similar operating conditions, there is relative 7.8%, 7.7% and 9.1% improvement in CO_2 absorption efficiency and consequential 7.0, 7.9% and 8.4% energy penalty reduction for DP-1, DP-2 and DP-3 packings respectively. Considering only a portion of Mellapak 250Y steel packing is replaced by 3D printing packings during the preliminary test, the improvement will be more prominent after the absorber column is filled with new packings. The testing results imply that the application of these superior packings can potentially decrease the required absorber column size to cut down capital expense. Under some circumstances, it also enables the system to be operated under lower absorbent circulation rate while acquiring richer CO_2 loading to reduce operating expense.

Acknowledgements

Financial support from US Department of Energy (DOE) National Energy Technology Laboratory (NETL) is gratefully acknowledged, DE-FE0031661.

References

1. Vega F, Baena-Moreno FM, Gallego Fernández LM, Portillo E, Navarrete B, Zhang Z. Current status of CO_2 chemical absorption research applied to CCS: Towards full deployment at industrial scale. *Applied Energy.* 2020/02/15/ 2020;260:114313.

2. Bui M, Adjiman CS, Bardow A, et al. Carbon capture and storage (CCS): the way forward. *Energy & Environmental Science.* 2018;11(5):1062-1176.

3. Jiang Y, Mathias PM, Freeman CJ, et al. Techno-economic comparison of various process configurations for post-combustion carbon capture using a single-component water-lean solvent. *International Journal of Greenhouse Gas Control.* 2021/03/01/ 2021;106:103279.

4. Ngo TD, Kashani A, Imbalzano G, Nguyen KTQ, Hui D. Additive manufacturing (3D printing): A review of materials, methods, applications and challenges. *Composites Part B: Engineering.* 2018/06/15/ 2018;143:172-196.

Figure 1. Steel and dynamic polarity structured packings (a), contact angle of 1A2P solution on nylon and HIPS surface (b) and system CO_2 capture performance evaluation (c, bubble width indicates heat duty)

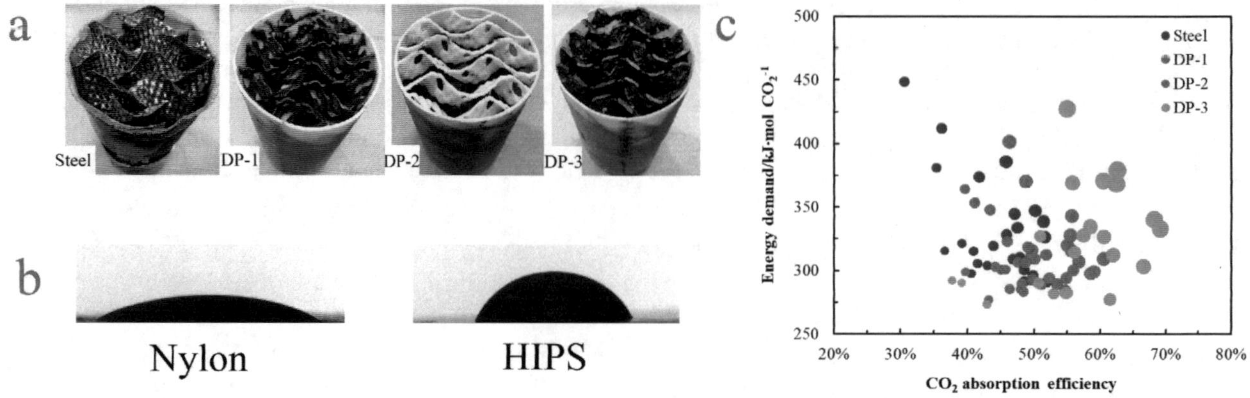
a
Steel
DP-1
DP-2
DP-3
b
Nylon
HIPS
c
Steel
DP-1
DP-2
DP-3
Energy demand/kJ·mol CO2⁻¹
CO2 absorption efficiency
500
450
400
350
300
250
20%
30%
40%
50%
60%
70%
80%

625517 Systematic Chem-Informatics and Machine Learning Studies for Gas Permeability and Selectivity in Polymers

Abstract:

Quickly and reliably predicting gas permeability and selectivity in polymers is important to develop effective and efficient polymers for specific gas separation applications, such as membrane development for post-combustion carbon capture. There are several challenges, such as building a polymer database with various different gases permeabilities, designing a scheme to encode the polymer repeat units, generating fingerprints and descriptors which could discern different polymer repeat unit structures, and building the machine learning (ML) models. We will address these problems in this presentation.

We extended the simplified molecular-input line-entry system (SMILES) for small molecules to encode the polymer repeat unit by utilizing the [*] ghost atom with different charges to specify different head and tail atoms for the repeat units of both simple and ladder homopolymers. Multiple-mers could be generated from this extended SMILE scheme. Starting from the Membrane Society of Australasia (MSA) polymer database, which contains different gases permeabilities, we built an in-house polymer database by carefully checking the data and references for accuracy, adding the SMILES for the polymer repeat units for homopolymers, random polymers, and polymer blends. We added more than 100 new datasets, and we also added other polymer properties, such as glass transition temperature, polymer density and free volume fraction. There are 1,674 sets of data in our in-house polymer database; each set corresponds to one polymer. There are 1,210 sets of data corresponding to homopolymers, out of which there are only 807 unique polymer repeat units. Five different fingerprints, that is, RDKfingerprint, MACCS keys, AtomPair fingerprint, Torsional fingerprint, and Morgan fingerprint were tested along with 189 different 2D descriptors. Although the RDKfingerprint has been used by another researcher group [2] to develop ML models to predict gas permeability in polymers, it was found in our work that the RDKfingerprint fails to describe 2.2% of the 807 unique repeat units. That is, for some different polymer repeat unit structures, the RDKfingerprint gives the same fingerprint values, which is unexpected. To fix this problem, we combined the RDKfingerprint with the 189 2D descriptors, which was found to be able to discern different polymer structures. In addition to the comparison between different fingerprints and descriptors to describe polymer repeat units, we will also show their ML performance by using a Gaussian process regression (GPR) algorithm and a different number of repeat units. Our work shows that the RDKfingerprint combined with 189 2D descriptors exhibits the best ML performance at a polymer repeat unit length of 10. Furthermore, shuffling of the data was found to significantly affect the ML performance. For example, using the same 1,071 sets of data for CO_2 permeability along with 70% training and 30% test data, shuffling data (leading to different assignment to training and test data) could give R^2 values (accuracy of determination) for the test data sets between 0.84 to 0.91 although the R^2 values for the training set are approximately 0.98. We will show strategies to alleviate this problem. We will also show our preliminary results of designing polymers by using the iQSPR method [3,4].

References:

1. https://docs.google.com/spreadsheets/d/1LXwkZfhrdLtLuG7WvrZBgjZ2u5QsATsby9yN_FW6kcw/edit#gid=1
2. *Sci. Adv.* 2020; **6** : eaaz4301
3. *J Comput Aided Mol Des* ***31****, 379–391*
4. *Mol Inf 2020, 39, 1900107*

630556 Costs and CO_2 Sequestration Potential of Near-Term and Long-Term Implementation of Bio-Energy with Carbon Capture and Storage (BECCS) in the U.S.

Abstract:

Bioenergy with carbon capture and storage (BECCS) can remove CO_2 from the atmosphere. This work assesses the potential scale of CO_2 removal and cost of implementing BECCS in the U.S. in the near term (now) and long term (2040) while considering supply chain emissions of CO_2, costs of biomass harvesting and transportation, and economics of power generation and carbon sequestration. This study includes two power generation technologies, pulverized coal (PC) and integrated gasification combined cycle (IGCC) power plants, and two biomass treatment scenarios (conventional and pelletized). Potential sites were determined by considering the proximity of saline aquifers, protected lands, and populated regions. Supply chain costs and emissions for the harvesting, pretreatment, and transportation of biomass feed were evaluated to determine accurate costs and associated CO_2 penalties. Performance and economics of PC and IGCC power plants were simulated using the Integrated Environmental Control Model (IECM) to determine parameters like net MWh of electricity produced, levelized cost of electricity (LCOE), emissions intensity, capital costs, and operation and maintenance (O&M) costs. Costs are presented in 2018 USD. Sensitivity analyses were performed to determine influential parameters in power plant performance. The LCOE ranges between \$140 and \$180 per MWh in the 2020 scenario, and between \$130 and \$180 per MWh in the 2040 scenario. The cost of BECCS ranges between \$35 and \$176 per tonne of CO_2, depending on the extent of biomass utilization. Under a near-term scenario using up to 206 million tonnes per year of biomass, roughly 181 million tonnes of CO_2 can be sequestered annually in the U.S. at CO_2 avoidance costs (CAC) ranging between \$62 - \$137 per tonne CO_2. Long-term scenarios have the potential to use 740 million tonnes of biomass and sequester 737 million tonnes of CO_2 per year at avoidance costs ranging between \$42 to \$92 per tonne CO_2, depending on power plant configuration and fuel type. The cost of removing CO_2 from the atmosphere, beyond post-combustion treatment, ranges from \$30 to \$140 per tonne CO_2. Lastly, the economics of co-firing coal with biomass was considered. For co-fired corn stover, the CAC ranges between \$58 and \$108 per tonne of CO_2 depending on degree of co-firing and power plant size. For poplar-fueled co-fired NGCC power plants, the LCOE ranges between \$117 and \$208 per MWh depending on extent of co-firing, plant size, and cost of fuel. Results presented in this work indicate that BECCS has the potential to remove CO_2 at costs competitive with other negative emissions technologies.

629204 Waste Activated Sludge Valorization for Short-Chain Fatty Acids Production By Integrated Electro-Alkaline Treatment

Abstract:

A significant amount of waste activated sludge (WAS) is generated during the wastewater biological treatment process, which needs to be properly treated before disposal. The disposal cost of WAS approximately accounts for 50–60% of the total operation cost of a wastewater treatment plant. Due to the increase in wastewater treatment capacity, the disposal of a huge volume of WAS has become a serious environmental and economic issue. On the other hand, WAS has been considered as a sustainable resource owing to a high organic matter content (>40%, consist of mostly proteins and carbohydrates) with an average calorific value of 6094 kcal/kg. Therefore, harnessing bioenergy from WAS could simultaneously address environmental and economic concerns imposed by the growing amount of WAS.

Short-chain fatty acids (SCFAs) (e.g., acetic acid, propionic acid, butyric acid, etc.) are the intermediate chemicals in anaerobic digestion. SCFAs are more valuable (600-3815$/ton) than biogas (150$/ton) and need shorter retention time (< 2d). The production of SCFAs from waste activated sludge has attracted growing attention in wastewater treatment plants (WWTPs) because SCFAs could be utilized for a wide range of applications including biohydrogen production, bioplastics (polyhydroxyalkanoate (PHA)), and biological removal of nutrients from wastewater.

It has been confirmed that electrochemical treatment is a powerful and environment-friendly pathway to disrupt sludge floc and sludge disintegration to produce SCFAs through oxidation of the complex organic compounds on the surface of the electrode. On the other hand, it has been reported that SCFAs production at alkaline pH was significantly higher than that at neutral or acidic pH because the alkaline medium is more favorable for protein and carbohydrate destruction. Moreover, it results in increased SCFA accumulation by mitigating the methanogenesis reactions.

In this study, the combined alkaline-electrochemical treatment was utilized in a single chamber electrolysis cell at low operating temperatures (22-55 °C) to investigate the synergetic effect of alkaline and electrochemical treatment on the solid reduction and SCFAs production. The effect of applied potential, electrode materials, operating temperature, and alkaline agent concentration on SCFAs production was evaluated. The sludge solid before and after treatment was characterized by thermogravimetric analysis (TGA), Scanning electron microscope (SEM), and elemental analysis techniques. The liquid filtrate after treatment was analyzed with gas chromatography- flame ionization detector (GC-FID) to evaluate the SCFAs concentration. In this presentation, the results from this work will be provided including volatile solid removal, SCFAs concentration, and solid residue characterization.

632485 pH Effects on Factors Influencing Nutrient Removal and Recovery from Synthetic Animal Waste Via Chemical and Electrochemical Techniques

Abstract:

It has been estimated that there may be as much as 2 billion tons of animal wastes produced in the United States annually. Application of excreta to soil brings benefits such as improved soil tilth, increased water-holding capacity, and some plant nutrients. Concentrated forms of excreta increase the eutrophication problems which has caused a lot of harm to the aquatic life and to water supply for industrial and domestic uses [1]. This eutrophication is due to the presence of phosphorus (P) and nitrogen (N), which are essential nutrients for plant cultivation and the human diet. Therefore, it is desirable that P and N are decreased from these waste streams before discharged to soil and the environment.

In a prior study, the efficacy of electrochemical method was examined with key operating parameters affecting struvite deposition identified as temperature, turbulence, applied potential and competing ion (Ca^{2+}) concentration [2]. In addition, a thermodynamic simulation was developed to further investigate the impact of Ca^{2+} in animal waste on the technology's performance [3]. High phosphorus recovery (83%) and low specific energy consumption (0.43kWh/kg P recovered) were discovered at optimum conditions in these previous studies. However, the effect of multi-factors interactions on electrochemical nutrients reduction is unknown. In this new study, the efficiency of nutrient (P and N) removal and struvite quality from animal waste will be assessed based on multivariate analyses including flow conditions, temperature, Mg^{2+}:Ca^{2+} ratio, N:P ratio and the cathodic potential. In addition, the effect of initial solution pH will be evaluated to delineate the difference between chemical and electrochemical driving forces. Initial results from this study indicates that under certain process conditions, a high initial pH results in significant solid dissolution under electrochemical control and decreases the overall nutrient recovery. This suggests chemical driving forces are more significant than electrochemical driving forces for nutrient recovery under certain conditions. These results along with solid characterization analyses and solution speciation will be discussed at the conference.

References

[1] Animal waste " Environmental Encylopedia" https://www.encyclopedia.com/environment/encyclopedias-almanacs-transcripts-and-maps/animal-waste

[2] Z. Belarbi and J. P. Trembly, "Electrochemical Processing to Capture Phosphorus from Simulated Concentrated Animal Feeding Operations Waste," J. Electrochem. Soc., vol. 165, no. 13, pp. E685–E693, Jan. 2018, doi: 10.1149/2.0891813jes.

[3] Z. Belarbi, D. A. Daramola and J. P. Trembly, "Bench-Scale Demonstration and Thermodynamic Simulations of Electrochemical Nutrient Reduction in Wastewater via Recovery as Struvite," J. Electrochem. Soc., vol. 167, no. 4, Nov. 2020, doi: 10.1149/1945-7111/abc58f.

625827 Systematic Evaluation of Emerging Wastewater Nutrient Removal and Recovery Technologies to Inform Practice and Advance Resource Efficiency

Abstract:

Background

Nutrient discharges from wastewater treatment plants to the environment cause widespread eutrophication (algae overgrowth), harming aquatic life and resulting in economic losses. Conventional nutrient removal from wastewater can prevent these negative impacts but requires substantial energy and chemical inputs. Meanwhile, synthesis of ammonia via the Haber-Bosch process and phosphorus mining have substantial environmental footprints and deplete natural resources. Nutrient (nitrogen and phosphorus) recovery technologies could offset chemical production and reduce impacts of the wastewater and chemical industries. Widespread implementation of novel, more efficient nutrient removal and recovery technologies typically takes decades. We systematically reviewed 157 nutrient technologies spanning diverse mechanisms and stages of development and proposed a framework for standardized reporting and comparison of nutrient technologies to accelerate adoption. Our specific objectives were to: (1) characterize and compare nutrient technologies by a common set of qualitative attributes and quantitative performance metrics; (2) identify mismatches between academic literature and the needs of practitioners; and (3) catalogue value propositions, barriers, and knowledge gaps to prioritize future research objectives.

Methods

We extracted information about 157 nutrient technologies from articles in peer-reviewed journals, conference proceedings, and industry reports using a data collection template devised to capture all data needed for our analysis. We recorded reported quantitative performance metrics, value propositions, barriers to adoption, and knowledge gaps. In addition, we conducted a survey of practitioners in the wastewater industry to understand practitioners' needs, collecting 54 responses predominantly from public wastewater utilities.

Results and Implications

Our review indicated diversity in mechanisms (biological, physicochemical, both), nutrient(s) addressed (nitrogen, phosphorus, both), and treatment objectives (removal, recovery) of nutrient technologies. To facilitate meaningful analysis, we created 38 process groups consisting of similar processes. The observed heterogeneity complicates comparison among technologies. Therefore, we proposed quantitative performance metrics (i.e., removal and recovery efficiency, removal and recovery rate, energy consumption, cost, greenhouse gas emissions, effluent concentration) and qualitative attributes (e.g., technology readiness level, adaptation level required). These metrics constitute a minimal set that is generalizable across all nutrient technologies and facilitates multi-faceted comparisons (e.g., using an overall treatment score based on multiple metrics). Comparing peer-reviewed literature with practitioner needs revealed limited reporting of energy consumption and cost, indicating misalignment between research and practice (Figure 1). Synthesizing barriers to adoption and knowledge gaps reported in literature and by practitioners, we propose a research agenda addressing the most reported gaps (e.g., underlying process mechanisms, scale-up, optimization of reactor configuration and operating conditions) and emphasizing rigorous investigations of systems-level impacts and product-market fit. This study's results will drive interdisciplinary research on nutrient technologies, guide technology development by academics and practitioners, and accelerate implementation for resource-efficient nutrient management, including recovery technologies.

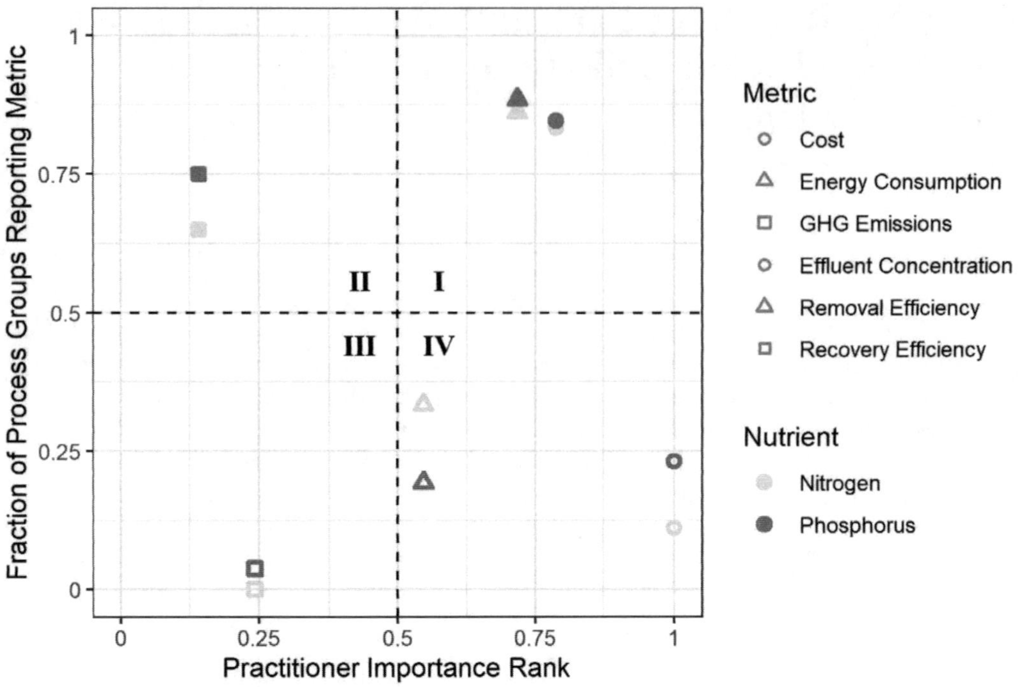

Figure 1. Standard Metric Reporting in Literature vs. Importance to Practitioners. Practitioner importance rank is calculated based on scores of importance that practitioners assigned to each metric.

632044 A Circular Economy Systems Engineering Framework for the Optimization of Food Supply Chains

Abstract:

Food supply chains rely heavily on the extraction of natural resources, including water, phosphorus, potassium, and fossil fuels. Population and welfare growth, along with the inflated standard of living increase the demand for food, leading to natural resource degradation, increased landfill wastes, water contamination, and greenhouse gas emissions. Circular Economy (CE) can be a solution for the transition of food supply chains to a more sustainable and regenerative future [1]. To achieve CE, we need to close material loops and connect different stages of the food supply chains that in a linear economy are discrete. These interconnections along with the multiple stakeholders connected with them make decision making for CE food supply chains very challenging. A holistic systems engineering approach is thus clearly needed to navigate the multi-scale, multi-faceted, and interconnected CE food supply chains, identify opportunities for synergistic benefits and systematically explore interactions and trade-offs [2].

In this work, we present the foundations of a systems engineering framework and decision-making tool for the analysis and trade-off optimization of interconnected CE food supply chains. The framework utilizes superstructure representations, along with mixed-integer modeling and optimization methods to establish the interconnections between different stages of the CE food supply chains. The analysis of the trade-offs is empowered by the introduction of composite metrics for CE that include waste, energy, and resource use minimization, as means to facilitate decision making and compare alternative processes, materials, resources, and technological options. To illustrate the applicability of the proposed framework we focus on the supply chain of coffee. A superstructure representation of the entire supply chain of coffee is presented, that involves alternative pathways for coffee harvesting and processing, waste utilization, product distribution, and new/alternative product introduction. Five different coffee-product demand scenarios are explored, showing that the production of instant coffee is the least energy and environmentally efficient scenario. On the contrary, the production of whole beans sets a hypothetical upper bound on the most circular coffee supply chain design and operation.

References:

[1] Ellen MacArthur Foundation, Intelligent assets: Unlocking the circular economy potential, 2016.

[2] Avraamidou, S., Baratsas, S.G., Tian, Y., Pistikopoulos, E.N., 2020. Circular economy - a challenge and an opportunity for process systems engineering. Computers & Chemical Engineering 133, 106629.

627949 Options for Reuse or Recycling of Wind Turbine Blades

Abstract:

The use of wind energy has expanded significantly in the last 20 years. It began the century as a niche energy source in Europe and the United States and ended up currently being a mainstream source of clean, cost-competitive energy around the world. In 2019, the global wind energy market reached a new milestone of 651 GW cumulative installed capacity, with strong continued growth foreseen across America, Asia, and Europe. And as onshore wind technology has matured, offshore wind has been embraced by governments and international institutions as one of the next game-changers in the energy transition.

With the increased use of wind turbines to support the energy transition to renewable sources comes the increasing challenge of the recycling or reuse of the turbine blades after decommissioning of turbines. Due to their materials of construction turbine blades cannot be disposed of easily. Dumping blades on a landfill is not a viable and environmentally acceptable solution. And other ideas currently applied may not be capable of handling the number of blades that will require disposal in the near future.

The main objective of this report is to provide information on possible solutions for the recycling of wind turbine blades, considering their components and materials of construction, the number of wind turbines in different geographical regions, and the turbine's expected lifetime. Several options for turbine blades recycling will be compared and their advantages and disadvantages will be discussed. The report concludes with our recommendation for a path forward including the identification of existing gaps in information and research that need to be closed.

628115 Use of Responsible R2 (Reuse and Recycling) Practices for Sustaiable Ewaste Supply Chain

Abstract:

The generation of eWaste in today's interconnected world has been posing many challenges especially for the large corporations and the entire eWaste supply chain. Some of the key challenges that the Industry is facing today include inter alia the following:

- The world produces as much as 50 million ton of electronic and electrical waste (e-waste) a year, weighing more than all of the commercial airliners ever made. This may more than double to 120 millions tons by 2050. Only 20% of this is formally recycled.
- The e-waste produced annually is worth over $62.5 billion, more than the GDP of many countries.
- There is 100 times more gold in a tonne of e-waste than in a tonne of gold ore.
- With an astounding growth and development in Consumer Electronics (especially Mobile Platforms) and Digital disruption in Technology Sector, Electronics Industry will continue to witness double digit growth.
- This means that from Sustainably standpoint, a circular electronics system - one in which resources are not extracted, used and wasted, but re-used in countless ways - creates decent, sustainable jobs and retains more value in the industry.

As we move towards circular economy in the interconnected world, we need to consider the increasing deployment of circular electronics ystem which offet the following key advantages:

–Lowering of Carbon Footprint

–Recovery potential of 100 times more gold in a ton of mobile phones than in a ton of gold ore.

–Realization of Full Life Cycle benefits when waste electronics are put to reuse

With tremendous potential of adoption of sustainable eWaste supply chain model, it is imperative to consider responsible Reuse and Recycling (R2) practices from the revised R2V3 (Version 3) standard which was released in July 2020 by 'Sustainable Electronics Recycling International', also known as "SERI". This standard is designed with the single most avoved objective of making eWaste supply chain greener and sustainable.

Based on the above backdrop, this presentation is aimed to cover the following learning and knowledge-sharing objectives and opportunities to the audience:

1) Introduction to responsibe Reuse and Recycling (R2) practices

2) Focus Materials in the eWaste supply chain

3) Introduction to R2 Equipment Categorization (REC)

4) Overview of R2V3 standard

5) eWaste Dowstream Vendors (DSVs) selection and qualification criteria

6) Importance of Data Sanitization, Security and Testing and Repair practices for electronic equipment

7) Auditing and Certification requirements

8) Roadmap for implementation of R2V3 standard

9) Benefits of adoption of R2 practices and R2V3 standard for circular eWaste management system

10) Case study discussion

11) Conclusion and Summary for the next steps

12) Q & A

631216 Capturing Spatial, Temporal and Technological Detail in Hydrogen Supply Chains By a Bilevel Optimization Strategy

Abstract:

Related background and motivation

Hydrogen is one of the key enablers of the energy transition and the decarbonization of our societies as confirmed by the most recent energy roadmaps [1] and national plans aiming for carbon neutrality by 2050. For hydrogen to play a significant role in the future demand mix, its costs should yet go down implying improved infrastructure design and operations. This paper presents a methodological framework for the optimal design of Hydrogen Supply Chains (HSC) to provide effective support for the study of deployment scenarios. According to the energy source used for its production, hydrogen can be labeled as: green, if produced through Power-to-Gas (PtG) systems, converting renewable electrical energy into green hydrogen and offering a more flexible power grid, better management of the intermittency of renewable energy sources, and a low-carbon fuel supply: as grey if obtained from fossil energy sources with Steam Methane Reforming (SMR) which is the most common process used today and finally as blue h if produced by the SMR process combined with carbon capture, utilization, and storage (CCUS).

Due to the characteristics of the HSC network, economic and environmental assessment is key to its optimal design and deployment that requires the consideration of the set of all possible raw materials, energy sources, technology maturity levels, potential sites for location, transportation and distribution options, storage, end uses, as well as the dynamic change of hydrogen demand over time.

Several works have been identified in the scientific literature to address this complex optimization problem [2]. Most studies involve mathematical models with mixed-integer linear programming (MILP) formulation, associated with an epsilon-constraint technique for tackling the multi-objective issue, mainly based on the total daily cost of hydrogen (TDC) and the Greenhouse Gas (GHG) emissions of the whole supply chain [3,4]. Yet for more than one objective, the construction of the Pareto front by mathematical programming techniques is usually a time-consuming task, since each Pareto solution requires at least one scalar optimization. An alternative to the solution of such problems is to use metaheuristics, and specifically, Multi-objective Evolutionary Algorithms (MOEAs). MOEAs are stochastic population-based nature-inspired algorithms, able of handling multiple objectives, continuous and discrete variables, nonlinear and multimodal functions. However, as these search algorithms are designed to work over unconstrained search spaces, they generally exhibit poor performance when solving highly constrained problems, even if sophisticated constraint-handling techniques have been developed. The scientific objective of this work is thus to propose a hybrid methodology to solve the HSC problem, which decomposes the global problem in such a way that some advantages can be taken of the strengths of every single approach, mitigating simultaneously their disadvantages.

Methods

The original HSC problem [4] is decomposed into two parts (see Figure 1): the (master) sub-problem and the continuous (slave) sub-problem. The master sub-problem is treated by the multi-objective evolutionary algorithm, in which the integer variables (number of production units) and their size (continuous variables) are encoded in every individual in the population. For each individual, a linear programming sub-problem is solved to determine the corresponding optimum production rates and transportation flows. This Linear Programming (LP) problem is constructed according to the values of the integer variables provided by the MOEA in such a way that it is almost a canonical transportation problem. The integer variables produced by the MOEA can be repaired to ensure that this problem is feasible. Subsequently, the MOEA recovers the continuous variables from the LP solver, to compute the *TDC* and *GHG* objectives. Once evaluated, each individual has its fitness function value assigned by the MOEA, which evolves the population for the next generation.

The proposed model takes the economy of scale for production and storage facilities into account with the "six-tenth-factor rule" [5]. The technological maturity of the various units that is reflected in their installation cost, power efficiency, maximal production capacity and the cost evolution of the different energy feedstocks over time is considered. The resulting model can thus be formulated as a mixed-integer nonlinear programming problem (MINLP) with MATLAB environment. An MOA algorithm is a tailored made version used to solve the upper-level problem, whereas the solution of the lower-level problem is performed using CPLEX v12.8.0.

Computational experiments

The proposed model is applied to a case study, i.e., Midi-Pyrénées part of the Occitanie Region in France [6]. The available resources are renewable energy sources (wind, solar and hydro) and natural gas, which must be exported. The set of options for hydrogen production accounts for green hydrogen (alkaline electrolysis cells (AEL) or polymer electrolyte membranes (PEM) with renewable energy sources), blue (SMR with CCUS), and grey (SMR without CCUS). Hydrogen demand is related to the transport sector for use in fuel cell electric vehicles (FCEV).

The region under study is divided into eight subregions (grids) and the time horizon (2020-2050) is divided into seven periods of five years each. Figure 2 presents the obtained Pareto front for this instance. It is composed of 1000 points yielded by MOEA. Each point represents a complete solution, that is, the result of strategic decisions, i.e., size, type, location of production and storage technologies, as well as logistics related to hydrogen transport between grids, for each period.

The Pareto front is divided into segments that are colored according to the shade of hydrogen mostly used to meet the hydrogen demand. Segment 1 comprises non-inferior solutions with Levelized Cost of Hydrogen (LCOH) from 1.64 - 1.90 USD/kgH_2 ; segment 2 from 1.90 - 1.95 USD/kgH_2 TDC; segment 3 from 1.95 - 2.13 USD/kgH_2 TDC; segment 4 from 2.13 - 2.49 USD/kgH_2 TDC; and segment 5 from 2.49 - 2.86 USD/kgH_2 TDC. In this way, the solutions in the left part of the front mainly involve the SMR technology, whereas the solutions at the right part of the front mainly use PEM electrolysis powered by renewable (wind) energy to meet the hydrogen demand.

For the sake of illustration, let us examine the composition of the segment that is mainly constituted of solutions using water electrolysis technology to satisfy the hydrogen demand. These solutions can be considered as the most expensive alternatives for deploying the HSC, but involve the best green options. Figure 3 shows a detailed solution from this segment. The squares indicate the number of production units, the arrows between grids indicate transport flows, and the pie charts indicate demand satisfaction by production type. It can be observed that the solution presents an LCOH of 2.62 USD/kgH_2, and GHG emissions of 1.86 kgCO_2-eq/kgH_2. Alkaline electrolysis (AEL) technology is employed using electricity from wind. In the first period, 33% and 67% of H_2 demand are fulfilled by AEL/wind and PEM/wind, respectively. Note that the AEL demand satisfaction drops from 33% to only 6%, from the first to the second period, due to the decrease in PEM technology cost compared to AEL. In periods 5 and 6, the network shows the installation of PEM electrolysis coupled with photovoltaic systems; nevertheless, most of the hydrogen is still produced by PEM/wind technology. In the last two periods, the contribution of demand satisfaction by technology remains the same, even if two new electrolysis PEM/wind plants are installed in the last period.

Conclusions

This work proposed a comprehensive methodological framework for the optimal design, deployment and management of Hydrogen Supply Chains HSC focusing on eco-mobility applications and their spatio-temporal integration considering the potential of Renewable Energy Sources. This framework can be viewed as a decision support system for scenario analysis. to achieve the desired step-up across the hydrogen value chain. An innovative point of the methodology is to explore the wide range of capacity sizes for a range of new technologies that may be involved gradually on the market.

The optimization strategy could now be extended to address other objective functions such as risk, the inclusion of dynamic aspects associated with the seasonal variability of energy sources, and the incorporation of other hydrogen demands and interactions with systems (ammonia and methanol production, buildings and industrial sector, etc.).

Reference

(1) IEA - International Energy Agency. *The Future of Hydrogen*; 2019.

(2) Li, L.; Manier, H.; Manier, M. A. Hydrogen Supply Chain Network Design: An Optimization-Oriented Review. *Renew. Sustain. Energy Rev.* **2019**, *103* (January), 342–360. https://doi.org/10.1016/j.rser.2018.12.060.

(3) Almansoori, A.; Shah, N. Design and Operation of a Future Hydrogen Supply Chain: Multi-Period Model. *Int. J. Hydrogen Energy* **2009**, *34* (19), 7883–7897. https://doi.org/10.1016/j.ijhydene.2009.07.109.

(4) De-León Almaraz, S.; Azzaro-Pantel, C.; Montastruc, L.; Boix, M. Deployment of a Hydrogen Supply Chain by Multi-Objective / Multi-Period Optimisation at Regional and National Scales. *Chem. Eng. Res. Des.* **2015**, *104*, 11–31. https://doi.org/10.1016/j.cherd.2015.07.005.

(5) Böhm, H.; Zauner, A.; Rosenfeld, D. C.; Tichler, R. Projecting Cost Development for Future Large-Scale Power-to-Gas Implementations by Scaling Effects. *Appl. Energy* **2020**, *264* (December 2019), 114780. https://doi.org/10.1016/j.apenergy.2020.114780.

(6) Région Occitanie / Pyrénées-Méditerranée. *Scénario Région à Énergie Positive de La Région Occitanie / Pyrénées-Méditerranée*; 2017.

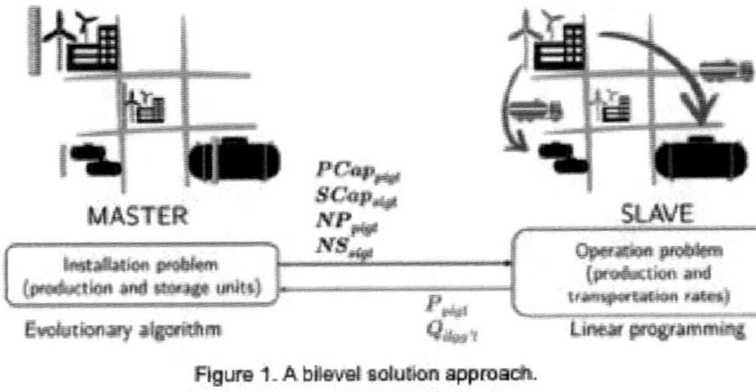

Figure 1. A bilevel solution approach.

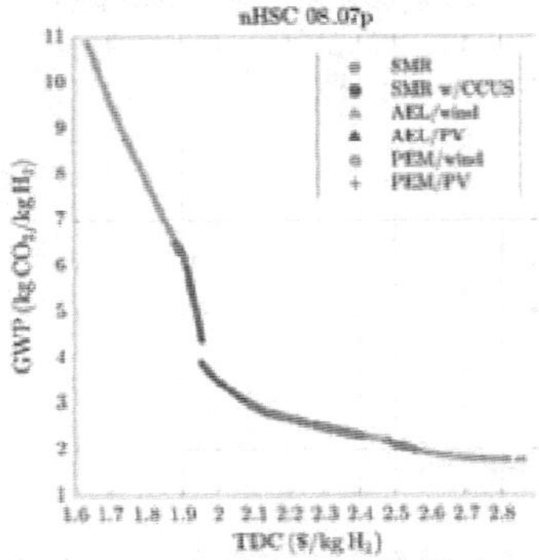

Figure 2. Pareto front approximation of HSC08g07p instance (8 grids-7 periods). The color code indicates the majority of production technology used for hydrogen demand.

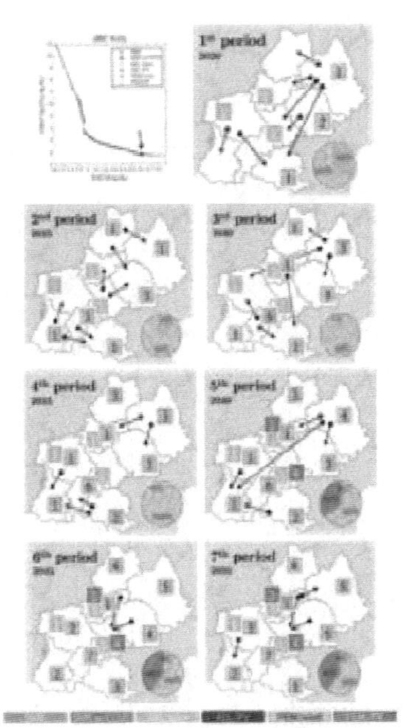

Figure 3. Presentation of a HSC solution from segment 5 for nHSC08g07p instance.

630396 Strength and Elastic Properties of Corn Stover Stalks

Abstract:

This poster overviews a series of experimental tests that measure the variability of strength and elastic properties of corn stover stalks to better understand the governing behavior during size reduction. Internodal sections of hand cut, and separated corn stalks were selected in three size regimes for this study: (**A**) 5 mm minor axis diameter x 80 mm stalk length, (**B**) 10 mm minor axis diameter x 160 mm stalk length, and (**C**) 15 mm minor axis diameter x 225 mm stalk length. Cyclic bending (0.1 mm increments at 1 mm/min), cyclic radial compression through the minor axis (0.1 mm increments at 1 mm/min), and shearing (1 mm/min) were performed with an electromechanical load frame. The data presented investigates the scale relationships for strength and elastic properties, as well as provides a calibration data set for a structure-informed discrete element modeling (DEM) framework to simulate corn stover comminution. Through a system energy balance, the incremental rotational moments about the stalk's center during the bending tests are equated to an effective pseudo-rotational spring constant for representation in the DEM approach. Experimentally, scaling relationships are explored for the proportionality limit, and the elasto-plastic transition to an ultimate strength point followed by strain-softening during bending, as well as similar properties under compression and shear forces. This fundamental data set provides a rich understanding for the damage and fracture of corn stover under different failure modes which will help inform and support further large-scale particle deconstruction population balance modelling studies to understand comminution behavior in pilot scale mills.

628889 Selective Oxidation of Glycerol to Glycolic Acid over Hignly Efficient Fe/TiO_2 Catalysts

Abstract:

As biomass energy continues to replace part of fossil fuels, it is of great significance to complete the cheap and efficient conversion of bio-polyols. Glycerol (GLY) is an important component of energy because the production of biodiesel could generate a lot of inexpensive glycerol as a by-product. However, developing efficient catalytic process of non-precious metals remains a challenge during the selective oxidation of glycerol. With different glycerol conversion pathways, oxidation in the liquid phase using metal catalysts has been studied mostly, targeting various value-added products (i.e., 1,3-dihydroxyacetone (DHA), glycolic acid (GLYOA), or tartronic acid (TA)), among which GLYOA has the greatest commercial value and the maximum demand. In particular, its polymer is a new degradable material and has a wide application prospect.

Herein, we prepared Fe catalysts supported by TiO_2 with different crystal forms for the direct oxidation of glycerol to glycolic acid. Through the optimization of system parameters, the Fe-TiO_2 catalyst exhibits excellent catalytic performance with more than 50% yields under mild conditions. It is found that $Fe^{\delta+}$-Ti_{5c}-O_{2c} active species could be generated on the interface of the TiO_2 (101) crystal plane due to the strong interaction of Fe-TiO_2, further promoting the oxidation of glycerol and the subsequent C-C cleavage of oxygen-containing intermediates. The insights here may pave the way to the rational design of non-noble metal catalysts for efficient conversion of biomass under mild conditions.

633162 Chemical Composition Study for Eggshells of Different Categories

Abstract:

Waste management has globally become a major concern due to the increased waste-load. Eggshells are still largely discarded as waste. Numerous studies have already been carried out to address the critical issues and eggshells are now used by pharmaceutical, cosmetic and food industries in various extents. Recent studies claim that calcium oxide (CaO) obtained from waste eggshells also has catalytic activity for trans-esterification of trygliceride. However, the study of chemical composition for eggshells is limited in literature. It is therefore assessed in the present study. The eggs were classified into three categories of hens, layer and ducks. The egg samples of each category were collected from various sources and compounded. The shells of each category was washed, air-dried and passed through calcination in a muffle furnace separately. The calcined samples were analyzed by X-ray Fluorescence (XRF) to determine the chemical composition. It was found that eggshells of all three categories contain approximately 99 wt% CaO. The chemical composition was found similar. It can therefore be concluded that the eggshells are not required to be screened into different categories for bulk uses.

623041 Metal Recovery Using Oxalate Chemistry: Recycling of Lithium-Ion Batteries and Refining of Bauxite Ore

Abstract:

Oxalate ion is a bidentate anionic ligand with excellent chelation and complexation properties. Its unique ability to precipitate divalent (M^{+2}) metal ions and dissolve monovalent (M^{+}) and trivalent (M^{+3}) metal ions in aqueous medium is advanatageous for separation of metals. In this poster, the use of oxalate chemistry will be described in detail for two commercially important metal recovery and separation processes. The first application is for the recovery and separation of lithium (Li) and cobalt (Co) from spent lithium-ion battery cathodes. In a single-step digestion process with an oxalate reagent, Li and Co from lithium cobalt oxide ($LiCoO_2$) can be separated in aqueous and solid phases, respectively. The Li can be precipitated into lithium carbonate (Li_2CO_3) and Co can be converted into cobalt oxide (Co_3O_4) or cobalt hydroxide ($Co(OH)_2$). To ensure sustainability and no waste is generated in the process, aqueous oxalate reagent used in the digestion was recovered and recycled. The oxalate chemistry was also efficiently used for refining of bauxite ore. Bauxite is a commercially important source for aluminum (Al) and also contains iron (Fe). In the bauxite refining process, closed-loop extraction and separation of Al and Fe from a variety of ores with varying alumina content has been evaluated. The experimental and mechanistic details of both of these processes will be described in detail and potential applications of oxalate chemistry for recovery of the critical metals involving titanium as well as rare earth elements such as samarium and neodymium will also be discussed.

631854 Overview of the Fcic

Abstract:

The FCIC is led by the Department of Energy's Bioenergy Technologies Office (BETO) and is a collaborative effort among researchers from 9 different National Laboratories to address the challenges posed by feedstock and process variability on biorefineries. Two key ideas underpin our work - biomass feedstock properties are variable and different from other commodities, and empirical approaches to address these issues have not been very successful.

FCIC researchers are developing first-principles based knowledge and tools to understand and mitigate the effects of biomass feedstock and process variability on biorefineries.

In this session and the companion session FCIC researchers will discuss their work addressing the impacts of feedstock variability across the bioenergy value chain.

631875 Good Enough for Biotech, Good Enough for Biofuels: A Risk-Based Approach to Managing Biomass Feedstock Variability

Abstract:

Quality-by-Design (QbD) is a decades-old framework and process for ensuring that end product specifications can be met as a function of the input feedstock and operating parameters through the process. In 2019, the DOE funded Feedstock Conversion Interface Consortium implemented a quality-by-design approach to managing the inherent variability that is found in lignocellulosic biomass. Historically, biorefineries have sought to employ equipment or processes from other industries such as grain processing or paper and pulp, and often to no avail. This talk will explore the progress made on employing this QbD process to biofuels derived from lignocellulosic feedstocks. Specifically, several advancements will be discussed.

First, progress made towards identifying, quantifying, and understanding the downstream impacts of critical material attributes will be discussed. Historically total ash, moisture content, and average particle size have been used to define feedstock 'quality', however, these fail to account for the variability imparted by environmental conditions (e.g. drought), storage conditions, and harvesting method. Work done by the Consortium has identified many additional critical material attributes that better characterize downstream performance, particularly in feeding operations and conversion performance. Second, the talk will discuss the use of risk-based methods for identifying and classifying material attributes. Through failure mode and effects analysis, the Consortium has developed a bias-free approach for identifying and classifying new feedstock attributes and processing parameters that must be monitored and mitigated through the process. Finally, the talk will explore several new tools and instruments that can readily be employed by industry to manage feedstock variability while simultaneously achieving economic and environmental sustainability targets.

632196 Impact of Material Attributes on Mill Throughput and Performance

Abstract:

After removal of feedstock moisture, comminution is the most expensive operation to biomass preprocessing. To better understand and optimize this processing stage, this work investigates linking material attributes and properties of Loblolly Pine to the fundamental stress-strain relationships under controlled deconstruction and pilot-scale hammermilling performance. In this work, fundamental strength properties and probabilities and variabilities for fracture and damage events are connected to a Population Balance Model (PBM) framework. This framework uses statistical and outcome distributions to predict the size reduction performance of a mill. From these fundamental damage properties, the resulting particle size distribution and relative hold-up time in the mill are predicted based on the original particle size, size distribution and moisture content as well as operation parameters of mill speed and retention screen size. In addition, this presentation reports pilot-scale milling throughput and energy consumption trials to relate the fundamental-scale testing and prediction performance to experimental data. These relationships and framework provide an opportunity to perform a small series of reduced-order lab-scale tests to predict at-scale mill operation. This operational knowledge and informatics for data-driven biomass preprocessing is housed under the BETO-funded Feedstock Conversion and Interface Consortium to investigate feedstock variability and properties that are presenting barriers to nascent biorefineries.

627541 Tool Wear Analysis and Mitigation for Various Biomass Comminution Systems

Abstract:

Feedstock size reduction is essential for efficient biomass conversion. However, a comminution system inevitably experiences tool wear issues especially in processing dirty feedstocks. DOE's Feedstock Conversion Interface Consortium (FCIC) uses integrated efforts of characterization, modeling, and testing to gain fundamental understanding of the wear mechanisms of biomass preprocessing equipment, develop analytical models to predict wear and establish material property specifications, and develop mitigations via feedstock modifications to reduce abrasiveness, cost-effective wear-resistant tool materials, and improved tool designs. Specifically, two conventional size reduction systems, hammer mill and knife mill, and the recently developed Crumbler® rotary shear were studied. Distinct tool wear mechanisms were identified for these three systems based on tribo-system analysis and worn component characterization. The content and composition of the biomass extrinsic and intrinsic inorganics were determined for various feedstocks and correlated to the tool wear performance. The effectiveness of several feedstock modification methods including air classification, size separation, and water washing on reducing tool wear was investigated. Candidate tool alloys, coatings, and surface treatments were selected and screened using bench-scale abrasion and/or erosion tests. Finite element analysis (FEA) was used to optimize the tool geometric design. Prototype tools were then fabricated using the top candidate materials and/or designs based on the laboratory wear testing results and FEA. The prototypes are being tested in actual comminution systems to demonstrate improved throughput and tool life. The knowledge and tools developed in this study enable rapid mitigation selection for sustainable performance and product quality during biomass preprocessing.

Research was sponsored by the Feedstock Conversion Interface Consortium (FCIC) of the Bioenergy Technologies Office, Office of Energy Efficiency and Renewable Energy, US Department of Energy (DOE).

632121 Advanced Analytical Methodologies Enable Feedstock Screening and Correlation to Pyrolysis Yields and Catalytic Upgrading

Abstract:

Pyrolysis of lignocellulosic biomass can serve as a powerful pathway for the generation of renewable fuels and industrial products. However, procurement of some biomass may be costly, especially when there is competition for alternative uses. Forestry and agricultural byproducts may serve as valuable pyrolysis feedstocks due to their high availability and low cost. Although waste products may fill demand for affordable pyrolysis feedstocks, they are often high in ash and extractives which contribute to unfavorable changes in pyrolysis vapor composition, as well as end products. The work presented here investigates the correlations between various pine anatomical fractions to understand their effect on pyrolysis oil composition for forest residues. Analytical pyrolysis coupled to direct analysis by molecular beam mass spectrometry (MBMS) provides real-time monitoring of pyrolysis vapor composition, even for compounds that are not amenable to gas chromatography. Statistical analysis of MBMS data from the fast pyrolysis (FP) of clean pine, cambium, twigs and branches, needles, bark, and a forest residue samples revealed that bark and needles made the greatest differences in the vapor composition of forest residues during pyrolysis and were enriched in resin acids, furfural derivatives, and ions that have been previously observed from catalysis corresponding to ash content. These results were also compared to catalytic fast pyrolysis (CFP) using a platinum on titania catalyst, and correlations between FP and CFP products were observed for the anatomical fractions. The forest residues and corresponding anatomical fractions from a stand of 23-year-old pine will be converted by fast pyrolysis in a bench-scale 2-inch fluidized-bed reactor, and statistical correlations between the bench-scale products and analytical-scale pyrolysis vapors will be shown. This work highlights the power of advanced analytical methodologies in determining the contributions of vapor components to catalytic pyrolysis outcomes for mixed feedstocks and demonstrates the use of analytical methodologies to better screen feedstocks prior to pyrolysis.

631280 Biomass Fast Pyrolysis Vapor and Bio-Oil Upgrading: Catalyst Deactivation and Its Correlation with Feedstock Properties

Abstract:

Extensive oxygen removal is required to upgrade bio-oils to fuel-range hydrocarbons, while hydrotreating is one of the most common and cost-effective processes already available in existing oil refineries. Long-term operation of the hydrotreater system with raw pyrolysis oil is not feasible because of catalyst fouling and severe reactor bed plugging mainly owing to instability of bio-oil and the fast condensation or polymerization of the active compounds in bio-oil. Therefore, bio-oil quality improvement, especially regarding stability enhancement, is required and still represents the greatest challenge in biomass fast pyrolysis and upgrading technology. The bio-oil stability improvement can be achieved by either vapor phase upgrading (a process called catalytic fast pyrolysis) or liquid phase stabilization by catalytic hydrogenation. Catalysts, especially reduced metal on oxide support, were primarily used in both processes. However, the stability of the upgrading catalyst was the biggest barrier due to the unique qualities of biomass (e.g., high oxygen content, high moisture content, highly reactive functional groups) which causes certain problematic catalyst deactivation. Sensitivity analyses performed for biomass conversion models routinely show that catalyst cost, especially related to catalyst lifetime, has a significant impact on process economics; therefore, improvement in the lifetime of catalyst used will have significant impact on the economic viability of a catalytic process. To this aspect, there is a critical need to develop a correlation of the critical material attributes of fast pyrolysis vapor/oil with upgrading catalyst life to guide the proper select/prioritize of value feedstock and further extend the upgrading catalyst life. This presentation will provide a brief introduction of current fast pyrolysis vapor/oil upgrading process and catalyst, the insight on catalyst deactivation mechanisms, and the establishment of critical material attributes, especially the sulfur content, inorganic content, and foulant/coke precursors in fast pyrolysis vapor/oil, influencing catalyst life.

624169 Applications of Phytoremediation for Catalytic Enhancement of Biocarbon Towards Supercapacitors

Abstract:

The U.S. generates approximately two billion tons of mining waste annually, leading to inorganic heavy metal contaminates such as Ni, deposited into surrounding ecosystems at toxic levels. Phytoremediation offers a low-cost solution for environmental clean-up, particularly in impoverished areas, however current phytoremediation strategies generate unusable biomass due to the toxic metal contents. A novel solution for this unused waste lies in the area of energy storage, especially for supercapacitor applications. This work focuses on the cultivation of water hyacinths, a known metal hyperaccumulator, in various concentrations of Ni-doped water followed by synthesis of harvested biomass into high quality biocarbon. Biomass was subjected to thermal annealing at 500 °C for 1 h under Argon environment to cleave functional groups. Subsequent biochar was improved through thermochemical/catalytic activation at 800 °C for 1 h with KOH providing additional enhancement during biocarbon synthesis. Electrochemical results demonstrate a specific capacitance of 541 F g^{-1} making it suitable for use in the supercapacitor for electricity storage. Physical and electrochemical characterizations, including ICP, SEM, BET, ATR-FTIR, Raman, and XRD, cyclic voltammetry, chronopotentiometry, and EIS will be discussed. This work provides a value-added product and in-depth insights into naturally embedded catalyst, while simultaneously remediating contaminated mining sites.

624185 Electrochemical Studies on Graphitized Biocarbon Derived from Hydrothermally Liquefied Low Ash Content Corn Stover

Abstract:

Biomass-derived biocarbon has gained considerable attention due to its abundance, low-cost, and chemical structures conducive for energy storage applications, particularly supercapacitors. Lignocellulosic components embedded within the cell wall of biomass substrates can provide a fine template for enhanced ion storage, transport, and rate capabilities, desirable for electrochemical storage. Due to its abundant availability of agricultural residues such as Corn Stover, in this research we have utilized the homogenized low ash content corn stover milled and sieved to desired specifications. As obtained corn stover is converted to hydrochar via catalytic hydrothermal liquefaction in the presence of $Ni(NO_3)_2$ catalyst at 275°C and its morphological properties such as surface area, pore volume and diameter were enhanced in the presence of $ZnCl_2$, followed by thermal activation at 450°C. Later as derived hydrochar is pH neutralized and acid washed with dilute HCl to extract metal contaminants. Biochar underwent thermal carbonization at 850 °C for 3 h to enhance pore structures and graphitization, conducive for electrochemical applications. Electrochemical results demonstrated specific capacitances in excess of 300 F g^{-1}, comparable to literature values for corn stover-derived biocarbons. Physical and electrochemical characterizations, such as BET, ATR-FTIR, Raman, XRD, cyclic voltammetry and chronopotentiometry, will be discussed to give fundamental insights into biocarbon enhancement. The results obtained during the electrochemical characterization and studies will be presented.

625756 Electrocatalyst Development for Making Ammonia from Air and Water at Ambient Condition

Abstract:

Renewable power generated from wave energy has faced technological and cost barriers that have hindered its penetration into utility-scale electricity markets. As an alternative, the production of chemical fuels—for example, ammonia (NH_3), which has a high energy density (11.5 MJ/L) and facile storage properties—may open wave energy to new markets including ocean exploration and transportation. We developed a unique catalyst and tested its performance toward nitrogen reduction reaction using the rotating round disk electrode (RRDE) method. The current density increased 0.25 mA/cm^2 after switching the gas from Ar to N_2 and the production rate of NH_3 and Faradaic efficiency (FE) are in the top list of reported results.

627567 Ammonium Formate As a Safe, Energy-Dense Electrochemical Fuel Ionic Liquid

Abstract:

Hydrogen has been investigated widely as an energy carrier for clean electrons, however its low energy density and large energy losses incurred in compression have limited its use. For these reasons, solid and liquid carriers of hydrogen fuel have been investigated, including liquid ammonia, formic acid/formate salts, and methanol, to name a few. While these proposed fuels have significant potential, many of them do not meet the energy efficiency requirements necessary to outperform compressed hydrogen and enable facile transport and storage of renewable energy. For example, ammonia is an appealing energy carrier because of its high hydrogen content: 110 kg H_2/m^3 for liquid ammonia compared to 40 kg H_2/m^3 for compressed hydrogen at 700 bar and 288 K. It is already produced, transported, and stored at scale, and there has been recent progress in reducing the carbon footprint of ammonia production technologies. Formic acid is also appealing due to its hydrogen content (53 kg H_2/m^3), its ease of transport and storage since it is a liquid, and its potential to be synthesized efficiently from carbon dioxide, water, and renewable electricity. However, both ammonia and formic acid suffer from safety issues because they are caustic and corrosive, as well as difficulties with efficient release of energy at scale.

In this work, we propose ammonium formate, a combination of ammonia and formic acid, as an energy carrier. Not only is its production as cheap and simple as its constituent molecules, ammonium formate is a benign solid at ambient conditions, making it safe to transport and store. We demonstrate an electrochemical cell that takes advantage of the fact that ammonium formate forms an intrinsically conductive ionic liquid at 116°C; as an ionic liquid, ammonium formate has negligible vapor pressure and is significantly safer than either ammonia or formic acid alone. In this system, hydrogen is evolved at the cathode with near 100% Faradaic efficiency and formate is oxidized to carbon dioxide at the anode with near 100% Faradaic efficiency. In theory, ammonia could be oxidized to nitrogen in the same cell, however we observe that the operating temperature makes it so that ammonia gas evaporates from the cell and can be oxidized using a second, modular ammonia fuel cell. Because molten ammonium formate is conductive, the electrolyte and feed can approach 100% ammonium formate fuel, resulting in a large thermodynamic driving force. Overall, this system represents a new class of electrochemical fuel ionic liquids and results in modular release of H_2 with the potential for zero net carbon emissions.

632405 Enhanced CO_2 Electroreduction to CH_4 and C_2H_4 Via Selective Proton Transfer

Abstract:

CO_2-derived fuels, synthesized using renewable energy, present an attractive route towards fuel formation from sustainable energy. The role of proton transfer in dictating the efficiency and selectivity of the two-electron, two-proton reduction of CO_2 to CO has been well-studied. Indeed, impressive progress has been achieved in the production of CO from CO_2. However, the role of proton transfer to surface-bound CO intermediates in governing the selective electrosynthesis of higher order products beyond CO, such as methane or ethylene, remains poorly understood.

Here, we investigate the electrochemical reduction of CO at Cu electrodes in nonaqueous electrolytes at low temperatures. The choice of the electrolyte and reaction conditions allows fine-tuned control of the proton donor and the CO binding strength, enabling for the first time activation-controlled kinetic studies over an extended parameter range. Our studies lead to the surprising finding that increasing the concentration of CO serves to suppress methane production, in addition to hydrogen. In contrast, ethylene production remains unaffected by CO concentration and depends predominantly on the electrochemical potential.

These observations provide unprecedented insight into the mechanism of electrochemical CO and CO_2 reduction. Among others, we are able to show that the rate of methane and hydrogen formation is governed by the competition of CO and H for surface sites and that ethylene formation is taking place from sites which are saturated with CO, indicating that simply increasing the supply of CO may not be a viable strategy for improving CO and CO_2 reduction systems. Based on our results, we subsequently demonstrate how knowledge of the reaction mechanism can be exploited to selectively suppress hydrogen evolution relative to CO reduction by rational tuning of the proton donor.

621275 Recent Progress on the Development of a Virtual Feedstock Preprocessing & Handling Laboratory

Abstract:

Computational modeling is an emergent approach that is aimed to enhance and expand the quality-by-design (QbD) approach through the development and deployment of experiment informed and/or validated multiphysics models and predictive simulation tools for the design of biomass feedstock preprocessing and material handling operation equipment. This presentation will exhibit the recent technical advancement in the research, development, validation, and deployment of a science-based, advanced computing-powered virtual preprocessing & handling laboratory as an integral part of the Feedstock-Conversion Interface Consortium (FCIC) funded by the U.S. Department of Energy Bioenergy Technologies Office (BETO). The over-arching objective of this virtual laboratory is a predictive numerical replica of the operation units (e.g., grinders, feeders, and hoppers) that can simulate not only the preprocessing and handling equipment in normal working conditions and generate the experiment-validated performance attributes (e.g., particle size distributions, feeding and discharge mass rates), but also expand the possible testing ranges of the material attributes and processing parameters to the extent that is too costly for physical experimentation in terms of material preparation, equipment wear, and power consumption. A number of first-in-its-kind particulate biomass flow simulation techniques based on discrete element method (DEM) models and continuum mechanics finite element method (FEM) and smoothed particle hydrodynamics (SPH) models, as well as their implementations in proprietary and open-source toolkits and application examples, will be presented. Novel DEM models that can predict the fracture mechanics of biomass materials, and that are coupled to computational fluid dynamics (CFD) for predicting biomass fractionation separation will also be introduced. The limitations and best practices of those numerical models regarding their fidelity and computational affordability will be discussed as well.

626770 Modelling a Compressible Packed Bed Flow-through Deacetylation Reactor for Corn Stover Pretreatment

Abstract:

The deacetylation process is a key step in the lignocellulosic biomass pretreatment process that currently is done in batch reactors. A flow-through diffuser type reactor is a promising approach to make the process more efficient and economically favorable. However, the highly compacted biomass particles can cause significant pressure-drop for fluid flow through the bed, which in the worst case can cause self-enhancing pressure-drop leading to reactor blocking. Detailed understanding of the fundamentals of flow through a compressible packed bed is necessary to scale up the washing and deacetylation process. This study closely examines the classical Kozeny Carman equation and proposes modifications to accurately predict the pressure-drop at different compaction states. A compressible packed bed model is further developed based on the interaction between solid mechanics and fluid dynamics. Experimental measurements of water flow through a packed bed of corn stover show good agreement with the developed model. Corn stover washing process scale up simulations show that smaller reactor diameter, higher friction lining material, lower process throughput, and higher process temperature is favorable in stabilizing the bed from extensive compression when the bed height is fixed.

625518 Multi-Scale Fast Pyrolysis Simulation Framework for Varied Biomass Feedstocks

Abstract:

This presentation summarizes the recent work of building an ***MFiX-based Multi-scale Biomass Fast Pyrolysis Simulation Framework*** for biomass feedstocks. The **MFiX Suite** is a family of multiphase CFD software developed at the National Energy Technology Laboratory and is designed for accurate simulation of multiphase reactors like biomass pyrolysis reactors. In simulating multiphase reactors, the aerodynamic drag related to particle density, shape, and size is a critical element of the biomass material attribute as it determines the particle residence time in the reactor. A hybrid drag model has been proposed to address the challenging problem of accurately capturing the drag of diverse biomass feedstocks. In this approach, a machine learning derived mesoscale drag model was used for small-sized biomass and biochar particles, and a macro-scale experiment measured drag model was used for sand and large biomass particles. The simulation accurately predicted the mixing of sand and biomass. Predicted biochar elutriation rate compared well with experimental data measured in a 2.5-inch fluidized bed using video (60 Hz camera) and high-speed differential pressure techniques with seven 100 Hz pressure differential transducers. A detailed biomass pyrolysis kinetic model with 32 heterogeneous reactions and 59 species was implemented in the MFiX suite of software. The reaction model was validated against two experimental pyrolysis data sets that provided detailed data describing chemical component yields. Sensitivity analysis using the simulation framework revealed the effects of critical material attributes on yields of pyrolysis oil. After validating the drag model and pyrolysis kinetics, the influence of particle shapes was considered using a glued-sphere approach where the composite shape of connected spheres was used to resolve the complex biomass shape, intra-particle heat conduction, and intra-particle species distribution. A simple rolling friction model was introduced to reflect the influence of particle shapes during collisions. To increase simulation speed, intra-particle heat and species distributions were then simulated using a machine learning model derived from detailed, particle-resolved simulations. Finally, the simulation framework was used to investigate the pyrolysis of varied biomass feedstocks over a range of operating conditions and scales to help guide pyrolysis reactor design and optimization.

634993 Technoeconomic and Sustainability Impacts of Feedstock Variability and Mitigation within the Biomass-to-Fuels Value Chains for Corn Stover and Forest Residues

Abstract:

The Crosscutting Analyses task of the Feedstock-Conversion Interface Consortium publishes case studies to quantify and communicate industrially relevant, system-level cost and environmental impacts for the discoveries and innovations of the FCIC. Analysis tools are used to evaluate how feedstock variability affects economics and sustainability throughout the biomass-to-fuel value chain. The scope of work encompasses the value chains for 1) high-temperature conversion of forest residues by ex situ catalytic fast pyrolysis and 2) low temperature corn stover deconstruction to cellulosic sugars and subsequent biological upgrading to multiple products. Case studies provide economic and sustainability information to help biorefinery stakeholders make informed decisions about feedstock collection, processing, and conversion processes to minimize or control the impacts of feedstock variability. An overview of the results, insights and lessons learned from across eighteen case studies will be presented.

631644 TEA Modeling to Quantify Economic Implications for Biorefinery Processing of Isolated Anatomical Fractions of Corn Stover

Abstract:

The Feedstock-Conversion Interface Consortium (FCIC; https://www.energy.gov/sites/prod/files/2020/01/f70/beto-fcic-overview-web.pdf), a collaboration of nine national laboratory partners, seeks to understand impacts of feedstock attributes on biorefinery performance. It is hypothesized that different individual anatomical fractions of corn stover vary in composition and recalcitrance, such that processing each fraction on its own through dedicated campaigns may enable better biorefinery economics overall relative to processing the whole stover material. This presentation focuses on techno-economic analysis (TEA) modeling to quantify the yield and cost ramifications for processing isolated anatomical fractions of corn stover through a low-temperature conversion biorefinery, reflecting a biochemical processing pathway consisting of biomass deconstruction through pretreatment and enzymatic hydrolysis, sugar fermentation and upgrading to hydrocarbon fuels, and lignin upgrading to value-added coproducts.

Commercial-scale process simulation and economic evaluation leveraged experimental and analytical data from FCIC researchers for conversion of whole corn stover plus three individual anatomical fractions (cobs, husks, and stalks) across key steps of the conversion process. Our assessment found encouraging potential for biorefinery economic gains that may be achieved through this approach. TEA results indicated fuel yields varying from 29–44 gallons gasoline equivalent (GGE)/dry ton for the individual anatomical fractions compared to whole stover at 34 GGE/ton, equating to minimum fuel selling prices (MFSPs) between \$6.37–\$10.18/GGE for the fractions versus \$8.76/GGE for whole stover (when lignin is burned), or \$9.15–\$15.19/GGE for the fractions versus \$13.11/GGE for whole stover (when lignin is upgraded to coproducts, based on current experimental performance levels). Cobs and husks demonstrated the ability to achieve the highest fuel yields and lowest MFSPs, outperforming whole stover, while stalks led to the opposite result, as a composite reflection of compositional differences and process convertibility.

Notably, even when taking the weighted average of the results reflecting each anatomical fraction weighted by its corresponding makeup of corn stover, this feasibility TEA screening supports feedstock cost allowances on the order of roughly \$22–\$29/ton as may reflect accommodating additional biomass fractionation equipment during feedstock pre-processing upstream of the conversion biorefinery gate to separate corn stover into such constituent fractions. Or viewed differently, the weighted average MFSP for the fractions was found to be \$0.31–\$0.32/GGE lower than the MFSP for the whole stover basis across either lignin scenario, when maintaining a fixed biomass feedstock cost. These findings highlight favorable implications for biorefinery economics as may be achieved by moving to a staged campaign approach for processing different corn stover fractions sequentially. Further opportunities exist for future work to fill in data gaps for remaining anatomical constituents (e.g. leaves) that were not included in the initial experimental studies, though are expected to maintain similar trends.

631273 Dynamic Life-Cycle Analysis of Fast Pyrolysis Conversion: Implications on Strategies to Manage Feedstock Variability and Improve Sustainability

Abstract:

Pyrolysis conversion is a main contributor to the GHG emissions for biofuels produced from fast pyrolysis. Feedstock composition and reaction temperature are important parameters of the conversion unit operation as they affect the yields of bio-oil, char, and gaseous products from fast pyrolysis. In this case study, we used the fluidized bed reactor experimental data to examine the impacts of feedstock properties and pyrolysis temperature on greenhouse gas (GHG) emissions of the conversion unit operation. We evaluated the trade-off between the main- and co-product, i.e., hydrocarbon fuels and surplus electricity, and its impact on the conversion GHG emissions. We also compared biorefinery-level GHG emissions reduction at different pyrolysis temperatures as a function of various feedstock properties to quantify the GHG emission reduction potential relative to petroleum fuel production and electricity generation at power plants. Results suggested that conversion GHG emissions increase as ash and lignin contents increase at all investigated temperatures (500 ^{o}C, 525 ^{o}C, and 550 ^{o}C). A trade-off between the organics yield and biochar and off-gas yields as a function of feedstock attributes such as ash content and the pyrolysis temperatures is observed, which influences the conversion GHG emissions. Biorefinery-level GHG emission reduction results show that for a low-ash feedstock, a lower pyrolysis temperature benefits greater biorefinery-level GHG emission reductions, while for a high-ash feedstock, a higher pyrolysis temperature benefits greater biorefinery-level GHG reduction. Biorefinery-level GHG emission reduction could vary from 0.35 to 0.42 tonne per tonne of feedstock converted to biofuels, depending on feedstock attributes and pyrolysis temperatures, which translates to $3 - $14 per ton of feedstock converted with a current carbon credit of $195/tonne of GHG reduction under California's Low Carbon Fuel Standard. Such results shed light on how the biorefinery can adjust the operating strategy based on the feedstock properties to achieve better sustainability and at the same time maximize its overall carbon credit, which could also help lowering the fuel MFSP.

630200 Resilience in the Face of COVID-19: Managing a User Program at the Molecular Foundry

Abstract:

Introduction

The Molecular Foundry is a Nanoscale Science Research Center, located at Lawrence Berkeley National Laboratory (Berkeley Lab) in Berkeley, California, sponsored by the Department of Energy (DOE). The Molecular Foundry is a knowledge-based User Facility, where leading-edge resources in instrumentation, analytical capabilities, and scientific expertise in nanoscale science are made available via a peer-reviewed proposal process. Roughly 1000 researchers or "users" come to the Foundry from around the world to conduct their work each year. The COVID-19 pandemic presented challenges and opportunities in conducting collaborative on site research. This presentation focuses on the operation of the user program amid the pandemic, with an emphasis on strategies for maintaining research productivity and user access. We also present an outlook on ways that gains in virtual collaboration and science during the pandemic can provide opportunities in a post-pandemic world.

The Molecular Foundry provides scientific expertise and capabilities distributed among seven facilities: Inorganic, Organic, and Biological facilities for synthesis, preparation, and assembly; Nanofabrication, for processing and integration; the National Center for Electron Microscopy and Imaging and Manipulation, for characterization; and Theory, for understanding and predicting material properties.

The Molecular Foundry maintains a highly collaborative research model. Staff scientists and technical staff at the Molecular Foundry interact with users through instrument training, consultation during experiments, support with data interpretation, and collaborative writing of journal articles. For each 12 month project, users stay in the local area for an average of three months while they work on site.

Spring 2020 Shelter-in-Place

In spring 2020, the scale and urgency of the COVID-19 pandemic in the United States became apparent, and the city of Berkeley enacted a Shelter-in-Place order effective March 17, 2020. The Molecular Foundry responded by pausing research operations on site while keeping research facilities safe and ready for reopening. This directly impacted every one of the hundreds of research projects that were active at the time, and presented a challenge to the Molecular Foundry's commitment to safety and user access.

The Foundry staff collaborated to respond to this challenge with a holistic approach to keep science moving forward safely. All user projects received a 12-week extension, to account for this pause on work. During this period, the Foundry developed protocols for safe work on site, aligned with requirements for Berkeley Lab and applicable regional, state, and national guidelines. These included the use of face coverings, hand-washing, distancing/density control, and surface disinfection.

While maintaining a very low density of people on site, a few high-priority projects were allowed to proceed, including research on COVID-19 and SARS-CoV-2 detection, treatment, and transmission mitigation. These early research activities served as a testing ground and model for the further expansion of access in the months to follow. Staff from across all seven facilities began to explore investment in virtual modes for training, user support, and instrument access amid the pandemic. Berkeley Lab developed new training modules for safe work on site during the COVID-19 pandemic, launched a lab density study to determine appropriate safe occupancy limits for all laboratories across the site, and developed a Health Pledge,symptom check system, and a specific COVID19 training module that was required for all on site personnel.

Summer 2020 and Beyond

The months that followed the onset of the pandemic were characterized by continual adaptation to the emerging pandemic. As of the writing of this abstract, this phase continues to evolve as vaccination campaigns are gaining momentum nationally. These are some of the strategies employed to manage the user program during this phase of the pandemic:

Communication

Communication with the user community and the staff of the Molecular Foundry was a central pillar of maintaining safe work at the Molecular Foundry. This included email messages to all users generally, frequent discussions among staff across all roles, and communication with Berkeley Laby leadership.

Shift Work to Manage Density

Upon emerging from the full shelter-in-place order, moderate numbers of researchers were allowed to work on site at the Molecular Foundry. The total number of persons accessing Berkeley Lab was managed centrally by the Laboratory, and the Molecular Foundry worked with an allotment of "slots" for site access throughout the duration of the pandemic restrictions. A three-shift model was deployed. Slots were distributed among the seven facilities, and researchers requested a slot ahead of time, subject to approval by Molecular Foundry leadership and compliance with the Health Pledge, weekly symptom check, and a specific COVID19 training requirements.

Prioritizing Research Projects

Facing limited access compared to normal operations, it was necessary to consider carefully which staff and user work could be accomplished on site. The Molecular Foundry values equity of access for users, held in balance with safety precautions. We aimed to carefully consider such factors as the potential to impact the fight against COVID-19, DOE and Berkeley Lab research priorities, inherently safe activities, experimental complexity, travel considerations, and user training.

Badged Entry for Approved Work

During the pandemic, the Molecular Foundry implemented badge-reader entry for all buildings. Coupled with weekly calendars and daily posting of all persons with prior approval to enter the buildings, this provided a layer of accountability around access to the spaces.

Regular Safety Meetings

Berkeley Lab implemented weekly virtual meetings focused on COVID-19 safe operations and other safety and space planning topics. These meetings require attendance by at least one staff representative from each of the seven facilities, and are open to all Molecular Foundry staff and users. Each facility hosted a daily safety check-in for work planning.

Carefully Considered Training

Training presented a particular challenge during the pandemic, as many users receive training on complex tasks at the Molecular Foundry, traditionally accomplished side-by-side. After several months, the Foundry formalized guidelines to consider for new user training, considering such factors as lab occupancy, the potential to develop tools to support distanced training, inherent hazards in the proposed work, and robustness of the instruments.

Development of New Remote and Virtual Training and Access

Molecular Foundry staff rallied around the call to support user access virtually. Specific responses are adapted to the needs and opportunities of each facility. A sampling of tools include: training videos, holographic augmented reality for training, audio headsets to support distancing, remote operation of select equipment, remote login for data processing, virtual presence via video conferencing during experiments, telepresence robots, and a variety of informal communication and collaboration tools.

Calls for Proposals Scope

Two of the standard calls for proposals have been fully within the pandemic. While we kept the calls open for the most exciting research across nanoscience, we balanced the practical considerations with some emphasis for these calls. We encouraged users to submit proposals that included COVID-19 research, theoretical and computational work, remote access to instruments, or data collection for remote analysis, and proposals requesting samples of materials developed at the Foundry.

Expanded Definition of Remote Users

The Molecular Foundry is one of five Nanoscience Research Centers sponsored by the DOE, and therefore shares certain reporting parameters for our user programs. Effective for FY 2020 reporting, the centers collectively adopted an expanded definition of a remote user, to recognize the increasing role of remote access for science at these knowledge-based user facilities.

Virtual Meetings and Conferences

Meetings have shifted online, supported by zoom video conferencing. This includes smaller team meetings, larger seminars spanning the Molecular Foundry, and our annual User Meeting--a nanoscale science research conference which drew 900 registrants in 2020.

Outcomes and Future Outlook

The pandemic presented challenges to maintaining a strong nanoscience research user program. Nevertheless, overall metrics reflect resilience in the face of these challenges. The User Program remains strong, serving 643 on-site and 97 remote users in FY 2020, according to formal reporting definitions, with a total of over 1500 named researchers on active projects during FY 2020. Demand for the user program remains strong, with around 225 standard proposals received during the most recent spring 2021 call for proposals, roughly two thirds our pre-COVID maximum. Research productivity as measured by journal articles has remained robust when comparing 2020 to 2019 and 2018.

The shift to increased virtual access presents opportunities to expand the geography of our user base to include more researchers outside our local area, due to reduced barriers in travel once post-pandemic travel resumes. We have seen an increase in sample-only proposals, enabling new ways of working together across institutions. We also have potential to include more people in the active experimental data collection phase, as undergraduate researchers and principal investigators increasingly log in remotely to participate in experiments that previously were completed by just one member of the research team. Furthermore, shifting our annual User Meeting online has facilitated a tripling of conference registrations, with potential to reach new communities with our science.

As we look ahead to 2022 with optimism, we hope to learn from the unique challenges presented by the pandemic. We will have new facets of this experience to manage, including a safe return to work for our users. The Molecular Foundry remains committed to advancing user science to bring solutions to the world.

629703 Global Supply Chain Optimization for COVID-19 Vaccines Under Covax Facility

Abstract:

With the pandemic in action, over 130 million people across 219 countries have been infected by COVID-19. The number of mortalities due to COVID-19 has climbed to around 3 million [1]. Though around 80% of the infected patients have recovered, several patients are becoming reinfected recently appeared. As a result, efforts for COVID-19 vaccines have accelerated around the world. The world's top pharmaceutical companies and research organizations have developed multiple vaccines. It is expected that the transmission can be slowed considerably if a sufficient population is vaccinated. Though a typical vaccine takes around ten years for approval, several COVID-19 vaccines have already been developed and approved in record times. As of 31 March 2021, around 750 million people worldwide have been vaccinated [2]. With world governments aggressively supporting vaccine developments and procurements, the vaccination drive is progressing rapidly.

However, to end the pandemic, we not only need COVID-19 vaccines, but the development of vaccines must be followed by a fair, fast, and efficient global distribution. The current challenge is the timely delivery of the vaccines. The expected volume of the vaccines per month adds up to around 850 tons and is more than the UNICEF suppliers' capacity combined [3]. Two main concerns regarding the global COVID-19 vaccine distribution are 1) Limited supply of resources for the delivery of the vaccines, e.g., cold chain equipment, and 2) Vaccine nationalism, i.e., governments signing agreements with companies to procure surplus vaccines. To overcome these challenges, the COVAX facility was launched by WHO for multiple vaccines to prevent vaccine nationalism. It aims to ensure fair access to COVID-19 vaccines for all countries, irrespective of their financial status. It allows countries to pool resources for effective vaccine development and distribution. Under COVAX, deals for vaccine procurement have already been made. However, vaccine distribution is still under development. It faces various challenges due to (1) the variety of vaccines, (2) person-specific doses, 2) unknown country-specific demands required for herd immunity threshold, especially low-middle income countries, (5) low-temperature storage protocols for most vaccines, and (6) unanticipated disruptions in both production and distribution.

Addressing these challenges requires a multi-prong approach that must go beyond straightforward pharmacological measures. The two primary goals are to (1) estimate the herd immunity threshold and (2) effect an efficient global vaccine distribution. In this work, we explore the time-varying nature of reproduction number during the first wave of the

COVID-19 pandemic using confirmed cases, recovered cases, fatalities, and timings of lockdown phases. We propose a mean-field epidemiological model for COVID-19 that extends the classical SEIR model [4]. We describe NEIASCFR, an 8-compartmental model with variable transmissibility over time, by considering the time-varying nature of parameter values over different lockdown phases. We estimate our model's parameters with public data on confirmed cases, recovered cases, and fatalities for each country. The parameter values are used to determine reproduction number and herd immunity threshold. The model is applied to study the herd immunity threshold of 9 countries (India, Nigeria, Iran, Pakistan, Thailand, Mexico, Democratic Republic of Congo, Egypt, Thailand), constituting 50% of demand under the COVAX facility.

Using this within the framework of COVAX, we develop a global supply chain model to optimize the production and allocation of vaccines to participating countries from various production centers. The supply chain was optimized considering potential disruptions, vaccine approval dynamics, and vaccine candidates' approval with different risk mitigation strategies. Given: 1) Network configuration consisting of production facilities and different regions. 2) Initial vaccine procurement quantities. 3) Capacity of different storage containers. 4) Costs of procurement, storage, transportation, and risk mitigation per unit. 5) Different scenarios, we determine 1) Production quantities of vaccines at different centers and vaccine allocation to each demand region. 2) Extra procurement of vaccines at each center. 3) Cost of procurement and transportation.

References:

1. Worldometer. COVID-19 CORONAVIRUS PANDEMIC. 2021; Available from: https://www.worldometers.info/coronavirus/.

2. Bloomberg. Covid-19 Tracker. 2021; Available from: https://www.bloomberg.com/graphics/covid-vaccine-tracker-global-distribution/.

3. UNICEF, UNICEF outlining plans to transport up to 850 tonnes of COVID-19 vaccines per month on behalf of COVAX. 2020.

4. Li, M.Y. and J.S. Muldowney, Global stability for the SEIR model in epidemiology. Mathematical biosciences, 1995. 125(2): p. 155-164.

632079 Sustainability and Resiliency Assessment for Industrial Synergies between Renewable Energy and Chemical Production

Abstract:

Photovoltaic (PV) is the fastest growing energy technology and is expected to maintain this growth in the coming years. Due to the durability of PV modules, there is a lengthy time gap between installation and decommissioning of the modules, and it is expected to create large scale waste worldwide in the near future. On the other hand, chemical manufacturing systems are undergoing significant transition towards bio-based production to meet the goal of sustainable low carbon feedstock. This work presents a potential synergy between renewable energy wastes and bio-based polymer production through extending a previously proposed system to circularize the PV lifecycle by diverting organic byproducts from PV recycling to agricultural production. We identify and introduce additional synergies as links in the network to close the loop between PV module disposal and manufacturing. The chemical outputs from the PV recycling process are used as fertilizers for crops that are potential feedstock for biopolymer production. These biopolymers are used in the manufacturing of PV backsheets, replacing the current petroleum-based products. We quantify the total life cycle environmental and economic benefits at scale of implementing this synergy along with total reduction in GHG emissions. To evaluate the design, we also use a network-based approach to understand the structure and topology of the proposed system by exploring various network measures as well as information theory-based measures, to assess resilience and long-term sustainability. We show quantitatively that network measures like diameter and density are ineffective at scoring the network resilience due to the nature of these measures in original network science literature, while centrality measures are useful for finding the relative importance of certain nodes, which may highlight network vulnerabilities that factor into the overall resilience. Likewise, ascendency and reserve values for the network appear useful in determining the overall resilience and sustainability of the network, however, based on their application to comparing two alternative network scenarios, these metric interpretations should be more rigorously defined for industrial networks before being incorporated for industrial network design.

633441 A Community-Level Energy Planning Model for the Dynamic Optimization of Waste-to-Gas and Waste-to-Power Anaerobic Digestion Supply Chains.

Abstract:

Society has growing concerns over the long-term environmental impacts of climate change and the accumulation of organic waste in local communities. Community energy planning is increasingly focused on the use of novel technologies to provide sustainable and cost-effective heat and power. Furthermore, changes in energy demand patterns support the development of dynamic optimization models of energy dispatch. This study focuses on the use of anaerobic digestion (AD) technology for converting human, agricultural, and industrial waste into renewable natural gas (RNG) and power. We develop a dynamic mixed integer linear programming (MILP) in Calliope to optimize the production and dispatch of RNG and power.
Anaerobic Digestion (AD) is a microbiological process that is used in processing organic, biodegradable wastes and side streams into methane-rich biogas and nutrient-rich digestate. Previous techno-economic (TEA) and lifecycle (LCA) analyses indicate that AD can be a commercially competitive and sustainable technology for communities to generate clean energy. In this study, we evaluate the use of AD in Ames, Iowa to convert human and animal waste, biomass, and wastewater into RNG and power. RNG and power are consumed locally to provide industrial, commercial, and residential heat and power demand. The AD process generates by-product digestate that is returned to Iowa farms. The model covers simulates 1 year of energy demand in hourly and monthly increments.

The MILP identified 13 locations in the vicinity of Ames, Iowa capable of producing power or RNG. The electricity price ranged between \$0.065 - \$0.18/kWh and the RNG price is estimated at \$0.026/kWh (\$7.62/mmbtu). Electricity production varied between 30 and 50 million kWh per month. Gas production varied between 15 and 40 million kWh per month. All the electricity demand in Ames, Iowa could be supplied from local waste resources. Summer gas demand can be supplied locally. However, the gas supply will need to be supplemented with gas from the external grid to meet winter demand. Technological improvements and public incentives could increase the adoption of AD in community energy planning systems. Further research will evaluate the impacts of variabilities in energy supply and demand to address grid disruptions.

625575 Asset Integrity Management (AIM) for Hydrogen: Major Incident Risk Reduction By Integrating a Risk-Based Safety Approach

Abstract:

Asset Integrity Management (AIM) is a well-understood process for optimizing the management of assets over their whole life-cycle. If applied correctly, AIM can offer asset owners and operators the ability to manage risk and enable the assertion, with confidence on an evidential basis, that their assets are safe and reliable. For AIM to be most effective, the assets' design must consider the whole life-cycle of the assets' operation. Adopting a risk based approach to safety and operations as early as possible in the design phase of a facility is demonstrated to improve long-term reliability, availability and maintainability. Combining a risk based approach at the design stage with AIM during operations has the potential to simplify and align processes. Additional benefits include the improvement of long-term reliability, availability and maintainability. By integrating a risk based approach with AIM the realization of common goals can be facilitated: the reduction of major accident hazard risks and consistent performance of assets. For a sector on which many emergent national decarbonization policies are reliant upon to deliver "net zero", an all-encompassing risk based safety and AIM approach offers asset owners and operators in the beginnings of the hydrogen economy boom a means by which to deliver business objectives profitably without major incident.

625565 More Haste, Less Speed: Proportionate Supply Chain Risk Management for the Accelerating Hydrogen Economy

Abstract:

The long awaited acceleration of the hydrogen economy is dawning. It is viewed as a necessity in the race to decarbonize industry, heat and transport, amongst other key sectors. Existing hydrogen economy supply chains serving the current markets face the challenge of ramping up to meet future demand. Aside from hydrogen growth, many organizations are increasingly concerned about supply chain risk, not only the traditional supplier issues of late delivery and poor quality but also other risks with security, environmental, social and ethical impacts. Furthermore, hydrogen as an energy vector requires significant safety management to mitigate its potential to cause major accidents. The confluence of increasing demand, major accident hazards associated with hydrogen and concerns of supply chain risks present a significant challenge for those organizations in the hydrogen supply chain. When viewed through the lens of rapid demand acceleration, supply chain risks are magnified by the potential for short-cuts and risk taking. Such risks are exacerbated by the potential set-back in the global decarbonization goal, should a major accident be realized. Formal Supply Chain Risk Management (SCRM) techniques adopting a risk based approach are explored to help mitigate supply chain risks in a proportionate way, reducing supply chain vulnerability, ensuring continuity if risks occur and increasing value within the supply chain through improved performance. Crucial attributes of successful SCRM are considered to be: application of good governance, early supplier engagement, proportionate assurance levels based on risk, supply chain mapping, monitoring of risk events/risk event pre-cursors and risk register maintenance.

633712 The Role of Carbon Capture in Meeting Net-Zero Carbon Goals

Abstract:

President Biden has laid out a bold and ambitious goal of achieving net-zero carbon emissions in the U.S. by 2050. The pathway to that target includes cutting total greenhouse gas emissions in half by 2030 and eliminating them entirely from the Nation's electricity sector by 2035. Research, , development, demonstration, and deployment (RDD&D) will be required to achieve the president's objectives, including investments in both point source carbon capture and carbon dioxide removal approaches that target the accumulated pool of carbon in the atmosphere. Both will be required to achieve net-zero carbon emissions in time and they will require increased deployment on an accelerated timeline in order to move down the cost curve. These efforts, combined with effective policy, will improve the economic viability of the implementation of these critical approaches.

Deployment of these technologies at the scale required will necessitate the use of resources including land, water, and low-carbon energy, while ensuring the secure and reliable storage of carbon dioxide (CO_2) on a timescale that impacts climate. Therefore, CCS and CDR deployment must be implemented strategically in terms of regional goals and requirements.

The Office of Fossil Energy and Carbon Management and the National Energy Technology Laboratory will play an important role in the transition to net-zero carbon emissions by reducing the environmental impacts of fossil energy production and use. This includes helping decarbonize other hard-to abate industrial sectors through investments in approaches like CCS, direct air capture, and technologies to produce low-carbon products and fuel - including hydrogen.

633330 Clean Energy and Decarbonization Research, Development and Demonstration at Oak Ridge National Laboratory

Abstract:

This talk provides the overview of Oak Ridge National Laboratory's (ORNL) transformational science and technologies towards the clean and decarbonized energy future with capabilities spanning the entire spectrum from clean energy generation to distribution, at-scale storage, and end-use. Being the largest DOE Office of Science Laboratory, ORNL is well-recognized for its energy efficiency research portfolio with the most comprehensive user facilities and capabilities in fundamental science and applied energy space among all the DOE national laboratories. We are accelerating decarbonization in the US with a three-pronged approach with integrated energy systems: 1.) Achieve deeper energy savings through enhanced energy efficiency efforts across all spectrums of the energy service sectors, i.e., buildings, industry, and transportation; 2.) Accelerate net-carbon neutrality by capturing and converting CO_2 into valuable materials, fuels, and feedstocks for downstream utilizations; and 3.) Enable bi-directional electron flows between the grid and end users with at-scale energy storage and transactive controls to enable a sustainable, resilient, and secure energy infrastructure. Exemplary ORNL RD&D advancements in all the areas above will be presented and future opportunities and challenges in the integrated energy systems approach will be discussed.

633329 Getting to Zero Emissions: The Need for Nuclear Energy

Abstract:

Sustainability demands a scalable source of clean and cheap electricity. Nuclear power can deliver that. It has the smallest environmental footprint and the best safety record but first public opposition to nuclear power must be overcome. I will address the concerns typically voiced in opposing nuclear power such plant safety, long-term storage of waste, fuel supply, and cost. Getting the public to embrace nuclear power is a Herculean task, but it must be undertaken. We have to (i) educate the public (ii) stop closing functional nuclear power plants; (iii) remove the barriers for financing nuclear projects; (iv) expand the fleet of nuclear power plants; and (v) develop and deploy the next generation of walk-away safe plants that can also use the spent fuel as a resource.

633349 High-Temperature Co-Electrolysis of CO_2 and H_2o on Lanthanum Ferrite-Type Perovskite Oxide Cathodes

Abstract:

High-temperature electrolysis of water and co-electrolysis of water and carbon dioxide are promising ways to produce clean hydrogen and syngas (a mixture of hydrogen and carbon monoxide) with pure oxygen being the only by-product. The co-electrolysis of H_2O and CO_2 also provides a potential CO_2 mitigation route by converting this greenhouse gas into syngas, an important chemical building block, which can be turned into valuable chemicals by Fischer-Tropsch synthesis. In this study, a high-temperature (700-850°C) solid electrolysis cell (SOEC) is used to co-electrolyze H_2O and CO_2 into syngas.

The SOEC employed in the present work consists of an yttria-stabilized zirconia (YSZ) solid oxide oxygen ion conducting electrolyte, sandwiched between two electrode layers. The counter electrode (anode) is a commercially available mixture of $La_{0.8}Sr_{0.2}MnO_3$ and YSZ (LSM-YSZ), whereas the working electrode (cathode) consists of an in-house developed lanthanum ferrite-type perovskite oxide. H_2O and CO_2 get electrolyzed at the cathode producing O^{2-} ions which travel through the YSZ electrolyte to the anode where they combine to form molecular oxygen.

The B-site doped lanthanum ferrite perovskite materials used in this study as the working electrode were synthesized via EDTA-citric acid complexation method. The synthesized materials were characterized ex-situ, in-situ and operando using X-ray diffraction (XRD), X-ray photoelectron spectroscopy (XPS) and X-ray absorption spectroscopy (XAS), Raman spectroscopy, transmission electron microscopy and four probe DC van der Paw techniques to investigate their morphology, bulk and surface structure and electrical conductivity and the changes in these properties during the electrochemical reaction under bias. The electrocatalytic activity tests showed that the synthesized perovskites are highly active in electrolysis of water and co-electrolysis of H_2O and CO_2, and H_2/CO ratio of the product can be tuned through modifications on the A- and B-site doping.

629894 Combined Alkali Treatment and Anaerobic Storage in Corn Stover Enhances Reactivity and Surface Energy Properties

Abstract:

The inherent recalcitrance of lignocellulosic biomass to physical and chemical depolymerization necessitates high severity thermal and chemical treatments to liberate carbohydrates and lignin for production of fuels and chemicals. The aim of this study was to examine the potential for process intensification within the feedstock supply chain operation of long-term storage. The aim was to use anaerobic storage to provide an environment conducive to slow transformations in corn stover following saponification of ester bonds in lignin and hemicellulose brought about by alkali addition. Results indicate the alkali-treated corn stover was well preserved in anaerobic storage over 2 weeks, showing only 4% dry matter loss. Monomerization of glucose and xylose in the presence of glycosidases confirmed the alkali-treated, stored biomass was more susceptible to enzymatic cleavage, suggesting reduced recalcitrance. Washing at room temperature was employed to remove any molecules that could be blocking enzyme assess. Washing increased enzymatic accessibility in terms of carbohydrate monomerization in the native sample and increased surface area in both native and alkali-stored samples. Surface energy confirmed the unwashed samples contained equivalent basic features, which primarily are attributed to lignin, yet washing removed the coalesced lignin in the alkali-treated stored samples. The alkali-stored, washed stover exhibited increased surface area, hydrophilicity, and work of cohesion and adhesion compared to the native, washed samples. This research drives the fundamental understanding that adding alkali prior to anaerobic storage not only improves enzyme accessibility to corn stover but, when combined with a room temperature wash, will result in a product that has favorable physical properties in downstream preprocessing.

632230 Air-Classification of Forest Residues for Beneficiated High Temperature Conversion Feedstock

Abstract:

Economic viability of current pathways to renewable fuels and materials relies on the utilization of all available biomass resources including residues and low-cost materials. These low-cost resources such as the tops, limbs, and other wood waste after lumber harvest or milling operations, construction and demolition debris, industrial and residential residues, etc. These wastes, however, often have deleterious composition/properties for subsequent upgrading to fuels or products. There are many potential treatments to beneficiate feedstock prior to downstream utilization, and this work investigates using air-classification and other density-based separations to separate the feedstock down to tissues with differential anatomical function. Separation of these materials can improve the product yield, quality, and overall process economics of thermochemical conversion of biomass. This work also presents how these fractionation methods impact and control the feedstock material attributes. Material attributes related to the classification efficiency include particle size, size distribution, particle density, morphology, and moisture content. These properties alter the drag and hydrodynamics of the particles and ultimately impact the separation efficiency. These results and properties are also investigated in a preliminary discrete element modeling effort to understand the separations from a first-principles perspective.

630331 Population Balance of Knife Milled Corn Stover

Abstract:

In this work we develop a predictive population balance model for knife milling of corn stover stalks, based on a probability and breakage function to predict the milled particle size distribution as a function of variation in mill tip speed, moisture content, initial particle size, and retention screen size. These probability and breakage parameters are dependent on material properties like the fracture resistance and minimum specific energy. Batch scale experiments were used to determine material and mill parameters within the breakage law. A breakage matrix was developed using fracture resistance and minimum specific energy, which were found from calibration tests with iterative regression to measured product size distributions as well as inputs for feed size, screen classification efficiency, amount of recirculated material and the number of impacts that a particle experiences. In the knife mill experiments, the initial particle size was obtained from analytical splits of the sample along with separations with an orbital screen. Screen opening was altered by interchangeable screen sizes while the mill speed was controlled with a VFD. These variables were associated with the impact energy, number of impacts and the size of the particles under impact. This work reports on the imparted control over resulting particle size distribution as a function of controllable operational parameters (e.g. tip speed, feed rate, initial material size and retention screen size). In this work, results for both a 'once-through' configuration in addition to a recirculating fractional milling technique was evaluated for degree of control over the comminution process, as well as potential energy savings and optimal configuration/recirculating rate. The model provides an estimate for daughter particle size distribution for a range of operational conditions which is then be validated by experimental results. Determining the highest throughput to achieve a specific particle size distribution will allow optimization on energy costs and milling configuration.

629793 Simulating the Operational Effect of Particle Scale Impacts on Deconstruction Energy of Pine Residues.

Abstract:

Feedstock supply systems are highly complex organizations of operations required to move and transform biomass from a raw form at the point of production into a formatted, on-spec feedstock meeting all conversion Critical Material Attributes at the throat of the reactor. Feedstock logistics can be broken down into subsystems, including harvest and collection, storage, transportation, preprocessing, and queuing and handling. Designing economic and environmentally sustainable feedstock supply systems, while providing necessary resource quantities at the appropriate quality, is critical to growth of the bioenergy industry. This study is focused on the feedstock preprocessing production cost impacts of variable moisture and ash on hammer mill throughput and energy consumption and on generation of fines that are not able to be fed to conversion. In addition, the quality of the feedstock, convertible carbon content (minimum carbon specification) and maximum ash content, was examined in terms of cost impacts of not being able to feed residue not meeting both specifications to the conversion reactor. This work compares a base system to a system where the rotary dryer is moved downstream of the disk screen in order to assess the cost impacts of performing size reduction before drying and then drying the smaller particles, by reducing fines and lowering the required energy. The modified system design achieved 27.6% higher throughput capacity than the base system due to the generation of fewer fines during wet versus dry grinding; the production cost adjusted for discarded fines were $137.46 and $72.04/dry ton (2016$) preprocessed for the base design and modified system, respectively. In the modified system, grinder energy consumption more than doubled going from 26.2 kWh/dry ton to 59.9 kWh/dry ton for grinding wet material. However, drying energy savings from the more efficient drying of smaller particles following wet grinding far outstripped that increase reducing the drying energy from 2,328 kWh/dry ton for drying wet chips to 1,238 kWh/dry ton for drying hammer milled residue. When it is assumed that on the material that meets the minimum compositional Critical Quality Attributes for the conversion process, the final delivered feedstock costs for the base system and modified system rose by $17.10 and $9.09/dry ton, respectively. Additionally, the modification of the system yielded a modeled reduction of at least $1.24/gge to the minimum fuel selling price.

625871 Optimal Depot Size and Location Selection in Biofuel Supply Chain Under Temporal and Spatial Variabilities over a 10-Year Period

Abstract:

Numerous studies have been conducted to quantify biomass availability, supply uncertainty, and to design biomass supply chains with minimized total logistics costs. These studies are mostly conducted relied on biomass yield and qualities at a single point in time and ignored variability that may arise simply due to variation in environmental factors. However, the most economical size and location for biomass depot should be defined based on temporal and spatial variability of feedstock supply and quality. Using static biomass quality attributes in biofuel supply chain design can lead to system designs with non-robust critical material attributes (CMAs) for preprocessing operation and non-consistent critical quality attributes for biofuel conversion process. To address this issue, an innovative computational model was developed to determine the optimal biofuel supply system configuration over a 10-year period (2010- 2019) considering the variabilities that temporally affect both biomass quantity and quality. In this analysis, the variability of biomass yield, quality parameters were linked to weather factors, especially precipitation and drought events occurred during growing season. Modeling results identified the optimal depot location and size in the selected supply region, under variability of herbaceous biomass sources and quality parameters encountered in the biomass sources over a 10-year period. The resulted total delivered feedstock cost of $83.78/dry ton was lower compared to the cost of $91.75/dry ton in the case where optimal depot location and size were determined from 1-year biomass sources and quality variability. In conclusion, this novel computation tool can be used to design the least cost supply chain configuration that delivers feedstock blend that meets or exceeds the CMAs for biofuel production, more importantly, with the ability to handle temporal and spatial variability of feedstock supply and quality.

631117 Nuclear Powered Biorefineries: A Nationwide Supply Chain Analysis

Abstract:

Advancement in biofuel and biopower technology is on the rise to mitigate climate change and to ensure constant energy supply to a growing population. Replacing all liquid fossil fuels with biofuel will require large-scale biorefineries and an extensive source of biomass. Lignocellulosic biomass is the largest biomass source in the world as it comes from numerous nonedible materials such as energy crops, forest residues, agricultural residues and waste materials. However, conversion of lignocellulosic biomass into liquid hydrocarbon biofuels requires intensive energy requirements. Using biomass for both the carbon source as well as the energy source for conversion will decrease the biofuel production by half of the biomass potential. This study analyzes the new concept of nuclear powered large-scale biorefineries (82 million dry tons per year) in which nuclear energy will provide the heat and hydrogen needed for biomass conversion. Nuclear powered biorefineries have the potential to increase the total biofuel yield and lower the carbon emissions of the biofuel conversion systems. We developed a Mixed Integer Linear Programming (MILP) model to analyze the nationwide on-spec delivery of biomass to nuclear powered biorefineries. Delivering on-spec biomass, includes feedstock with ash content less than or equal to 5% (dry basis), moisture content greater than or equal to 20% and carbohydrate content greater than or equal to 59%. Approximately 164, 246 and 328 million dry tons of corn stover and switchgrass could be delivered nationwide at less than $110 per dry ton to the biorefinery gate with the establishment of two, three and four nuclear powered biorefineries in years 2022, 2030, and 2040 respectively.

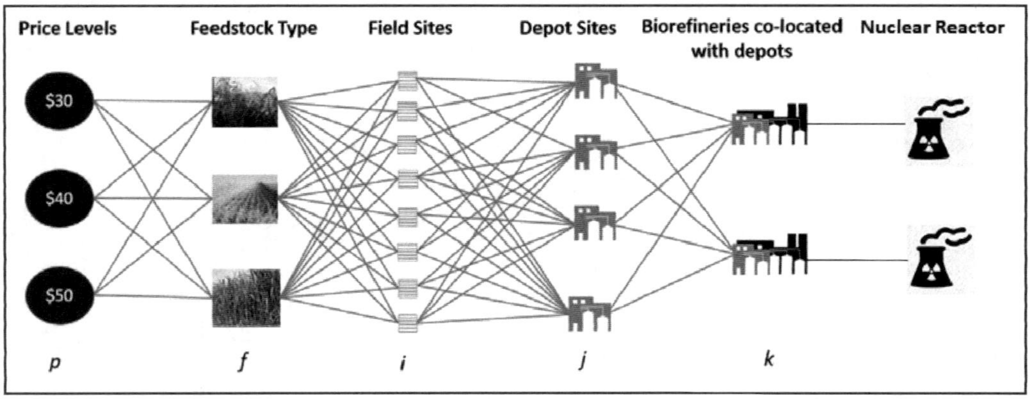

622570 Optimal Design of Controlled Environment Agricultural Systems Under Market Uncertainty

Abstract:

Global population growth and increased climate variability have posed major challenges to traditional agricultural production [2,6]. As these trends persist, meeting increased consumer demand under heightened uncertainty will further stress the capabilities of our existing food production infrastructure [4,5]. For this reason, alternative methods are being sought to augment conventional food production. Controlled environment agriculture (CEA) is an alternative strategy that decouples crop production from environmental conditions through indoor cultivation. CEA offers a more sustainable alternative to conventional agriculture and presents opportunities for improved food security by increasing public access to nutritious fresh produce. Startup costs for CEA systems, however, are significantly higher than those of traditional growing operations, and therefore present a larger barrier to adoption. To accelerate the adoption of this potentially disruptive technology for next generation food production, a better understanding of the long-term economic viability of CEA technology is desired. To provide this valuable insight, we have developed the first economics-based optimal design and planning model for CEA systems.

To date, the majority of agricultural planning models have focused on traditional outdoor food production. Many of the existing models have been tailored to highly specific operations, few have employed mathematically rigorous optimization techniques, and only a handful have accounted for uncertainty, which is desired to realistically assess the long-term economic viability of the system [1,3,7]. Since existing models are not applicable to CEA systems, we have developed an optimal design and planning model that can be used to assess the long-term economic viability of a broad range of CEA systems. Our proposed model employs robust optimization to account for uncertainty in the design and planning stages of CEA production, which enables risk-averse investment decision making for long-term economic viability.

In this work, we present a novel methodology for the simultaneous robust design and planning of CEA systems under multi-period market price uncertainty/risk. This problem is cast as a bilevel program with a nonconvex outer program pertaining to the design objective and multiple semidefinite inner programs pertaining to mean-variance risk exposure over each period considered in the planning horizon. The bilevel program is reformulated as a nonconvex semi-infinite program with several semi-infinite constraints pertaining to the multi-period risk exposure over the horizon. The problem is solved to global optimality using a custom branch-and-bound strategy constructed with EAGO.jl [10] that exploits a combination of McCormick-based envelopes and Gurobi's nonconvex solver [8], to solve the relaxed (upper-bounding in the maximization sense) subproblems, and IPOPT [9] for local solution of the lower-bounding (in the maximization sense) subproblems. A solution to the model represents a design and operating schedule that is robust to worst-case market uncertainty and therefore provides a conservative basis for CEA engineering and investment decision-making.

Our proposed model has been formulated generally and can be used to optimize large-scale CEA systems for the cultivation of any crop portfolio via single or multi-mode cultivation. The optimization objective employed is to maximize the net present value of the CEA system over its expected lifespan. Constraints limiting risk exposure and production capacity are also included to ensure that the optimal design is both robust and practical. Capital and operating expenses, including those associated with system construction, cultivation equipment, resources, labor, and distribution, are accounted for in the economic objective through sub-models that capture the location, capacity, and crop allocation cost dependencies of the specific CEA system being modeled. These sub-models can be easily adjusted for different CEA configurations without structural changes to the main optimization algorithm. This not only enables efficient determination of the optimal design for a given CEA configuration, but also enables efficient comparison between multiple candidate systems for strategic adoption of CEA technology.

To evaluate the efficacy of our approach, the economic performance of the robust design was compared against that of naïve design and operating strategies through two case studies. In each of the presented case studies, an economically feasible design was identified, even when the naïve design was economically infeasible, and the improvement in system performance attributed to the implementation of a robust design was quantified. This approach represents, to our knowledge, the first robust optimization approach to CEA systems and is demonstrated to effectively increase the robustness of CEA systems to market uncertainty, improve the long-term economics of CEA systems over naïve operating strategies, and validate the economic viability of single and multi-mode CEA production of distinct crop portfolios. This work establishes a means of evaluating the economic viability of a wide array of CEA systems and serves as a baseline for future development of CEA design and planning models to accelerate the adoption of CEA technology.

References:

[1] Ahumada, O., Villalobos, J.R., 2009. Application of planning models in the agri-food supply chain: a review. Eur. J. Oper. Res. 196 (1), 1–20. doi: 10.1016/j.ejor.2008.02.014 .

[2] Deschênes, O., Greenstone, M., 2007. The economic impacts of climate change: evidence from agricultural output and random fluctuations in weather. Am. Econ. Rev. 97 (1), 354–385. doi: 10.1257/aer.97.1.354 .

[3] Dury, J., Schaller, N., Garcia, F., Reynaud, A., Bergez, J.E., 2012. Models to support cropping plan and crop rotation decisions. A review. Agron. Sustain. Dev. 32 (2), 567–580. doi: 10.1007/s13593- 011- 0037- x .

[4] Fedoroff, N.V., 2015. Food in a future of 10 billion. Agric. Food Secur. 4 (1), 11. doi: 10.1186/s40066- 015- 0031- 7 .

[5] Field, C.B., Barros, V., Stocker, T.F., Dahe, Q., 2012. Managing the Risks of Extreme 1083 Events and Disasters to Advance Climate Change adaptation: Special Report of the Intergovernmental Panel on Climate Change (IPCC). Cambridge University Press.

[6] Food and Agriculture Organization of the United Nations (FAO), 2018. The state of agricultural commodity markets 2018: Agricultural Trade, Climate Change and Food Security. Technical Report. FAO. URL http://www.fao.org/3/I9542EN/i9542en.pdf .

[7] Glen, J.J., 1987. Mathematical models in farm planning: a survey. Oper. Res. 35 (5), 641–666. doi: 10.1287/opre.35.5.641 .

[8] Gurobi Optimization, LLC, 2020. Gurobi optimizer reference manual. URL http://www.gurobi.com .

[9] Wächter, A., Biegler, L.T., 2006. On the implementation of an interior-point filter line-search algorithm for large-scale nonlinear programming. Math. Program. 106 (1), 25–57. doi:

10.1007/s10107- 004- 0559- y.

[10] Wilhelm, M.E., Stuber, M.D., 2020. EAGO.jl: Easy advanced global optimization in Julia. Optim. Methods Softw. doi: 10.1080/10556788.2020.1786566 .

632232 Reproducible and Fast Mapping of Material Flows in Industrial Networks Using Piot Hub: A Novel Cloud Based Computational Tool

Abstract:

An important goal of industrial ecology is to ensure a balance is maintained between economic and environmental sustainability. Facilitation of this balance requires an understanding of the structure of the material economy and the material flows/waste generation in industrial networks in order to minimize the environmental consequences of these industrial systems. Physical Input-Output Tables (PIOTs) are uniquely suited for this purpose by mapping physical flows between industrial sectors. In previous work, PIOT-Hub was developed to automate the generation of both Physical Supply Tables (PSTs) and Physical Use Tables (PUTs) as well as their conversion into PIOTs by integrating the strength of mechanistic models and macroeconomic Input-Output framework [1]. Hence, PIOT-Hub, a cloud-based tool can be used to generate the PST and PUT for a region using a bottom-up approach where relevant industrial processes are simulated via mechanistic models to determine steady-state material input and output requirements. Thus, the tool establishes a connection between engineering and economic model in such a way that it is insightful for transitioning to a low carbon and zero waste circular economy. In past work, PIOT-Hub was applied to 11 agro-based economic sectors in Illinois, thus capturing elemental flows across the economy, demonstrating the automation aspect. Here we evaluate the reproducibility and time aspect of PIOT-Hub to map the material flows of equivalent economic sectors in Indiana. We develop simulation models for industries in Indiana and utilize PIOT-Hub for mapping the industrial network, thus evaluating the functionality of this novel computational platform. This work not only demonstrates the reproducibility of the tool's utilization to other geographic regions, but also serves as a case study of how the fast generation of PSTs, PUTs, and PIOTs by PIOT-Hub can be used to compare the material economies of multiple regions. We perform a comparative analysis of material flows in Illinois and Indiana using our work and the PIOT generated for Illinois in previous work. We will also demonstrate the collaborative aspect of this computational tool by live demo of the tool during this talk, thus highlighting the scaling impact of collaboration through cloud-based platform in making sustainable manufacturing decisions.

References

[1] Shweta Singh; Lan Zhao; Jaewoo Shin; Venkata Sai Gargeya Vunnava (2021), "PIOT hub for Automated Generation of Regional PIOTs," *https://mygeohub.org/resources/piotpy*.

625295 Solid Oxide Fuel Cell – Thermal Energy Storage (SOFC-TES) Integrated System Modeling and Energy Efficiency Optimization.

Abstract:

Concerns over greenhouse gas emissions and their associated environmental challenges continue to increase. As a result, cleaner energy generation technologies, such as fuel cells, are receiving attention as the next generation's energy sources. One of these attractive energy technologies is the solid oxide fuel cells (SOFCs) due to its several advantages such as 1) reduced greenhouse gas emissions, 2) higher efficiency, and 3) flexible fuel selection compared to other types of fuel cells. However, the temperature of flue gas from SOFCs is significantly high due to its operating temperature between 600°C and 1000°C. Therefore, it is necessary to recover this waste heat so that we can maximize the process energy efficiency.

Many studies have been conducted to utilize the high-temperature exhaust gases from SOFCs so far. However, they only focused on the direct utilization of the heat through the electricity generation, using processes such as gas turbine[1], Rankine cycle[2], and combined heat and power (CHP)[3] in order to maximize the process energy efficiency. However, the aforementioned utilization methods result in losses where the generated electricity is further converted to other forms of energy such as thermal energy for space heating. The energy losses during this process are often significant.

Motivated by this, we propose the SOFC-integrated thermal energy storage (TES) method as a more efficient utilization technique for recovering waste heat from SOFCs. We employed a 4 kW-SOFC and combined it with the thermal energy storage system for storing the excess energy from the high-temperature exhaust gas. In order to integrate both processes, SOFC and TES mathematical models were developed using MATLAB. Next, the developed models are integrated, simulated, and the overall efficiency of the SOFC-TES model is compared to the general utilization methods. To further improve the proposed model's thermal efficiency, we developed a heat exchanger network using Aspen Energy Analyzer. Through the developed heat exchanger network, the integrated model was optimized by pinch analysis.

The optimization results revealed that the process efficiency of the integrated system, that is, SOFC-TES, showed superiority compared with the traditional methods of utilizing the waste heat from SOFCs. By this, excess heat from the exhaust gases can be reliably stored during low-demand seasons and reutilized during high-demand seasons to supplement residential and industrial energy demand or deficits. We concluded that the SOFC-TES could be widely used for various purposes due to its inherent characteristics, i.e., storing heat, such as a greenhouse in the smart-farm and human life, compared with the traditional method of waste heat utilization from the SOFCs.

References

[1] S.H. Chan, H.K. Ho, Y. Tian, **Modeling of simple hybrid solid oxide fuel cell and gas turbine power plant**, Journal of Power Sources, 109 (1) (2002), pp. 111-120

[2] Masoud Rokni, **Thermodynamic analysis of an integrated solid oxide fuel cell cycle with a Rankine cycle**, Energy Conversion and Management, 51 (2010), pp. 2724-2732

[3] S.H. Chan, C.F. Low, O.L. Ding, **Energy and exergy analysis of simple solid-oxide fuel-cell power systems**, Journal of Power Sources, 103 (2002), pp. 188-200

630767 Upcycling Food Waste for Hydrogen Storage: Technoeconomic Assessment

Abstract:

While annually 1.6 billion tons of food waste is estimated to be discarded globally by 2030, according to U.S. Environmental Protection Agency (EPA) the corresponding figure is projected forecast to reach approximately 66 million tons for U.S. With the monumental challenge of food waste management, researchers are seeking avenues of upcycling FW into higher value materials such as hydrogen storage medium. Therefore, the objective of this research was to evaluate the technoeconomic prospect of activated carbon production from food waste for the ultimate purpose of hydrogen storage through the application of hydrothermal carbonization of food waste followed by chemical activation of the hydrochar using KOH. The process was simulated using Aspen Plus and the process flow diagram (PFD) was constructed such that it mainly constituted of 2 rotary kilns and filter press, 1 rotary cooler, dryer, homogenizer, as well as other necessary equipment - pumps, heat exchangers, storage tanks. The facility processes 96 ton of food waste per day to be converted to activated carbon powder with a carbon yield of 20.4%. Also, it was assumed that the activated carbon manufacturing facility obtains food waste of Brevard County, Florida and the facility operates continuously 330 days a year. Several parameters were incorporated in the study including equipment sizing, capital costs and operating costs, such as labor, utilities, maintenance and equipment depreciation. Break-even cost from the economic analysis was evaluated and sensitivity analysis was carried out to reveal the effect of 7 different process variables on the break-even cost where raw materials' (KOH and food waste) and wastewater treatment cost significantly affected it. The unprecedented analysis shed light on the economic feasibility of producing activated carbon from food waste as a cost-effective route of upcycling it.

625694 Economic and Environmental Impact of Biochar Addition on Anaerobic Digestion Process

Abstract:

Anaerobic digestion (AD) technology is becoming a promising pathway for producing clean and renewable natural gas from waste resources. Despite growing industrial adoption, AD technologies still suffer from inefficiency due to process instability and microbial inhibitions. Scientists have demonstrated that biochar addition has positive impacts on AD. Biochar addition increases biogas yield and methane concentration and reduces gas contaminants. This influence is attributed to several mechanisms: providing support for microbial colonization, adsorbing excess inhibitory products, and improving electron transfer between syntrophic microorganisms. However, due to the cost and energy-intensive conversion process, the additional costs for biochar addition could offset the resulting energy benefits. This study investigates an integrated biorefinery where corn stover biochar is added to the AD of animal manure to produce biogas-derived Renewable Natural Gas (RNG) for pipeline injection.

A kinetic model of AD, biogas upgrading, and pipeline injection is simulated in Aspen Plus ^{TM} for a mid-size farm with 5000 cattle heads. The process simulation model estimates a 28% increase in biogas yield for 3% (wt. % of wet mass) of biochar addition. High pressurized water scrubbing process is used for biogas upgrading into RNG. Techno-economic analysis (TEA) determines the minimum RNG selling price for a 30-year plant lifetime with a 10% Internal Rate of Return (IRR). TEA estimates the minimum RNG selling price as \$42/MMBTU and \$35/MMBTU for the without biochar and with biochar scenario, respectively. Biochar addition lowers the RNG selling price by 16% whereas adding revenue from digestate selling and renewable identification number (RIN) credits result in up to 32% reduction. Biochar addition increases the capital and operating costs by 8% and 13%, respectively. Cradle-to-gate life cycle analysis (LCA) is performed in SimaPro ^{TM} to evaluate the GHG emissions of biochar amended AD for RNG production. GHG abatement for the without biochar and with biochar scenario is 0.038 kg $CO_{2\ eq}$/MJ of RNG and 0.123 $CO_{2\ eq}$ /MJ of RNG, respectively. LCA results show biochar addition leads to 2.18 times GHGs abatement compared to the without biochar scenario. Environmental benefits mostly come from the additional carbon sequestration with biochar land use and fossil fuel displacement by syngas from pyrolysis. Sensitivity analysis shows the influence of different major processes and economic parameters on the RNG selling price. The process is cost-prohibitive for small-scale farms due to the significant economy of scale penalty for the biogas upgrading and RNG injection facility. Increased biogas yield reduces the manure input leading to less digestate production. This would decrease the environmental benefits of emissions avoidance from conventional manure management and carbon sequestration per MJ of RNG while reducing the emissions from digestate transportation and applications. Uncertainty analysis considers the variations of key technical and economic parameters within a reasonable range and gives an estimated range of fuel selling prices and GHG abatement. Minimum RNG selling prices vary between \$8.82/MMBTU and \$16.28/MMBTU and GHG abatement vary between 0.06 kg $CO_{2\ eq}$/MJ and 0.16 kg $CO_{2\ eq}$/MJ within 10-90% confidence intervals.

627985 Optimizing Syngas Production Process By Downdraft Gasifier Using Response Surface Methodology

Abstract:

The gasification process of biomass to produce combustible gases has an attracted interest recently. The downdraft gasifier has an advantage over other types of gasifiers because of its small size, high quality gas production, and low tar yield. One of the challenges in the gasification process is the production of low-quality syngas (low concentration of H2 and CO) and large amount of tar. To overcome this problem, optimization of the gasifier operating conditions was employed to achieve high-quality product gases and hence high gasification process efficiency. In this work, the influence of initial gas temperature, airflow rate, and mole fraction of oxygen in primary air, on the gasification efficiency and product quality was optimized by response surface methodology. The obtained results showed that the reduction in the temperature caused a low concentration of carbon monoxide and a high amount of tar in the product, thereby there was a decline in the efficiency of the gasification process.

623925 Techno-Economic Evaluation of Nano-Porous Silica Production from Rice Husk and Sand

Abstract:

Global rice production in 2019 was estimated at about 500 million tons. Rice husk is a by-product from rice, including about 20 wt% of rice and is needed to be disposed as waste. Rice husk contains approximately 14-25% of ash. The silica content in rice husk ash (RHA) is above 90%. Nanoporous silica (NPS) used in industrial fields such as reinforced rubber additives (tires), concrete mixture (cement binder), nano-fibers, nano-filters, and membranescan be produced from RHA. Rice hisk is combusted at a relatively low temperature (650-800°C) and RHA is produced, whereas the silica sand is operated at 1400-1500°C to produce silicate. In addition, compared to the conventional NPS production from sand, The NPS production from RHA has an advantage in lowering the production cost due to low material cost and energy consumption.

In this study, the techno-economic feasibility for the sodium silicate ($Na_2O{\cdot}xSiO_2$)-based nano-porous silica (SS-NPS) production from sand and rice husk was performed. The return on investment (ROI) and payback period (PBP) were evaluated according to the plant size for the two cases. The effects of raw material and product prices on ROI and PBP were analyzed and economic conditions to be economically viable were suggested.

AIChE
The Global Home of Chemical Engineers

A CAPITAL AND ENERGY EFFICIENT ALTERNATIVE EMISSIONS CONTROL TECHNOLOGY FOR CONTROLLING EMISSIONS USING BEAD ACTIVATED CARBON

Kim K. Tutin, Captis Aire LLC, East Point, GA United States
Harold R. Cowles, Environmental C&C Inc., Scottsdale, AZ United States
John M. Berger P.E., Environmental C&C Inc., Fishers, IN United States

TABLE OF CONTENTS:

I. INTRODUCTION:

In manufacturing, process exhaust gas (air) streams released may contain volatile organic compounds (VOCs), Hazardous Air Pollutants (HAPs), particulate, etc. These are considered pollutants/emissions and must be removed from the air prior to release to avoid air pollution emissions into our communities. Add-on controls either collect the VOCs for recovery and reuse or destroy the VOCs. A number of control options are available including thermal oxidizers, catalytic oxidizers, and biofiltration which destroy pollutants. Other control systems including condensation, scrubbing, and adsorption systems may recover pollutants.

This paper provides an introduction to an adsorption technology, the Fluidized Bed Concentrator (FBC) with Bead Activated Carbon (BAC) media. It describes 1) the FBC technology, 2) the BAC media used in these systems, 3) the performance characteristics, 3) typical applications of this technology, and 4) a case study review of the FBC technology demonstration for controlling emissions and collecting terpenes from wood drying.

II. DESCRIPTION OF FBC TECHNOLOGY: FOUR PRIMARY PARTS

1. Adsorber:

The adsorber section has trays filled with very adsorbent BAC media. Contaminated process exhaust enters from the bottom, passes upward through the adsorption trays, and fluidizes the Bead Activated Carbon (BAC) adsorbent media. When the process air interacts with the BAC, it adsorbs (captures) the terpenes Volatile Organic Compounds (VOCs) in its pores. VOC saturated BAC flows to the bottom of the adsorber vessel and is transferred using fans to pneumatically transfer it to the desorber. Cleaned (reusable) BAC from the desorber feeds continuously into the top of the adsorber, providing counter-current VOC removal. Cleaned air exits out the top of the adsorber.

2. Desorber:

In the desorber, the BAC media is heated (regenerated), causing it to release the VOC vapors into a low volume, gas stream. This gas stream is then circulated continuously to the next stage in the process. The regenerated (cleaned) BAC is then returned continuously to top of the adsorber for reuse. Returning this regenerated BAC to the top of the adsorber provides a final "polishing" of emissions to remove residual emissions prior to release of the process exhaust air.

Desorber Recovery System: In the recovery system, a low volume inert carrier (nitrogen) gas is introduced into the bottom of the desorber. Gas flows up through the desorber as the BAC flows downward. The system is heated via either hot oil or electric heating. The heat causes the VOCs to volatilize off the BAC. The VOC vapors are liberated resulting in a concentrated waste flue gas stream that moves from the desorber to the condenser. This system may eliminate the on-site use of natural gas entirely. Thus, it offers an "All Electric" control technology.

"All Electric" Option

Desorber Oxidation System: In the oxidation system, clean hot stripping air at 100-700°F is introduced to the bottom of the desorber. It flows upward across the BAC media to both volatilize the VOCs and fluidize the BAC. The VOC vapors are liberated resulting in a concentrated waste flue gas stream that moves from the desorber to the oxidizer.

3. Condenser or Oxidizer:

The gas stream from the desorber is circulated continuously to this system for further treatment to either capture and recover VOCs or oxidize them.

> **"All Electric" Option**

Condenser Recovery System: In the recovery system, nitrogen gas circulates from the desorber to the condenser continuously. The VOC vapors are sufficiently concentrated such that the vapors can be cooled via a tube and shell condenser converting the vapor to liquid. The liquid drains continuously into a storage system. The liquid VOCs can then be reused or sold. The nitrogen recirculates back to the desorber where it entrains the VOCs vapors again and the cycle continues.

Oxidizer Oxidation System: In the oxidation system, the vapors are instead sent to a small thermal oxidizer to treat the VOCs. While this system does not provide recovered (collected) VOCs, it nevertheless retains FBC technology benefits including significantly reducing natural gas usage. The burner component of the thermal oxidizer is selected to provide enough energy to increase the thermal oxidizer to the 1400°F operating temperature. The burner is designed to minimize the "fire rate and thus consume negligible fuel. Optionally, a "low NO_x" design can be selected to minimize production of NO and NO_2 (oxides of nitrogen). Thus, the system minimizes energy usage and GHG emissions. [1]

4. Side Stream Reactivator (SSR) [2]:

A small portion of the BAC media exiting the desorber may be transferred to an optional SSR. The SSR reactivates the BAC by heating it to 1450°F with the introduction of a small amount of water which reactivates the BAC continuously to regenerate it and thus maintain its ability to adsorb VOCs. Reactivated BAC from the SSR is transferred back to the top of the adsorber for reuse.

III. ADSORBENT MEDIA:

The BAC adsorbent media utilized in this system is an essential element of the FBC technology. Specialized Bead Activated Carbon (BAC) efficiently adsorbs VOCs (i.e. terpenes) from the process air. The BAC has high-fill (adsorption) capacity, is a small sphere-shaped particle that is less than 1 mm in diameter with a narrow particle size distribution, has low dust, has high purity, has high flowability, and has high strength and wear resistance as measured by ASTM D 3802 – 79 (2005) Standard Test Method for Ball-Pan Hardness of Activated Carbon.[3] The spherical shape enables the BAC to fluidize and flow.

The BAC media can be reactivated and reused. This is accomplished either by either continuous or batch processes.

---

[1] "VOC Removal and Destruction System and Method". U.S. Patent and Trademark Office. (USPTO) https://appft.uspto.gov/netacgi/nph-Parser?Sect1=PTO2&Sect2=HITOFF&p=1&u=%2Fnetahtml%2FPTO%2Fsearch-bool.html&r=1&f=G&l=50&co1=AND&d=PG01&s1=20190022579.PGNR.&OS=DN/20190022579&RS=DN/20190022579 (Accessed October 4, 2021).

[2] "GASTAK® Solvent Recovery and Odor Removal by Fluidized Bed Process". Kureha Ecology Management Co. , LTD website. https://www.kurekan.co.jp/en/download/gastak.pdf (Accessed September 18, 2021).

[3] "ASTM D3802-79 (2005) Standard Test Method For Ball-Pan Hardness Of Activated Carbon. https://webstore.ansi.org/standards/astm/astmd3802792005 (Accessed September 18, 2021).

Continuous Reactivation: In the continuous process, a small amount of BAC media is diverted from the BAC exiting the desorber and sent to an in-line side stream reactivator (SSR).[4] It is then returned to the top of adsorber. Returning this reactivated BAC to the top of the adsorber provides a final "polishing" of emissions to remove residual emissions prior to release of the process air.

Batch Reactivation: In the batch process, for operators that do not have an SSR, the BAC media is unloaded periodically, shipped offsite for reactivation, and then reloaded once returned. In the interim, a second load of BAC is utilized. Reactivation is typically performed by California Carbon who has been providing reactivation services since 1975.[5] This service includes a reactivation warranty *"Carbon will be reactivated to 95% of Virgin Activity per ASTM D 3467 and Carbon will lose no more than 10% of incoming volume of the specified mesh size in the reactivation process."*[6] BAC lifetime is site and application specific. Some operators have been reusing their BAC for years.

Bead Activated Carbon (BAC) Adsorbent Media

Bead Activated Carbon (BAC), the Adsorbent Media, Fluidized in Tray

IV. FBC OPTIONS: RECOVERY AND OXIDATION

The Fluidized Bed Concentrator can be utilized to either recover the VOCs or oxidize them as shown in the figures. Operators typically select the desired option based on a number of factors especially including the anticipated value and reusability of the recovered VOCs. The value of the VOCs depends on a number of factors including the anticipated quantity recovered, purity, salability, and/or reusability.

Recovery: When the value of the recovered VOCS is high and if the cost of recovery is less than the cost of purchasing new material, then recovery makes sense.

---

[4] "GASTAK® Solvent Recovery and Odor Removal by Fluidized Bed Process". Kureha Ecology Management Co. , LTD website. https://www.kurekan.co.jp/en/download/gastak.pdf (Accessed September 18, 2021).

[5] "California Carbon Co. Inc. Welcome". California Carbon website. http://www.californiacarbon.com/index.html (Accessed September 18, 2021).

[6] "Custom Spent Carbon Reactivation". California Carbon website. http://www.californiacarbon.com/pdfs/Custom%20Reactivation.pdf (Accessed September 18, 2021).

Oxidation: If the VOC has low or no recovery value, as, for example, when it is a mixture, or if it poses disposal concerns, such as for toxic compounds, then destruction probably makes the most sense. Excluding value derived from VOC recovery, the oxidation option retains FBC system performance characteristics including reduction in energy usage.

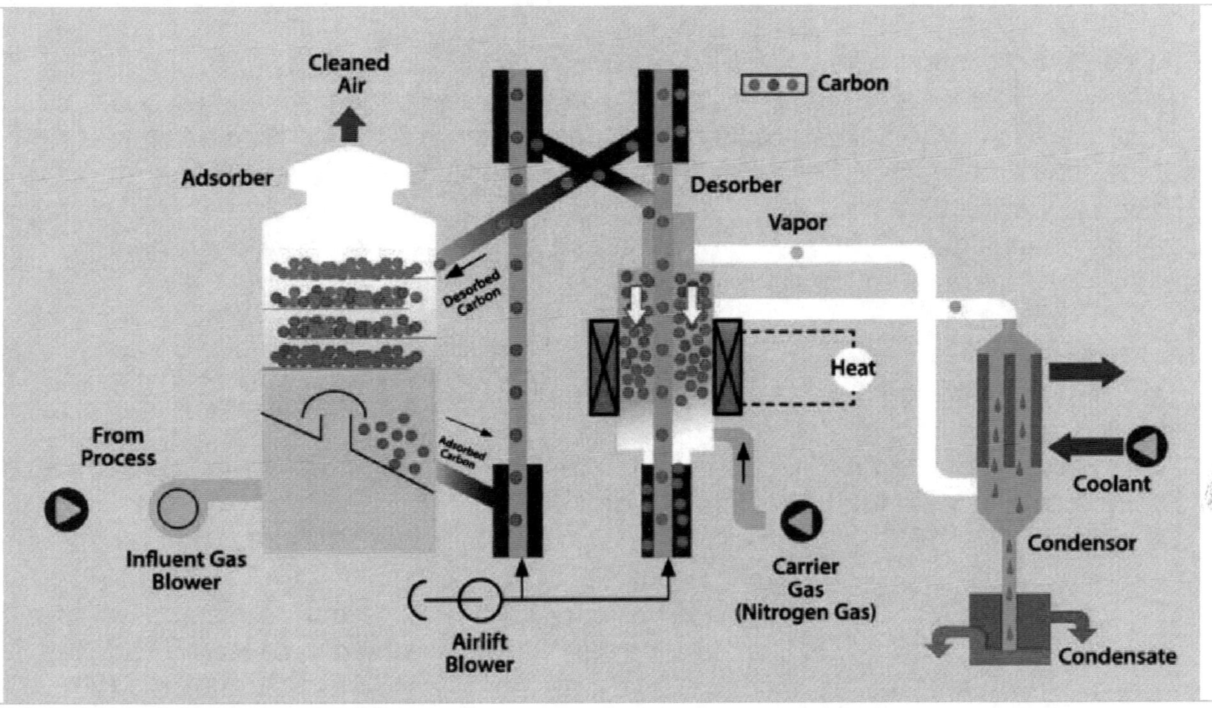

RECOVERY SYSTEM: Adsorbs, Desorbs, and Condenses VOCs

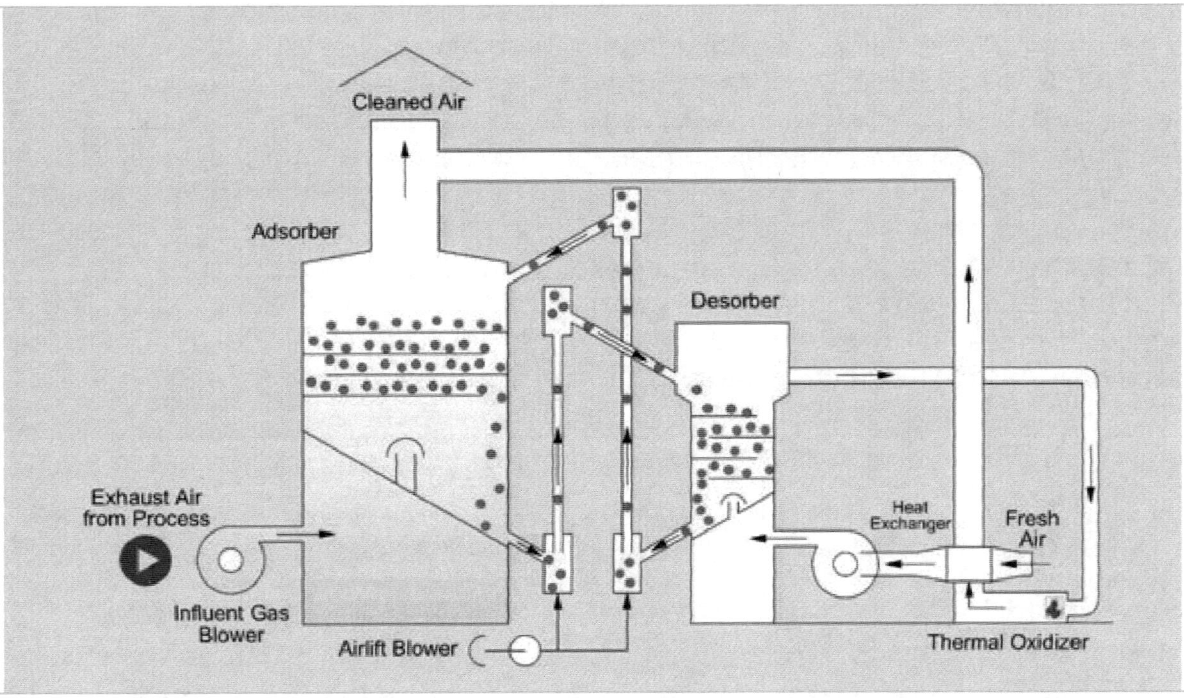

OXIDATION SYSTEM: Adsorbs, Desorbs, and Oxidizes VOCs

V. PROCESS PERFORMANCE CHARACTERISTICS:

1. The FBC reduces energy usage (i.e. natural gas) versus oxidation systems due to reduction or elimination of energy used to heat the air going through thermal oxidizers:

Baseline Oxidation Systems:

Thermal oxidizers burn VOCs by heating them in enclosed combustion chambers in the presence of excess oxygen. The precise temperature and residence time needed for effective reduction of VOC emissions varies with the nature of the VOC. However, temperatures of 1400-2000°F and residence times of 0.5-2 seconds are typical. Since auxiliary fuel is necessary to maintain combustion temperatures, most thermal oxidizers incorporate some form of heat recovery to lower auxiliary fuel costs.

FBC Technology:

The air going through the adsorber is not heated. Thus, the natural gas usage is reduced or eliminated.

Recovery – "All Electric" Option: For a recovery system, the control device can be "All Electric". It could eliminate the use of natural gas entirely.

Oxidation: The FBC oxidation system significantly reduces natural gas usage but does not eliminate it.

Energy Reduction Calculation: The amount of energy that is eliminated, that would have been used by the thermal oxidizer, can be calculated utilizing one of many credible sources. One of the best sources is the Institute of Clean Air Companies' Guidance Methodology[7]. Calculators are also available on-line including those provided by the EPA[8], Cycle Therm[9], and Captis Aire[10]. With these calculators, users can input their specific data and the calculators will calculate the estimated amount of energy required to run the thermal oxidizer. All these sources rely on a single primary factor to calculate energy required to heat the air. This factor is the specific heat of air which is the number of British Thermal Units (BTUs) required to heat one pound of air by one °F. This factor along with the system specific factors including the standard cubic feet per minute (sCFM) of air, the temperature increase, the density of air, and conversion factors provide the number of BTUs required to heat the air going through the device. These factors are used together with the energy efficiency and the heat generated by the oxidation of the emissions to calculate the BTUs of energy required to heat the air going through the RTO. Users can then multiply BTUs times the forecasted long-term cost of natural gas to estimate the long-term cost of natural gas required over the 20+ year life of the thermal oxidizer. This calculation provides the value of the natural gas elimination.

---

[7] "Institute of Clean Air Companies (ICAC) Guidance Method for Estimation of Gas Consumption in a RTO". Institute of Clean Air Companies website. https://cdn.ymaws.com/www.icac.com/resource/resmgr/RTO-F1.pdf (Accessed September 18, 2021)

[8] "Air Pollution Control Cost Estimation Spreadsheet". Environmental Protection Agency (EPA) website. https://www.epa.gov/sites/production/files/2018-01/oxidizers_calc_sheet_finalversion_1-16-2018.xlsm (Accessed Feb 3, 2021).

[9] "Operating Cost Calculator". Cycle Therm website. http://www.cycletherm.com/resources/operating-cost-calculator.aspx (Accessed Feb 3, 2021).

[10] "Calculate Engineering Estimate of Natural Gas Usage and Greenhouse Gas Generation". Captis Aire website. https://www.captisaire.com/Value-Calculator/ (Accessed Feb 3, 2021).

2. The FBC generates revenue from sales of recovered organics:

When using a recovery system, including the FBC, operators have the option to collect and sell or reuse the volatile organic compounds recovered. The liquid VOCs are collected from the FBC condenser for reuse or sale. The value of recovered VOCs will depend on the type, amount, and purity of VOCs collected.

3. The FBC reduces Greenhouse Gases (GHG) emissions versus oxidation systems due to the reduction or elimination of burning natural gas:

Due to the reduction or elimination of natural gas usage and reduction in electrical usage by the FBC, greenhouse gas emissions are also reduced. The greenhouse gas emissions reduction can be calculated based on the amount of carbon in natural gas eliminated and its stoichiometric conversion to carbon dioxide (CO_2). The Environmental Protection Agency (EPA) has an on-line calculator[11] that enables the user to calculate greenhouse gas emissions that are eliminated by eliminating natural gas used to heat the air in the thermal oxidizer.

4. The FBC enables creation of carbon offsets versus oxidation systems:

In conversations with American Carbon Registry[12] and ClimeCo[13], it may be possible to measure, register, and verify carbon offsets for a site converting to the FBC technology from an oxidation system due to the reduction in energy usage.[14]

The quantification methodology and value of carbon offsets is evolving quickly. But these carbon offsets can be sold or traded externally. Or they may be utilized for internal credit accounting purposes. Potential values for these carbon offsets vary widely.[15] However, as of early 2021, American Carbon Registry and ClimeCo estimated the value of carbon offsets for an FBC project at $5/ton.[16]

5. The FBC reduces electricity usage versus oxidation systems:

The horsepower required to run the fans to move the process exhaust air through the BAC media is reduced for the FBC due to the fact that it's BAC is fluidized (moving) rather than packed (non-moving). The pressure drop across the system is reduced from a typical 10-20" to 3.5". Thus, the horsepower required to run the fans for the FBC may be ~ ¼ of that required for the RTO.

To estimate the electrical savings, the sum of electrical demands from the FBC is subtracted from the sum of electrical demands of the thermal oxidizer. Net kW·h along with site-specific electricity costs and operating hours per year are used in combination to estimate electrical savings.

---

[11] "Greenhouse Gases Equivalencies Calculator - Calculations and References". EPA website. https://www.epa.gov/energy/greenhouse-gases-equivalencies-calculator-calculations-and-references (Accessed December 28, 2020).

[12] "Harnessing the Power of the Markets to Improve the Environment". American Carbon Registry website. https://americancarbonregistry.org/ (Accessed February 3, 2021).

[13] "Scalable greenhouse gas projects to earn real financial returns for your business." ClimeCo website. https://climeco.com/ (Accessed February 3, 2021).

[14] "AM0044: Energy efficiency improvement projects - boiler rehabilitation or replacement in industrial and district heating sectors --- Version 2.0.0". Clean Development Mechanism (CDM) website. https://cdm.unfccc.int/methodologies/DB/3HZ4USHZ2W449HMAXZN420E5PJB1QF (Accessed Sept 18, 2021)

[15] "Allowance Price Explorer". International Carbon Action Partnership website. https://icapcarbonaction.com/en/ets-prices (Accessed Jan 20, 2021).

[16] Eric Ripley (Director Industrial Programs, American Carbon Registry), in a phone conversation with Kim Tutin November 17, 2020.

6. <u>The FBC enables superior mass transfer of VOCs through counter current flow fluidized BAC media that is continuously regenerated:</u>

Baseline Adsorption Systems:

Adsorption: In sorbent treatment systems, process exhaust air moves across the sorbent. VOCs are removed via attraction (adsorption) of the VOCs on carbon or zeolite sorbent media in packed (non-moving) beds. This may be accomplished by concentrators that have wheels with sectors containing activated carbon or zeolite media. In this arrangement, most of the space on the wheel functions in adsorption mode, with one small sector being regenerated at any given time. Or the process exhaust air may be forced through large vessels containing sorbent media. In this arrangement, the VOCs are treated via a once-through process.

Regeneration / Reactivation: Regeneration is accomplished by heating the sorbent media with heaters, heated gases, steam, etc. If using steam, water may be a byproduct. This can be accomplished either via off-site regeneration or by alternating the flow between multiple vessels where one vessel functions in adsorption mode while the other is being regenerated at any given time. In both these arrangements, the sorbent media is being regenerated in batch or semi-continuous fashion. Reactivation may be accomplished by sending the sorbent media off site for reactivation. Or the sorbent media may be discarded when it becomes saturated.

Effect of water vapor: Especially for process exhaust air heavily laden with water vapor, i.e. 10% B_{ws} moisture by volume[17] or more, the sorbent media may become partially or fully saturated with water. At high moisture content, water molecules begin to compete with the hydrocarbon molecules for active adsorption sites. This reduces the capacity and the efficiency of the adsorption system.

FBC Technology:

Adsorption: The BAC media in the fluidized bed concentrator adsorber is fluidized (moving) by the process air moving opposite (counter current) to the BAC as it moves downward through the trays in the adsorber. This provides excellent mixing of the BAC and contact with the air. As this occurs, VOCs or odors are transferred from the process air to the BAC. The cleaned air exits through the top of the adsorber.

Regeneration / Reactivation: The BAC media is regenerated continuously via continuous recirculation to and from the desorber. It is reactivated continuously if system includes the Side Stream Reactivator (SSR). This ensures the BAC retains the capacity to maintain adsorption efficiency. This is done at temperatures well above the boiling point of water, often around 400°F or more. This minimizes or eliminates accumulation of water and/or organics on the BAC.

Effect of water vapor: Continuous regeneration and reactivation at temperatures above the boiling point of water, i.e. 400°F, volatilizes the water off the BAC media and prevents the BAC from becoming saturated with water. Thus, it maintains adsorption efficiency even when treating process exhaust streams with high moisture contents, as are typical for wood drying applications, i.e. ≥20% B_{ws} moisture by volume[18]. This system minimizes or eliminates need for dehumidification systems.

---

[17] "METHOD 4—DETERMINATION OF MOISTURE CONTENT IN STACK GASES". EPA website. https://www.epa.gov/sites/default/files/2016-06/documents/m-04.pdf. (Accessed October 17, 2021).
[18] "METHOD 4—DETERMINATION OF MOISTURE CONTENT IN STACK GASES". EPA website. https://www.epa.gov/sites/default/files/2016-06/documents/m-04.pdf. (Accessed October 17, 2021).

Effect of Fluidized (moving) vs Packed (non-moving) media: BAC media fluidization prevents pockets of high concentration VOCs, provides higher dry particulate tolerance (reduced plugging of interstitial spaces between media), and enables continuous on-site regeneration and reactivation.

TABLE I
PHYSICAL AND OPERATIONAL COMPARISON OF THE FBC VS THE RTO

Parameters:	RTO	FBC	Advantages of FBC
Treatment Temperature, °F	>1400°F	175°F	• Adsorption at process air temperature • No natural gas required for heating
Media Desorption Temperature, °F	>1400°F	400°F	• Much less energy required for desorption • Continuous media regeneration, above the boiling point of water, prevents the media from getting saturated with water. Thus, it maintains its ability to adsorb pollutants.
Media depth, feet (ft)	16 ft	1 ft	• Less Electricity (fan HP) required to move air through less media
Media Configuration	Packed (Not Moving)	Fluidized (Moving)	• Less Electricity (fan HP) required to move process air through fluidized media
Pressure Drop, Inches (")	10-20"	3.5"	• Less Electricity required to move process air against less pressure
Fan size, Horsepower, (HP)	2400 HP	600 HP	• Less HP due to media that is fluidized. • Less susceptible to plugging from particles
Emissions Treatment	Oxidation to CO^2	Adsorption (Capture)	• Emissions are NOT oxidized to CO_2, so reduces GHG emissions • Emissions collected can be sold or reused
Media Reuse	Discarded	Reused	• Media desorption with 'in unit' continuous regeneration. • Flowable bead shaped media can be removed and reused
Safety		Inherently Safer	• Lower operating temperatures • No large heavy moving parts • Easier cleaning and maintenance

VI. AWARDS:

The FBC has been recognized for excellence in environmental performance. Honda Motors of Alabama was recognized as *"Air Conservationist of the Year"* when they implemented the FBC to control emissions when they increased production.[19] Ford's "Fumes to Fuels" work was awarded the EPA's *"Clean Air Excellence"* award.[20] In this work, they utilized the FBC to capture solvent-laden air, strip the solvent from the air, concentrate the solvent up to 2,000 times, and send it to a fuel reformer. The fuel reformer used the concentrate to generate hydrogen-rich fuel, which was directed into a fuel cell to generate electricity.

VII. APPLICATIONS:

Concentration systems, including the FBC, are advantageously used in front of control devices when treating very high volumes of very dilute exhaust gas to reduce energy usage. Recovery systems are beneficial when the value of the recovered VOCs is high. The following are a few examples where the FBC technology has been utilized historically.

TABLE II
FBC APPLICATIONS

Application	Chemistry
Adhesive Tape Mfg	Toluene
Agricultural Products	Organic Acid & Ammonia Odor
Aluminum Casting	Phenol, Formaldehyde, Ammonia
Animal Lab	Odor Control
Atomic Power Plant	Styrene
Auto Parts Degreasing	Trichloroethylene
Bakery	Odor Control
Brewery	Odor Control
Car Wash	Kerosene
Cellophane Coating	Toluene, Ethyl Acetate, Butyl Acetate
Ceramic Condenser	Terpineol
Confectionery	Odor control
Chemical Production	Toluene, Higher Molecular Weight Alcohols, Carbon Tetrachloride

---

19 "HMA Honored as Air Conservationist of the Year". Honda website. https://www.hondaalabama.com/article/hma-honored-as-air-conservationist-of-the-year (Accessed September 21, 2021).

20 "Clean Air Excellence Award Recipients". Environmental Protection Agency (EPA) website. https://www.epa.gov/sites/default/files/2015-06/documents/clean_air_excellence_award_recipients_year_2003.pdf (Accessed September 21, 2021).

TABLE II (CONTINUED)
FBC APPLICATIONS

Application	Chemistry
Dye Production	Perchloroethylene
Electronics Mfg Plant	DMF, MEK, etc.
Electronics Parts Cleaning	Trichloroethylene, Toluene, 1,1,1-Trichloroethane
Fabric Washing	Perchloroethylene
Feed Processing	Odor Control
Film Coating	MEK, Methyl Cellosolve, Toluene, Tetrahydrofuran
Film Laminating	Toluene, Ethyl Acetate, n-Hexane
Gravure Printing	Toluene, Xylene, n-Propanol, n-Propyl Acetate, mixed
Iron Casting	Phenol, Formaldehyde, Ammonia
Lacquering of Film	Toluene, Ethyl Acetate, MEK
Landfill	Odor Control
LSI Mfg Plant	Phenol, Acetone, Methanol, Mixed Solvents
Lubrication Mfg	Mixed Solvent Odor Control Magnetic Tape, Chlorinated Hydrocarbon
Medicine Mfg Plant	Butanol, Methylene Chloride
Paint Booth	Thinner
Printing Ink Mfg	Mixed Solvents
PVC Resin Mfg Plant	Vinyl Chloride
Resin Mfg	N-Hexane, Methylene Chloride
Rubber Coating	Toluene
Rubber Vulcanizing	Odor Control
Sandpaper Mfg	Toluene, Xylene, Ethyl Acetate
Semiconductor Mfg	IPA, HMDS, NMP, Ethanol, Methanol, Ethyl lactate
Silk Screen Printing	Xylene, Mixed Solvents
Teflon Tubing Mfg	Perchloroethylene
Wastewater Aeration	Odor Control

TABLE III
EXAMPLES OF FBC SYSTEMS INSTALLED

FLUIDIZED BED CONCENTRATOR SYSTEMS INSTALLED: [21]	
Industry **Location** **Inlet Stream** **VOCs** **Performance Achieved** **Notes**	Semiconductor Chip Manufacturing Israel 20,000 CFM Semiconductor solvents, 50 ppmv 95% Reduction Efficiency System located in populated area of a major city has eliminated nuisance odor
Industry **Location** **Inlet Stream** **VOCs** **Performance Achieved** **Notes**	Semiconductor Chip Manufacturing Arizona 36,000 CFM Semiconductor solvents, 25-100 ppmv 90%+ Reduction Efficiency System met strict local compliance requirements while achieving Zero Unscheduled Downtime (ZUD)
Industry **Location** **Inlet Stream** **VOCs** **Performance Achieved** **Notes**	Teflon Tubing Manufacturing Massachusetts 1,800 CFM Perchloroethylene, 1500 ppmv 98% Reduction Efficiency Recovered solvent re-used in process. System payback in 1 year.
Industry **Location** **Inlet Stream** **VOCs** **Performance Achieved** **Notes**	Furniture Painting California 50,000 CFM Paint Solvents, 25-50 ppmv 90%+ Reduction Efficiency Recovered solvent and reformulated with paint
Industry **Location** **Inlet Stream** **VOCs** **Performance Achieved** **Notes**	Military Vehicles Michigan 150,000 CFM Paint Solvents 95% Reduction Efficiency Controlled multiple different interchangeable VOC sources.

[21] "Environmental C&C Inc". Environmental C&C website. https://www.environmentalcc.com/ (Accessed September 18, 2021).

Automotive Site, Alabama, 100,000 CFM
Paint Solvents, >90% Reduction Efficiency
Two adsorbers and one desorber

Image shows three adsorbers one desorber	
Image shows ductwork that enabled control of any of the 3 sources to any of the 3 adsorbers. **Each adsorber has its own process fan.** **System provided control of multiple different interchangeable VOC sources**	

Military Vehicles, Michigan, 150,000 CFM
Paint Solvents, 95% Reduction Efficiency

TABLE IV: VALUE CREATION CALCULATION EXAMPLE

Fluidized Bed Concentrator (FBC) vs Regenerative Thermal Oxidizer (RTO)	**Value per Year**
1) Reduce Energy Usage (Natural Gas) due to Elimination of Energy Used to Heat the Air in the RTO: [22] [23] a. $\frac{mmBTU}{Year}\ to\ heat\ air\ in\ RTO = (\Delta\ \mathrm{T})\ (sCFM)(1.08)(\%Heat\ Loss)(\frac{Operating\ Hours}{Year})/1000000$ b. $(1500°F - 175°F) * (300{,}000sCFM) * \left(\frac{1.08\ BTUs}{sCFM \cdot Hour \cdot Degree°F}\right) * (7.5\%) * (8000\frac{Hrs}{Yr})/1000000$ c. $= \left(257{,}580\frac{mmBTU}{Yr}\right) - [\left(250\frac{lbs\ VOC}{Hour}\right) * \left(15{,}000\frac{BTUs}{lb\ VOCs}\right) * \left(\frac{8000\ Hrs}{Year}\right)] = \mathbf{227,580}\frac{\mathbf{mmBTUs}}{\mathbf{Year}}$ d. $\left(\frac{227{,}580\ mmBTUs}{Year}\right)(\$3.5\frac{Long\ Term\ Cost\ of\ Natural\ Gas}{mmBTU}) = \mathbf{\$796,530}\ \boldsymbol{Savings/Year}$	Save by Eliminating Natural Gas **$796,530**
2) Reduce Greenhouse Gases (GHG) Generated from Burning Natural Gas: [24] a. $Tons\ GHG = (mmBTU) * \left(\frac{117\ lbs\ CO2}{mmBTU}\right) * (\frac{1\ Metric\ Ton}{2205\ lbs})$ b. $\left(227{,}580\frac{mmBTUs}{Yr}\right) * \left(117\ lbs\frac{CO2}{mmBTU}\right) * \left(1\frac{Metric\ Ton}{2205\ lbs}\right) = \mathbf{12,076}\frac{\boldsymbol{Tons\ GHG\ Reduction}}{\boldsymbol{Year}}$	Sell Carbon Offsets **$60,378**
3) Sell Carbon Offsets (Carbon Credits) Utilizing Carbon Accounting Service Providers, i.e. ClimeCo: [25] a. $\left(\frac{12{,}076\ Tons\ GHG}{Year}\right) * \left(\frac{\$5}{Ton\ GHG}\right) = \mathbf{\$60,378}/\boldsymbol{Year\ Selling\ Carbon\ Offsets}$	
4) Reduce Electricity Usage by Reducing Fan Power to Move Process Air through Fluidized Media: $(2400RTO - 600FBC\ HP\ fans - 800\ HP) * \left(\frac{0.75\ kW \cdot h}{HP}\right)\left(\frac{\$0.07}{kW \cdot h}\right)\left(\frac{8000\ Hrs}{Year}\right) = \$420{,}000\frac{\boldsymbol{Savings}}{\boldsymbol{Year}}$	Save by Reducing Electrical Usage **$420,000**
5) Sell Terpenes [26] **Yielded from Green Pine Wood** [27]**:** a. $\left(\frac{1\ Gallon\ Terpenes}{Oven\ Dry\ Ton\ Wood}\right) * \left(\frac{300{,}0000\ Dry\ Tons\ Wood}{Year}\right) * (75\%Yield) * \left(\frac{\$3}{Gal}\right) = \mathbf{\$675,000}/\boldsymbol{Yr}$ **Selling Terpenes**	Sell Terpenes **$675,000**
6) Reduce Maintenance Costs:[28] Media changeouts, large heavy moving part replacement, corrosion repairs, etc.	Save up to **~$1 Million**
Total Value Creation per Year per Site:	**Up to $3 Million**

---

22 ICAC Guidance Method for Estimation of Gas Consumption in a RTO. Institute of Clean Air Companies website. https://cdn.ymaws.com/www.icac.com/resource/resmgr/RTO-F1.pdf (Accessed September 18, 2021)

23 "Operating Cost Calculator". Cycle Therm website. http://www.cycletherm.com/resources/operating-cost-calculator.aspx (Accessed November 21, 2021).

24 "Greenhouse Gases Equivalencies Calculator - Calculations and References". Environmental Protection Agency (EPA) website. https://www.epa.gov/energy/greenhouse-gases-equivalencies-calculator-calculations-and-references (Accessed December 28, 2020).

25 "Allowance Price Explorer". International Carbon Action Partnership website. https://icapcarbonaction.com/en/ets-prices (Accessed Jan 20, 2021).

26 "Price Chart for Gum Turpentine". PineChem website. http://www.pinechem.net/images/gall/ChartGNE.GIF Accessed Jan 3, 2021).

27 "Sulfate Turpentine Recovery". Pulp Chemicals Association. Pages 9-12, 127. https://www.pinechemicals.org/store/viewproduct.aspx?id=1579188 (Accessed Jan 2, 2021).

28 "Air Pollution Control Technology Fact Sheet". EPA website. https://www3.epa.gov/ttncatc1/dir1/fregen.pdf (Accessed October 17, 2021).

VIII. CASE STUDY: FBC TECHNOLOGY SPECIFIC TO WOOD DRYING

DEMONSTRATION DESCRIPTION:

Captis Aire LLC founder and Environmental C&C Inc. worked on a project together from 2015-2021 with a commercial oriented strand board (OSB) manufacturer to demonstrate the ability to utilize the FBC to control emissions from wood drying.

In wood products manufacturing, process gas exhaust streams released from drying wood contain VOCs, including turpentine (terpenes). These VOCs are considered pollutants/emissions and must be removed from the air, prior to release, to avoid air pollution emissions into our communities. Wood drying operations utilized in the manufacture of oriented strand board (OSB), lumber, wood to biofuels, i.e. wood pellets, plywood, particleboard, etc. are few examples of these processes that have process exhaust streams containing VOCs.

Work on this patent[29] pending[30] technology was executed in three phases. In the first phase, laboratory screening work was done to explore whether emissions from wood drying, primarily terpenes, could be adsorbed by Bead Activated Carbon (BAC) media at the expected operating temperatures. In the second phase, a small-scale pilot adsorber was installed at a commercial Oriented Strand Board (OSB) manufacturing site to demonstrate ability to adsorb emissions from wood drying under commercial manufacturing conditions. In the third phase, the pilot unit capabilities were expanded to include the adsorber, desorber, and condenser. The SSR performance was simulated by semi-continuous additions of virgin BAC during air emissions compliance testing.

In the third testing phase, the pilot unit ran for 6 months with over 1700 hours on-line. Ability to meet air emissions control compliance, per the Plywood and Composite Wood Products Maximum Achievable Control Technology (PCWP MACT) standard, was determined by two independent certified air emissions[31] stack testing companies[32] during two separate testing events.

---

[29] "METHODS AND SYSTEMS FOR RECOVERING TERPENE COMPOSITIONS FROM WOOD DRYING EXHAUST". World International Patent Office (WIPO) Website. https://patentscope.wipo.int/search/en/detail.jsf?docId=WO2019148208 (Accessed September 18, 2021).

[30] "METHODS AND SYSTEMS FOR CONTROLLING EMISSIONS FROM WOOD DRYING PROCESSES". WIPO Website. https://patentscope.wipo.int/search/en/detail.jsf?docId=WO2019148207 (Accessed September 18, 2021).

[31] "AIR Advanced Industrial Resources". AIR website. https://airtest1.com/ (Accessed April 16, 2020).

[32] "Testing and Modeling". John Zink Hamworthy Combustion website. https://www.johnzinkhamworthy.com/testing-modeling/ (Accessed Feb 3, 2021).

TABLE V
RESULTS OF AIR POLLUTION CONTROL COMPLIANCE TESTING (FBC UNIT):

Emissions Measured:	FBC %Reduction Efficiency[33]
Total Hydrocarbons (THC), as carbon including Methane	92%
Volatile Organic Compounds (VOCs), as carbon. (Total non-methane hydrocarbons – THC minus Methane)	92%
Formaldehyde	83%
Methanol	75%
GHC - Nitrogen Oxides (NOx)	55%
Particulate Matter (PM), total filterable	50%
Condensable Particulate Matter	59%
Total Particulate Matter, Filterable + Condensable	51%
Total PM2.5, Particle Size Analyses (PSA) Filterable + Condensable	60%
Total PM10, Particle Size Analyses (PSA) Filterable + Condensable	64%
Carbon Monoxide (CO)	7%
Sulfur Dioxide (SO2)	99%

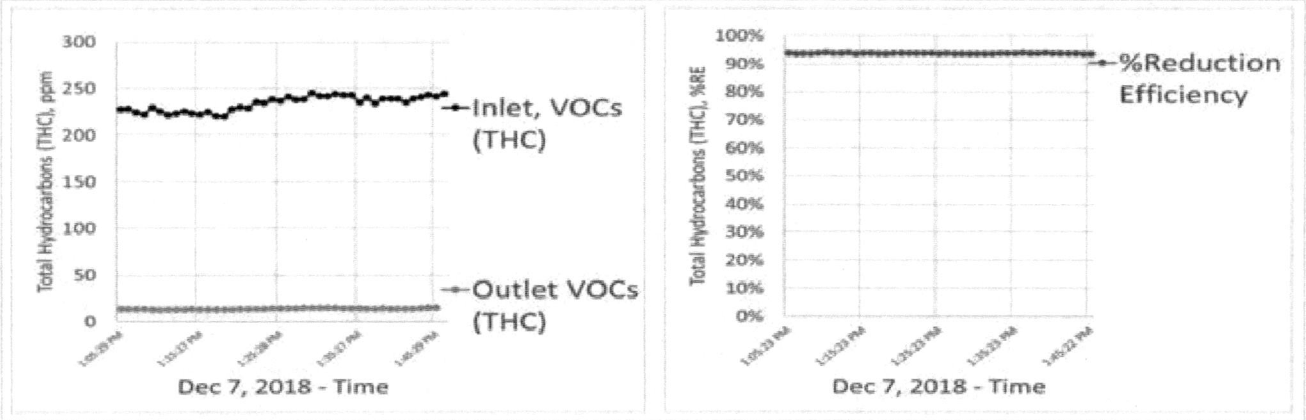

RESULTS OF VOC REDUCTION EFFICIENCY PERFORMANCE
MEASURED BY FLAME IONIZATION DETECTOR (FBC DEMONSTRATION UNIT):
Achieved >90% Reduction Efficiency meeting PCWP MACT Standard[34]

33 Emissions results FBC 2018 at commercial wood drying site done by Advanced Industrial Resources, Inc. Results were demonstrated during three 1-hour testing intervals. https://airtest1.com/ . (Accessed September 18, 2021)

34 "Electronic Code of Federal Regulations (eCFR)". eCFR website. See https://www.ecfr.gov/cgi-bin/text-idx?node=sp40.13.63.dddd#ap40.14.63_12292.1 (Accessed April 16, 2020).

RESULTS OF VOLATILE ORGANIC COMPOUNDS CAPTURE:

Terpenes were collected from the FBC demonstration unit. These terpenes are a mixture of terpene hydrocarbons, primarily Alpha and Beta pinenes. These terpenes are responsibly-recovered, biobased materials. Potential applications for these terpenes include use in fuels, fragrances, solvents, adjuvants, cleaners, and polyterpene resins.

TERPENES COLLECTED FROM FBC DEMONSTRATION UNIT

VALUE OF TERPENES:

The Pine Chemicals Association has documented the gallons of terpenes per oven dry ton of wood in a variety of tree species. Typically, there is ~1 gallon of turpentine (terpenes)[35] per oven dry ton of wood product produced from green pine wood from the SE part of the United States, i.e. Loblolly Pine, Jack Pine, Slash Pine, Shortleaf Pine, Ponderosa Pine, Pitch Pine, Pond Pine, Red Pine, and Spruce Pine. If 75% of this is evaporated and captured from the drying process, the value of the terpenes per year can be calculated by multiplying this yield by the oven dry tons (ODT) of wood produced and a typical price per gallon[36] for terpenes. For example:

$$\left(\frac{1\ Gallon\ Terpenes}{ODT\ Wood}\right) * \left(\frac{ODT\ Wood}{Year}\right) * (\%Yield) * \left(\frac{\$\ Dollars}{Gallon\ Terpenes}\right) = \$\ /\ Year$$

TABLE VI
TERPENES TYPICAL PROPERTIES:

Property	Result
Appearance	Clear Liquid
Moisture	<0.1%
Weight per Gallon	7.2 pounds per gallon
Specific Gravity	0.86
Sulfur Content	< 10 ppm

---

[35] "Sulfate Turpentine Recovery". Pulp Chemicals Association. Pages 9-12, 127. https://www.pinechemicals.org/store/viewproduct.aspx?id=1579188 (Accessed Jan 2, 2021).
[36] "Price Chart for Gum Turpentine". PineChem website. http://www.pinechem.net/images/gall/ChartGNE.GIF Accessed Jan 3, 2021).

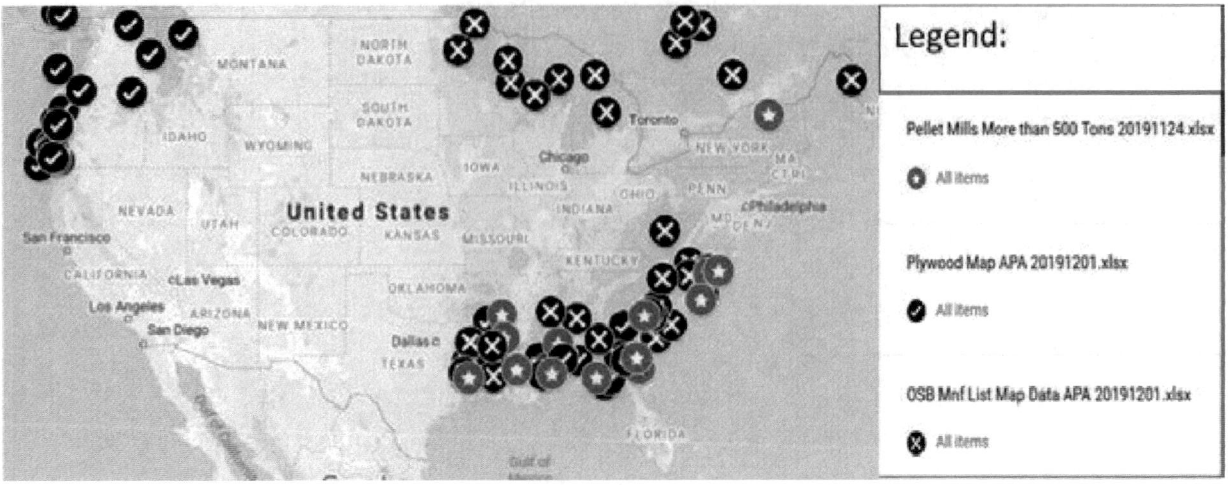

MAP OF POTENTIAL WOOD DRYING SITES FOR FBC USE:
This map shows potential sites where various wood products are produced where the FBC may be used to control emissions. Potential sites include but are not limited to Oriented Strand Board (OSB), lumber (sawmills)[37], plywood[38], wood to biofuels, i.e. wood pellets[39], particleboard, Medium Density Fiberboard (MDF), etc.

RESULTS OF DISCUSSIONS WITH ENVIRONMENTAL AGENCIES:

Official stack testing results were discussed with two state environmental agencies and two federal environmental protection agency (EPA) personnel. In all discussions, environmental agency personnel were supportive of providing a permit for full size commercial FBC unit.

FINAL RESULTS FROM WOOD DRYING DEMONSTRATION. THE UNIT:

- Provided salable terpenes, composed primarily of Alpha and Beta Pinenes.
- Reduced greenhouse gas (GHG) emissions.
- Reduced energy usage. It eliminated natural gas usage and reduced electrical usage.
- Met the PCWP MACT requirement of >90% Reduction Efficiency[40] in both tests. In the second test, the system met the compliance requirements three times in a row, over each of the three one-hour testing intervals.
- Demonstrated good operational reliability.
- Enabled State and Federal environmental personnel that reviewed these results to support permitting for a full commercial FBC system.

---

[37] "The mill map allows you to visually locate and drill down to find the mill you're interested in." Primary Forest Products Network website. https://primary.forestproductslocator.org/mill-map (Accessed September 18, 2021).
[38] "Manufacturer Directory". Engineered Wood Association website. https://www.apawood.org/manufacturer-directory (Accessed September 18, 2021).
[39] "Biomass Magazine". Biomass Magazine website. See http://biomassmagazine.com/plants/map/pellet/ (Accessed September 18, 2021).
[40] "Electronic Code of Federal Regulations (eCFR)". eCFR website. See https://www.ecfr.gov/cgi-bin/text-idx?node=sp40.13.63.dddd#ap40.14.63_12292.1 (Accessed April 16, 2020).

IX. CONCLUSIONS:

The FBC provides an affordable, reliable, and accessible clean technology for controlling VOCS, HAPS, and even particulate in some applications. This technology is well established in many applications including automotive and electronics manufacturing where it has been used successfully for many years. It is an emerging technology for use in biomass (i.e. wood drying) applications.

Especially for process exhaust streams with low VOC concentrations, the FBC technology is a patent[41] pending[42], cost competitive, industrial air pollution control technology that enables operators to advantageously:

1. Generate revenue via sales of recovered organic products (including biobased products such as terpenes in biomass applications) and sales of carbon offsets
2. Improve sustainability by reducing energy usage and thus greenhouse gas (GHG) emissions
3. Reduce operational costs by reducing energy usage (natural gas and electricity) and maintenance costs.
4. Enhance VOC mass transfer to the BAC media
5. Provide continuous in-situ BAC media regeneration and reactivation. This enables the media to continue to be reused.
6. Minimize water adsorbed on the BAC media thus maintaining its ability to adsorb VOCs, even in high moisture content process exhaust streams

Like any air pollution control technology installation, careful consideration of many design parameters, the unit's specific operating profile, and the operator's goals need to be taken into account during the design, installation, and startup/optimization processes. There are excellent resources, including Environmental C&C Inc.[43], Captis Aire LLC[44], and the Institute of Clean Air Companies (ICAC)[45] personnel, to assist with evaluating, implementing, and optimizing FBC applications.

---

[41] "METHODS AND SYSTEMS FOR RECOVERING TERPENE COMPOSITIONS FROM WOOD DRYING EXHAUST". World International Patent Office (WIPO) Website. https://patentscope.wipo.int/search/en/detail.jsf?docId=WO2019148208 (Accessed September 18, 2021).

[42] "METHODS AND SYSTEMS FOR CONTROLLING EMISSIONS FROM WOOD DRYING PROCESSES". WIPO Website. https://patentscope.wipo.int/search/en/detail.jsf?docId=WO2019148207 (Accessed September 18, 2021).

[43] "Environmental C&C Inc". Environmental C&C website. https://www.environmentalcc.com/ (Accessed September 18, 2021).

[44] "Emissions Control Technology Captures Emissions to Clean the Air". Captis Aire website. https://www.captisaire.com/ (Accessed November 21, 2021).

[45] "The Trusted Voice of the Clean Air Technology Industry". Institute of Clean Air Companies (ICAC) website. https://www.icac.com/ (Accessed September 18, 2021)

X. ACKNOWLEDGEMENTS:

The author recognizes inputs from a broad spectrum of personnel at the commercial OSB manufacturer and at Environmental C&C[46] including Harold Cowles, John Berger P.E., and Jim Starek.

CAPTIS AIRE CAPTURE-CONCENTRATE-CLEAN	Environmental C&C, Inc. Solvent Recovery & VOC Abatement Systems
Kim TUTIN Captis Aire LLC [47] 1612 W. Cleveland Ave, Suite 001 East Point, GA 30344-3213 Tel: +01 404 5802795, E-mail: kim@captisaire.com Website: www.captisaire.com	Harold COWLES Environmental C&C Inc. 16705 E. Calle de las Estrellas Road Scottsdale, AZ 85262-5982 Tel: +01 518 2250315, E-mail: halc@ecnc.com Website: https://www.environmentalcc.com/

[46] "Environmental C&C Inc". Environmental C&C website. https://www.environmentalcc.com/ (Accessed September 18, 2021).
[47] "Emissions Control Technology Captures Emissions to Clean the Air". Captis Aire website. https://www.captisaire.com/ (Accessed November 21, 2021).

623593 Process Optimization of Ion Exchange Processes for CO_2 Mineralization Utilizing Industrial Waste Streams

Abstract:

Large coal-fired power plants (>500 MW) account for 30% of global CO_2 emissions, and long-term management of this CO_2 to is urgently needed mitigate global temperature increases. Sequestration of CO_2 within stable mineral carbonates (e.g., $CaCO_3$) represents an attractive emission reduction strategy because it offers a leakage-free alternative to geological storage of CO_2 in an environmentally friendly form. We have previously described a mineralization process in which divalent cations are sourced from various waste streams (e.g., produced water and brackish water) and alkalinity is induced via regenerable ion-exchange materials (Bustillos et. al. *Frontiers in Energy Research.* 2020, 8, 352). In our process, aqueous carbonate-bearing streams with pH > 8 are produced by contacting fresh water and carbon dioxide with various ion-exchange materials (e.g., Na form zeolites or ion exchange resins). These streams are mixed with produced water containing varying concentrations (~0.01 – 1.0 M) of Ca^{2+} leading to the precipitation of solid calcium carbonate (PCC). This process has the advantages of using regenerable solids in a simple and continuous process to increase the pH of water by ion exchange instead of relying on the consumption of costly and unsustainable sources of alkalinity (e.g., sodium hydroxide). While once-through column experiments showed the above benefits, the same were yet to established in a steady-state process with recycle streams. In this work, we set up a process simulation to quantify the energy requirements and CO_2 emissions associated with the process and seek optimal produced water compositions and CO_2 concentrations (5 – 20 vol%).

The process simulation was set up in ASPEN Plus using eRNTL as the thermodynamic property method and sequential modular strategy. Ion exchange alkaline solution was simulated using sodium hydroxide and validated against the experimental data obtained from once-through kinetic experiments. Nanofiltration and reverse osmosis membrane steps were also implemented for the separation of divalent cations and production of fresh water and a regeneration stream following mineralization. Sensitivity analysis was carried out using a range of produced water compositions (0.01 – 1.0 M Ca^{2+}, 0.001 – 0.15 M Mg^{2+}, 0.5 – 3.5 M Na^{+} and 0.0004 – 0.002 M Fe^{2+}) according to the United States Geological Survey (USGS) database.

Calcium carbonate yields increased with increasing CO_2 concentrations and were maximized using produced water compositions with larger Ca^{2+} concentrations. Maximum calcium carbonate yields produced at 5 vol%, 12 vol% and 20 vol% CO_2 were 2.3 mmol/L, 5.5 mmol/L, and 9.3 mmol/L, respectively, with the formation of brucite (a magnesium hydroxide phase, $Mg(OH)_2$) and goethite (an iron hydroxide phase, FeOOH) as the primary contaminant phases (99% calcite, 0.6% brucite, 0.4% goethite), which agree with phases detected by XRD experimentally. These results indicate high purity calcium carbonate can be precipitated using industrial waste streams.

Consequentially, energy consumption and net CO_2 emissions were minimized where precipitated calcium carbonate was maximized for all produced water compositions and CO_2 concentrations. Minimum energy consumptions were 0.21 kWh/ton CO_2 processed, with 98% of the energy input required coming from the membrane filtration steps. Produced water compositions with large Na concentrations (> 0.5 M) were effective at reducing energy consumptions due to faster regeneration time of ion exchange materials. Additionally, calculated net CO_2 emissions were negative for the process and ranged from -0.02 kg/ton CO_2 to -0.15 kg/ton CO_2 processed, indicating a low emission process. We will also present techno-economic assessment showing the economic benefits of the current process as an alternative to the addition of stoichiometric bases to induce alkalinity for the precipitation of $CaCO_3$.

625931 Sustainable Processing of Composite Materials

Abstract:

Utilization of captured carbon dioxide (CO_2) by converting it into valuable products, such as fuels, chemicals, plastics, and building materials is needed to offset the >36 GT/y of anthropogenic emissions. This presentation will describe two breakthrough processes invented and patented by Rutgers University: Hydrothermal Vapor Synthesis (HVS) and Gas-assisted Reactive Hydrothermal Liquid Phase Densification (g-rHLPD). These processes collectively enable the production of ceramics with a greatly reduced carbon footprint, which in many cases can be carbon negative. HVS enables the production of feedstock anhydrous inorganic oxides, an essential precursor for carbonate ceramics, at temperatures <500˚C, which is up to 1000˚C lower than the temperatures of traditional pyrothermal reactions. g-rHLPD is a low temperature (<100˚C) densification process using CO_2 as a reactant with anhydrous oxides for creating strong, dense carbonate-bound ceramics with outstanding physical and chemical properties. Materials produced from HVS and g-rHLPD have utility for both commodity-based concrete and advanced composite materials that can substitute for wood, fiber-reinforced resins and even steel. Materials produced with these technologies will be presented and compared with the incumbent materials to demonstrate the outstanding future awaiting carbonate-based materials.

622440 Field Demonstration of the Reversa^{TM} Mineral Carbonation Process Using Coal and Natural Gas Flue Gas Streams

Abstract:

Concrete, a mixture composed of a cementation agent, mineral aggregates, and water has the potential to serve as a gigaton-scale sink for carbon dioxide (CO_2). This could make concrete the world's largest CO_2 utilization opportunity. CarbonBuilt's Reversa^{TM} process, developed at UCLA's Institute for Carbon Management exploits simple acid-base chemistry to mineralize CO_2-dilute flue gas emissions into mineral carbonate-based cementation agents at ambient pressure, at flue gas temperatures, and without a need for carbon capture. The approach leverages innovations in the use of portlandite ($Ca(OH)_2$: calcium hydroxide, or slaked lime) which carbonates readily, and produces limestone

($CaCO_3$: calcium carbonate) – a potent cementation agent – upon its carbonation. Within the scope of a project sponsored by the US Department of Energy's Office of Fossil Energy, the Reversa technology was upscaled and demonstrated using a modularized pilot-plant at the Integrated Test Center (Gillette, WY) and National Carbon Capture Center (Wilsonville, AL) using coal- (~12 vol. % CO_2) and natural gas (~4 vol. % CO_2) flue gas streams. The field demonstration led to the production of over 15,000 concrete masonry units (CMUs, also known as concrete blocks) and achieved: (1) a CO_2 utilization efficiency in excess of 75%, and (2) greater than 250 kg of CO_2 utilization per production run.

Importantly, based on rigorous 3rd-party validation, the CMUs produced were confirmed to be compliant with all relevant industry specifications (ASTM C90). The success of this demonstration suggests that the pioneering Reversa technology is ready for commercialization.

625647 High-Efficiency Capture and Utilization of CO_2 from Flue Gas Using Algae-Based Amino Acid Salt Solution

Abstract:

The power industry in the U.S. emits more than 1.8 gigaton of CO2 (30% of the total CO2 emissions) per year. New solutions are urgently needed to capture and utilize the CO2. It has been reported that amino acids neutralized with a base compound as alternatives of alkanolamines to capture carbon dioxide (CO2) from flue gas. The study develops a new integrated system based on algal amino acids. The system applies four consecutive steps of preparation of algal amino acid salt solution, algal amino acid salt absorption, spent amino acid utilization of polyurethane production, and conversion of collected CO2 for organic acid production to realize CO2 capture and utilization. A bench-scale system was used to evaluate performance of CO2 capture and utilization. A detailed techno-economic analysis and life-cycle assessment are conducted based on the results from the bench-scale studies. A technically feasible and economically sound CO2 capture and utilization process is concluded from this study.

632855 Premium Animal Feed from Filamentous Algae Grown on Florida Power Plant CO_2

Abstract:

Microalgae cultivation in large-scale raceway ponds is being developed by various organizations for production of commodities, in particular for animal feeds—a billion ton market, valued at over US$500 billion world-wide. Microalgae production requires concentrated CO_2 sources, with CO_2 supply being a major cost and engineering challenge. A coal- and natural gas-fired power plant owned by the Orlando Utilities Commission (Florida), is the focus of experimental and regional scale-up planning studies – 4,000 ha (10,000 acres) of CO_2 distribution and algae cultivation – all within 4 km of the power plant. At this power plant, the Stanton Energy Center (SEC), MicroBio Engineering Inc. is operating an algae carbon capture and utilization (ACCU) research facility, with a slipstream of scrubbed flue gas. The facility is being used to develop cultivation techniques for various easy-to-harvest filamentous algal strains suitable for production in Florida of biomass rich in omega-3 and -6 fatty acids and other phytonutrients valued in animal feeds. Poultry feeding trials have been conducted with laying hens at Cal Poly State University. Here we model and compare alternatives technologies for the capture, concentration, transport and utilization for algae cultivation using flue gas CO_2 from both coal- and natural gas-fired power plants at the SEC. Algae cultivation data are used as inputs to scale-up engineering designs, techno-economic analysis, and environmental life cycle assessments for the capture, concentration, and transport up to 40 km to large-scale algae cultivation facilities. The outputs are the projected costs of CO_2 utilization and algal biomass production for commodity animal feeds, at 400,000 tons/year. Although the cost of algae biomass will be higher than soybeans, it is justified by a greater content of phytonutrients, as well as the overall reductions in CO_2 emissions and other environmental benefits.

Supported by DOE NETL (National Energy Technology Laboratory) Award DE-FE-0031717

625497 Combined Techno-Economic Analysis and Life Cycle Assessment of an Integrated Direct Air Capture System with Advanced Algal Biofuel Production

Abstract:

The continuous increase in carbon dioxide (CO_2) concentration in the atmosphere since the First Industrial Revolution correlates convincingly with the ongoing rise of the Earth's global average temperature contributing to climate change. As a response, carbon capture and sequestration (CCS) technologies are being implemented to mitigate anthropogenic CO_2 emissions by capturing CO_2 from high emitting point sources, such as power plants, refineries, and cement factories, and subsequently buried in geological formations underground for long term storage. Recently direct air capture (DAC) has emerged as a promising alternative technology that captures CO_2 directly from the atmosphere for use or sequestration. This study investigates the potential synergistic benefits of integrating a solid amine-based DAC system with advanced algal biofuel production in photobioreactors (PBRs). DAC utilization allows the removal of atmospheric CO_2 while also decoupling algae production facilities from anthropogenic point CO_2 sources and avoiding the cost and logistics challenges of transporting CO_2 long distances to remote facilities. Techno-economic analysis and life cycle assessment are performed to assess the economic and environmental benefits of heat and mass integration between the DAC and the PBR for biofuel production. The DAC-PBR system integration also considers on-site flue gas handling options, DAC capital utilization, the tradeoff between centralization and decentralization of key unit operations, and the PBR array. This presentation will discuss how optimizing DAC-PBR process integration can enhance the algal biofuel's economic and environmental sustainability and the prospect of DAC enabling the circular carbon economy.

631638 Conversion of Plastic Waste to Fuel: Assessment of Technologies and Economics

Abstract:

Plastics are extremely valuable to our society and are widely used across industries such as food, beverage, textile, healthcare, automotive, aerospace, construction, agriculture, and oil. The global production of plastics, accumulative from 1950 to 2017, is 8.3 billion metric tons, with 6.3 billion tons waste generated, of which only 9% has been recycled, 12% incinerated, with the remaining 79% landfilled or released into the environment. Mechanical recycling is used to recycle about 95% of plastic waste today, however, this method works well only for homogeneous and non-contaminated plastic waste, and for easily identifiable objects such as bottles and containers made of PET and PE. Chemical recycling can offer a solution in recycling many of the plastics which mechanical recycling fails to process. Chemical recycling involves the deconstruction of polymers to monomers for conversion back into polymers, or to intermediates that can be used as fuel or feedstock by the chemical industry. A number of chemical recycling technologies are being developed, and the processes are being tailored to handle specific feedstocks and/or produce desirable products. In this project we assess companies that are involved in the chemical recycling of plastic waste: their technology and/or key innovations, the feedstocks and product(s), and the current scale of operation. Further, we evaluate in this project the economics of chemical recycling with a specific interest on plastic-to-fuel conversion.

623136 Continuous Back to Monomer Recycling of PET from Composite Materials Using a Twin-Screw Extruder

Abstract:

The current Packaging Act requires a 63% quota for the mechanical recycling of plastics from 2022 in Germany [1]. In order to meet this requirement, new efficient recycling technologies are necessary. Many plastics, such as polyethylene terephthalate (PET), are used in combination with other polymers. Apart from energy recovery, no recycling process for such multilayer material is available and it is therefore largely lost to the resource cycle.

The research project revolPET® specifically addresses this issue through the development of a continuous process for recovering the monomers terephthalic acid (TA) and ethylene glycol (EG) in primary raw material quality from mixed PET-containing waste [2]. While literature mainly reports batchwise depolymerization of PET in stirred tank reactors [3, 4], the revolPET® process uses an extruder as a continuous chemical reactor.

This contribution presents experimental results on the successful depolymerization of PET in composite materials into a TA salt and EG. The high flexibility of the extruder offers a wide range of options for process design in terms of thermal, mechanical and chemical conditions, so that suitable process windows are identified depending on the feed material. Key to this technology, the twin-screw extruder allows the two reagents to be mixed as solids. This innovative process intensification enables high degrees of depolymerization at average residence times below 2 minutes. In the experiments, a PET depolymerization degree of ≥ 97 % was achieved. Systematic parameter studies reveal the influence of temperature, filling degree of the screw chamber and residence time on the successful PET depolymerization. Figure 1 presents typical results for degree of depolymerization and residence time at varied screw revolutions.

In addition to the technical aspects, the contribution provides insight into the ecological assessment of this new recycling process. In particular, the challenge of the exothermic reaction to identify an optimal process window is shown. This topic is examined in more detail against the background of scaling. The presented experimental results will be discussed in detail and placed in the context of the current political debate.

Literature

[1] § 16 Absatz 2 VerpackG

[2] www.revolpet.eu

[3] A.M. Al-Sabagh, F.Z. Yehia, Gh. Eshaq, A.M. Rabie, A.E. ElMetwally, *Egyptian Journal of Petroleum* (2016) 25, 53-64, doi: 10.1016/j.ejpe.2015.03.001

[4] A. Oku, L.-C. Hu, E. Yamada, *J. Appl. Polym. Sci.* (1997) *63*, 595-601.

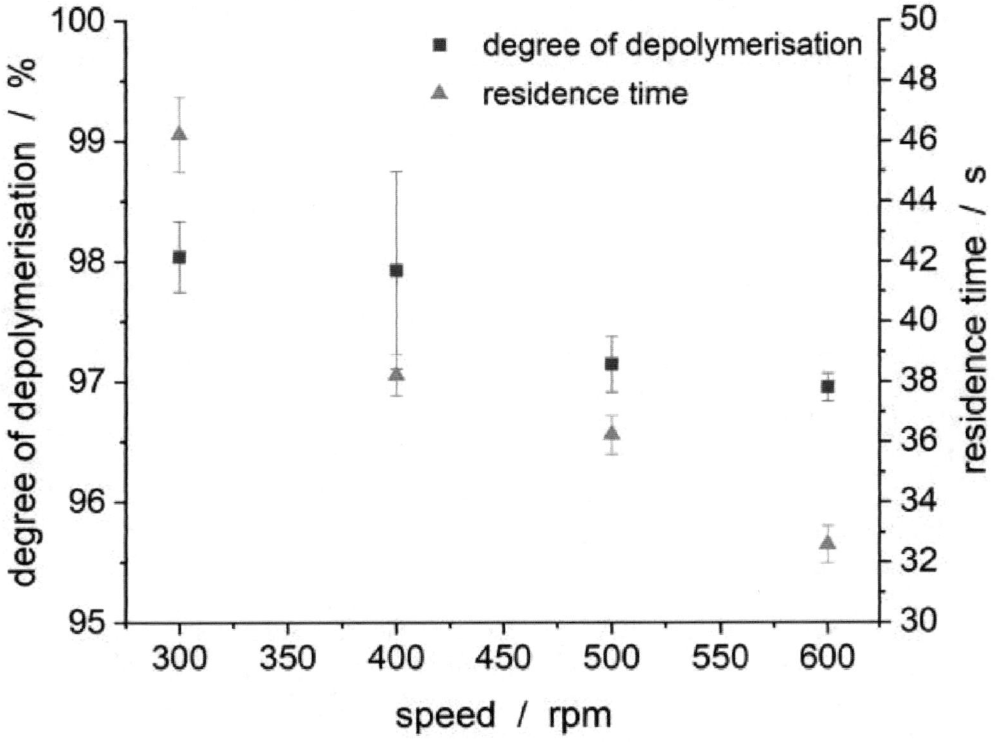

Figure 1: Degree of PET depolymerization in relation to the residence time at different speed

629642 The Effect of Residence Time and Heterogeneous Catalysis on Continuous Hydrothermal Liquefaction of Food Waste

Abstract:

The United States releases a significant amount of methane through the decomposition of waste in landfills; methane is a greenhouse gas with a global warming potential 25 times greater than carbon dioxide [1]. According to the EPA, landfilling accounts for approximately 17% of the total US methane emissions [1], and food waste is the largest non-recyclable component of municipal waste [2]. Although food waste currently harms the environment, it also presents an opportunity to solve another looming threat: the US dependency on non-renewable fossil fuels. Mainstream Engineering, in collaboration with Worcester Polytechnic Institute, is developing technology to convert food waste to energy through hydrothermal liquefaction (HTL) that simultaneously addresses our reliance on fossil fuels and the environmental impact of food waste.

HTL converts food waste, and other wet wastes, into an energy dense biocrude by reacting the food mixture of carbohydrates, lipids, and proteins in near supercritical water. Traditionally, HTL reactions have been uncatalyzed or used a homogeneous catalyst (i.e., Na_2CO_3). We have shown previously that using a solid oxide, heterogeneous catalyst, $CeZrO_x$, in a batch reactor improves biocrude yield and energy recovery [3]. We have subsequently studied the effects HTL residence time and catalyst addition in a continuous stirred tank reactor (CSTR). A model food waste slurry was fed to the CSTR via a dual high pressure syringe pump system. The pressure of the system was maintained using a back pressure regulator rated to 300 °C and 5,000 PSI. Commercially available $CeZrO_x$ powder was compressed into tablets and immobilized in the CSTR using a catalyst cage. The catalyst tablets were able to withstand the hydrothermal environment without physically disintegrating. The stability of the catalyst tablets was characterized using XRD and BET before and after HTL. The energy recovery, product yields, and biocrude properties (HHV, TAN, and CHNO) were determined to characterize the effectiveness of $CeZrO_x$ in a continuous HTL reactor and evaluate the impact of residence time on continuous HTL of food waste.

References

[1] EPA. *Greenhouse Gas Emissions*. [cited 2021; Available from: https://www.epa.gov/ghgemissions/overview-greenhouse-gases.

[2] USDA. *Why should we care about food waste?* [cited 2021; Available from: https://www.usda.gov/foodlossandwaste/why.

[3] Maag, A.R., et al., *Catalytic hydrothermal liquefaction of food waste using CeZrOx*. Energies, 2018. **11**(3): p. 564.

629415 Continuous Hydrotreatment of Sewage Sludge and *Spirulina* Algae Biocrudes from Hydrothermal Liquefaction: Different Catalysts for Different Organic Contaminants

Abstract:

Unprecedented global warming and raising public awareness are the driving factors to decarbonize long-haul transportation sector. In this regard, biomass could be a potential renewable source for the production of sustainable, carbon-neutral liquid fuels. Hydrothermal liquefaction (HTL) followed by catalytic hydrotreatment is one of the most promising pathways to covert varying feedstocks into drop-in fuels. The energetically dense black viscous product from HTL ("biocrude"), contains a diverse organic pool with considerable amount of oxygen (~5-18%), nitrogen (~1-8%) and metallic content (~0.1-0.6%). The presence of these organic contaminant results in poor miscibility with fossil fuels and poses new and exciting challenges e.g. thermal instability of biocrude at high temperatures which are indispensable for complete hydrodenitrogenation. This is why special attention is needed not only to realize this complex biocrude mixture but also to select different catalysts which performs best to saturate and remove certain organic contaminants under optimized conditions.

In this work, we utilized advanced analytical tools (i.e. FT-ICR MS etc.) and observed different classes of organometallics and nitrogen compounds from sewage sludge and *Spirulina* algae biocrude. From these results, we utilized three different types hydrotreating catalysts based on their porosity and nickel loading on γ-Al_2O_3 support and successfully demonstrate smooth continuous hydroprocessing operations (Figure 1) for hundreds of hours without coke formation, catalyst deactivation, and reactor plugging. Finally, 100% deoxygenation and ~96% denitrogenation were achieved. Furthermore, hydrotreated products underwent true-boiling point fractional distillation and showed on-spec metal content, density, pour point, cloud point and HHV for diesel and jet fuel.

.

Figure 1: Hydrotreated samples from the continuous hydrotreatment of sewage sludge (a) and *Spirulina* algae (b) biocrude.

Acknowledgement: This research has received funding from the European Union's Horizon 2020 research and innovation programme under grant agreement No 764734.

627420 Review and Modeling of Organic Removal in Anaerobic Sludge Blanket Bioreactors for Methane Production and Energy Recovery from Domestic Wastewater

Abstract:

The current domestic wastewater treatment paradigm centers on aerobic wastewater treatment technologies such as activated sludge. While effective in meeting U.S. EPA regulated discharge standards for organics and suspended solids, providing oxygen to aerobic microorganisms is energy-intensive, typically accounting for ~50% of wastewater treatment facility costs. In total, wastewater conveyance and treatment accounts for ~3% of our nation's electrical energy usage. Direct anaerobic treatment of dilute wastewater is a viable methane-generating alternative to aerobic wastewater treatment. Anaerobic microorganisms convert wastewater organics (fats, carbohydrates, proteins, etc.) to methane-rich biogas, which can be harnessed to create energy via technologies such as combined heat & power. If more energy is generated from the methane produced than is used to run the process itself, then the process can be considered "energy-positive" (or a net energy producer). Anaerobic sludge blanket bioreactors, most notably upflow anaerobic sludge blanket (UASB) bioreactors and anaerobic baffled reactors (ABRs), have been studied at bench- and pilot-scale for several decades. UASBs are also currently operated at full-scale in several locations worldwide, mostly in warmer climates. Despite the number of completed studies, no current review critically examines performance data from different sized sludge blanket bioreactors collected under varying conditions (e.g., varying wastewater temperatures and hydraulic residence times). Further, no current review study conducts predictive modeling (i.e., Monte Carlo Simulation) on aggregated sludge blanket bioreactor data to better understand the operation of such bioreactors under varying conditions. This study aggregates and examines data from 145 UASBs studies and 56 ABRs studies to understand bioreactor performance (e.g., chemical oxygen demand and total suspended solids removal) and methane generation. This study also employs Monte Carlo Simulation using @Risk to better predict bioreactor performance under varying conditions. Additionally, this study identifies critical gaps in literature (e.g., a lack of consistently reported methane generation) that may impact full-scale implementation in certain climates (e.g., cold weather climates). Last, results from this study can help inform decision makers to better understand of anaerobic sludge blanket bioreactors can meet regulatory standards or generate methane for renewable energy at their wastewater treatment facilities.

629993 Utilization of Marble Dust As a Concrete Additive to Enhance Its Properties

Abstract:

With the advancement in the building technology, each and every step to alter the cement properties for making it environmentally beneficial and equally strong is important. In this advancement, one always opts for an alternative which provides the strength to the building by reducing its cost and weight.

Marble often known as calcium carbonate ($CaCO_3$) has been a well known material used as a building material since the ancient times. Marble is often available in combination with Silicon Oxide (SiO_2), Aluminum Oxide (Al_2O_3), Iron Oxide (Fe_2O_3), Magnesium Oxide (MgO) and Calcium Oxide (CaO).

Marble is utilized at a number of operations and the waste obtained requires an extra effort for the adequate environmental disposal. The waste marble left in the environment can be harmful therefore must be utilized for an alternative process to reduce its harmful effect. The addition of marble dust to concrete is proving to be an effective alternative for the process. The process of marble addition is participating in the hydraulic reactions contributing significantly to the composition and microstructure of hydrated products. Another way of utilizing the marble dust is to utilize it in the geo-polymer concrete which will be strengthening the concrete properties.

The entire paper is evaluating the utilization of the marble dust in concrete so that the unsafe waste of marble is used and fewer efforts is to be taken for its disposal and also to accomplish viability, economy and environmental conservation.

624910 Ionic Liquid Initiated New CO2 Reactions for the Orginc Synthesis and Conversion

Abstract:

Ionic liquids typically are considered as stable media for electrochemical and chemical reactions. Currently a serial of reactions have been discovered from the ionic liquids due to its unique chemical properties and chemical environment offered, Here, we are presenting highly efficient CO_2 reduction and utilization method derived from ionic liquids performed in aerobic environments. CO_2 reduction to CO is observed at a very low over-potential, achieving >90% faradic efficiency and near 100% carbon selectivity on three conventional disk electrodes (Pt, Au and glass carbon), in the presence of O_2 at an over-potential that is approximately 1.6V lower than that required for reduction of pure CO_2. Furthermore, the facile electrochemical approach to the formation of several urea compounds from CO2 and the associated amine compounds (R-NH2) at room temperature in ionic liquid (IL) environments. These reactions were carried out at very low potentials under mild conditions without addition of catalysts or other agents, other than those available from the atmospheric environment. The new processes have potential to be transformative, in that it can lead to new pathways that can be developed to utilize atmospheric CO2 as a starting material. A process that begins with atmospheric CO2 as a reactant and results in production of urea fertilizer has very exciting possibilities, in terms of taking an unwanted substance and converting it to an essential one.

625542 Life Cycle Analysis Tools to Evaluate Environmental Footprint of Carbon Utilization Projects

Abstract:

Utilizing captured carbon dioxide (CO_2) emitted from power and industrial sources to make useful products is an emerging area of research. Life cycle assessment/analysis (LCA) can support to evaluate the potential environmental impacts and net life cycle greenhouse gas (GHG) reductions of CO_2 utilization technologies compared to existing alternatives in the marketplace from a consistent and unbiased viewpoint. CO_2 utilization systems are complex based on methodological perspective due to the intrinsic links established between the power sector and the utilization sector (e.g., biofuels, cement, chemicals, etc.). In their 2019 report, "Gaseous Carbon Waste Streams: Utilization Status and Research Needs," The National Academies identified LCA benchmarking as a key research priority to "facilitate consistent and transparent assessments of the net greenhouse gas emissions of carbon utilization technologies." A robust and consistent guideline that establishes best practices to perform LCA of CO_2 utilization technologies could be a great benefit to technology developers and LCA analysts to evaluate potential environmental performances and opportunities for further research.

The holistic approach of LCA, which forces a cradle-to grave analysis, is uniquely suited for appraisal of CO_2 utilization technologies. Since there is important link between the source of the CO_2 and the use of the CO_2 in the overall environmental impact, robust treatment of the upstream CO_2 sources is imperative in LCA of CO_2 utilization technologies. For emerging CO_2 utilization technologies, conducting LCA at early stages of development is important, as it allows flexibility to implement changes while technologies are still nascent. Additionally, the U.S. Department of Energy (DOE) and the federal government are increasingly requiring for LCA as part of funding for primary research and tax incentives such as "45Q" for CO_2 capture, utilization, and storage projects (H.R. 1892, 2018).

In the interest of supporting the creation of useful and consistent LCAs of CO_2 utilization projects, the DOE National Energy Technology Laboratory has developed a CO_2 utilization LCA toolkit. The primary audience for the toolkit is the group of principal investigators that have been awarded a grant through a DOE Carbon Use and Reuse funding opportunity announcement. However, we believe that there is wider application beyond those grant recipients. The toolkit comprises a guidance document, reporting templates, open LCA contribution analysis tools, LCI data, and subject matter expertise. The guidance document should be considered as a companion to the requirements established in ISO 14040/14044, providing specific guidance on methodological issues for scoping and completing the LCA. The guidance, data, and tools are publicly available and free. This presentation will provide an overview of each part of this first-of-a-kind toolkit, its utility to the LCA community, and its value in assessing emerging technologies.

References

H.R. 1892, 115 Cong. (2018) (enacted). Retrieved from https://www.congress.gov/bill/115th-congress/house-bill/1892

National Academies of Sciences, Engineering, Medicine. (2019). Gaseous Carbon Waste Streams Utilization: Status and Research Needs. Washington, DC: The National Academies Press.

DISCLAIMER

This project was funded by the United States Department of Energy, National Energy Technology Laboratory, in part, through a site support contract. Neither the United States Government nor any agency thereof, nor any of their employees, nor the support contractor, nor any of their employees, makes any warranty, express or implied, or assumes any legal liability or responsibility for the accuracy, completeness, or usefulness of any information, apparatus, product, or process disclosed, or represents that its use would not infringe privately owned rights. Reference herein to any specific commercial product, process, or service by trade name, trademark, manufacturer, or otherwise does not necessarily constitute or imply its endorsement, recommendation, or favoring by the United States Government or any agency thereof. The views and opinions of authors expressed herein do not necessarily state or reflect those of the United States Government or any agency thereof.

626846 CO_2 Electrolyzer to Produce Formic Acid at Industry Relevant Current Density

Abstract:

Electrochemical CO_2 reduction reaction (CO_2RR) to produce as formic acid has attracted great interest recently since formic acid can be used as a feedstock for biorefineries. Over the past decade, much effort has been devoted to the prospects of electrochemical CO_2 conversion to formic acid. However, issues including low current density, low selectivity, low energy efficiency, poor long-term stability, and production of formate salts instead of formic acid remain as the great challenges for the development and commercialization of this technology. Recently, Dioxide Materials (DM) has developed an electrolyzer technology that can directly convert CO_2 to pure formic acid with a three-compartment CO_2 electrolyzer. In this work, DM will present our recent effort on improving the electrolyzer performance operating at industry relevant current density (200 mA/cm^2). We have demonstrated 200 mA/cm^2 with 80% Faradaic efficiency for 1000 hours of operation The effects of CO_2 concentration and impurities in the flue gas on the electrolyzer performance will be discussed, as will ongoing efforts to test the device at a coal fired power plant. The long-term (1000 h) performance of the electrolyzer will also be presented.

628598 Electrochemical Reduction of Carbon Dioxide By Means of Transition Metal Infiltrated Cathodes in Solid Oxide Electrolyzer Cells

Abstract:

Population and economic growth have led to a substantial increase in worldwide energy demand and CO_2 emissions. Although the world is transitioning towards renewable energy sources such as wind and solar, these sources will not satisfy increasing energy demand requiring continued use of fossil fuels. The continued use of fossil fuels makes development of cost-effective carbon capture, utilization, and storage technologies essential to sustainable worldwide growth. Over the past two years, Ohio University (OHIO) has been investigating conversion of carbon dioxide (CO_2) into more valuable industrial feedstocks such as carbon monoxide (CO). CO plays a significant role in various industrial applications including the production of chemical compounds as well as the production of synthetic fuels.

Commercially, CO is produced either by steam reforming of gaseous hydrocarbons or by gasification of heavy hydrocarbons (liquids and solids), where it is produced along with hydrogen (H_2) to form synthetic gas mixture (syngas) [1]. Pure CO is obtained by separating it from the syngas mixture using different techniques such as cryogenic separation, pressure swing adsorption (PSA), and membrane separation [1,2]. The aforementioned techniques are energy intensive requiring large, centralized facilities to achieve favorable economics. CO_2 electrolysis has shown to be a viable option for CO production due to its simplicity and low operating cost. High temperature CO_2 electrolysis utilizing solid oxide fuel cell (SOFC) technology offers enhanced reaction kinetics and high cell efficiencies at high temperatures, potentially offering a cost-effective modular alternative which can be integrated into industrial settings [3].

OHIO, with support from the U.S. Department of Energy [DE-FE0031709], is developing a solid oxide electrolyzer cell (SOEC) to convert CO_2 into more valuable products such as CO. In this study, SOEC cathodes were fabricated using various transition metal electrocatalysts (cobalt, copper, and nickel) infiltrated into a porous $Gd_{0.10}Ce_{0.90}O_{1.95}$ (GDC-10) scaffold, while $(La_{0.80}Sr_{0.20})_{0.95}MnO_3$ (LSM20) was used for the anode fabrication. The electrochemical performance of the transition metal-based electrocatalysts were evaluated at different temperatures (750, 800, and 850 °C) by supplying a blend of CO_2, CO, and N_2 (81% CO_2, 10% CO, 9% N_2) on the cathode side and air on the anode side. The fabrication techniques as well as the electrochemical performance of the transition metal electrocatalysts will be discussed in this presentation.

[1] Wilbur, S.; Williams, M.; Williams, R.; Scinicariello, F.; Klotzbach, J. M.; Diamond, G. L.; Citra, M. Toxicological Profile for Carbon Monoxide. *Agency for Toxic Substances and Disease Registry (ATSDR) Toxicological Profiles; Agency for Toxic Substances and Disease Registry* (US): Atlanta (GA), **2012**.

[2] Poudel, J.; Choi, J.; Oh, S. Process Design Characteristics of Syngas (CO/H2) Separation Using Composite Membrane. *Sustainability* **2019**, *11* (3), 703.

[3] Ebbesen, S. D.; Mogensen, M. Electrolysis of Carbon Dioxide in Solid Oxide Electrolysis Cells. *J.*

Power Sources **2009**, *193* (1), 349–358.

625154 Carbon Dioxide Electrolyzer Scale up and Commercial Opportunity

Abstract:

Opus 12 is developing an efficient polymer-electrolyte membrane (PEM) CO_2 electrolyzer that can couple to sources of renewable electricity and recycle carbon dioxide emissions into useful chemicals and fuels. CO_2 electrolysis has the potential to monetize and reduce greenhouse gas emissions by providing a profitable alternative to releasing carbon dioxide into the atmosphere. Large-scale deployment of this technology is key to achieving these benefits.

Our first target product is carbon monoxide, where an electrochemical approach can leverage renewable electricity generation on the grid to make low-cost, carbon neutral fuels, materials, and chemicals. This talk will describe progress and lessons learned in scaling up our electrolyzer to reduce the cost and smaller scale projects completed with commercial partners that demonstrate the commercial opportunity for CO2-derived products at an industrial scale.

625532 Microwave-Assisted Dry Reforming for Energy Efficient CO_2 and CH_4 Conversion

Abstract:

Microwave-assisted catalysis is an emerging concept for promoting thermally-challenging reactions. This talk describes NETL's efforts to develop microwave-active catalysts for the dry reforming of methane (DRM: $CO_2 + CH_4$ à $2CO + 2H_2$). DRM is an appealing reaction because it converts CO_2 and CH_4 into a carbon neutral industrial commodity (syngas), but high reaction temperatures currently make it impractical in most cases. Microwave-assisted catalysis can overcome this limitation by selectively heating the catalyst bed to high temperatures (900 ^oC) at extremely fast rates (200-300 ^0C/min). This rapid on/off cycling will allow utilization of excess renewable energy and reduces heat management issues compared with traditional (always on) thermal reactors. We based our catalyst system on the perovskite oxide $La_{0.8}Sr_{0.2}CoO_3$ (LSC). This catalyst shows good microwave absorptivity and high temperature stability. We found that incorporating first row-transition metals (Mn, Fe, Ni and Cu) into the LSC structure dramatically impacted catalytic activity and stability. *Ex situ* O K-edge x-ray absorption spectroscopy, temperature programmed reduction, and density functional theory correlated the dopant electronegativity to catalyst activity and overall stability. Incorporating dopants that were less reducible than Co (Mn and Fe) improved LSC performance by strengthening metal-oxygen bonding and stabilizing the catalyst structure. On the other hand, incorporating more reducible dopants (Ni and Cu) led to severe phase segregation during DRM and rapid catalyst deactivation. Our best in class catalyst, Mn-doped LSC, demonstrated over 90% single-pass CO_2 and CH_4 conversion with good long-term stability. *In situ* synchrotron-based X-ray diffraction evaluated catalyst structure during DRM and identified the formation of small Co nanoparticles at the catalyst surface as key reaction centers. The increased performance of LSC-Mn was directly related to Mn-dopants stabilizing the perovskite structure and preventing sintering of catalytically active Co nanoparticles. Scaling studies show improved performance at higher flow rates and energy efficiencies of 10 kWh/kg_{CO2} that approach the efficiency of electrochemical CO_2 conversion technologies. Our results demonstrate the potential for microwave-assisted DRM. The ability to use intermittent renewable energy to promote thermally-demanding catalytic processes may allow the development of new reactor configurations and reconsideration of currently impractical reactions. The results were published in *Applied Catalysis B*, **2021**, *284*, article: 119711

623574 Electrochemical CO_2 Conversion to Formic Acid: Optimization of Production & Efficiency Via Operating Voltage Tuning and pH Regulation

Abstract:

Electrochemical utilization of CO_2 has been a hot topic to target high-value, cost-effective products. Conventional processes for CO_2 conversion to fuel sources, such as the Kemira processes or hydrogenation, are challenging and expensive due to their requirement for high temperatures and pressures. Electrochemically, a catalyst can be used at a fixed applied potential to reduce CO_2 to form a desired product with less competition at modest operating conditions. Formic acid (FA) has been a product of interest due to its niche market in livestock feeds as well as for hydrogen storage. Furthermore, FA has the added benefits of low thermodynamic input and a favorable atom economy. Electrochemically, FA is produced through a direct two-electron transfer process involving CO_2 with a proton source, requiring less energy input and fewer reaction steps than traditional processes. However, there remain obstacles for the broader-scale implementation of CO_2 on the market, including competition with unwanted CO and H_2, the presence of molecular oxygen, and stability of the electrocatalyst . Additionally, electrode efficiencies with conventional catalysts for FA have rarely been more than 50%.

The unique assets of the UK CAER electrochemical CO_2 reduction process include an organic charge carrier to shuttle electrons to the catalyst for more efficient FA production, and novel electrode materials to improve efficiency and stability. The flow system comprises two separate cells to decouple electrochemical reduction of the charge carrier with the FA production via an engineered catalyst. The design and material considerations iterate from previous studies based on appropriate choice of electrode material and electrolyte concentrations.

Recent results show that with appropriate design changes and operating conditions to the flow system, FA production above 25 mM can be achieved along with efficiencies above 50%. Peak FA production is found to occur at an applied voltage of approximately -0.85 V vs. Ag/AgCl reference electrode, which aligns with the expected voltage for peak charge carrier production. Due to the mixing with the bulk vessel solution in the flow cell and transit time delays, a gradient is present which results in a 20% higher measured FA production at the immediate production cell exit, relative to within the mixing vessel. This will indicate a further increase in extracted FA concentration when it is siphoned out from the production cell during system scale-up. One major finding from this study is that there is a tradeoff between faster kinetics to produce FA and system efficiency with the engineered catalyst. System design, selection of operating conditions and their inter-dependence will be discussed.

1. Spurgeon JM, Kumar B. A comparative technoeconomic analysis of pathways for commercial electrochemical CO 2 reduction to liquid products. Energy & Environmental Science. 2018;11(6):1536-51.
2. Kortlever R, Balemans C, Kwon Y, Koper MT. Electrochemical CO2 reduction to formic acid on a Pd-based formic acid oxidation catalyst. Catalysis Today. 2015 Apr 15;244:58-62.
3. Omosebi, A., Landon, J., Thompson, J., and Liu, K. "Process Intensification for Electrochemical Utilization of CO2." Proceedings of AIChE 2019 Annual Meeting. 11 Nov 2019.

4. Moreno D, Omosebi A, Abad K, Thompson J, Liu K. Electrochemical CO2 Utilization: Scalable System Operation for Formic Acid Production. University of Kentucky Center for Applied Energy Research; 2020 Nov 17.
5. Moreno, D., Omosebi, A., Abad, K., Jeon, B.W., Landon, J., Liu, K., Kim, Y. H., and Thompson, J. "Electrochemical utilization of CO2 from coal power plants" 15th International Conference on Greenhouse Gas Control Technologies, GHGT-15, 1-18 Mar 2021. Submitted 15 Feb. 2021.

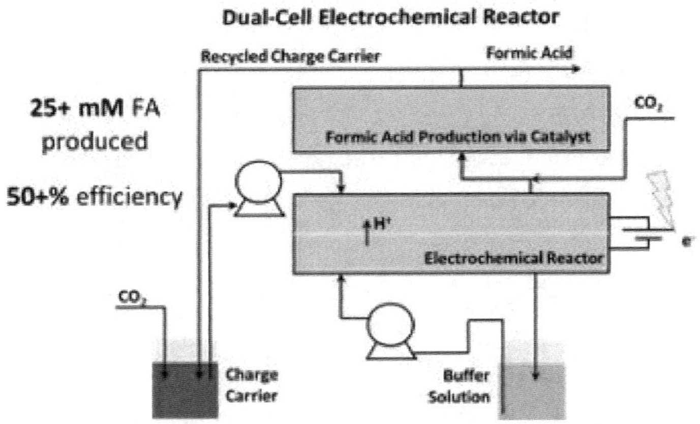

623134 Selective Conversion of CO_2 to CO at Low Temperatures Using Cu and Ni Catalysts Prepared By Combustion Synthesis Technique

Abstract:

Utilization of fossil fuels accounts for a large portion of the total energy required resulting in an exponential growth in CO_2 emission into the atmosphere. In this regard, the Carbon Capture and Utilization (CCU) strategy, a catalytic conversion of CO_2 into CO through reverse water gas shift (RWGS) reaction can be considered an attractive approach to CO_2 valorization. Herein, we investigate Cu and Ni-based catalysts prepared by the combustion synthesis method for RWGS reaction. The CO_2 thermal hydrogenation reaction was carried out successfully using a tubular flow reactor in the temperature range of 100 to 600 °C with a CO_2/H_2 ratio of 1 to 4. The catalysts showed ~70% conversion at 600 °C stable for 17h [1,2]. The copper catalysts show exceptional selectivity for CO, even at low temperature, whereas addition of nickel results in better activity with a slight trade off in CO selectivity to CH_4. We also employed electrocatalytic routes to convert CO_2 on Cu, Ni electrocatalysts to understand the reaction pathway [3-6]. The catalysts (both fresh and used ones) were characterized by different techniques such as: XRD, TPR and SEM/EDX, TEM, and Raman; to study the morphological properties as well as the impact of reaction exposure on the catalysts.

References:

[1] Mohammed AAA, Saad MAHS, Kumar A, Al-Marri MJ.; Greenh Gases Sci Technol 2020;10:715–24.

[2] Kumar A, Mohammed AAA, Saad MAHS, Al-Marri MJ.; Int J Energy Res 2021:er.6586.https://doi.org/10.1002/er.6586.

[3] Nazir R, Khalfani A, Abdelfattah O, Kumar A, Saleh Saad MA, Ali S.; Langmuir 2020;36:12760-12771

[4] Nazir R, Kumar A, Ali Saleh Saad M, Ali S.; J Colloid Interface Sci 2020;578:726–37.

[5] Nazir R, Kumar A, Saleh Saad MA, Ashok A.;Colloids Surfaces A Physicochem Eng Asp 2020;598:124835.

[6] Nazir R, Kumar A, Ali S, Saad MAS, Al-Marri MJ.; Catalysts 2019;9:860

620938 Tandem Carbon Dioxide Electrolysis Process for Multicarbon Chemical Production

Abstract:

Electrochemical conversion of carbon dioxide (CO_2) using renewable electricity is an attractive approach for sustainable production of fuels and chemicals. Copper (Cu) has a unique capability of catalyzing carbon-carbon (C-C) bond formation to form high-value multi-carbon (C_{2+}) products. While highly alkaline electrolytes are often used to enhance C_{2+} selectivity, the inevitable reaction of hydroxide ions with CO_2 to form undesired carbonates at the electrode-electrolyte interface disrupts the electrolysis process. This fundamental challenge can be solved by decoupling the CO_2 electrolysis into a two-step process, where CO_2 is first electrochemically reduced to carbon monoxide (CO) at neutral conditions, followed by CO electroreduction to produce C_{2+} chemicals in alkaline environments. The feasibility of two-step tandem CO_2 electrolysis has been demonstrated experimentally by Romero Cuellar and co-workers [1], who reported a total CO_2 reduction current density of 300 mA/cm^2 with a cumulative Faradaic efficiency of 62% towards C_{2+} products. However, the system is far from optimized when compared to standalone CO_2 and CO electrolysis technologies. In this talk, we will present the latest development of the tandem CO_2 electrolysis process for multicarbon chemical production, including the electrocatalyst design, reactor engineering, process optimization, and techno-economic analysis. Additionally, we will discuss the future opportunities for carbon dioxide utilization through tandem and hybrid processes [2].

Reference:

[1] Romero Cuellar, N.S., Scherer, C., Kaçkar, B., Eisenreich, W., Huber, C., Wiesner-Fleischer, K., Fleischer, M., and Hinrichsen, O. Two-step electrochemical reduction of CO_2 towards multi-carbon products at high current densities. *J. CO_2 Util.* 2020; **36**: 263-275.

[2] Overa, S., Feric, T. G., Park, A. H. A.* & Jiao, F.* Tandem and Hybrid Processes for Carbon Dioxide Utilization. Joule 5, 8-13 (2021). doi: 10.1016/j.joule.2020.12.004

624710 Techno-Economic Evaluation of H_2 Production Using Chemical Looping Dry Reforming

Abstract:

Chemical looping dry reforming (CLDR) is a novel process for production of inherently separated syngas (H_2 and CO) streams. In CLDR, CH_4 is first converted to a pure hydrogen stream and solid carbon via thermocatalytic cracking over a metal catalyst in a "cracker" reactor. The deposited carbon is then selectively removed in an oxidizer reactor using CO_2 as oxidant, resulting in the production of high-purity CO and regenerating the metal which is recycled back to the cracker reactor. The metal thus acts as a "carbon carrier" as opposed to an "oxygen carrier" as in typical chemical looping processes. Overall, CLDR results in the formation of inherently separated syngas streams, while at the same time addressing some of the key shortcomings of conventional methane cracking, such as low CH_4 conversion, rapid catalyst deactivation, and large temperature excursions during carbon burn-off in the oxidation half cycle.

In the present study, we perform a techno-economic evaluation of a CLDR process that is designed to maximize H_2 production by combining CLDR with water gas shift (WGS) in order to utilize the produced CO for further H_2 production. Different configurations are evaluated for supplying the heat required for the endothermic dry reforming reaction via combustion of additional CH_4 or combustion of a fraction of the carbon produced in the cracker reactor using either pure oxygen or air. All configurations include CO_2 capture and H_2 purification via pressure swing adsorption. Building on our previous fixed-bed experimental results, reactor-level mass and energy balance calculations are performed and coupled with a systems-level model that incorporates thermodynamic calculations of individual process components such as water gas shift reactors, blowers, feed compressors, CO_2 capture, PSA, and heat recovery equipment as well as steam turbines. Capital cost models for the major equipment and operating costs (catalysts, chemicals, fuel etc.) are also developed using data in the open literature. The results from the performance model are combined with the cost models to comparatively evaluate the three configurations based on performance and cost metrics such as energy efficiency, chemical efficiency, CO_2 emissions, capital costs, and cost of H_2 production. The results are compared with conventional H_2 production pathways using steam reforming and dry reforming of methane. The results are used to identify areas where CLDR systems have technical and cost advantages over conventional processes and highlight factors which need improvement.

631031 Chemical Looping CO_2 splitting and Methane Reforming for Low-Carbon-Footprint Chemical Synthesis

Abstract:

Conversion of CO_2 to CO for chemical production is one of the most promising routs for widescale carbon utilization. However, traditional CO_2 to CO technologies pose challenges. Dry reforming of methane (DRM) which co-produces hydrogen is one of the most widely studied but has several drawbacks, including high energy demand, high temperatures (>800 °C), and coke formation. The 1:1 H_2:CO syngas product of dry reforming is useful for several chemicals, but few commercial processes can directly convert a 1:1 syngas without costly syngas separation, and/or water-gas shift (WGS). Due to the challenges of DRM, industrial syngas producers typically use methane partial-oxidation/autothermal reforming, followed by energy intensive separation to obtain high concentration CO feedstocks for synthesis (which offers no CO_2 utilization benefit). We present chemical looping catalyst for a hybrid redox process (HRP) that produces separate CO and syngas streams. The syngas is produced by partial oxidation of methane by the lattice oxygen of the HRP catalyst. In a subsequent step the lattice oxygen is regenerated by CO_2 splitting. Through choosing mixed metal oxides with catalytically active surfaces and/or modifying the particle surfaces with catalytically active metals, HRP offers high conversions in both reaction steps. Compared to dry reforming, HRP permits the efficient utilization of heat directly produces concentrated CO and 2:1 H_2:CO streams without parasitic separations/WGS reactions.

We present experimental results that show over 95 % CO yield in the CO_2-splitting step and >95% selectivity towards syngas with a near 2:1 H2:CO in the partial oxidation step. The activity and selectivity of the redox catalysts during both the reduction and oxidation steps were tested. The performance of both platinum group metal (PGM)-containing and PGM-free redox catalyst are presented. PGM-free catalyst are shown to maintain excellent activity and stability in HRP operation, greatly enhancing the economic feasibility of the approach. Stable performance is observed over 500 cycles/125 hours with the phases intact. Simulations based on cxperimental HRP results are briefly presented indicating large cost and net CO_2 savings for production. HRP is an ideal choice for sustainable CO_2 fixation due to the wide industrial use for CO in chemical production.

628906 Experimental Evaluation of a Solar Carbonation-Calcination Reactor Under Simulated High-Flux Solar Irradiation

Abstract:

Two-step calcium oxide carbonation–calcination chemical looping is studied for carbon dioxide (CO_2) capture and high-temperature thermochemical energy storage applications [1]. The process involves solar-driven endothermic calcination reaction of calcium carbonate ($CaCO_3$) and non-solar exothermic carbonation reaction of calcium oxide (CaO) with CO_2:

Calcination (endothermic): $CaCO_3 \rightarrow CaO + CO_2$

Carbonation (exothermic): $CaO + CO_2 \rightarrow CaCO_3$

Previous pertinent studies of the process include investigations of reaction chemistry, transport phenomena, solar reactor design, and process integration [1–4].

In this work, we experimentally evaluate the thermochemical performance of a sorbent material and a solar thermochemical reactor. The reaction kinetics are investigated using non-isothermal thermogravimetric analysis (TGA) with raw $CaCO_3$ pellets of size in the range of 2–5 mm. Calcination and carbonation reaction rates and extents are measured as a function of temperature, CO_2 flow rate and sample heating rate. Under a heating rate of 10 K min^{-1} and the atmosphere of 100 mL min^{-1} CO_2 and 20 mL min^{-1} Ar, the calcination is found to occur at about 910°C and is completed in about 18 min. For the same conditions, the carbonation occurs at about 860°C and its extent reaches 47.3% after 34 min.

The solar reactor experimental setup consists of a high-flux solar simulator as a source of concentrated solar radiation [5], a 1 kW dual-cavity solar-driven packed-bed reactor [4], and auxiliary equipment. The reactor consists of two concentric cylindrical cavities, an irradiated inner cavity surrounded by a packed bed of reacting particles, as previously described in [4]. The reactor has an easy-to-assemble and modular design, allowing for future upgrades and studies of other processes. In a typical experiment, the inlet gas is preheated to an intermediate temperature for simulating heat recuperation, and the product gas composition is monitored using a mass spectrometer. The composition and internal morphology of the sorbent particles are analyzed using XRD and SEM techniques before and after the reactor experiments. The effects of calcination and carbonation temperatures, reactor heating rate, inlet gas temperature, CO_2 flow rate, and duration of the calcination and carbonation steps on the reactor solar-to-chemical conversion efficiency are investigated. The experimental results will be used to validate a transient heat transfer model of the reactor and to optimize the solar reactor and process design.

References:

[1] Reich, L., Yue, L., Bader, R. and Lipiński, W., 2014. Towards solar thermochemical carbon dioxide capture via calcium oxide looping: A review. *Aerosol and Air Quality Research*, 14(2), pp.500–514.

[2] Yue, L., Reich, L., Simon, T., Bader, R. and Lipiński, W., 2017. Progress in thermal transport modelling of carbonate-based reacting systems. *International Journal of Numerical Methods for Heat*, 27(5), pp.1098–1107.

[3] Bayon, A., Bader, R., Jafarian, M., Fedunik-Hofman, L., Sun, Y., Hinkley, J., Miller, S. and Lipiński, W., 2018. Techno-economic assessment of solid–gas thermochemical energy storage systems for solar thermal power applications. *Energy*, 149, pp.473–484.

[4] Reich, L., Melmoth, L., Yue, L., Bader, R., Gresham, R., Simon, T. and Lipiński, W., 2017. A solar reactor design for research on calcium oxide-based carbon dioxide capture. *Journal of Solar Energy Engineering*, 139(5).

[5] Bader, R., Haussener, S. and Lipiński, W., 2015. Optical design of multisource high-flux solar simulators. *Journal of Solar Energy Engineering*, 137(2).

628142 Ultra Accelerated Aging and Characterization of Thin-Film Materials with Concentrating Light.

Abstract:

The materials' sectors have experienced the most advancement in years to provide solutions to our societies' ever-increasing demands. Therefore, significant engineering tests for products and materials have been developed, like thermal-stability and predicted lifetime as a function of exposure to temperature, humidity, and UV radiation. However, there is still a deficiency in the understanding and testing when it comes to the degradation effects driven by the combined effect of solar light intensity and exposure duration. To cover this gap, a small number of accelerated aging methods based on concentrating natural or artificial light have been developed with the mature ones to have reached the standardization level and the rest still in low readiness levels. The former methods utilize concentration ratios of a few tenths of suns and focus on paints, plastics and other commercial materials for outdoor applications. The latter methods focus on solar energy-absorbing materials with concentration ratios that might exceed 1,500 suns. Our group has produced published work in both directions.

Noteworthy, all the above methods require specialized equipment, weeks or months of testing, and a large number of samples to produce reliable results. Additionally, many methods are based on indirect degradation mechanisms by inducing very large variations of temperature and humidity. However, they neglect the impact of solar radiation and that most materials and products will operate between low temperature and humidity ranges. Moreover, most of them provide a single degradation data point (often just a pass or fail) per sample and exposure test, which implies using many samples to extract reliable and long-term degradation predictions. A few of the materials that depend on such accelerated aging methods are paints, plastics, coatings, window films, and solar cells with very large market sizes and predicted growths. For example, the global paints and coatings market size was around USD 150 billion (2019), around USD 570 billion for plastics, and USD 10 billion for window films; all three of these markets are expected to grow by ~4-5% per year from 2020-2027.

To address the above technical limitations, we developed a method which combines in one apparatus concentrating artificial light with continuous optical characterization. The apparatus can be used to investigate the light-induced degradation of (research and commercial-grade materials) by conducting accelerated aging tests efficiently under fully controlled conditions (acceleration ratio, temperature, humidity, pressure, and atmosphere composition). In parallel, it continuously collects and analyzes the light transmitted through the sample and its temperature (optionally the evolved gases). The transmitted light provides online insights as simple as the material's aggregated light properties and as complex as its composition. Until now, successful aging tests, using our High Flux Solar Simulator and several lab-components, on new and standard materials proved the concept and the above claims. Here we present our results from the first two campaigns with materials used in organic photovoltaics (OPV) and alkyd paints.

Poly-3-hexylthiophene (P3HT) is one of the many polymers in the field of OPV and has the largest set of degradation data available in the literature; thus, one of the model systems for degradation benchmarking. Combination of P3HT with phenyl-C61-Butyric-Acid-Methyl Ester (PCBM) in a bulk heterojunction OPV cell (P3HT:PCBM) has dominated the research for a decade. Qualitatively, two degradation mechanisms were observed on the continuous UV-Vis spectrograms for both materials when exposed to various intensities of concentrated light. For P3HT a chemical transformation captured as the blue-shift at ~554nm and the overall decomposition of the material captured with the gradual decrease of the absorbance. These results agree with the observations in literature suggesting that both mechanisms could be attributed to the photo-oxidation of P3HT. Indeed, conducting the experiments in an inert Argon atmosphere showed a similar behavior but with significantly lower rate. In comparison with P3HT, P3HT:PCBM proved significantly more stable when exposed in the same conditions; as expected.

Coating materials such as paints and varnishes are mixtures of several (often numerous) components (binding media, pigments, solvents and fillers). Since binding media are not simple mixtures but rather complex polymers, degradation mechanisms are more complicated, because the initial components have reacted to form new compounds (polymers). Past studies on the photo degradation kinetics of Alkyd paints were able to simulate a year of sunlight exposure in 1,008 hours with the use of a commercial light aging device. The estimated degradations were estimated with the use of FTIR-ATR at around 48% to 65%. On the other hand, we were able to simulate the same degradation (~50% also checked with FTIR-ATR) in just 0.5 hours i.e. circa 2,000 times reduction in the experimental time. In addition, the degradation mechanisms were monitored continuously using the transmitted light which yielded a similar overall degradation (55% - 60%).

632029 Preliminary Reactor Design for High Temperature Discharge of Solid-State Thermochemical Energy Storage Particles

Abstract:

The increased interest in harvesting and storing solar energy has led to the development of multiple Thermochemical Energy Storage (TCES) materials that enable the accumulation of solar energy in the form of chemical energy. In addition to the TCES materials, suitable reactors for the charging and discharging processes must be developed. The use of Concentrating Solar Power (CSP) enables the use of heat transfer fluids (HTF) at high temperatures, which then can be used to run thermal power plants or can also be used to charge TCES materials by means of endothermic reduction reactions, allowing the conversion and storage of thermal energy into chemical energy. Moreover, the discharge of TCES can be carried out as high temperature exothermic oxidation reactions, returning the stored chemical energy in the form of thermal energy. Higher temperature discharge allows for higher power generation efficiency, at the expense of more difficult process conditions. Presented here is a preliminary concept development of a reactor for high temperature (≥900 °C) oxidation of solid particles used for thermochemical storage of solar energy. An analytical hierarchy process was performed to ponder the main design criteria, such as flowability, heat transfer, fabricability and reliability, among others. Reactor concepts such as packed bed, moving bed, fluidized bed, and combinations thereof were then evaluated based on the criteria. The 'hybrid' design presented here consists of a top moving bed, a middle fluidized bed with heat exchanger and a bottom moving bed. The top and bottom moving beds will serve as heat recuperation devices, and the middle fluidized bed with heat exchanger will be used to carry out the oxidation reaction and to transfer the heat of reaction to the HTF. The results indicate that the hybrid concept surpasses its contenders in flowability and heat transfer parameters, but it had low scores in fabricability, control and operation aspects. Additionally, a preliminary design of a hybrid reactor for the oxidation of relatively large TCES particles (1 mm or greater) is presented with a focus on enhancing the heat transfer between the particles and the walls, as well as improving the solids flow within the reactor for better residence time distributions. A heat output of 1 kW is used as basis to design the oxidation reactor and heat exchanger and moving beds. The preliminary modelling results indicate that pressure has a significant effect on heat transfer and fluidization. Also, under specific conditions, particle preheating is feasible in the top and bottom moving beds, and a working fluid outlet temperature of 800 °C or greater is achievable.

632874 Thermal Energy Storage Using Sorption-Assisted Boudouard Processes

Abstract:

This study presents current results on the development of a long term storage of solar heat via chemical conversion of simple carbon-based chemicals using Sorption-Assisted Boudouard Processes [1-3].

$2CO + MO \leftrightarrow MCO_3 + C$ $\Delta H = -350$ kJ/mol

where M represents a Group 2A element.

This approach enhances and simplifies our prior design [4] by enabling operation at higher temperatures, and bypassing the need to store CO_2 in addition to CO. Because energy storage occurs at ambient temperature, this approach allows for very long-term storage of the heat without degradation, enabling heat storage at time scales from intra-day to seasonal displacement (i.e., indefinitely) as desired. The system is being examined to enable continuous solar thermal energy storage, as well as for the development of solar thermal industrial process heat.

Current results reflecting system performance, modulating charge and discharge temperatures, and exergetic efficiency will be presented.

References:

[1] Boudouard, O. Influence De La Vapeur D'eau Sur La Réduction De L'anhydride Carbonique Par Le Charbon *C. R. Hebd. Acad. Sci.* **1905**, *141*, 252– 253.

[2] Hunt, J.; Ferrari, A.; Lita, A.; Crosswhite, M.; Ashley, B.; Stiegman, A.E. Microwave-Specific Enhancement of the Carbon–Carbon Dioxide (Boudouard) Reaction *J. Phys. Chem. C* **2013**, *117*, 26871–26880.

[3] Rout, K.R.; Gilband, M. V.; Chen, D. Highly Selective CO Removal by Sorption Enhanced Boudouard Reaction for Hydrogen Production *Catal. Sci. Technol.* **2019**, *9*, 4100–4107.

[4] Miao, Y.; Yokochi, A; AuYeung, N.; von Jouanne, A Thermal Energy Storage Using the Boudouard Reaction *2020 Annual Meeting of the American Institute of Chemical Engineers*, **Virtual, Nov 16-20 2020**, Paper 559f.

624168 Concentrated Solar–Plasma Chemical Processing: Reactors Design and Implementations for CO_2 Decomposition

Abstract:

The production of chemicals from low-value feedstock using renewable energy by environmentally benign methods is prone to have an increasingly important role towards achieving sustainable development. Particularly, the conversion of carbon dioxide (CO_2) using renewable energy can fulfill the growing need for fuels and chemicals, leading to economic growth while mitigating environmental emissions. Solar thermochemical processes, given their direct use of solar irradiation, have among the greatest sustainability potential; nonetheless, the inherent intermittency of the solar resource prevents them from continuous operation, which limits their viability. The combination of solar thermochemical with non-solar approaches powered by renewable electricity can lead to continuous-operation sustainable processes with improved viability. As an example of a solar – non-solar chemical synthesis process, the integrated use of solar thermochemical and plasmachemical approaches applied to the decomposition of CO_2 is presented.

Plasmachemical processes utilize electricity to power an electrical discharge to drive feedstock gas to a plasma (i.e. partially ionized gas) state. Solar-plasma chemical processing is motivated by the potential for synergistic interactions among solar photons and plasma species (free electrons, ions, excited molecules and atoms) to lead to enhanced process performance. Computational studies [1] have shown that CO_2 in non-local thermodynamic equilibrium state (i.e. significantly greater equivalent temperature of free electrons, typically > 1 eV ~ 11600 K, compared to the gas species temperature, typically of a few hundred ^oC) absorbs drastically more solar radiation than CO_2 in thermodynamic equilibrium, which can potentially lead to greater CO_2 decomposition.

The design of reactors for solar-plasma chemical conversion present distinct challenges, such as ensuring appropriate interaction between incident solar radiation and plasma, maintaining the integrity of the optical aperture in the proximity of plasma, and integrating catalytic monoliths that exploit solar photon- and plasma-driven reactivity. Sola-plasma reactor design rationale and representative design concepts are reviewed and discussed.

Solar-plasma chemical conversion processes can be categorized based on the ratio of solar input power (P_s) to electrical input power (P_e). If $P_s > P_e$, then the process is denoted plasma-enhanced solar thermochemical (PES); whereas if $P_s < P_e$, then it is denoted solar-enhanced plasmachemical (SEP). A solar-gliding arc plasma reactor is implemented for a representative PES process [2, 3] and a solar-microwave plasma reactor for a SEP process [4, 5]. Both reactors are evaluated for the decomposition of CO_2 at (near) atmospheric pressure conditions using up to 525 W of radiative energy flux from a high-flux solar simulator. The solar-gliding arc reactor is powered by up to 240 W from two high-voltage AC power supplies and the solar-microwave plasma reactor by up to 900 W of net power from a 2.45 GHz magnetron. The studies did not include the incorporation of catalytic monoliths in order to discern the level of solar- or plasma-enhancement. CO_2 conversion and energy efficiency as function of solar and electrical input power, flow rate, and gas composition (100% CO_2 and CO_2-N_2/Ar/H_2O/CH_4 mixtures) are presented. Process characterization included optical and high-speed imaging, electrical diagnostics, optical emission spectroscopy, and gas chromatography of products. The results show that greater CO_2 conversion and energy efficiency are obtained by the combined use of concentrated solar and plasma as compared to plasma or solar energy alone. Current efforts encompass the incorporation of different catalytic monoliths into the reactors and detailed computational multiphysics analyses of the solar-plasma CO_2 conversion processes.

Acknowledgements: The author gratefully acknowledges the work of Dr. Sina Mohsenian, Dr. Dassou

Nagassou, Rasool Elahi, Benard Tabu, and Visal Veng, from which this work is based. This work has been supported by the U.S. National Science Foundation through award CBET-1552037.

References:

[1] R. Elahi, D. Nagassou, S. Mohsenian, J. P. Trelles (2020). Enhanced Solar Radiation Absorption by Carbon Dioxide in Thermodynamic Nonequilibrium: A Computational Study, *Solar Energy*, 195, 369-381.

[2] Nagassou, D., Mohsenian, S., Bhatta, S., Elahi, R., & Trelles, J. P. (2019). Solar–gliding arc plasma reactor for carbon dioxide decomposition: Design and characterization. *Solar Energy*, 180, 678-689.

[3] Nagassou, D., Mohsenian, S., Nallar, M., Yu, P., Wong, H. W., & Trelles, J. P. (2020). Decomposition of CO2 in a solar-gliding arc plasma reactor: Effects of water, nitrogen, methane, and process optimization. *Journal of CO2 Utilization*, 38, 39-48.

[4] Mohsenian, S., Nagassou, D., Bhatta, S., Elahi, R., & Trelles J. P. (2019). Design and Characterization of a Solar-Enhanced Microwave Plasma Reactor for Atmospheric Pressure Carbon Dioxide Decomposition. *Plasma Sources Science and Technology*, 28, 065001.

[5] Mohsenian, S., Nagassou, D., Elahi, R., Yu, P., Nallar, M., Wong, H.-W., & Trelles, J. P. (2019). Carbon Dioxide Conversion by Solar-Enhanced Microwave Plasma: Effect of Specific Power and Argon/Nitrogen Carrier Gases, *Journal of CO2 Utilization*, 34, 725-732.

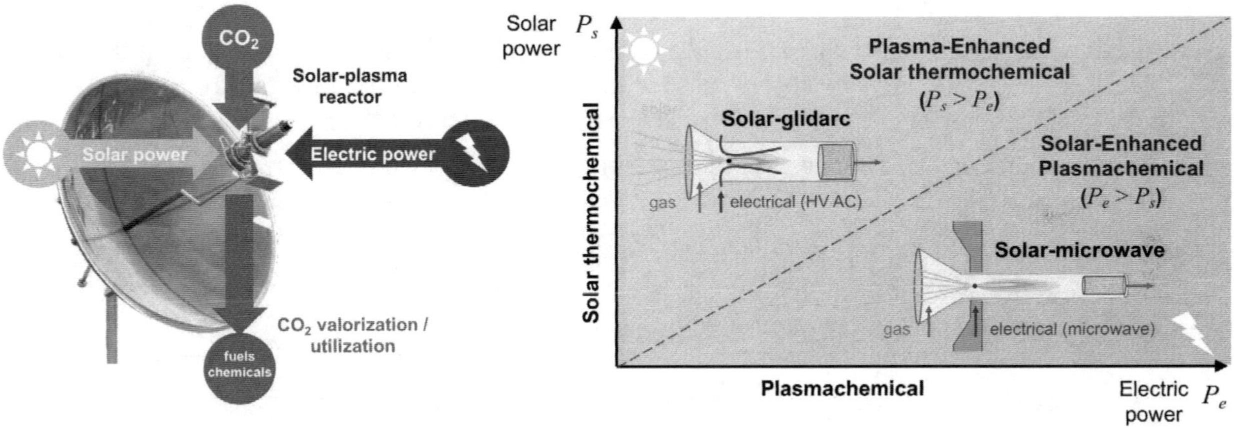

627463 North Dakota Carbonsafe – Early Successes

Abstract:

As part of the U.S. Department of Energy's (DOE's) CarbonSAFE program, the Energy & Environmental Research Center (EERC) is conducting commercial-scale site characterization and permitting of up to two geologic horizons for the geologic storage of anthropogenic CO_2 emissions. These storage horizons would collectively and permanently store nearly 4 million metric tons (MMt) of CO_2 per year captured as part of Minnkota Power Cooperative's (Minnkota's) Project Tundra. Minnkota's ambitious initiative to build the world's largest postcombustion carbon capture and storage facility in central North Dakota demonstrates the existence of a business case for carbon capture, and storage (CCS). The EERC and project team is building upon the success, learnings, and momentum of the North Dakota CarbonSAFE Phase II feasibility study. Minnkota's primary generating resource is the two-unit Milton R. Young Station (MRYS), a minemouth lignite coal-fired power plant. The mine which provides the lignite coal for MRYS is owned and operated by BNI Coal, Inc. (BNI) and is adjacent to the MRYS facility. The lignite used as the fuel for electrical generation also serves as the primary source of the captured CO_2 that will be securely stored by Tundra SGS. The storage project together with the carbon capture project are commonly referred to as Project Tundra.

The proposed Tundra SGS injection site in Oliver County, North Dakota, will include up to three injection wells, one dedicated monitoring well for the lowest USDW (underground source of drinking water), and associated surface facility infrastructure that will accept CO_2 transported via a CO_2 flowline. In addition, one deep subsurface monitoring well is proposed to be installed. Two wells are being evaluated for the injection of CO_2 into the Broom Creek Formation with the option of either a third Broom Creek well or a well for injection of CO_2 into the Black Island–Deadwood Formation. Project partners anticipate submitting applications and supporting documents for CO_2 storage facility permits to the North Dakota Department of Mineral Resources in 2021.

To acquire the necessary near-surface and subsurface information to fully characterize the site and meet the needs of the storage facility permits, two stratigraphic test wells were recently drilled, cored (>1300 feet of core), and logged. One of the wells was drilled outside of the CarbonSAFE project. In addition, approximately 21 miles of 2D seismic data and 12 square miles of 3D seismic data were acquired in the project area near MRYS and integrated into 6 square miles of existing 3D seismic data and the geophysical logs collected from the stratigraphic test wells.

Geophysical well logs were used to pick formation top depths and interpret lithology and petrophysical properties. The site-specific data collected were used to validate the presence of upper and lower seals and inform the construction of geologic models, reservoir simulations, geochemical modeling, and geomechanical analysis, with the goal of demonstrating the suitability of the formations for CO_2 storage and to complete the UIC (underground injection control) Class VI permitting assessments. Geologic modeling efforts incorporated the characterization data, and the resulting model was used in reservoir simulation to delineate the extent of the necessary pore space to store the CO_2 and support a risk-based area of review. Both large domain and small domain versions of these geomodels were created to accurately simulate small-scale CO_2 plume location and migration as well as large-scale pressure plume effects in each potential storage formation.

The EERC and the project team leveraged North Dakota's Class VI Primacy status to prepare and submit applications for two storage facility permits and associated monitoring well and UIC Class VI injection wells. As such, the applications and their supporting documents were prepared in accordance with the North Dakota Century Code and the North Dakota Administrative Code and demonstrate that 1) the proposed Tundra SGS site comprises injection zones of sufficient areal extent, thickness, porosity, and permeability to receive up to 80 MMt of CO_2 over 20 years and

2) the confining and secondary confining zones are free of faults and transmissive fractures.

625628 Effects of Carbonated Brine-Rock Interactions on Multiphase Flow Properties in Heterogeneous Sandstone

Abstract:

With the injection of CO_2 into a deep saline aquifer, CO_2 can dissolve in the in-situ brine to form carbonic acid, which drives fluid-rock geochemical reactions. The impacts of these fluid-rock geochemical reactions on porosity, permeability, and multiphase flow responses in rock are important to evaluate CO_2 storage capacity in deep saline aquifers. However, such investigations on heterogeneous eolian sandstone, consisting of poorly sorted, quartz-rich sand with laminated bedding, have been rare.

In this research, carbonated brine injection experiments were performed on core samples from the Upper Minnelusa Formation at the Wyoming CarbonSAFE project site located in Northeast Wyoming. The core samples exhibit heterogeneity consisting of poorly sorted sand with laminated bedding. Complementary pre- and post-injection porosity, permeability, SEM, thin section, Mercury Intrusion Capillary Pressure (MICP), and Brunauer-Emmett-Teller (BET) surface area were measured to study the petrophysical properties changes. Overall, porosity and permeability of the core sample both increased significantly after experiencing 7-day carbonic acid injection, from 6.2% to 8.4% and 1.6mD to 3.7mD, respectively. We attributed these changes to mineral dissolution, which was confirmed by the effluent brine geochemistry, pore throat size distribution resulting from MICP, and BET surface area. To be more specific, the more permeable section of core sample owes larger pore size, the permeability increment of this section is apparent due to dolomite cement dissolution, as shown by SEM and thin section. However, within the less permeable rock section, the smaller pore size and possible mineral precipitation lessened dissolution resulting insignificant petrophysical properties changes. Consequently, the observed heterogeneous carbonated brine-rock interactions resulted in alterations of CO_2/brine relative permeability, i.e. a decrease from 0.29 to 0.2 in CO_2 saturation at irreducible water saturation, while CO_2 relative permeability increased as a whole. The work in this study provides a fundamental understanding of the effect of fluid-rock reactions on the changes in static and multiphase flow properties, which lays the foundation for accurate prediction/simulation of CO_2 injection into deep saline aquifers. This work is funded under the Department of Energy CarbonSAFE program (awards DE-FE0031624 DE0031891).

624439 Experimental and Simulation Study of CO_2 EOR and Associated Storage in a Naturally Fractured Reservoir

Abstract:

Injecting carbon dioxide (CO_2) within deep geologic formations offers a potential means to reduce greenhouse gas emissions. Currently, dedicated CO_2 storage in saline aquifers and associated CO_2 storage in oil reservoirs with enhanced oil recovery (EOR) are the two most widely reported methods for long-term CO_2 storage. The target formations are usually characterized by good porosity and permeability to ensure sufficient injectivity and storage capacity. However, recent unconventional reservoir development gives an indication that tight formations could also be possible opportunities for CO_2 EOR and storage despite relatively low matrix porosity and permeability. Abundant fractures in the reservoir provide flow paths for CO_2 migration and greater surface area for penetration into the tight rock.

A series of experimental and simulation activities were conducted in this study to investigate CO_2 EOR and the associated storage potential of the Dickinson Lodgepole Mounds (DLM) located near Dickinson, North Dakota. These mound structures have abundant natural fractures but relatively low matrix porosity (2%–6%) and permeability (0.1–10 md). Ten individual DLM fields have been discovered and developed with prolific oil production since 1993. The fields contain some of the best oil-producing wells in North Dakota history. Based on the analysis of the production and injection history in the DLM, Stadium Field was selected for an in-depth study. The field is relatively small, with five active wells.

Core samples from the field were used in laboratory investigations. Contact angle measurements showed the reservoir is strongly oil wet. Core samples were saturated with oil and then flooded with water to mimic the production mechanisms in the reservoir. Both immiscible and miscible CO_2 flooding tests were then conducted to observe the EOR response under reservoir conditions. Experimental results showed that water could only displace minor amounts of oil from the rock while CO_2 could recover as much as 44% of the remaining oil. Also, miscible CO_2 flooding was more effective than immiscible flooding for EOR.

A compositional reservoir simulation model integrating a ten-component equation-of-state was developed to predict the CO_2 EOR and associated storage performance in the field. An innovative fracture simulation method, embedded discrete fracture modeling (EDFM), was adopted to simulate natural fractures efficiently using SimTech's EDGS Suite. EDFM enables modeling of complex fracture geometries using structured grids. History-matching results showed the model could reproduce historical data satisfactorily.

Case studies were conducted to assess the CO_2 EOR and associated storage potential in both immiscible and miscible flooding modes. Simulation results indicated injectivity was sufficient because of the relatively high degree of fracturing in the reservoir. The results also suggested CO_2 flooding in Stadium Field could recover an additional 11%–14% of oil from the reservoir, with associated CO_2 storage potential of up to 2.05 million tonnes within 10 years.

Both experimental and simulation results showed that a fractured reservoir with relatively low matrix porosity and permeability could host an effective EOR operation and have the potential to store a significant quantity of CO_2. The findings from this study may have implications for CO_2 EOR and associated storage assessments in large-scale unconventional reservoirs.

632734 Increasing the Temperature and pH Range of Urease to Enable Biomineralization Sealing of Leaky Wells

Abstract:

Urease Induced Calcium Carbonate Precipitation (UICP) has been shown to seal sub-surface fractures around leaky wells (Cunningham et al., 2011; Kirkland et al., 2021, in review; Phillips et al., 2016). UICP could be useful in carbon capture and storage (CCS) to seal leaking wells. CCS sites tend to have a low pH due to the mixing of supercritical CO_2 and water and sometimes elevated reservoir temperatures. These conditions may be inhibitory to the urease enzyme that is involved in the UICP process, which could limit the application of sealing leakage pathways in CCS wells. We investigated enzyme immobilization as a strategy to increase the stability of urease at low pH and increased temperatures.

Jack Bean Meal (JBM) suspensions as the source of urease were prepared from JBM powder in deionized water (DI) and mixed for 16 hours at 22°C. To test the potential increase in enzyme stability, immobilized JBM was prepared by sorption to a porous ceramic proppant. After 24 hours, the liquid was decanted, retaining the ceramic proppant in the flask. Phosphate buffered saline solution (PBS) was added to rinse the ceramic proppant before being decanted again. This rinse procedure was repeated twice to remove residual suspended enzyme.

Four different pH buffers were prepared, three with acetic acid and one with phosphate. After preparation of these buffers, pH of JBM solutions ranged from 3.7 to 7.1. First, 5 mL of each buffer was transferred to triplicate glass vials. Second, 5 mL of 10 g/L suspended JBM solution was added. In the case of immobilized enzyme samples, solid proppant particles were added as the enzyme sources and 5 mL of sterile water were added to keep buffer concentrations identical with suspended enzyme experiments. The suspended or immobilized enzyme was exposed to the pH conditions at 22°C or 60°C. After one hour the pH was adjusted to 7, urea was added, and the glass vials were placed in a water bath at 30°C and agitated at 120 RPM. Samples were taken for urea analysis after 0, 0.5, 1, 3, 8, 24, and 48 hours.

A first order rate model was used to describe the ureolytic activity.

$$-dA/dt=kA$$

Using the integrated rate form, a plot of ln(A) versus time was made.

$$\ln(A)=\ln(A_0) - kt$$

Using linear regression, the first order rate constant, k, was found. The first order rate coefficient, k, was taken as a measure for the inactivation that occurred during the 1-hour pH and/or temperature exposure. Residual activities were normalized to the 1-hour exposure to pH 7. Suspended and immobilized JBM were exposed to pH conditions 3.7, 4.1, 4.7, and 7.1 at 22°C and 60°C.

Urease appeared to be protected from inactivation by immobilization (Figure 1). The immobilized JBM exposed to pH values of 3.7, 4.1, 4.7, and 7.1 at 22°C for one hour remained active. This is an improvement in comparison to the suspended JBM exposure, where suspended enzyme became fully inactivated at pH 3.7 (Figure 1 and 2). The immobilized JBM urease showed activity after exposures to 60°C for 1-hour and pH conditions 4.1, 4.7, and 7.1, however higher enzyme inactivation was observed in combination with high temperature. There was no remaining activity observed for the immobilized enzyme exposed to pH condition 3.7 at 60 °C for 1-hour. The immobilization allowed the enzyme to retain activity when exposed to pH 4.1 at 60°C, however less activity than after exposure to higher pH values.

Attached as Abstract.PNG

Figure 1: Urea concentrations over time during incubation at 30°C at pH 7, after 1-hour exposure of suspended and immobilized urease to various pH conditions and temperatures.

Attached as Abstract.PNG

Figure 2: First order reaction rate coefficients determined from urea hydrolysis experiments at 30°C at pH 7, after 1-hour exposure of urease to pH values and temperatures indicated on the x-axis and the legend, respectively.

After a 1-hour exposure of urease to pH 3.7, 4.1, and 4.7 at 22°C the activity of the suspended enzyme decreased by 100% (complete inactivation), 57% and -7% (negative number indicates increase), while the activity of immobilized enzyme decreased by only 13.5%, -8%, and -30%, respectively. When the pH exposures were performed at 60°C, to mimic potential conditions in the deep subsurface, the extent of irreversible enzyme inactivation increased. After a 1-hour exposure of suspended urease to pH 3.7, 4.1, and 4.7 at 60°C, the activity decreased by 100%, 100% and 72%, while the immobilized enzyme activity decreased by 100%, 98%, and 42% respectively.

The pH range, for which the enzyme retained significant activity after exposure, was narrower during 60°C exposure than during 22°C exposure. The combination of heat and low pH completely inactivated the suspended enzyme exposed to pH 3.7 and 4.1 at 60°C and decreased the activity when exposed to pH 4.1 at 22°C. Meanwhile, immobilized urease remained active after exposure to pH values as low as 3.7 at 22°C and as low as 4.1 at 60°C.

Inactivation appeared to be irreversible after the 1-hour exposure to the low pH buffers since the subsequent activity assessments were performed at pH values and temperatures ideal for urease activity (pH 7 and 30°C). The extent of inactivation was increased significantly for the suspended enzyme between pH 3.7-4.7 at room temperature and elevated temperature exposure; however immobilized enzyme was able to remain active even at pH 3.7 during exposure at 22°C.

Immobilization decreased enzyme inactivation due to, both, high temperature and low pH exposure. Immobilized urease retained ureolytic activity even after exposure to pH 3.7 at 22°C. The ability of urease to remain active after exposure to low pH conditions and elevated temperatures indicates its potential usefulness in sealing leakage pathways in geologic CO_2 sequestration scenarios.

Ureolysis can occur under low pH conditions. This study suggests that using enzyme- catalyzed UICP in CCS reservoirs may be possible even if the enzyme is exposed to high temperatures, acidity. This study demonstrates the possibility for ureolysis to occur even after exposure to harsh subsurface conditions. The exposure to pH 3.7 did not fully inactivate the urease at room temperature when immobilized; at 60°C urease was completely inactivated after 1-hour long exposure at pH 3.7 even if immobilized. The combination of temperature and low pH appears to cause faster (and thus more complete) inactivation.

References:

Cunningham, A. B., Gerlach, R., Spangler, L., Mitchell, A. C., Parks, S., & Phillips, A. (2011). Reducing the risk of well bore leakage of CO2 using engineered biomineralization barriers. *Energy Procedia, 4*, 5178-5185. doi:10.1016/j.egypro.2011.02.495

Kirkland, C., Akyel, A., Hiebert, R., McCloskey, J., Kirksey, J., Cunningham, A. B., Phillips, A. (2021, in review). Ureolysis-induced calcium carbonate precipitation (UICP) in the presence of CO2-affected brine: a field demonstration. *International Journal of Greenhouse Gas Control.*

Phillips, A. J., Cunningham, A. B., Gerlach, R., Hiebert, R., Hwang, C. C., Lomans, B. P., Spangler, L. (2016). Fracture Sealing with Microbially-Induced Calcium Carbonate Precipitation: A Field Study. *Environmental Science & Technology, 50*(7), 4111-4117. doi:10.1021/acs.est.5b05559

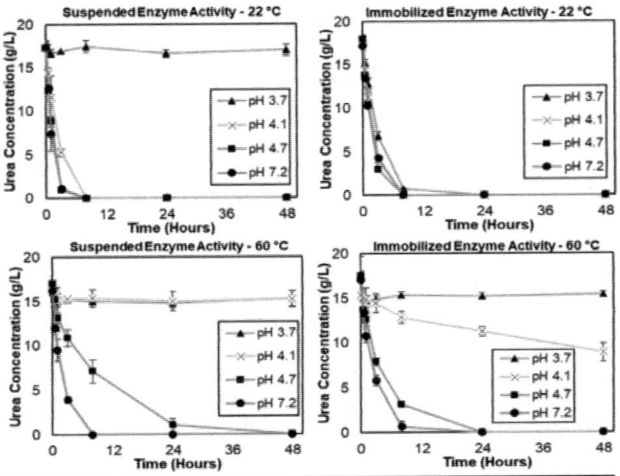

Figure 1: Urea concentrations over time during incubation at 30°C at pH 7, after 1-hour exposure of suspended and immobilized urease to various pH conditions and temperatures.

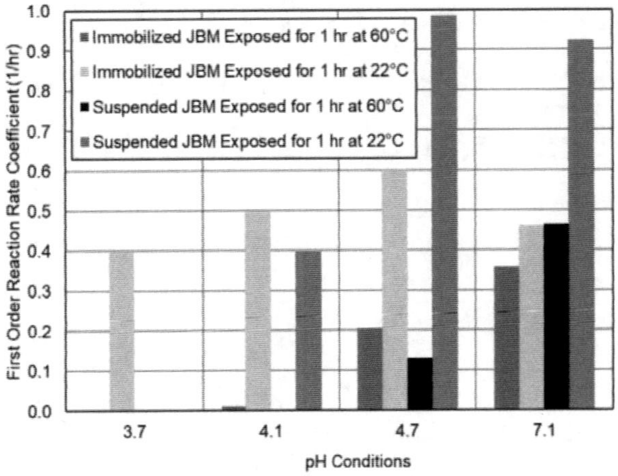

Figure 2: First order reaction rate coefficients determined from urea hydrolysis experiments at 30°C at pH 7, after 1-hour exposure of urease to pH values and temperatures indicated on the x-axis and the legend, respectively.

631835 Chemo-Mechanical Impacts on Morrow B Sandstone Reservoir and Caprock in an Active CO_2 Injection Project, Farnsworth Field Unit, Texas.

Abstract:

Background

Carbon capture, utilization, and storage (CCUS) is considered as one of the most effective technologies used to avert global warming. This technology uses CO_2 miscible and immiscible injections to enhance residual oil recovery after primary and secondary recovery phases. As such CCUS technology provides a solution to utilize and store CO_2 in underground reservoirs simultaneously. Unlike primary and secondary recovery stages, the usage of CO_2 as EOR and/or CO_2 sequestration medium introduces more complex geochemical and thermodynamical effects to the reservoirs. At the minimum miscibility pressure (MMP), the injected CO_2 dissolves in the oil phase, and this results in the alteration of fluid properties such as viscosity, density, compressibility, etc. The injected CO_2 may also undergo chemical reactions with the resident brine to form acidized brine. This acidized brine may react with the formation minerals leading to dissolution and/or precipitation of minerals that could ultimately alternate petrophysical properties such as porosity and permeability. Not only could these properties being changed by chemical reactions, but they may be changed due to stress changes during injection and production processes. More importantly, the stability of the caprock is analyzed to ensure that CO_2 is contained for a very long time without leaking. The injection of CO_2 may lead to an increase in the reservoir pore pressure. Variations in the pore pressure are always accompanied by changes in stress and strain; thus, the need to determine whether the mechanical damage is irreversible.

Objective

The main objective of this work is to investigate the chemo-mechanical impacts of CO_2 injection to a depleted oil reservoir and its caprock. A matured oil reservoir, Morrow B, from Farnsworth Unit, Texas will be studied through CO_2-water-alternate injection process.

Motivation

Even though current research works have brought on tremendous knowledge and insights, especially in the area of CO_2 storage, but there are still some aspects of numerical simulation of CO_2 sequestration operation that need further investigation, and that have served as the motivation for this work, and the novelties of this work are:

Novelty

1. CCUS operations involve three disciplines: hydrodynamics, geochemistry, and geomechanics. However, most of the published works have only coupled hydrodynamics with geochemistry or geomechanics, without considering the impacts of geochemistry and geomechanics concurrently. Therefore, this paper seeks to investigate the impacts of coupling all three disciplines.

2. To our best knowledge, most published simulation works lacked field data but used hypothetical cases to investigate the fundamental processes of CO_2 injection. In using a history-matched model, the simulation results could possess more confidence than an absolute hypothetical case.
3. While most CO_2 storage studies have concentrated on injection in deep saline aquifers, this work investigates the process in depleted oil reservoirs. It will expand the existing knowledge of CO_2 sequestration with comprehensive chemo-mechanical effects in depleted or matured oil reservoirs.

Methodology

The simulation model was constructed by an advanced multi-mechanism simulator that considered petrophysical and geological characterizations of the field. Laboratory experiments that investigated five hydraulic flow units (HFU) from the field were employed to populate the permeability and the relative permeability curves within the simulation model. The volume of CO_2 dissolved in the aqueous phase was modeled with Henry's law under reservoir pressure, temperature, and salinity. The solubility of CO_2 in the oleic phase was governed by the CO_2 injection above the MMP. The relative gas hysteresis effect was employed to estimate the amount of residual CO_2 trapped in the porous media displacing the in-situ fluids. An Elemental Analysis (ELAN) was performed to determine the initial formation mineral compositions, and water samples collected from the field were analyzed to determine the initial formation brine ion composition. Further, the rock mechanical properties were acquired from core analysis and well logs. A numerical simulation model was then constructed with the afore mentioned input data and used to determine the amounts of CO_2 trapped by structural, residual, solubility, and mineral trapping mechanisms. The chemo-mechanical impacts on rock properties such as permeability and porosity were then investigated.

To couple the hydrodynamic and geomechanics, a two-way simulation approach was used. In this approach, the geomechanics model estimates stress and strain distributions and displacement using the pressure perturbations output from the hydrodynamic model within a prescribed time interval. The hydrodynamic model then updates the petrophysical properties using the output of the geomechanics model and estimates the pressure and saturation responses for the next time step interval. The numerical reservoir model was validated by a rigorous history matching process using 10-year of ternary recovery data. Four models were constructed from the history matched model: hydrodynamic-only model, coupled hydro-geochemical (HGC) model, coupled hydro-geomechanical (HGM) model, and coupled hydro-geochemical-mechanical (HGCM) model. These models were then used to forecast production for the next 20 years. Finally, the model ran for 1000 years post-injection period to monitor the CO_2 plume migration and its effects on the reservoir and to establish the long-term storage capacity of the field. The caprock integrity evaluation was based on multiple scenarios. The coupled hydro-geomechanical model simulation results with the current field operation strategies, that include production, and CO_2-water-alternate injection strategies, were analyzed for effects of pore pressure on group uplift, deformation, and CO_2 leakage into the overburden caprock. Other cases to be considered are to convert all CO_2-water-alternate wells into CO_2 injection-only wells, injecting at variable rates, and bottomhole pressure constraints without production. This will help ascertain the maximum allowable operating conditions in order to store CO_2 safely for a long time post-EOR operations.

Results

The preliminary results of this work showed that the geochemical reactions of CO_2 with the formation minerals did not have significant effects on porosity and permeability. The maximum change in porosity is 0.05%. However, after accounting for stresses within the reservoir, it exhibited relatively more impacts on porosity and permeability. The maximum changes in porosity and permeability are 0.6% and 2% respectively. Also, most CO_2 is stored as free gas in the reservoir, this is followed by residual gas, dissolution in oil, and aqueous phases. Considering the coupled hydro-geochemical-mechanical model, 48% of the stored CO_2 is free gas, 34% is stored as residual gas, 10% dissolved in the oleic phase and 8% dissolved in the aqueous phase. However, there is no CO_2 stored as carbonate mineral because calcite being the only carbonate mineral dissolved faster than it precipitated. Also, based on the preliminary results, the long-term fate analysis of the CO_2 storage shows that CO_2 is trapped within the reservoir and there is no leakage within the time frame of 1030 years considering the current field operating strategies. Other operating strategies would be considered to determine the maximum allowable CO_2 injection strategies for the Morrow B reservoir. The addition of geomechanics in CO_2 sequestration processes presents more realistic responses of the CO_2 injection to the depleted oil reservoir.

This study presents the evaluation of chemo-mechanical impacts during CO_2 injection in depleted oil reservoirs. The numerical simulation model successfully coupled hydrodynamic, geochemical, and geomechanical models by utilizing history-matched numerical simulation models to fill current gaps in CO_2 sequestration studies pertaining to oil reservoirs.

626504 Greenhouse Gas Emissions Life Cycle Analysis of Carbon Capture and Storage for Industrial Sources in the Midwest-Northeast United States

Abstract:

Greenhouse gases generated for carbon dioxide point source facilities in the Midwest-Northeast region of the United States were evaluated to account for "cradle to grave" CO_2 equivalent emissions for carbon capture, compression, transport, and storage in relation to volumes of CO_2 stored underground. Seven scenarios were modeled: ethanol plant, natural gas power plant, direct air capture plant, hydrogen plant, petroleum refinery, cement plant, and fertilizer/ammonia plant. The analysis integrated specific factors for CO_2 sources in the region, including geologic storage setting, geographic location, CO_2 emission volumes, capture requirements, compression, CO_2 transport, and injection. A greenhouse gas life cycle analysis model was applied for CCUS operations that may contribute to greenhouse gas emissions. The model included general capture parameters with more detailed parameters for CO_2 compression, transport, and injection. The model did not account for combustion of fuel products or displaced electricity. Results of the greenhouse gas life cycle analysis provide insight into developing CCUS in the region in terms of net CO_2 storage versus the emissions related to facility operations. Emissions generated by CO_2 compression, transport, and injection may counter the carbon storage mass for smaller CO_2 sources, especially in areas where geologic CO_2 storage options are not readily accessible. Therefore, greenhouse gas emissions life cycle analysis is useful in ensuring maximum, net CO_2 storage for development of CCUS facilities in the Midwest-Northeast United States.

The work was supported by the Midwest Regional Carbon Initiative (U.S. Department of Energy Award No.: DE-FE0031836).

626957 Machine Learning Applications for Geologic Data Integration and Operational Data Analysis in Geologic Carbon Dioxide Storage Systems

Abstract:

Problem:

Data-driven models that are built using machine learning (ML) algorithms are becoming increasingly common-place in subsurface science and engineering applications. The impetus for adopting this emerging technology comes from its success in multiple fields such as consumer marketing, finance, design and manufacturing, health care, etc. The use of ML is particularly well-suited for characterizing, describing, and forecasting the behavior of geologic carbon dioxide storage systems where typical data analysis challenges include: (a) Incomplete data, (b) unreliable physics-based models (if they exist), and (c) data-driven models using conventional statistical methods are not robust.

In this presentation, we will describe the application of ML for two specific problems: (1) geologic data integration, i.e., identification and prediction of electrofacies from well-log data, and (2) operational data analysis, i.e., prediction of bottomhole pressure and temperature in CO2 injection wells from injection rate and wellhead pressure and temperature measurements.

Methods:

Our systematic ML workflow for building data-driven models involves the following steps: (a) exploratory data analysis to visually understand patterns, trends and outliers in the multivariate datasets, (b) statistical imputation to fill-in missing values (if any), (c) unsupervised learning to identify natural groupings (statistically homogeneous subsets) across the space of independent variables (predictors), and (d) supervised learning to fit predictive models between known predictors and responses (dependent variables). Unsupervised learning is typically carried out using principal component analysis (PCA), k-means clustering (kMC), hierarchical clustering (HC), etc. Supervised learning can be formulated either as a classification problem where the response is categorical, or a regression problem where the response is continuous. This is typically carried out using algorithms such as k-nearest neighbors (kNN), random forest (RF), artificial neural network (ANN), etc.

Results:

For the geologic data integration problem in a CO2 enhanced oil recovery project, well-logs from 250+ oil wells in the Albion-Scipio field in Southern Michigan were collected in a database. The database was checked for outlier values which were either rectified or eliminated. Missing values for several well logs were then imputed using a Random Forest algorithm to create a full database. Cluster analysis using both kMC and HC were used to identify the presence of 6 natural groups (or electrofacies) within the dataset. Finally, highly accurate models to predict the electrofacies based on well-log attributes were built and validated using both traditional statistical approaches (i.e., logistic regression) and machine learning (i.e., RF) approaches, as shown below in Figure – Part A.

For the bottom-hole pressure and temperature prediction problem, data from 3 different CO2 injection wells in different pinnacle reefs in Northern Michigan were collected. The dataset includes hourly values for wellhead pressure, wellhead temperature, wellhead density, injection rate, bottomhole pressure and bottomhole temperature. For all three wells, examination of the data revealed a bifurcation of the data around a wellhead density of 25 lb/ft^3. Therefore, different models were built for high- and low-density subsets of the data corresponding to this threshold. As in the previous case, missing values were filled in (imputed) using a random forest regression approach. Separate predictive models were built for bottom hole pressure and temperature as a function of the surface conditions for each well. The baseline model was a multivariate linear regression model with quadratic and cross terms. Machine learning options included a kNN model, an RF model, and an ANN model. The models were validated using three replicates of randomized 80-20 split sample testing (i.e., 80% training and 20% test data). The machine learning models were generally more successful, as demonstrated from the performance on the held-out data for one of the wells in the figure below in Figure – Part B.

Implications:

A systematic workflow for machine learning applications has been demonstrated for two representative subsurface problems. The identification of electrofacies helps build robust predictive models between well-log attributes and dynamic reservoir properties such as permeability. Also, bottom-hole gauges are not present in all wells (or they tend to malfunction from time to time), so the ability to predict pressure and temperature downhole from surface measurements is a valuable capability. Successful application of machine learning for these two different types of problems (one for geologic reservoir characterization side and one for operational data analysis) demonstrates the added value from these workflows. We are currently investigating the transferability of such models to other wells in the vicinity, as well as across geologic basins.

KEYWORDS: geologic storage; carbon dioxide; machine learning; classification; regression

Figure – Part A

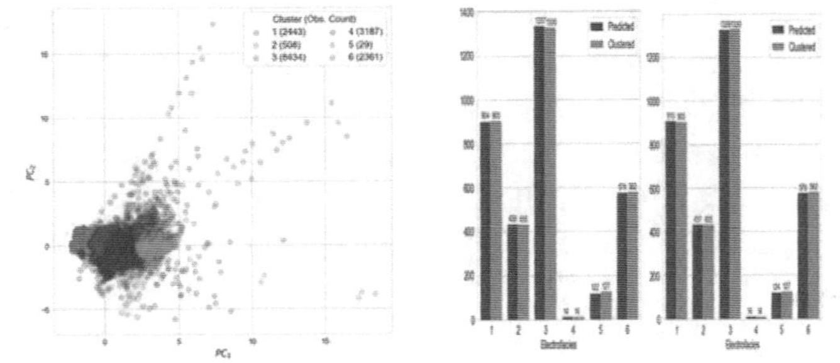

Figure – Part B

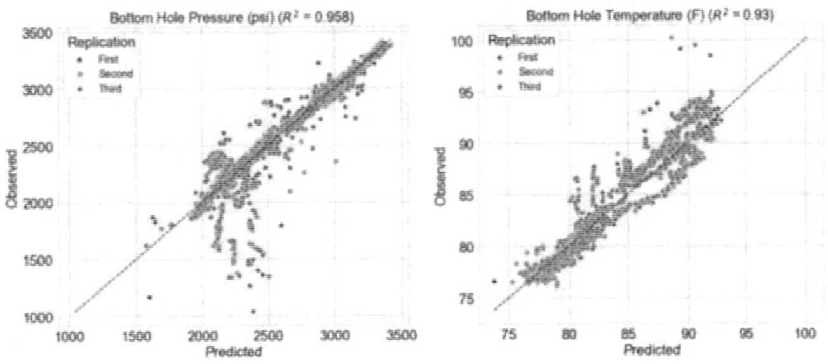

625586 Carbon Capture and Storage – the Risk and Rewards

Abstract:

Carbon Capture and Storage (CCS) is expected to play a vital role in the global fight against climate change. This presentation will introduce the overall concept of CCS, along with some of its nuances, before describing the rewards it can deliver and the risks and challenges it presents. It will outline the key health and safety risks, focusing in particular on the issues around reservoir integrity; widely seen as the greatest safety and ethical challenge. Geological security is clearly paramount to the success of CCS. Unless the long-term integrity of underground storage can be ensured, the environmental benefits of CCS will not be achieved and public health could be compromised. Using practical case studies, this presentation will explore the factors that can reduce storage integrity by weakening the cap rock or creating leak paths that bypass the sea, namely; acidic fluid action, existing geological faults, stress of injection, diffusion and lateral migration for example, before describing how existing knowledge, experience tools and techniques can be innovatively reworked to provide an effective means of managing these risks.

This presentation will also touch on what we consider to be the greatest risk to the sustainable proliferation of CCS - treating it as something entirely new and innovative. Whilst clearly a novel concept presenting significant new challenges, it will rely on many existing systems and processes. Granted these will often be used in reverse (e.g. using an offshore platform and subsea pipeline in reverse to inject CO_2 rather than extract oil and gas) and will accommodate a different medium, but many of the tools and techniques required to assess and manage these risks will be very similar. Also deep knowledge and experience of these facilities will exist which will be of huge benefit to this repurposing.

CCS can play a major role in our fight against climate change but it comes with a range of technical, geological, health, safety, environmental, and ethical challenges. However, by repurposing the old, borrowing the proven, and innovating the new, CCS can meet these challenges and achieve its full potential in a safe and sustainable way. This presentation will focus on our experiences in meeting these challenges using practical case studies from projects we have completed.

623388 Exploration and Forecasting of Water Usage for Treatment Process Development in Small Disadvantaged Agricultural Communities

Abstract:

Remote and disadvantaged communities in agricultural regions are confronted with impairment of their local potable groundwater supplies due to contamination and rising salinity associated with agricultural activities. Nitrate contamination of community potable groundwater sources has been reported in various areas throughout the United States. In California, for example, more than 205 community well water systems have been found (during 2002- 2010) to exceed the maximum nitrate contaminant level (MCL) of 10 mg/L as N and a significant number of communities are also confronted with high salinity above the recommended level of about 500 mg/L total dissolved solids (TDS). These communities rely on local water supplies which must be treated in order to provide safe drinking water.

In order to develop the required local wellhead water treatment and water distribution infrastructure for small remote communities, forecasting of water use information is needed to quantify the temporal (from hourly to seasonal) variability of was use and thus the treatment and delivery capacity. Moreover, information regarding water storage capacity needs and handling of sanitary water in community septic systems must be assessed. Accordingly, the present work demonstrates a data-driven modeling approach to describe and forecast water use patterns in small communities that rely on local well water for their potable water supply. The study included three small communities having population in the range of 16 – 36 located in the midst of agricultural fields in the Salinas Valley (California). These communities rely on their local wells for domestic potable water supply and manage their sanitary wastewater in their local septic systems as they have no access to a centralized sewer system. Groundwater supplies of these communities have been impaired due to nitrate contamination and wellhead treatment via membrane water treatment technology is a viable option. However, the design and operation (which invariably is expected to be intermittent) of such systems has to consider temporal variability of water use patterns. In particular, daily and peak water use data are critical to assess needed water storage capacity, and the volumetric rate of treatment and/or water delivery, as well as establishment of operational protocols. Accordingly, the present study focused on developing data-driven model of water use patterns that is suitable for small agricultural communities. The model was based on water use data compiled for the three study communities, via smart water meters, over a period of six years.

Exploratory data analysis was first conducted to quantify water use patterns at hourly, daily, weekly, monthly and seasonal levels. Self-organizing maps (SOMs) was initially conducted to assess the overall patterns of average daily water use throughout the different months. Subsequently, autoregressive moving average (ARMA) models were developed for water use patterns at different temporal scales (hourly, daily). Climate metrics (i.e., daily and monthly low/high temperature (oC) and rainfall (inches/day) for the region were included in the analysis to assess their relevance to the correlation of water use patterns. Data outliers were retained in the ARMA models given that these models can handle outliers. Water usage patterns for the study communities revealed temporal irregularity (i.e., the variance was non-stationary for the time-series data). Thus, water use data for the study sites followed non-stationary stochastic patterns with non-uniform variance. Accordingly, for ARMA model development for water consumption, non-stationarity was removed via the 2nd order differencing. The ARMA models were based on the combination of auto-regressive (AR) and moving average (MA) polynomials. Combination of the AR and MA polynomials was achieved considering both the variable linear relationship and linear dependence of the forecast errors. The performance of the ARMA models was quantified by the Root Mean Squared Error (RMSE), R-squared, and Trends.

Analyses via ARMA models with good predictive performance (i.e., $R2 > 0.85$) demonstrated that the predicted changes of water consumption can provide critical information for water supply scheduling, sizing the required water treatment system capacity, the required level of product (treated) water recovery, operation frequency and required local water storage as a buffering capacity of needed water supply. Self-organizing maps (SOM) clustering and Kullback-Leibler (*KL*) divergence approaches identified similarities among the remote communities in terms of their water use at seasonal, monthly, weekly, and daily levels. This community similarity assessment facilitated the use of transfer learning (i.e., the use of community data for predicting water use for other communities) and data imputation (I.e., filling data gaps). Additionally, in order to detect and report outliers in real-time data (e.g., unexpected spikes in water use due to community garden irrigation or leaks), a Gaussian Mixture Model (GMM) with 2 Gaussian distributions was trained and tested. The outlier detection approach can be useful for operational decision support (i.e., to distinguish between normal and atypical conditions) as well as the development of robust and adaptive predictive model to automatically predict water use with high predictive accuracy and to detect atypical water use. Finally, it is noted that the inclusion of temperature and rainfall as model input attributes improved the model predictive performance in terms of *R2* by as high as 29% (from *R2* of 0.61 to 0.9) and 27% (from *R2* of 0.54 to 0.81) for average daily and hourly water use, respectively. The association of water consumption with meteorological parameters is consistent with the expectation of higher water usage during hotter summer period and lower usage during the colder winter rainy period.

The present study suggests that the ARMA type models can be a useful tool in support of planning regarding water source management, and the design and deployment of local water systems and needed water treatment and wastewater handling. However, development of water use patterns models for causal analysis would require additional model parameters. For example, such parameters could include, but are not limited to, community descriptors such as occupation, income, residents per household, size and number of residential units.

629736 Quantitative Assessment of the Impact of COVID 19 in the Indian Power Sector Via Machine Learning

Abstract:

The COVID-19 pandemic has spread rapidly around the globe and caused significant challenges to the energy industry. India is one of the most affected countries by COVID-19, with over 13 million cases and 170K deaths [1]. Besides, India is also experiencing the second wave in 2021, which is much worse than the first one. In 2020, the pandemic forced governments worldwide to impose restrictions to prevent the spread of the virus. In India, a nationwide lockdown was imposed from March to May 2020 [2]. With a series of nationwide lockdown and social-distancing measures, the country gradually resumed economic activity after the lockdown in 2020. Policymakers and power system operators must take a scientific approach to understand and evaluate the impact of COVID-19 on the electricity sector and prepare for the second wave in India. The Indian electricity distribution companies (DISCOMs) have already been witnessing financial losses for the past decade (due to India's growing energy demand as well as demand-supply imbalances). With the steep dip in the overall electricity consumption across various Indian states (as we have statistically quantified), the incurring losses of DISCOMs will only exacerbate. This could threaten the energy security of India going forward.

Several reports have been published, both peer-reviewed and non-peer reviewed articles have analyzed the impact of COVID-19 on electricity consumption [3-5]. However, these assessments have not calibrated a baseline electricity consumption profile in the absence of the pandemic. Also, no studies have quantified the effect of COVID-19 dependent factors on electricity consumption.

In this work, we have developed a backcast model used to determine the electricity consumption profile if no COVID-19 occurred. The model gives us a fair estimation of the baseline and is used to determine the percentage of deviation in the electricity consumption. After the backcast model, we developed a regression model to capture the effect of various factors on electricity consumption in 2020. We analyze the impact of the three lockdown phases (partial lockdown, strict lockdown, lifted lockdown) on India's power sector for seven states (i.e. Delhi, Maharashtra, Gujarat, West Bengal, Uttar Pradesh, Tamil Nadu, and Karnataka), constituting 65% of the total electricity consumption. Such state-level quantifications as analyzed in this study, can be directed to National Government to assist in the relevant state level policy making and tariff related support to state DISCOMs to ensure the nation-wide energy security in coming years.

The backcast model is developed to forecast the energy consumption of 2020, based on historical data for the past three years (2017-2019). It is a classical time-series model to capture the trends and seasonality. The model is expressed as a function of the factors that affect the power consumption, including weather parameters (temperature, humidity, and wind speed) and economic parameters, i.e., GDP. Using the forecast for 2020 by the backcast model as our baseline, the analysis shows that the various states experienced anywhere between 17-44% reduction in energy consumed than the same time period from the last two years and baseline 2020. The state-wise dip in electricity consumption eased during phases 2 and 3, ranging between 6%-21%, and gradually reaching the regular norms by the end of August 2020.

In the second model, we developed a support vector regression (SVR) model to map the state-wise energy consumption as a function of weather parameters (e.g. temperature, humidity, wind speed) and social distancing factors (reduction in work population, increase in homestay). To estimate the power consumption during the second wave of COVID-19, we then devised energy consumption

scenarios for each state for the months from May – August 2021, by varying the lockdown parameters. This model enables us to use the SVR model to create May onward scenarios to estimate energy consumption across states under varying degrees of lockdown. Certain states in India are going through the second lockdown (partial). A stricter lockdown measure in Delhi, from May onwards, may inflict a reduced energy consumption for the period from May to mid-June.

We also studied the contribution of renewable sources in the grid mix. India managed to increase the % RES (Renewables) in the grid mix by 4.83 % compared to 2019, despite generating 2.37% less energy for the same year. Therefore, India managed to keep the RE mix of the grid at 10-11% in 2020 compared to 8-9% in the preceding years of 2018 and 2019. The current crisis and studies such as these present an opportunity for policy analysts to convene and lend a common voice to long-pending structural changes needed in the power sector in India.

References:

1. Worldometer. Worldometer-India. 2021; Available from: https://www.worldometers.info/coronavirus/country/india/.

2. Lancet, T., India under COVID-19 lockdown. Lancet (London, England), 2020. 395(10233): p. 1315.

3. IEA, Covid-19 impact on electricity. 2021.

4. EnergyWorld. Coronavirus impact: Within ten days, 26 per cent fall in India's energy consumption. 2020; Available from: https://energy.economictimes.indiatimes.com/news/power/coronavirus-impact-within-ten-days-26-per-cent-fall-in-indias-energy-consumption/74854825.

5. Aruga, K., M. Islam, and A. Jannat, Effects of COVID-19 on Indian energy consumption. Sustainability, 2020. 12(14): p. 5616.

625540 Energy Efficiency and GHG Emission of Manufacturing Zones: A Data-Driven Modeling and Analysis

Abstract:

Continuously improved industrial practice over the past decades has led to very significant improvement of energy sustainability. This is evidenced by tremendous energy savings, waste reduction, and productivity gains. However, according to the recent data reported to the DOE Industrial Technologies Program, energy loss in manufacturing sectors is still very high (~50% of the total consumption), and the total combustion emissions have reached ~1,200 MMT CO2e. It is shown that small and medium sized industrial and manufacturing facilities have a lower energy use efficiency than large ones; the lack of resources and expertise to address these inefficiencies has made their business less sustainable, especially in global economic competition. Hence, how to improve energy efficiency has been continuously a main focus in industries. Numerous energy consumption/GHG emission analysis methods have been developed for in-depth analysis for industrial systems. However, a further methodological study is needed for systematic analysis of process and non-process energy consumption and loss.

In this work, we will introduce a general, systematic data-driven modeling and analysis method to study energy consumption and loss in a manufacturing zone that usually consists of various types of manufacturing sectors. Huge amounts of energy consumption data in different regions are available from the U.S. Census Bureau and U.S. Energy Information Administration, many of which have been reported by various DOE supported industrial assessment centers. In the presentation, we will present a model-based analysis of the data to characterize the effectiveness of known strategies for energy efficiency improvement. A case study on the energy efficiency of the automotive-centered manufacturing sector, where various chemical and material supply companies exist, will be presented to demonstrate methodological attractiveness.

625492 Methodological Study on Sustainability Assessment and Selection of New Technologies for Process Performance Improvement

Abstract:

Technology assessment is of utmost importance when choosing new technologies. Traditionally, technology assessment focuses on mostly the technological performance and economic benefits. With a great push towards creating a more environmentally friendly future and the rapid advancement of technologies, it is imperative to have a systematic methodology for evaluating comprehensive sustainability performance improvement of new technologies. It is of particular importance to stakeholders that the technology used is economically feasible, to the environment that waste in any form could be minimized in the first place, and to the workforce and people that it does not create problems such as health and safety risks.

In this paper, we will introduce a methodology for assessing new technologies in their early development stage. The methodology consists of three major functions that are developed for (1) integrated sustainability and technical performance assessment, (2) human-machine interacted technology analysis, and (3) goal-bearing technology improvement recommendation and impact analysis. The identified technologies will be comprehensively analyzed for their capacities in the improvement of process design and operational performance, as compared with known technologies. We will present a mathematical framework for each of the three major functions and discuss uncertainty impact and processing methods. We will demonstrate the methodological efficacy through an applied study on designing and operating a plant-wide chemical and water use network system in an electroplating plant with multiple plating lines for different surface coating applications.

631952 Catalyst Choice Impacts Aromatic Monomer Yields and Selectivity during Hydrogen-Free Reductive Catalytic Fractionation

Abstract:

Lignin is the only large volume feedstock of renewable aromatics, and its efficient conversion into fuels and chemicals is expected to play an important role in future biorefinery viability. Reductive Catalytic Fractionation (RCF) is a promising method to extract and depolymerize lignin from native biomass into aromatic monomers using an alcohol solvent and heterogeneous supported metal catalyst. However, the use of externally supplied hydrogen gas in conventional RCF leads to high capital costs due to the need for catalytic reactors capable of withstanding high operating pressures. Hydrogen-free RCF processes utilize alternative hydrogen sources to stabilize reactive lignin fragments, and thus avoid the use of hydrogen gas directly in the RCF reactor. To understand the role of the catalytic metal in hydrogen-free RCF, we compared four common carbon supported RCF catalysts (Ru, Ni, Pd, Pt) for their ability to stabilize lignin from whole poplar to aromatic monomers with and without externally supplied hydrogen gas. Significant differences in selectivity and total monomer yield were observed: namely, Pd/C and Pt/C exhibited the highest monomer yields during hydrogen-free RCF, indicating that catalyst choice has a significant impact on process performance. Hydrogen gas produced from reforming of the methanol solvent ($CH_3OH \rightarrow CO + H_2$) is often cited as the hydrogen source. However, hydrogen gas production measured during hydrogen-free RCF did not correlate with monomer yield, indicating that the stabilization process may go through hydrogen transfer instead of needing to first make H_2 gas. Hydrogen-free monomer yield and selectivity trends from whole biomass were mirrored in model compound experiments with a monomer substrate, but dimer model compounds gave conflicting trends, potentially indicating that the reactive species during whole biomass reactions is monomeric. Further understanding of hydrogen-free RCF, including identification of the hydrogen donor, will be important for further process improvements.

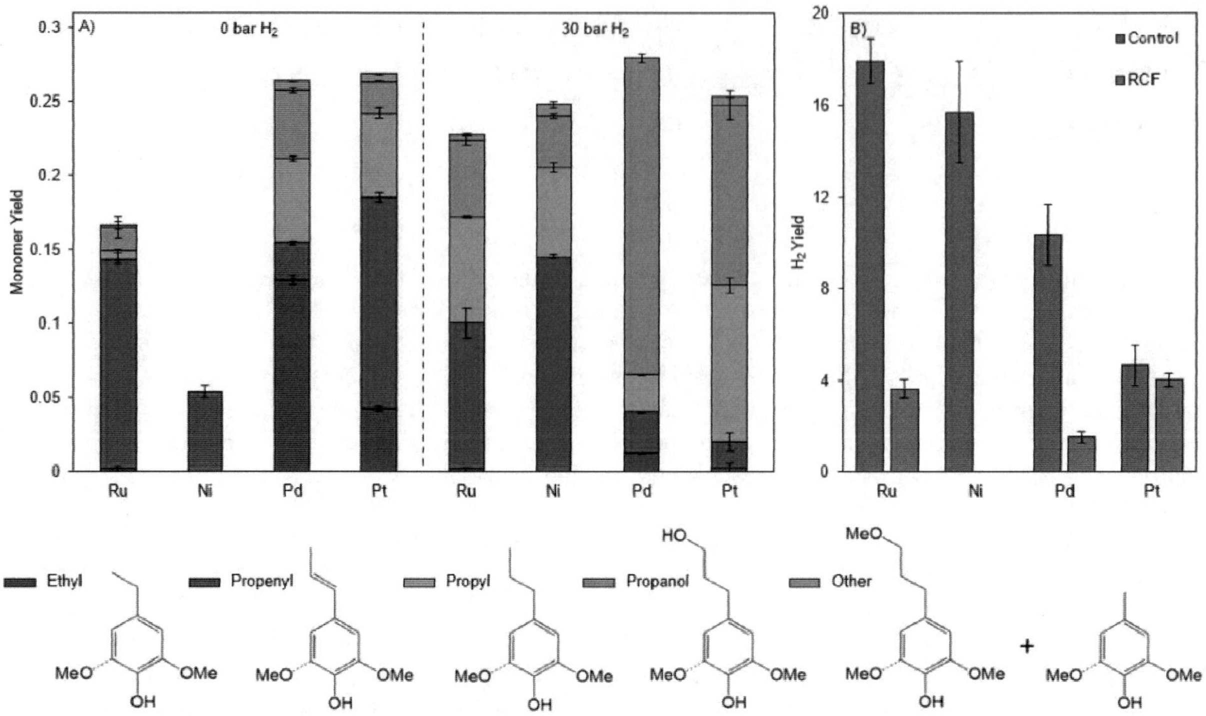

A) Monomer yields from 3 hr batch RCF reactions with 100 mg of catalyst, 2g poplar, 30 mL methanol, 225°C. Catalysts achieved similar monomer yields when 30 bars of external hydrogen is loaded, but clear differences are observed for hydrogen free reactions. B) Hydrogen gas yields from hydrogen-free RCF (grey) and control reforming reactions performed without poplar biomass (green) show clear suppression of the reforming pathway when biomass is present.

629510 Terpene Oxidation Routes to Value-Added Bio-Based Chemicals Utilizing Heterogeneous Catalysis

Abstract:

Terpene compounds generated through biosynthesis have the potential to be further upgraded through chemical catalysis to valuable products. Among these terpene-enabled molecules, oxidized products are particularly interesting, motivating the development of terpene oxidation reactions that can lead to enhanced product selectivity. In this work, we focused on limonene oxidation to limonene oxide and limonene diol through liquid phase epoxidation using H_2O_2. The selectivity to the desired limonene oxide product were significantly influenced by the reaction conditions. We found that the presence of water in the system particularly needed to be carefully controlled to enhance selectivity. To access more compounds via oxidation, we further extended our target to limonene diol generated through the ring-opening reaction of limonene oxide. The overall goal for the work is to examine the potential of integrating biological products from the terpeniod pathway with subsequent chemical conversion to valuable products not readily accessible when using only biological or chemical catalysis.

621593 Kinetics and Reactor Modeling of Volatile Fatty Acid Ketonization for Sustainable Aviation Fuel Production

Abstract:

Ketonization of volatile fatty acids (VFAs) produced via arrested methanogenesis of wet waste represents the only unit operation of wet waste upgrading to sustainable (net-zero or negative life cycle CO_2 emissions) aviation fuel not currently in industrial practice. Ketone product yields of close to 100% have been obtained during gas-phase reactions of VFAs over oxide catalysts at the laboratory scale, but design of an industrial ketonization reactor requires understanding of the impacts of reactant and product partial pressures, deactivation, and heat and mass transport phenomena on observed ketonization rates.

Our work leverages rigorous kinetic analysis of ketonization processes to inform reactor scale-up efforts through packed-bed reactor modeling. We first present results of a kinetic study of ketonization of a model VFA, hexanoic acid, over an industrial ZrO_2 catalyst performed in a packed-bed microreactor in conditions free of significant heat or mass transfer gradients. Major findings of the analysis include: (i) hexanoic acid saturates all catalyst active sites at relatively low partial pressure (~10 kPa) and (ii) ketonization products 6-undecanone, water, and CO_2 inhibit reaction rates. Kinetic data are used to fit a rate expression quantifying the functional dependence of ketonization rate on partial pressures of VFA reactants and ketone, water, and CO_2 products. A packed-bed reactor model describing vapor-phase hexanoic acid ketonization over ZrO_2 extrudate catalysts is developed based on the kinetic model. The effects of (i) bed- and pellet-scale mass- and heat-transfer limitations and (ii) axial pressure drop guide our development of recommendations for optimal sizing, temperature, and influent flow composition of an industrial-scale ketonization reactor. The quantitative understanding of VFA ketonization developed in this study represents an advance toward derisking the VFA ketonization step of wet waste upgrading to sustainable aviation fuel.

623493 Mechanocatalytic Hydrogenolysis of a Model Lignin Ether

Abstract:

Converting lignin to value-added chemicals will be essential as the world shifts to renewable sources of carbon. Recalcitrance of the lignin polymer during fractionation further limits the feasibility of many of these strategies, so unconventional valorization methods are necessary to overcome these issues. Mechanocatalysis is a promising novel approach to lignin upgrading because costly and environmentally harmful solvents can be avoided. These reactions use mechanical energy, typically in the form of collisions or impacts, to drive catalytic reactions in place of thermal energy. Here, the mechanocatalytic hydrogenolysis of benzyl phenyl ether (BPE), a lignin model compound, is studied over supported nickel catalysts.

The milling of BPE in a hydrogen environment with a commercial nickel (53 wt%) on silica-alumina catalyst resulted in 99% conversion within 120 min. Hydrogenolysis occurred at the benzyl ether bond, resulting in the primary products of toluene, phenol, and cyclohexanol. A maximum in phenol yield around 60 min indicated aromatic ring hydrogenation occurred after hydrogenolysis. The toluene volatilized and left the reactor before significant hydrogenation occurred. The mechanical action during also liberated fresh nickel surfaces, avoiding a high-temperature pretreatment step. Recycle experiments show that polyaromatic coke forms during milling, resulting in a steady deactivation.

Hydrogenolysis reactions were also conducted using three 5 wt% nickel catalysts to explore the influence of the support properties. These catalysts with less nickel were synthesized by dry impregnation using high surface area amorphous silica-alumina, high surface area silica gel, and low surface area silica as supports. Contrary to expectation, the low surface area catalyst had the best performance with the highest toluene yield and the lowest carbon loss. The high surface area silica supported catalyst did not show evidence of polyaromatic coke, despite having large carbon losses. Instead, the milling creates reactive sites on the silica that strongly binds the products.

630120 Seawater-Air-Sunlight Based Industrial Clusters for Carbon Dioxide Utilization

Abstract:

Climate change is an increasingly important crisis, and efforts must be directed towards it in order to lessen its threat and achieve sustainability on a global scale. One method of addressing the climate change crisis is by applying circular economy, in which resources are conserved and emissions may be reduced. Together, the power and industry sectors are responsible for most global carbon dioxide emissions, which significantly promote climate change. Carbon capture and sequestration (CCUS) systems can be used to reduce carbon dioxide emissions through reusing carbon dioxide and utilizing it to produce value added products. A resource integration framework (Ahmed et al. 2020) identifies materials for material exchange and assesses the impact of energy reuse and renewable energy on the system. This allows for an eco-industrial park to be designed, in which emissions are captured and utilized through resource integration, and sustainability and circularity are achieved.

A three-stage approach was developed in this work to design a carbon dioxide negative eco-industrial park that can convert carbon dioxide into value-added products. The first stage is process identification, where candidate plants are selected based on the cluster's objective and inputs. The objective here is to convert carbon dioxide using seawater, air, and sunlight. Therefore, the cluster must incorporate processes that can utilize carbon dioxide and process the input natural resources into functional forms. The inputs are converted into these useful forms through desalination, air separation, and solar electrical technologies. A reverse osmosis process produces water and brine from seawater, an air separation process separates air into oxygen and nitrogen, and a photovoltaic system and a concentrated solar power system use sunlight to produce the electricity needed to power the cluster. Therefore, identification of these processes is critical and performed through a stage-gate selection process. In the second stage, called the data collection stage, information related to each of the selected processes, such as its resource balances and associated economics, are collected. In the third and final stage, resource network optimization, information collected in the previous stage helps identify the cluster's optimal configurations to meet the specified economic and environmental objectives using a mixed-integer linear programming model. An illustrative example demonstrates how the model synthesizes carbon negative networks through optimization for different objectives while achieving a profit. The conversion of carbon dioxide into value-added products through natural resource inputs and renewable energy sources in the clusters designed, therefore, achieve overall economic and environmental benefits. The outcomes of this work reinforce industrial parks as a means to transition to circular economies.

Ahmed R, Shehab S, Al-Mohannadi DM, Linke P. Synthesis of integrated processing clusters. Chem Eng Sci. 2020. doi:10.1016/j.ces.2020.115922

631272 The Effect of Cultivation History on the Growth Phenotype of a Type I Methanotroph

Abstract:

Methane (CH_4) is the second most abundant greenhouse gas (GHG), whose global warming potential is 72 times that of CO_2 within a 20-year period. At the same time, CH_4 is a low-cost, rich source for carbon and energy, and an essential component of the global carbon cycle. Because of its low volumetric energy density, it is desirable to convert CH_4 into other forms of liquid fuels, and microbial conversion of CH_4 at ambient temperature and pressure is an attractive alternative to the thermochemical gas-to-liquid conversion, particularly for small scale application [1]. Among different methanotroph species, *Methylomicrobium buryatense* 5GB1 has been identified as a promising biocatalyst for industrial CH_4 conversion, and recently several key results have been published on the strain, including some baseline data of *M. buryatense* 5GB1 on its bioreactor performance as well as transcriptomic and ^{13}C labeling-based analyses have been reported recently [2-5].

Despite these advancements towards a better understanding of the type I methanotroph's cellular metabolism; available understanding on the difference of methane metabolism between oxygen-limited and methane-limited growth is still lacking. In our prior work, we have attempted to examine the different growth phenotypes, i.e., oxygen-limited vs. methane-limited, using batch experiment. However, we found that changing the composition of feed gas alone cannot drive the cells into different phenotypes. Instead, cell growth rate must be controlled to achieve oxygen-limited and methane-limited phenotypes [6].

In this work, we use continuous cultures to achieve oxygen-limited and methane-limited growth phenotypes by controlling both the cell growth rate and the feed gas composition. Additionally, we compare the differences between the two phenotypes through yield distribution.

Totally 6 steady-state growth conditions were examined, which consist of three different cell growth rates (controlled via dilution rate) and two feed gas compositions. The feed gas that contains 14% CH_4 and 23.3% O_2 is denoted as the methane-limited condition; while the feed gas that contains 14% CH_4 and 11.7% O_2 is denoted as the oxygen-limited condition. Two independent continuous runs were performed. The conditions tested in different runs were the same 6 conditions, but these conditions were carried out in different orders to test the hypothesis that the cells' prior growth condition has a noticeable effect on their current physiological state. Figure 1 depicts the order of the different conditions being tested in the first run (CE1) in (a) and the second run (CE2) in (b). For each growth condition, once a steady-state was achieved, it was maintained for at least three days to allow multiple data points to be collected. If a variation of larger than 10% was observed in the data, the steady-state was maintained for an additional day, after which the culture condition was switched to examine a different condition.

By adjusting dilution rate and feed gas composition, different cell growth rates can be obtained with different methane and oxygen uptake rates, enabling further examination of the differences between the methane-limited and the oxygen-limited growth phenotypes. It is interesting that both continuous runs showed that the methane limited conditions produce more organic carbon while the oxygen-limited conditions produce more biomass and less organic compounds, contrary to the common belief that oxygen-limited condition would result in higher yield for the organic compounds.

To better understand the differences between the oxygen-limited and methane-limited growth phenotypes of *M. buryatense* 5GB1, a modified GEM was used to conduct *in silico* analysis by incorporating the measurements obtained from the continuous culture. One major application of metabolic network models, particularly GEMs, is to predict different growth phenotypes (*e.g.*, how fast cells grow, what products are excreted) in various genetic and environmental conditions. Developed by the Palsson Lab, phenotype phase plane (PhPP) analysis is a powerful tool that uses FBA with the GEM to provide a global perspective on the genotype-phenotype relationship, and to help characterize different metabolic phenotypes [7]. In PhPP analysis, FBA is performed along two flux constraint dimensions (for this paper methane uptake rate and oxygen uptake rate), and the FBA results generated are used to construct a 2D or 3D (if growth rate is also considered) phase diagram.

In order to conduct FBA with the modified model, growth and non-growth associated ATP maintenance values are needed. The growth associated ATP maintenance value within the biomass equation was set at 23 ATP g DCW $^{-1}$, which is considered as the low mode [3]. The non-growth associated ATP maintenance (NGAM) was estimated using experimental data, as listed in Table 1 for both CE1 and CE2

Table 1 clearly shows that the methane-limited phenotype requires higher NGAM, indicating higher metabolic burden to the cells. This is consistent with the experimental results of increased production of organic compounds under methane-limited conditions. In addition, the NGAM estimated for CE1 have higher values than that for CE2, for both oxygen-limited and methane-limited phenotypes. This makes sense as in CE1, cells were driven to different phenotypes when moving to a new condition, which resulted in heavier metabolic burden by expressing the enzymes needed for the different phenotype. However, in CE2, cells stayed in the same growth phenotype and only switched once when transition from condition 6 (methane limited) to condition 5 (oxygen-limited). Less frequent changes of phenotype resulted in lower NGAM. Figure 1 (c), (d), (e), and (f) plotted the phenotype planes obtained based on the NGAM estimated for the different runs. It clearly demonstrates the effect of the cultivation history on the current phenotype: For condition 6 (methane-limited condition), in CE1, it was located in the oxygen-limited phenotype, because the two conditions tested before it were both oxygen-limited (condition 4 and 5); while in CE2, it was located in the methane-limited phenotype, because both culture conditions preceding it were methane-limited (condition 2 & 3).

References

1. K.A. Stone, M.V. Hilliard, Q.P. He, J. Wang, A mini review on bioreactor configurations and gas transfer enhancements for biochemical methane conversion, Biochem. Eng. J. 107 (2017) 52–60

2. Gilman, L.M. Laurens, A.W. Puri, F. Chu, P.T. Pienkos, M.E. Lidstrom, Bioreactor performance parameters for an industrially-promising methanotroph Methylomicrobium buryatense 5GB1, Microb. Cell Fact. 14 (2015) 1 http:// microbialcellfactories.biomedcentral.com/articles/10.1186/s12934-015-0372-8

3. Torre, A. Metivier, F. Chu, L.M.L. Laurens, D.A.C. Beck, P.T. Pienkos, M.E. Lidstrom, M.G. Kalyuzhnaya, Genome-scale metabolic reconstructions and theoretical investigation of methane conversion in Methylomicrobium buryatense strain 5G (B1), Microb. Cell Fact. 14 (2015) 1.

4. Gilman, Y. Fu, M. Hendershott, F. Chu, A.W. Puri, A.L. Smith, M. Pesesky, R. Lieberman, D.A.C. Beck, M.E. Lidstrom, Oxygen-limited metabolism in the methanotroph Methylomicrobium buryatense 5GB1C, PeerJ. 5 (2017) e3945, https:// doi.org/10.7717/peerj.3945

5. Fu, Y. Li, M. Lidstrom, The oxidative TCA cycle operates during methanotrophic growth of the Type I methanotroph Methylomicrobium buryatense 5GB1, Metab. Eng. 42 (2017) 43–51, https://doi.org/10.1016/J.YMBEN.2017.05.003

6. K.A. Stone, M.V. Hilliard, K. Badr, A.D. Bradford, Q.P. He, and J. Wang, Comparative study of oxygen-limited and methane-limited growth phenotypes of Methylomicrobium buryatense 5GB1, Biochemical Engineering Journal, 161 (2020), p.107707.

7. S. Edwards, R. Ramakrishna, B.O. Palsson, Characterizing the metabolic phenotype: a phenotype phase plane analysis, Biotechnol. Bioeng. 77 (2002) 27–36.

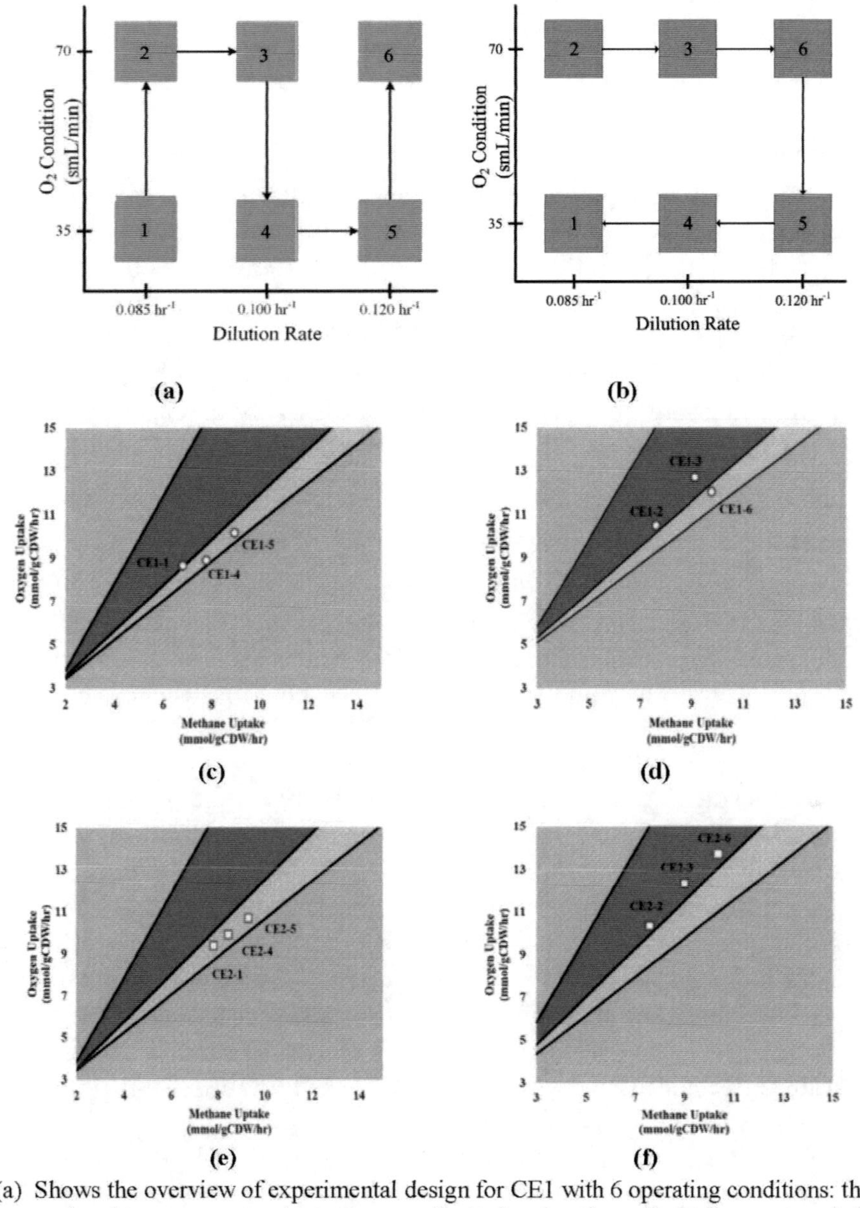

Figure 1. (a) Shows the overview of experimental design for CE1 with 6 operating conditions: three growth rates (dilution rates) and two gas compositions (oxygen-limited and methane-limited) were tested. (b) Shows the overview of experimental design for CE2 with 6 operating conditions: three growth rates (dilution rates) and two gas compositions (oxygen-limited and methane-limited) were tested. (c) and (d) Compares the 2D PhPP diagram for CE1. (e) and (f) Compares the 2D PhPP diagram for CE2.

Table 1 Non-growth associated maintenance energy from experimental data

NGAM	*Oxygen limited*	*methane limited*
CE1	10.6	15.53
CE2	9.89	10.93

631276 Improving Global Nitrogen Cycle Management Via Mechanochemical High Efficiency Urea Cocrystal Fertilizer Material Synthesis and Utilization

Abstract:

The world is experiencing unprecedented economic growth and increase in human population, thereby requiring more sustainable utilization of natural resources. Fertilizer production and usage is strongly correlated with food output and food security. While more than 200 million fewer people are undernourished than in 1990, 795 million people still remained undernourished in 2015. With the world population expected to grow by 35% to 9 billion by 2050, providing people around the world with nutritious supply of food at reasonable cost, is one of the greatest challenges. Fertilizer demand hence is expected to grow to about 200.5 million metric tons of ($N+P_2O_5+K_2O$) in 2018. Urea, $CO(NH_2)_2$, has been the most prominent N fertilizer making up ~60% of global nitrogen fertilizer use. Since the process to synthesize ammonia (NH_3), a reactant used to make urea, remains energy intensive and uses up to 1 % of the global energy and ~4 % of natural gas, it is critical that the urea nitrogen applied to soils is fixated in plants and not released in the form of gaseous NH_3 or otherwise lost to the environment. Unfortunately, only about 50% of the nitrogen fertilizer applied is absorbed by the crops and conceptually new chemical engineering processes are needed to avoid large exergy loss in the overall N-fertilizer production process as well as the emissions of reactive nitrogen into the environment.

The presented work will describe an approach to pre-emptively mitigate N-losses to the environment by describing mechanochemical synthesis of urea ionic cocrystals in combination with natural minerals. This will be done by converting urea powder and Ca- and Mg-containing natural minerals into value-added high volume sustainable fertilizers. In particular, mechanochemical methods applied to urea-CaSO4*2H2O and urea-Ca(H2PO4)2 were utilized and the corresponding reaction kinetics as well as the resulting product crystal structure and properties were monitored using in situ methods including XRD, TGA-DSC and Raman spectroscopy. Further, phase diagrams of the resulting cocrystals were described in silico by accounting for the environmental conditions of relative humidity and temperature. Resulting cocrystals possessed remarkable stability under moist conditions with decreased N-emissions. Finally, sustainable synthesis of CaSO4:4urea cocrystal was performed using widely available industrial gypsum drywall waste and the resulting enhanced efficiency fertilizer materials were tested in the laboratory as well as field experiments.

628475 European Path Towards Carbon Neutrality By 2050

Abstract:

With the European Green deal initiatives, Europe advances towards becoming the first carbon neutral continent by 2050. Using policy initiatives set by the European Commission, each member state is committed towards this 2050 goal. Along this commitment, the European Green deal also emphasises on the importance of interconnecting energy systems between member states. This is to ensure reaching the 2050 goal with a better energy infrastructure in the continent. To establish a net-zero carbon infrastructure in the European continent, many studies suggested relying on intermittent renewables. However, this raises the concern related to grid stability as energy production from intermittent renewables is not a stable process[1]. Hence, this brings the question of to what extent the EU can integrate these intermittent renewables and what better alternative technologies can be deployed instead. In terms of better alternatives, many studies have proved the effectiveness of deploying carbon capture and storage (CCS) and negative emissions technologies (NETs) for securing a stable grid[2], especially when an energy mix contains intermittent renewables. With this, a key question rises about how the energy system mix should evolve in the EU and what pathway is most economic, and how the pathway for each individual member state will vary. Thus, the objective of this work is to study the evolution of the electricity system in the EU to identify the economic pathways towards carbon-neutrality. To perform this case study, we chose the electricity sector as a representative and collected the necessary data. Such data consisted of installed capacity, demand profile, electricity prices, intermittent renewables availability, grid features, interconnections...etc. for each member state. These parameters were used as an input to the Electricity Systems Optimisation (ESO)[3] tool that optimises the system in terms of system cost. In this tool NETs and CCS are considered as technologies, and each member state is treated as a node connected to the other. The unique feature of this modelling tool to other EU electricity systems literature, is the consideration of grid features and grid expansion in terms of interconnections for each member state. The novelty of this work is shown through considering various factors in the model. First of which is the new carbon target; net-zero, where most studies in the literature considered the old carbon target (85% - 95% reduction) with a reference point of 2015. In addition to this, country level policies such as early decarbonisation and nuclear retirement were not included in most of the modelling work. Hence, this work addresses these common literature gaps by considering the most recent available data, country level policies and third countries that EU member states are connected to.

To conduct our assessment, we first assessed the current status of the EU in terms of previous carbon targets. This is to gain an image of how each member state's energy mix has evolved and whether previous carbon targets were met or not. From this, it was strongly observed that each member state is moving in a different direction as some were becoming more fossil fuel intense. This in return showed us how EU member states were not working collaboratively to meet their targets and hence, we next studied whether this was to their advantage or not. To quantify this in the ESO tool, we used grid constrains and emissions target to build our scenarios. In our first scenario, we implemented member states' grids connection in a way that they form one EU grid and each member state can contribute to the other's grid requirements. With this scenario, we also dedicated the net-zero target to the continent itself, (e.g. some countries can deploy NETs to offset emissions from other member states). In the second scenario, we separated the grids of each member state and connected them via interconnection lines. This is to ensure that expansion of the grid occurs within the member state itself and grid requirements are met solely by the member state. In terms of emissions, the member states in the second scenario were to target net-zero emissions separately. To asses which grid configuration is better, the results from these scenarios were compared based on total system cost and carbon intensity. The conclusion drawn from this analysis is the importance of EU member states to move forward by working together to reach the 2050 net-zero target. This is due to the benefits gained from sharing resources, as it was observed within the model how these resources vary based on geographical location.

References:

1. Heuberger CF, Mac Dowell N. (2018) ''Real-world challenges with a rapid transition to 100% renewable power systems,'' Joule, March, pp. 367 – 370.
2. Fajardy M, Mac Dowell N. (2017) ''Can BECCS deliver sustainable and resource efficient negative emissions?,'' Energy & Environmental Science, 1389 – 1426.
3. Heuberger CF, Bains PK, Mac Dowell N. (2020) ''The EV-olution of the power system: A spatio-temporal optimisation model to investigate the impact of electric vehicle deployment,'' Applied Energy, January, pp. 113715.

623737 Application of Membrane CO_2 Absorber in NH_4oh Looping Process for Enhanced Algae Growth

Abstract:

An energy-efficient and durable way to capture and store CO_2 for utilization is absorbing CO_2 into an aqueous solution for use as an in-situ carbon feedstock and using solar energy to produce to chemicals or biomass, such as via biofixation using algae. This is a sustainable process for CO_2 capture and utilization [1, 2]. Currently, most algae utilization systems utilize the CO_2 product stream after it has been separated from the CO_2-containing gas (flue gas). Unfortunately, this approach is not cost-effective on a large scale.

Dissolved NH_3 is attractive for both CO_2 capture and as an algae nutrient. For CO_2 capture, the ammonium hydroxide (NH_4OH) solvent has several advantages including low cost, no degradation in the presence of O_2, SO_2 or heat, high capture capacity, low regeneration energy, as well as the potential for capturing multiple acid gases (NO_x, SO_x and CO_2) in the flue gas [3,4]. Studies have shown that the scrubbing capacity of NH_3 is approximately 0.9-1.2 kg of CO_2/kg of NH_3, with a CO_2 removal efficiency of ~99% and half the solvent regeneration energy than that of 30 wt% MEA [5-7]. Its low cost makes NH_3 attractive as an algae nutrient. A practical, cost-effective, integrated CO_2 capture and utilization technology has been developed, constructed and tested at the University of Kentucky Center for Applied Energy Research (UK CAER) for algae production with three unique features. First, an aqueous NH_3 solution functions as both CO_2 capture absorbent and as an algae nutrient. Second, a downward flow, gas-liquid, indirect contact membrane CO_2 absorber is used, where the capture and biofixation is decoupled and NH_3 slip is minimized by including <1 wt% chelating compound in the solvent. Condensate from the saturated flue gas continually washes the gas side of the absorber membrane to ensure long-term operation. Third, solar energy-powered, distributed regenerators are installed near the algae bioreactor modules to provide local, just-in-time distribution of CO_2 and NH_3 at the appropriate ratio.

The UK CAER integrated capture and utilization system uses 10 cfm flue gas produced from either coal or natural gas combustion to simulate an industrial CO_2 source. The bench scale process consists of the membrane CO_2 absorber, solar powered flash stripper, algae bioreactors, and balance of plant, including NH_3 and water make-up systems. Commercially available polymeric membranes, whose surfaces are rendered hydrophobic, are used in the membrane absorber without the direct contact between flue gas and aqueous capture solution [8, 9]. During operation, flue gas is fed into the membrane tube-side followed by a water-wash column to remove any NH_3 slip prior to stack emission. The NH_4OH solvent flows through the shell side of the absorber and the exiting CO_2-rich solution is pumped to a pressurized solar energy heater and a solvent regenerator. Pressure is controlled in the regenerator to produce a product stream with a CO_2:NH_3 ratio of ~7, which is most appropriate as feed to the algae bioreactors. A modularized bioreactor, ~10 m^2, is used for performance evaluation.

Experimental results will be presented from operation of the UK CAER bench scale process detailing CO_2 capture and mass transfer enhancement, NH_3 slip and operation stability due to NH_3 salt formation and algae production. Control of NH_3 slip by solvent formulation optimization will be presented. A 2 M NH_4OH solvent is employed to keep the NH_3 vapor pressure low and a chelating agent (Zn^{2+}, amine and ionic liquid) is added to further reduce NH_3 volatility. Development of the membrane absorber will also be presented. Configuration of the membrane absorber utilizes condensed water from the flue gas to continually wash the gas-side to reduce fouling and recapture NH_3 slip. Hydrophobic coatings, such as with Teflon AF2400®, of the membrane are also considered. Preliminary bench scale studies will be detailed showing that NH_3 slip is manageable and controllable. Finally, *Scenedesmus acutus (*UTEX B72) algae productivity results will be presented when continuously fed with the appropriate CO_2:NH_3 ratio, optimized based on pH. Algae productivity is determined based on regular gravimetric culture density measurements and the algae yield at harvest. Culturing and harvesting follow standard practices. Measured productivity is compared with modeled productivity for the same period/climatic conditions, using an existing model.

References

[1] Klok, A. J., Lamers, P. P., Martens, D. E., Draaisma, R. B., & Wijffels, R. H. *Trends in biotechnology*, 2014, 32, 521-528.

[2] Shah, M., Mahfuzur, R., Liang, Y., Cheng, J. J., & Daroch, M., *Frontiers in plant science*, 2016, 7, 531.

[3] Hanak, D. P., Biliyok, C., & Manovic, V., *Applied Energy*, 2015, 151, 258-272.

[4] Li, K., Yu, H., Qi, G., Feron, P., Tade, M., Yu, J., & Wang, S., *Applied Energy*, 2015, 148, 66-77.

[5] Diao, N., Li, Q., & Fang, Z., *International Journal of Thermal Sciences*. 2004, 43, 1203-1211.

[6] He, Q., Chen, M., Meng, L., Liu, K., & Pan, W. P., *Institute for Combustion Science and Environmental Technology*, 2004.

[7] Yeh, A. C., & Bai, H., *The Science of the Total Environment*. 1999, 228, 121-133.

[8] Villeneuve, K., D. Roizard, J.C. Remigy, M. Iacono, and S. Rode., *Separation and Purification Technology*, 2018, 199, 189-197.

[9] Molina, C. T., & Bouallou, C., *Clean Techn Environ Policy*, 2016, 18, 2133-2146.

626976 Co-Product Recovery in Wet Anaerobic Storage to Reduce the Cost of Algae Biomass for Fuels and Chemicals Production

Abstract:

Seasonal variation in algae biomass productivity requires storage to ensure a constant feedstock supply to a biorefinery for year-round fuels production. Wet anaerobic storage instead of drying as a means of preserving algae biomass has been shown to reduce the cost of algae biofuels by as much as \$0.32/GGE. The low oxygen environment of wet anaerobic storage encourages the growth of lactic acid bacteria, which produce lactic acid, reducing the pH and inhibiting the growth of other microorganisms and thus preserving the biomass. These organic acids (lactic, succinic, and acetic acids) could potentially be recovered as a co-product to reduce the cost of algae biomass production. A two-stage membrane filtration process was developed to recover succinic acid from *Scenedesmus* biomass that accumulated 40% of its dry weight as succinic acid during storage. The first membrane consisted of a microfiltration membrane to separate soluble organic acids from the microalgae biomass after which a nanofiltration membrane (MW cutoff of 90 Da) was used to concentrate organic acids. The nanofiltration membrane achieved a succinic acid rejection rate of 95% and a transmembrane flux of 10 L m^{-2} h^{-1}. Techno-economic analysis was utilized to assess the impact storage byproducts could have on the cost of algae biomass when succinic acid was accumulated to 40% (w/w, dry basis) in storage through two different scenarios. The first applied the value of succinic acid as a credit toward the cost of algae biomass and provided high-acid biomass directly to a biorefinery. The second scenario utilized the two-stage filtration process to first recover organic acids from stored algae before providing low-acid biomass to a biorefinery. The minimum biomass selling price could be reduced by 20% when acid-rich biomass was provided directly to the biorefinery, while recovering storage co-products first could reduce MBSP by 9%. This study demonstrates that wet anaerobic storage, in addition to enabling the year-round operation of algae biorefineries, reduces the cost of producing fuels and chemicals from algae biomass by increasing its value.

630922 Simulations of Mixing and Radiative Interactions in Large-Scale Raceway Ponds for Algae Cultivation.

Abstract:

Understanding algae cultivation in raceway ponds is important for scaling up the growth of algae for many applications [1]. Among all the relevant variables affecting biomass growth in ponds, the degree of mixing and light penetration place at the top in terms of relative importance [2].

In this work, we simulate a 500 m^2 raceway pond for algae cultivation taking into account the dynamic interaction of algal cells with fluid flow and sunlight radiance. The fluid dynamics of the flows is simulated using the volume of fluid method (VOF), where the interface between the air and liquid phase is tracked and a single momentum equation to obtain the phase averaged velocity is solved [3]. This multiphase system is simulated using the open-source CFD software OpenFOAM [4]. Pond mixing is provided by a paddlewheel rotating between 10 – 15 RPM, which is simulated using a multi-reference frame method that imposes non-inertial frame momentum source terms in the vicinity of the paddlewheel. We validate our simulations with velocity measurements taken at different locations in the pond. We also study light availability as a function of depth which is one of the decisive factors in the overall algae growth and productivity. A set of radiation transport simulations are run in order to determine light availability as a function of concentration and depth. The objective is to obtain reduced order models (surrogate models) for light penetration, that along with the fluid dynamics will be coupled with an algae growth model.

[1] C. Inostroza, A. Solimeno, J. García, J. M. Fernández-Sevilla, F. G. Acién. *Improvement of real-scale raceway bioreactors for microalgae production using Computational Fluid Dynamics (CFD)* Algal Research 54 (2021) 102207

[2] J. Huang, X. Qu, M. Wan, J. Ying, Y. Li, F. Zhu, J. Wang, G. Shen, J. Chen, W. Li. *Investigation on the performance of raceway ponds with internal structures by the means of CFD simulations and experiments*. Algal Research 10 (2015) 64–71

[3] J. A. Heyns, and O. F. Oxtoby. *Modelling surface tension dominated multiphase flows using the VOF approach.* 6th European Conference on Computational Fluid Dynamics. 2014.

[4] S. S. Deshpande, L. Anumolu, M. F. Trujillo. *Evaluating the performance of the two-phase flow solver interFoam.* Computational science & discovery, 5(1), (2012) 014016.

630825 A Fed-Batch Mixotrophic Cultivation Strategy for Enhanced Biomass Growth and Metabolite Formation

Abstract:

Microalgal biomass - rich in industrially important biomolecules (e.g. carbohydrates, lipids, proteins) - has emerged as a suitable biorefinery platform for high value-added fuels and chemicals [1]. However, despite its promising potential, the large-scale production of microalgae for commercial applications is restricted by the low biomass productivities typically attained by the conventional batch cultivation of phototrophic species (i.e. those that assimilate inorganic carbon dioxide in the presence of light). Low biomass productivities, in turn, lead to a low production of biomolecules, making the overall production and conversion processes economically unattractive [2,3].

Studies have shown that nutrient-limited cultivation influences microalgae's intracellular composition, generally leading to increased accumulation of intracellular metabolites such as lipids and/or starch. Nevertheless, the benefits of nutrient limitation are often outweighed by a well-known trade-off between metabolite accumulation and cellular growth [4]. Instead, the establishment of a better-suited strategy for induced metabolite formation which does not simultaneously hinder biomass density can be approached through: i) the use of mixotrophic species (i.e. those capable of assimilating both inorganic and organic carbon) which exhibit higher growth rates [5], and ii) fed-batch systems which target prolonged cell life through adequate nutrient feeding regimes [6].

In this work we present a highly productive fed-batch mixotrophic cultivation strategy, consisting of intermittent nutrient pulses, leading to increased biomass growth whilst inducing enhanced metabolite formation. The fed-batch strategy was evaluated at laboratory-scale with the model species *Chlamydomonas reinhardtii*, grown mixotrophically in standard Tris-Acetate-Phosphate (TAP) medium. Furthermore, an optimal pulse feeding regime (i.e. pulse concentration, pulse feeding time) for maximal metabolite formation, specifically starch and lipid molecules, was identified by exploiting and further enhancing an *in-house* developed multiparametric kinetic model capable of simulating various fed-batch cultivation scenarios. The optimal pulse-feeding strategy was validated experimentally, and yielded significantly higher concentrations of biomass, starch, and lipids, compared to a typical batch system, making it a promising cultivation strategy fit for biorefinery applications.

References

[1] G.M. Figueroa-Torres, J.K. Pittman, C. Theodoropoulos, Optimisation of microalgal cultivation via nutrient-enhanced strategies: the biorefinery paradigm, Biotechnol. Biofuels. 14 (2021) 64.

[2] Y.-K. Oh, K.-R. Hwang, C. Kim, J.R. Kim, J.-S. Lee, Recent developments and key barriers to advanced biofuels: A short review, Bioresour. Technol. 257 (2018) 320–333.

[3] E.S. Shuba, D. Kifle, Microalgae to biofuels: 'Promising' alternative and renewable energy, review, Renew. Sustain. Energy Rev. 81 (2018) 743–755.

[4] G. Markou, I. Angelidaki, D. Georgakakis, Microalgal carbohydrates: an overview of the factors influencing carbohydrates production, and of main bioconversion technologies for production of biofuels, Appl. Microbiol. Biotechnol. 96 (2012) 631–645.

[5] M. Bekirogullari, J.K. Pittman, C. Theodoropoulos, Multi-factor kinetic modelling of microalgal biomass cultivation for optimised lipid production, Bioresour. Technol. 269 (2018) 417–425.

[6] F.J. Fields, J.T. Ostrand, S.P. Mayfield, Fed-batch mixotrophic cultivation of Chlamydomonas reinhardtii for high-density cultures, Algal Res. 33 (2018) 109–117.

631828 Integrated Conversion of Algal Biomass to Fusel Alcohols and Optimization of Alcohol Mixture Composition for Maximum Fuel Economy/Efficiency in Light Duty Vehicles

Abstract:

A key challenge for use of algae biomass for renewable fuels is efficient utilization of all of the major biochemical fractions of the biomass, including carbohydrates, proteins, and lipids. Development of a combined algae processing biorefinery would facilitate co-production of petroleum displacing chemicals with the intermediate to high value products that are currently produced from algae. To overcome issues involved with highly variable feedstock composition, our group is developing means for single-pot bioconversion of amino acid and sugar oligomers from algae hydrolysates to generate a variety of petroleum-displacing end products. Through these efforts we have recently demonstrated pretreatment and bioconversion of a variety whole algae hydrolysates, including raceway-cultivated *Nannochloropsis* sp., *Scenedesmus* obliquus, filamentous periphyton obtained from algae turf flow ways, and others. To enable high efficiency utilization of multiple fermentation substrates, we have developed an engineered *E. coli* biocatalyst consortium with each cohort optimized for utilization of either sugar or amino acid oligomers. Using this strategy, we have demonstrated production of fusel alcohols from diverse algae biomass and found that retention of algal lipids in the bioprocessing provides process intensification benefits by alleviating product inhibition and allowing co-extraction of the algae derived products. Furthermore, we have developed an optimization platform to evaluate the optimal alcohol mixture to maximize fuel performance and better inform further biocatalyst development. Our results demonstrate that an optimized solution has the potential to increase the change in total fuel economy of a light duty vehicle by more than 10% at high blend volumes compared to an unoptimized mixture. Optimal solutions can be tailored toward greatest increasing in vehicle fuel economy, or greatest thermodynamic efficiency gain/greatest CO_2 reductions. This work discusses these trade-offs and their implications for algae cultivation and conversation technologies at large.

623044 Efficient Resource Recovery to Enhance Biomass Conversion

Abstract:

Nitrogen is an essential nutrient for biological growth vital to the persistence of individuals and society in various aspects, from crop and livestock husbandry to microalgae cultivation for biofuel. Nitrogen is crucial for amino acids and protein synthesis, and it is commonly taken up in the inorganic forms of NO_3^- or NH_4^+. In the US alone, 14 million metric tons (MMT) of NH_3 are utilized directly and indirectly as fertilizer; the volume of N nutrients demand is substantial.

The ammonia demand is 2.7 kg NH_3 per million BTU of renewable diesel (RD) for microalgae hydrothermal liquefaction (HTL), which implies that 3.3 million metric tons of ammonia are required to produce 10 billion gallons per year of RD by HTL. HTL of algae feedstock at 300°C led to the nitrogen distribution that roughly 30 % of the feedstock nitrogen in the oil and solid phases, and 70 % in the HTL aqueous phase (HTL-AP). HTL-AP is generally characterized by high concentrations of ammonium, phosphates, and organic carbon. Besides from the prospective recovery of NH4-N, the separation of other high-value organic compounds, i.e., acids, further improves the economic upside of algae biofuel production. HTL oil has a high heating value but requires significant upgrading before it can be used as RD for transportation, because of high oxygen, nitrogen, and sulfur contents, together with acidity and viscosity issues. Depending on the upgrading process of HTL oil, nitrogen incorporated into HTL oil, which roughly accounts 6 wt.%, maybe dissolved in the process wastewater, and can be further recovered.

Controlling energy and operating costs are the key to economic viability and commercialization of ammonia recovery from wastewater. The unit energy consumption for conventional stripping and production of $(NH_4)_2SO_4$ from wastewater is approximately 9 kWh/kg-N, while ammonia fertilizer production with the Haber–Bosch process is 10 kWh/kg-N. At a 80% removal ratio, the addition of chemicals for pH adjustment has a profound impact on operating costs. Batch operations of electrochemical separation processes, e.g. electrodialysis (ED) and electrodeionization (EDI), has reported removal ratios over 90% with energy consumptions around 5 kWh/kg-N. Furthermore, ED and EDI may eliminate chemical additives for pH control during ammonia removal/recovery because of water splitting reactions that produce proton and hydroxyl ions and enable in-situ pH adjustment under the electric field. Resin wafer electrodeionization (RW-EDI) immobilizes the loose resin beads of conventional EDI into a porous matrix material and improves unstable process performance, stack leakage, and flow shortcut that increase operating and maintenance costs.

>90% recovery of both ammonia and organic acids from the waste algae HTL aqueous phase was achieved using RW-EDI technology. >75% of TOC was removed from the treated waste aqueous phase. TEA for these resource recovery and potential economic benefits from their utilization will be discussed.

AUTHOR INDEX

AIChE
120 Wall Street, FL 23
New York, NY 10005-4020

ISBN 978-1-7138-5716-7

CURRAN ASSOCIATES INC.
proceedings.com